E. Van der Li Keller

Developments in Petroleum Science, 16

petroleum geology

FURTHER TITLES IN THIS SERIES

1   A. GENE COLLINS
GEOCHEMISTRY OF OILFIELD WATERS

2   W.H. FERTL
ABNORMAL FORMATION PRESSURES

3   A.P. SZILAS
PRODUCTION AND TRANSPORT OF OIL AND GAS

4   C.E.B. CONYBEARE
GEOMORPHOLOGY OF OIL AND GAS FIELDS IN SANDSTONE BODIES

5   T.F. YEN and G.C. CHILINGARIAN (Editors)
OIL SHALE

6   D.W. PEACEMAN
FUNDAMENTALS OF NUMERICAL RESERVOIR SIMULATION

7   G.V. CHILINGARIAN and T.F. YEN (Editors)
BITUMENS, ASPHALTS AND TAR SANDS

8   L.P. DAKE
FUNDAMENTALS OF RESERVOIR ENGINEERING

9   K. MAGARA
COMPACTION AND FLUID MIGRATION

10   M.T. SILVIA and E.A. ROBINSON
DECONVOLUTION OF GEOPHYSICAL TIME SERIES IN THE EXPLORATION FOR OIL AND NATURAL GAS

11   G.V. CHILINGARIAN and P. VORABUTR
DRILLING AND DRILLING FLUIDS

12   T.D. VAN GOLF-RACHT
FUNDAMENTALS OF FRACTURED RESERVOIR ENGINEERING

13   J. FAYERS (Editor)
ENHANCED OIL RECOVERY

14   G. MOZES (Editor)
PARAFFIN PRODUCTS

15A   O. SERRA
FUNDAMENTALS OF WELL-LOG INTERPRETATION. I. THE ACQUISITION OF LOGGING DATA

Developments in Petroleum Science, 16

petroleum geology

R.E. CHAPMAN

Dept. of Geology and Mineralogy, University of Queensland,
St. Lucia, Qld. 4067, Australia

ELSEVIER — Amsterdam — Oxford — New York 1983

ELSEVIER SCIENCE PUBLISHERS B.V.
1 Molenwerf
P.O. Box 211, 1000 AE Amsterdam, The Netherlands

Distributors for the United States and Canada:

ELSEVIER SCIENCE PUBLISHING COMPANY INC.
52, Vanderbilt Avenue
New York, NY 10017

Library of Congress Cataloging in Publication Data

Chapman, Richard E.
  Petroleum geology.

    (Developments in petroleum science ; 16)
    Includes bibliographical references and index.
    1. Petroleum--Geology. I. Title. II. Series.
TN870.5.C445 1983   553.2'8   83-1549
ISBN 0-444-42165-3

ISBN 0-444-42165-3 (Vol. 16)
ISBN 0-444-41625-0 (Series)

© Elsevier Science Publishers B.V., 1983
All rights reserved. No part of this publication may be reproduced, stored in a retrieval system or transmitted in any form or by any means, electronic, mechanical, photocopying, recording or otherwise, without the prior written permission of the publisher, Elsevier Science Publishers B.V., P.O. Box 330, 1000 AH Amsterdam, The Netherlands

Printed in The Netherlands

To my wife, June

"Pour voir les choses, il faut les croire possibles."

(Marcel Bertrand, 1891)

PREFACE

The fascination of petroleum geology lies both in its complexity and in its importance to society. There is still much that we do not understand; and there is much to learn if remaining undiscovered reserves of oil and gas are to be found economically. It is also good geology with a healthy practical component.

The great advances in geological thought and understanding in the 19th Century were based largely on the construction of coal mines, railways and canals. But this was almost two-dimensional geology of the land, bounded by the low-tide line. In the last 30 years or so, geology has moved offshore onto the continental shelves and ocean floors, largely under the stimulus of petroleum exploration, and with it has grown a great wealth of geological information.

In "Petroleum Geology: a Concise Study" (published by Elsevier in 1973, with the paperback edition in 1976) I attempted to focus on those elements of petroleum geology that seemed to be amenable to synthesis and to provide a broader understanding of some significant processes in petroleum geology. Since then, there has been an even more spectacular growth in the quality and quantity of geological information. We are still being buried under a mountain of empirical data.

I remarked then, as many others had before me, that petroleum geology embraces more disciplines of science than one mortal can master. The same is true today, of course, and it is also true that in many of our fundamental topics, no true consensus has emerged. This is not through lack of information (although this is certainly retarding our progress in the microbiological aspects of petroleum geology). The notable exception is the consensus reached on the geochemical aspects of the origin and generation of petroleum.

Since writing my first book, books have been written on petroleum geochemistry, abnormal pore-fluid pressures, and empirical approaches to petroleum migration (among others), whereas only papers in journals or chapters of books had appeared before. These were all valuable contributions to petroleum geology; but there has grown from these and other works a need for another treatment of petroleum geology that will help the individual to get a grasp of the whole subject and the interactions between the specialist topics. This is particularly important for the student because, once an active career in industry begins, little time will be found to keep up with the literature.

VIII

This book is the child of the first. It was no longer possible, or even desirable, to follow the format of the first book, although I have no reason to wish to change the main conclusions in it. Once again, I have tried to present the subject in a way that will also interest the student who does not intend to follow a career in the petroleum industry. I hope also that, like its predecessor, this book will also interest those with some experience in the industry. My purpose here is to present a view of petroleum geology that may also contribute something to our understanding of wider aspects of geology. I have only paraphrased the works of others in essential outline. References to topics not considered here in detail are given at the end of each chapter, as are references to works that present a different view or interpretation. The reader is encouraged to delve into the literature because it is exciting.

The most worrying aspect of the developments of the last two or three decades concerns the eternal problem of scientific rigour in what is essentially an applied science. It is quite certain that many of us are in error in our ideas and assertions: what is not certain is who they are, and which ideas and assertions are in error. It is not the purpose of science to avoid error, but to avoid its propagation. In our branch of science, which depends almost exclusively on industrial operations for data, much of which is confidential to the company acquiring the data, there is little control in the scientific sense. This is not to impute dishonesty to anyone. The pressures within the petroleum industry simply do not allow much time for *thought*, and it is not necessary to *prove* a theory or hypothesis before it is put to practical use. But this can lead us into errors that may have important practical consequences. For example, vitrinite reflectance was found some years ago to have real practical value in assessing the prospects of exploratory wells in some areas (not all). There was danger of forgetting the logic of the association, and some came to believe that there was a causal association. There are many areas of the world where, if such a dogmatic approach had been taken, important reserves would not have been discovered. The danger here is that that information can be obtained from a single well. Fortunately, other techniques for assessing maturity of sedimentary rocks were developed, and vitrinite reflectance is but one of the methods used. If scientific proof had been required before its use, this technique would never have developed to usefulness. But there is still a danger that we have misunderstood the nature of maturity.

This raises the question of parochialism. Some years ago I suggested to the author of a paper on abnormally high pore pressures in mudstones that his reliance on clay-mineral diagenesis could lead him into difficulties if confronted with abnormal pressures at depths known to be shallower than the depths of this diagenesis. His reply was that his company had "proof" that the cause of abnormal pressures is clay-mineral diagenesis. A few months later, in the research laboratory of another company, I was shown "proof" that clay-mineral diagenesis is not the cause of abnormal pressures.

The point is this: if your perspective is limited to one part of the world, you are more likely to be led into erroneous ideas because the evidence that would distinguish cause and coincidence might be lacking. A geologist who has spent his career in the Western Canada basin would probably have totally different ideas about the generation, migration and entrapment of petroleum from one whose career had been spent in the U.S. Gulf Coast. Indeed, they would probably have different ideas about the nature of geology in general. Not the least of these contrasts would be the lack of deformation in the Western Canada basin where, from the well-head, the Rocky Mountains can be seen; while the Gulf Coast is deformed under the continental shelf with no land in sight, let alone mountains. But our Canadian geologist would feel quite at home in Mexico and Libya, while our Gulf Coast geologist would feel quite at home in Nigeria and South-East Asia.

Geology, I believe, still suffers from one important, but unavoidable, fact: it grew from studies of outcrop, which are necessarily confined to the land areas, with the third dimension limited to the depths of mines and the heights of mountains, and it is still practised by a majority of geologists within these dimensions. The geology of what we can see and touch is the geology of sedimentary basins that are no longer accumulating sediment, and the geology of orogeny. Petroleum geology gives us a glimpse of sedimentary basins that are still actively accumulating sediment, and are still being deformed in spite of the fact that they have not yet suffered orogeny. Petroleum geology, although not giving us a complete three-dimensional picture, has given us a three-dimensional picture of some areas in great detail to depths of two, three, and four kilometres. Who would have imagined that there could be Mesozoic thrust faults beneath the horizontal Tertiary of the north German plains? . . . or folds and faults in young Tertiary sediments in many continental shelves? Our conception of an unconformity seems to have been dominated by Hutton's unconformity on the east coast of Scotland, and an assumption of subaerial erosion; yet there are extensive and important unconformities in the continental shelves that were never subaerially formed, so far as we can determine, and the deformation, erosion and subsequent sediment accumulation were entirely submarine.

These matters affect our understanding of geology. There are others that affect our understanding of petroleum geology. When people peer into our science from another discipline, and speak with confidence, we tend to accept what they say. Lord Kelvin poured scorn on geologists of the last century, and few rose to defend geology against him. A century later, we can say with confidence that most of what he said about geology and geologists was wrong. During the last 20 years or so, chemists have spoken with increasing confidence about the generation of oil, and geologists have tended to mould their concepts to fit the hypotheses of chemists. Some seem to have forgotten that geology is also a science — even if an imprecise one compared to chemistry, physics and mathematics, but nevertheless a science with its own logic. No geochemical hypothesis can be satisfactory unless it is also satisfactory from a geological point of view. This is not to assert that the geochemists are wrong, but rather to assert that the search for the truth about oil generation and migration must be a search for hypotheses that are geologically as well as chemically satisfying. It is not at all clear that this happy state has been reached in any major petroleum province.

Finally, it must be remembered that whether an hypothesis is correct or erroneous cannot be determined by voting. The majority is not always right. A vote taken in 1950 on the validity of continental drift would have had a very different result from one taken in 1980. My friend S.W. Carey, who would probably have been in the minority on *both* votes, may yet turn out to be more correct than the rest of us!

It is therefore my earnest hope that readers will read this book not with their eyes but with their minds, to try and get at the fundamentals of the various problems and, above all, to develop an independent assessment of the nature of petroleum geology. The answers to our problems lie in the geology of our petroleum provinces and the geology of provinces without petroleum, not in books.

Brisbane, 3 October 1982   RICHARD E. CHAPMAN

CONTENTS

Preface	VII
Acknowledgement	XV
Postscript	XVII

Part 1. General

CHAPTER 1. CONCEPTS OF SEDIMENTARY BASINS 1

Summary	2
Introduction	2
Accumulation of sediment	5
Transgressions and regressions	8
Eustatic sea-level changes	11
Lithological associations in sedimentary basins	15
Classification of sedimentary basins	18
Sedimentary-basin geometry	19
References	21

CHAPTER 2. EARLY DEFORMATION OF SEDIMENTARY BASINS: GROWTH STRUCTURES 23

Summary	23
Introduction	24
Definitions and terminology	24
Growth faults	25
The nature of growth faults	28
Growth anticlines	34
References	38

CHAPTER 3. COMPACTION OF SEDIMENT AND SEDIMENTARY ROCKS, AND ITS CONSEQUENCES 41

Summary	41
Compaction of sediment and sedimentary rocks	41
Pore-water expulsion	53
Interstitial water pressures	54
Mechanical aspects of pore water in sedimentary rocks	57
Fluid migration from abnormally high pore pressures	61
References	63

CHAPTER 4. A SYNOPSIS OF PETROLEUM GEOLOGY 67

Introduction	67
Geological inference	68
Petroleum geology	72
Petroleum	72
The Alkane (Paraffin) Series	73

The Cyclo-Alkane Series	74
The Aromatic Series	74
Water	76
Generation of petroleum	78
Migration of petroleum	80
Entrapment of petroleum	81
Geophysics	85
Seismic stratigraphy	89
References	93

CHAPTER 5. DRILLING ... 95

Introduction	95
Cable-tool drilling	95
Rotary drilling	99
Deviated or directional drilling	103
References	105

CHAPTER 6. THE LOGGING OF BOREHOLES ... 107

Summary	107
General	107
Principles of electrical logging	110
Resistivity	110
Spontaneous potential (SP) log	125
Radioactivity logs	127
Sonic log	130
Dipmeter	132
Temperature	133
References	136

CHAPTER 7. THE NATURE OF OIL AND GAS FIELDS ... 139

Summary	139
Giant oil fields	142
Zipf's law	143
The oil-rich countries	149
References	153

CHAPTER 8. THE NATURE OF PETROLEUM RESERVOIRS ... 155

Summary	155
Water saturation	156
Pressures	161
Permeability and relative permeability	161
Mechanics of production	169
Secondary recovery	176
References	177

CHAPTER 9. ORIGIN AND MIGRATION OF PETROLEUM: GEOLOGICAL AND PHYSICAL ASPECTS ... 179

Summary	179

XIII

Introduction	179
Hydrodynamic aspects of migration	181
Secondary migration	185
Migration in moving water	189
Rates of secondary migration	198
Accumulation of petroleum in a trap	199
Origin and migration of petroleum in carbonate rocks	200
References	206

CHAPTER 10. ORIGIN AND MIGRATION OF PETROLEUM: GEOLOGICAL AND GEOCHEMICAL ASPECTS ... 209

Summary	209
Introduction	209
Coal and petroleum	213
Primary migration	215
Crude oils with high wax content	220
Alteration of crude oil after generation: water-washing and biodegradation	221
References	226

CHAPTER 11. ORIGIN, MIGRATION AND ACCUMULATION OF PETROLEUM: DISCUSSION ... 231

Summary	231
Evidence of stratigraphic position of source for accumulations	231
Petroleum migration and faults	245
References	251

Part 2. Petroleum geology of transgressive sequences

CHAPTER 12. FOSSIL CORAL REEFS ... 255

Summary	255
Introduction	255
Silurian reefs of the Great Lakes area	259
Devonian reefs of western Canada	262
Cretaceous reefs of Mexico	269
Paleocene reefs of Libya	272
References	276

CHAPTER 13. PALAEOGEOMORPHIC AND UNCONFORMITY TRAPS ... 279

Summary	279
Palaeogeomorphic, palaeotopographic traps	279
Unconformity traps	284
Prudhoe Bay oil field	286
The North Sea	291
Bass Strait oil and gas fields, Australia	296
References	301

Part 3. Petroleum geology of regressive sequences

CHAPTER 14. ABNORMAL PRESSURES ... 303

Summary	303
Observations	303
Interpretation of observations	306
The nature of the transition zone	310
Discussion	313
Other possible causes of abnormal pressures	318
Reservoir geometry	318
Clay-mineral diagenesis	318
Tectonic	319
Osmosis	319
Petroleum generation	320
References	321

CHAPTER 15. DIAPIRS, DIAPIRISM AND GROWTH STRUCTURES — 325

Summary	325
Diapirs	325
Salt diapirs	327
Mudstone (shale) diapirs	329
Generalizations	330
Diapirism	330
Discussion	336
Growth structures and incipient diapirism	340
Gesa anticline, Waropen coast, Irian Jaya	341
Seria field, Brunei	342
Miri field, Sarawak	344
References	346

CHAPTER 16. SYNTHESIS — 349

Summary	349
Introduction	350
Deformation of sedimentary basins	351
Regressive sequences	352
Rift basins	361
References	371

APPENDIX 1. CONSTRUCTION AND USE OF SHALE-TRANSIT-TIME PLOTS — 375

Operational uses of the method	378
Geological interpretation	378
Example of pressure estimation and interpretation	380
References	381

GLOSSARY	383
AUTHOR INDEX	399
SUBJECT INDEX	409

ACKNOWLEDGEMENTS

One cannot write a book after 30 years of practising geology and acknowledge adequately the great influence many people have had on one's development as a geologist. To all my past and present colleagues I owe more than I find it comfortable to admit. I thank particularly Pierre Freymond, a friend for more than 25 years and a colleague for several, who has greatly influenced my geological thinking; Hollis Hedberg, whose friendship I have greatly valued and who has been a constant source of enlightenment and encouragement; and John Webb, who has been a valued colleague, friend and critic for the last 12 years or so.

The authors of published works are, of course, the sowers of seeds and the cultivators of their progeny without whom there would be no science. In particular, I have benefited from the works of M.K. Hubbert, J.M. Hunt, P.E. Kent, W.W. Rubey, K. Terzaghi, R.W. van Bemmelen, L.G. Weeks, and, of course, H.D. Hedberg. None would agree with all I have written.

To my colleagues and students at this University I owe a special debt of gratitude. Margaret Eva, in the Geology Library of this University, has shown tenacity, tolerance and good humour in tracing and acquiring numerous works from other libraries. I am also most grateful to Irene Lenneberg, who drew almost all the figures that are not otherwise acknowledged.

No-one but myself is responsible for the shortcomings in this book.

I am greatly indebted to the following organizations for permission to use material for which they hold the copyright:

AAR Ltd, Brisbane: for Fig. 6-1.

American Association of Petroleum Geologists:

From their Bulletins; Figs. 9-19, 9-20, 9-21, 11-5b, 12-3, 12-4, 12-10, 15-3 and 16-5.

From Memoir 8 (1968); Figs. 15-8—15-11 inclusive.

From Memoir 14 (1970); Figs. 1-6, 8-18, 12-5—12-8 inclusive.

From Memoir 16 (1972); Figs. 12-11, 13-1 and 13-2.

From Memoir 30 (1980); Figs. 12-11, 12-12, 13-3—13-9 inclusive.

From Habitat of Oil (1958); Figs. 11-2, 11-3, 15-13, 15-14, 16-2 and 16-4.

Also, for permission to quote verbatim a section from Hedberg, Sass and Funkhouser, 1947.

Australian Petroleum Exploration Association:

From their Journal; Figs. 4-7, 4-10, 4-11, 12-2, 16-8, 16-9, 16-10, and Plates 8-1 and 8-2.

Canadian Society of Petroleum Geologists:

From their Memoir 3 (1974); Fig. 16-12.

Esso Australia Ltd:

For providing photographs of their original illustrations, Figs. 4-8—4-12 inclusive, 13-10—13-14 inclusive, and 16-11.

Institute of Petroleum, London:

From: Petroleum Geology of the Continental Shelf of North-West Europe (1981); Fig. 16-12.

Koninklijk Nederlands Geologisch Mijnbouwkundig Genootschap:

Fig. 11-5a.

NASA/Goddard Space Flight Center:

For Figs. 1-4 and 1-5.

Oil and Gas Journal:

For the data of Fig. 7-2 and Tables 7-7—7-10 inclusive.

Royal Dutch/Shell Group:

For the data from which Figs. 3-14, 14-3, 14-8 and 15-7 were drawn.

Schlumberger SEACO Inc., Sydney:

Figs. 6-7, 6-11 and 6-12.

Springer, Berlin:

From Petroleum Formation and Occurrence; Figs. 4-3 and 10-1.

Duckworth, London:

For "The Microbe" of Hilaire Belloc.

I am indebted to the authors of these works for their permission to use these figures, and in particular to the following who also provided me with clean copies of the originals: P.M. Barber and Phillips Australian Oil Company, D.L. Barss, I.R. Campbell and West Australian Petroleum Pty Ltd., G.A. Day and the Director of the Institute of Geological Sciences, D. Gill, C.R. Hemphill, H.C. Jamison and Atlantic Richfield Company, R.H. Kirk and Mobil Oil Corporation, and P.E. Playford.

If any name has been inadvertently omitted, I apologize.

POSTSCRIPT

While this book has been in press, two papers have been published that I would have referred to. There has been a stimulating discussion of Ekofisk by Minturn (1982) concerning the geometrical interpretation of seismic reflection surveys. There has also been an interesting account of petroleum in basement rocks by P'an Chung-Hsiang (1982) that includes descriptions of several fields not mentioned in this book. There has also been a new edition of the *Geothermal Gradient Map of Southeast Asia*, which I have not yet seen.

REFERENCES

Minturn, L.W., 1982. Ekofisk: first of the giant oil fields in western Europe: discussion. *Bull. Am. Ass. Petrol. Geol.*, 66: 1408—1411.
P'an Chung-Hsiang, 1982. Petroleum in basement rocks. *Bull. Am. Ass. Petrol. Geol.*, 66: 1597—1643.

Note added in proof

Readers will find the following recent papers of interest in connexion with the topics of petroleum generation and migration:

Goldstein, T.P., 1983. Geocatalytic reactions in formation and maturation of petroleum. *Bull. Am. Ass. Petrol. Geol.*, 67: 152—159.
Wilson, H.H., 1982. Hydrocarbon habitat in main producing areas, Saudi Arabia: discussion. *Bull. Am. Ass. Petrol. Geol.*, 66: 2688—2691.

PART 1. GENERAL

CHAPTER 1

CONCEPTS OF SEDIMENTARY BASINS

SUMMARY

(1) Sedimentary basins are areas in which sediment accumulated at a significantly greater rate than sediments of the same age in neighbouring areas, so accumulating a greater thickness. The sediments accumulate by virtue of subsidence.

(2) The geological concept of a sedimentary basin is distinct from the geographical concept of a physiographic basin. Sediment accumulation is also distinct from sedimentation and deposition, because not all sediment deposited accumulates for a significant period of time in the geological record.

(3) The nature of the sediments that accumulate in a sedimentary basin is related to the environments of the physiographic basin from which the sediments were derived and in which they were deposited.

(4) Sedimentary basins typically begin with a transgressive sequence and end with a regressive sequence, but they may have a long and complicated history. Transgressive sequences record a general deepening of the sea, with reduction of the land area and migration of the facies towards the land. Regressive sequences record a general shallowing of the sea, with extension of the land and migration of the facies seaward.

(5) Most significant carbonate sequences are transgressive: arenaceous sequences may be transgressive or regressive. All important regressive sequences are arenaceous.

(6) Eustatic changes of sea level leave a record in all active sedimentary basins that are accumulating sediment. Changes of sea level due to changes in the shape of the geoid may lead to transgressive sequences in some parts of the world contemporaneously with regressive sequences in others.

(7) Sedimentary basins are deformed by faults and folds while they accumulate sediment and the sedimentary column is still subsiding.

INTRODUCTION

The geology of petroleum is largely the geology of sedimentary basins because it is in sedimentary basins that the commercial accumulations of petroleum occur. It is therefore essential to have a clear idea from the outset of what a sedimentary basin is.

A sedimentary basin may be defined as "an area in which sediments accumulate during a particular time span at a significantly greater rate, and so to a significantly greater thickness, than in the surrounding areas". This is not entirely satisfactory because of the vagueness about thickness — yet this vagueness exists. The essential part of any definition must be the *accumulation* of sediment relative to neighbouring areas, and its relative rather than absolute thickness.

The surface of the Earth can be divided into three broad categories: areas of erosion, areas of sediment accumulation, and neutral areas in which neither erosion nor accumulation is dominant. Active sedimentary basins are areas of accumulation. They may be large or small, deep (thick) or shallow (thin). They may persist for a significant span of geological time measured usually in tens of millions of years, and they may persist in one place, or migrate to some extent. The age of the sedimentary basin is the age of the sediments that accumulated in it. Sedimentary basins are of infinite variety, and so individually unique; but they share some fundamental attributes.

The concept of a sedimentary basin is distinct from that of a physiographic basin. A physiographic basin is a depression in the surface of the land or sea floor that may or may not fill with sediment. Part of the area of a physiographic basin is an area of erosion, providing the materials that will form sediment in other areas of the physiographic basin. In those areas in which sediment accumulates, the upper surface of the sediment does not *necessarily* form a depression everywhere: it may be physiographically indistinguishable from the neighbouring areas that are not accumulating sediment.

For example, there is a large physiographic basin occupying the mid-continent region of the United States of America, drained largely by the Mississippi and Missouri rivers into the Gulf of Mexico (Fig. 1-1). The physiographic basin occupies a significant area of North America and includes the Gulf of Mexico. Within this area, sediment is derived from the peripheral areas, mainly the Rocky Mountains and the Appalachians, and transported towards the sea. Some of this sediment accumulates on and in the flood plains of the rivers, some accumulates along the Gulf Coast, and some is carried out into the Gulf where it accumulates. One of the important sedimentary basins of the world lies under the general area of the coast; and here sediments have accumulated to a far greater thickness than the contemporary sediments in the deeper parts of the Gulf of Mexico. The surface of this sedimentary basin does not form a depression, and the area of obvious depression in the Gulf of Mexico is not the main area of sediment accumulation.

The nature of the sediments in the physiographic basin is determined by the geology of the peripheral areas of weathering and erosion, and by the

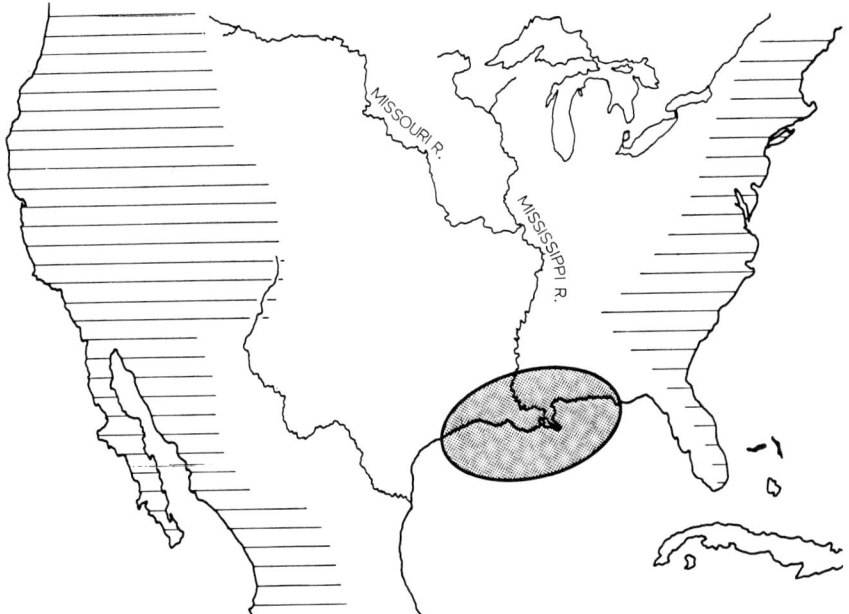

Fig. 1-1. Sediment transported in the central physiographic basin of North America accumulated mainly in the coastal area, not the deep part of the Gulf of Mexico.

physiography and climate of the basin. The nature of the sediments that accumulate depends on these factors, the processes of transportation within the basin, and the position of the sedimentary basin within the physiographic basin. If the sedimentary basin is in the coastal region, paralic sediments will accumulate; if in the coastal plain, fluvial sediments will accumulate; if offshore, marine sediments will accumulate.

Sedimentation is one thing: sediment accumulation is another. The mouth of a river may well be the site of heavy sedimentation; but if the energy of the environment in which the sediment is deposited (the energy of the waves and currents) is sufficient, the sediment will be transported elsewhere and the *accumulation* of sediment near the river mouth may be nil or very little. The redistribution of sediment depends on its physical properties — the density, size and shape of the particles — and the energy available to move it. Sediment will be moved along the sea floor under the influence of waves and currents until it arrives in a position where the available energy is insufficient to move it further. There it will accumulate.

But if the accumulation of sediment raises the depositional surface to a level of higher energy, where the waves and currents are stronger in the shallower water, it will only accumulate to the level at which the energy is sufficient to move the sediment elsewhere. This is the concept of *baselevel* in the marine environment that was developed by Barrell (1917) in one of the more important papers on sedimentary geology.

Baselevel is the level at which sediment neither accumulates nor is eroded: it is also the conceptual level at which this equilibrium would be achieved. The energy of the environment is one side of the equation, as it were, while the physical properties of the sediment are the other. Baselevel is used in the singular, and must be thought of with reference to a particular, perfectly sorted material. This is a device to simplify the more complex reality, which is that there is a baselevel for each grade of sediment, each grain of sediment. A poorly sorted sediment in a given position has a range of baselevels, and sediment accumulates or passes on according to whether the baselevel of each grain lies above or below the depositional surface.

Baselevel may be viewed another way. In each position on the sea floor there is a grade that just (but only just) cannot be moved by the available energy, and the baselevel for that grade coincides with the sea floor in that position. Finer grades cannot accumulate there.

It is this general process that leads to the accumulation of sand in one area and mud in another. It is a process that has been operating since the material became incorporated in the stream load (but baselevel is a more difficult and less useful concept outside the marine environment). The sediment that accumulates in a place is that fraction of the total sediment in that place that cannot be transported further. Therefore, one must not think of a rock unit as a *representative* sample of the sediment brought to the place.

Baselevel fluctuates with changing tides, seasons and weather. On the continental shelves, fluctuations of baselevel will accompany fluctuations of current intensity, and wave and swell *lengths* (because the energy of waves decreases exponentially with depth as a function of wave length rather than of wave amplitude). It seems reasonable to suppose that few areas on the continental shelves are permanently below the lowest baselevel induced by exceptional circumstances once in a hundred or thousand years. If the shape of the continental shelf and the level of the sea relative to it were to remain constant, and the mean storm intensity, current strength, and the variations about the mean, were also to remain constant, there would be a tendency for sediment to accumulate evenly and thinly over the shelf with the bulk of the total sediment eventually passing off the shelf into deeper water. The increasing depth of water away from the land, and hence the decreasing mean or maximum energy at the depositional surface, would result in a general sorting of clastic material from coarse to fine seawards, dense to less dense, spherical to angular, from the coast towards the margin of the continental shelf. The geological record does not show more than a general tendency for this to happen, and it does show significant thicknesses of accumulated sediment in some areas, not in others. There is another influence: *subsidence* of the depositional surfaces relative to baselevel.

ACCUMULATION OF SEDIMENT

The essence of Barrell's concept of sediment accumulation is embodied in his diagram, redrawn here as Fig. 1-2 (Barrell, 1917, p. 796, fig. 5). It depicts the consequences of a generally rising baselevel that includes two minor orders of fluctuation. The diagram is valid for a rising baselevel or a subsiding sedimentary column: it is the relative change that is important. Sediment accumulates during periods of rising baselevel, and is removed during periods of falling baselevel*. The process is both additive and subtractive, and only the sediment that is left permanently below the fluctuations of baselevel accumulates permanently. *Diastems* result from the smaller oscillations: disconformities from the larger. The actual sediment in a sequence may represent only a small fraction of the total time that elapsed for the accumulation of the sequence because the sequence on the left of the diagram accumulated during the black intervals along the time scale at the top.

Subsidence of a depositional surface, which we must now understand as a surface over which sediment is or was being transported, provides a potential to accumulate sediment because there is a tendency for part of the sedi-

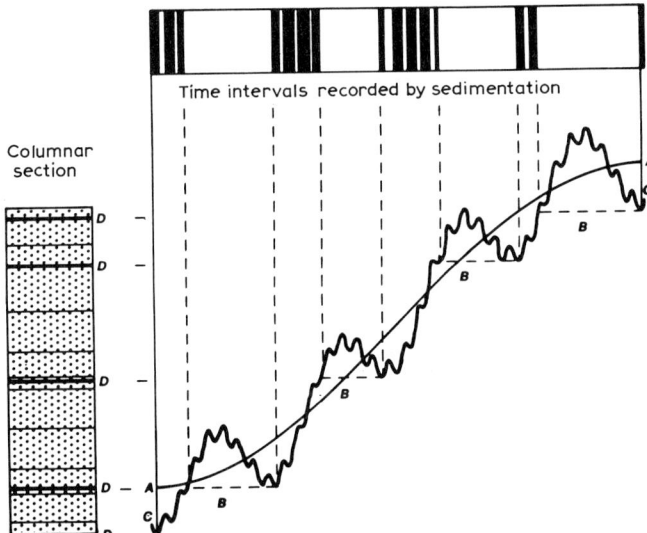

Fig. 1-2. Sedimentary record resulting from oscillations of baselevel. AA is the primary curve of rising baselevel; B are second-order oscillations giving rise to disconformities; $C-C$ contains third-order oscillations giving rise to diastems. (Redrawn from Barrell, 1917, p. 796, fig. 5.)

---

* A single, perfectly sorted material is implied in the diagram as a simplification.

ment to be taken permanently beyond the reach of fluctuations of baselevel. The more rapid the subsidence, the larger the proportion of sediment accumulated permanently (up to the limit, of course, of total accumulation of the sediment supplied).

Sediment accumulation may therefore be viewed as the difference between the supply of sediment and the capacity of the environment to remove it. Sedimentary basins are areas that, over a span of time, have had a supply of sediment and a capacity to retain all or part of it.

It is clearly erroneous to consider sediment accumulation generally as resulting from sediment falling directly from suspension. Sediment accumulates in this way only when the depositional surface lies below baselevel, and then the sediment usually is of very fine grade. In general, sediment is redistributed on the sea floor of the continental shelf, and this generalization applies to muds and other fine-grained sediment because few areas of the continental shelves are below the baselevel of muds, which lies deeper than those for coarser grades.

The work of Barrell forms the basis for an understanding of sedimentary basins. The concept of fluctuations of baselevel, leading to discontinuous sediment accumulation, reconciles the long-standing observation that the maximum net rate of sediment accumulation over major intervals of geological time (Eras, Periods) is very much slower than that suggested by the sediments themselves. The maximum net rate of accumulation, obtained by dividing the period of time in years by the maximum known thickness of sediment accumulated during that time, is given for the Phanerozoic in Table 1-1. In contrast with the rates in this table, Holocene sediment of the U.S. Gulf Coast accumulated at about 1 m in 500 years, that of the Orinoco delta at Pedernales accumulated at about 1 m in 100 years (Kidwell and Hunt, 1958) and, as Barrell pointed out, the preservation of tree trunks in the geological record of many parts of the world, in rocks of various ages,

TABLE 1-1

Phanerozoic rates of sediment accumulation (based on data of Holmes, 1960)

Phanerozoic:	1 m per 4400 yrs
	1 ft per 1300 yrs
Palaeozoic:	1 m per 5700 yrs
	1 ft per 1700 yrs
Mesozoic:	1 m per 4100 yrs
	1 ft per 1200 yrs
Cenozoic:	1 m per 2100 yrs
	1 ft per 600 yrs

The apparent increase in accumulation rates with time must be interpreted with caution. It may be real, but there has been less opportunity for younger rocks to be destroyed and more opportunity for them to be measured.

indicates that they were buried before the wood rotted, implying a rapid rate of sediment accumulation to be measured in tens of years per metre rather than thousands.

Barrell's concept also requires us to exercise extreme caution when interpreting the geological record in terms of present-day environments. Only a small proportion of the sediment at present in transport along the sea floor will normally accumulate near where it is at present, and not all of that that accumulates will accumulate for a significant time-span geologically. Moreover, the post-Pleistocene has been a time of generally rising sealevel and rising baselevel, so it has been a period of active, but not necessarily permanent, sediment accumulation at an average rate of about 1 m in 1000 years perhaps (if we accept the figure Gutenberg, 1941, p. 729, derived from tide gauge records as being representative of the Holocene).

In general, it is erroneous to consider rock units in the stratigraphic record as analogous to sand banks, shoals, bars, tidal flats, etc., of the seafloor today — although there are clear exceptions to be found in thin units with ripple marks and animal tracks and burrows. It is generally much more accurate to regard rock units as the incomplete record of the passage of sediment of different compositions with bedding planes representing diastems. A clean sand did not necessarily accumulate from a clean sandy sea floor because the mud with it might have been winnowed out, to accumulate elsewhere.

The migration of a large sand bank over the sea floor, for example, may be recorded in the stratigraphic sequence by a thin, laterally discontinuous sand unit that consists only of that portion that came to be permanently below baselevel. The dune shape will not be apparent, and there may even be no trace of current bedding. Likewise, on a smaller scale, the preservation of ripple marks requires that a surface that was close above the baselevel of that material on one tide be buried under a protective layer of sediment by the next tide, and for baselevel to be elevated permanently above the surface of the ripple marks. Only in areas of extremely rapid subsidence are ripple-marked surfaces likely to be common, and it is hard to escape the feeling that they are a common feature preserved by uncommon events — that the ordinary ripple-marked surface in shallow water and between the tide lines is unlikely to be preserved. Worm burrows are sometimes interpreted as evidence of slow sediment accumulation, but consideration of baselevel suggests that it is more likely to have been very rapid.

While the concept of a physiographic basin is quite distinct from that of a sedimentary basin, a sedimentary basin is necessarily situated within a physiographic basin because it is the latter that is the dominant influence on sediment supply. The combined concept is dynamic: changes in the physiographic basin with time affect the type and character of the sediment that accumulates in the sedimentary basin.

Sediment supply is often apparently in very close balance with subsidence, because we find considerable thicknesses of sediment of a particular environment. The balance is more apparent than real, and true balance is probably rare. It is commonly said that the rate of subsidence was equal to the rate of sediment accumulation, and this leads to the idea of balance. More correctly, and more usefully, one should think of the rate of accumulation being equal to the rate of subsidence. The accumulation of a considerable thickness of sediment of the same facies merely reflects the constancy of the physiographic environments over the area of the sedimentary basin, the surplus sediment being removed.

The primary control on sediment accumulation is subsidence (assuming sediment supply) because sediment accumulation without subsidence is vulnerable to subsequent dispersal. The primary controls on the nature of the sediment that accumulates are the sediment sources and the environments, and energy of the environments, of the physiographic basin on the sedimentary basin. The matter of lateral continuity of a particular rock type involves considerations of the three dimensions of space and that of time. A rock unit may be discontinuous over an area for a number of reasons:

— The sediment was not distributed over the whole area.
— The sediment was distributed but did not accumulate over the whole area.
— The sediment was distributed and accumulated over the whole area, but accumulation was only temporary over parts of the area due to changing energy patterns.

One conclusion is clear: the margin of a discontinuous rock unit in the sedimentary record is not necessarily the margin of the environment it represents in the physiographic basin. The sequences of rock types and their areal distribution in a sedimentary basin constitute a variable and very incomplete record in space of the variations of the environments in the physiographic basin over that area with time.

TRANSGRESSIONS AND REGRESSIONS

When subsidence relative to baselevel in a physiographic basin exceeds the supply of sediment, that is, the area has a capacity to accumulate a larger volume of sediment than is supplied, the sea tends to deepen over the depositional surface, and facies tend to migrate towards the land (Fig. 1-3). This is a *transgressive* phase of sedimentary basin development, leading to the accumulation of a transgressive sequence of sediments. Where this process raises baselevel above the depositional surface, pelagic sediments may accumulate (and may accumulate to a considerable thickness). Sedimentary basins tend to be enlarged during transgressive phases, but the enlargement is only permanent if the subsidence relative to baselevel is permanent.

If the sea over part of a physiographic basin becomes shallower, and the

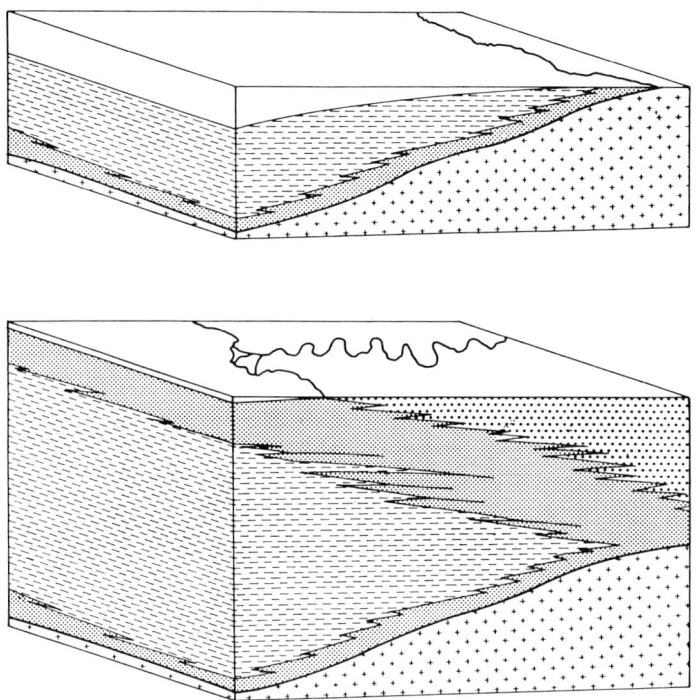

Fig. 1-3. Block diagram showing accumulation of transgressive sequence (above) and transgressive sequence followed by regressive sequence (below). Neritic facies, dashes; neritic-paralic, fine dots; paralic-terrestrial, coarse dots.

facies migrate seawards, the development is *regressive*. Let us clarify the terms and concepts involved in transgressions and regressions because they are vital to an understanding of geology in general and petroleum geology in particular. There are advantages in taking regressions first.

"Regression" is *defined* as a lowering of sea level relative to the land. When this happens, the shoreline and the associated environments and facies migrate in a seaward direction, with an extension of the land area. This is the basic concept. A second concept is derived from the first in the following manner. If the mass of sediment supplied to an area requires more energy for its dispersal than is available, baselevel rises and sediment accumulates commensurately. The accumulating sediment tends to extend the land area, and the facies also tend to migrate seawards, or *prograde*. There is thus a subtle difference between the two aspects of regression, but a difference of considerable importance geologically. Both can result stratigraphically in a sequence that shows sediments of shallower-water facies overlying those of deeper-water facies, or terrestrial over marine, *provided sediment accumulates*. The important difference between the two types of regression is that the relative lowering of sea level results in a *lowering of baselevel* while regression of the second type results from a *rising baselevel* — more sediment being supplied than can be removed.

Regressions of the first type are important for the erosion and redistribution of sediment that had previously accumulated in the sedimentary basin. This will accumulate elsewhere, and there will be a hiatus where the erosion took place. Regressions of the second type are important for the accumulation of major regressive sequences that persisted for much of the life of a sedimentary basin.

Deltas are examples of the second type of regression (which we regard as the normal type). They result from the volume of sediment supplied by the river exceeding the volume that can be removed by the sea, with a consequent shallowing of the sea and extension of the land area. Delta growth is not uniform. There may be strong seasonal influences, with periods of high river level and massive sediment transport. Distributary channels become unstable as they build up their levees, and change their courses from time to time, so changing the shape of the delta. The building of a delta is episodic, but its very existence indicates that more sediment is supplied to it than can be removed, and the net effect is regressive. The sequence that accumulates as the delta progrades has paralic sediments overlying the deeper-water, prodelta sediments of finer grade.

"Transgression" has two aspects analogous to those of regressions. The basic concept is that of a rising sea level relative to the land, with a consequent migration of the strand line with its associated environments and facies towards the land, and a reduction of land area. The derived concept is this: when the energy of the environment is greater than that required to remove the sediment supplied to the area, the surplus energy may erode and redistribute sediment that accumulated previously (e.g. coastal erosion). This also leads to the migration of environments and their facies landwards. Both types of transgression may result in accumulation of sediment in a stratigraphic sequence in which deeper-water facies overlie shallower-water facies (or marine overlie terrestrial). Transgressions of the first type result in rising baselevel and so a potential to accumulate sediment. Transgressions of the second type result in a lowering of baselevel and the erosion and redistribution of sediment that will accumulate again elsewhere. (Note in passing that a break in the stratigraphic sequence is not necessarily regressive.)

Petroleum geology is concerned with the accumulation of sediment as well as the accumulation of petroleum. We are therefore concerned with transgressions and regressions that tend to accumulate thick sedimentary sequences, rather than those that tend to erode and redistribute sediment, in sedimentary basins. Sedimentary basins typically begin with a transgressive sequence and end with a regressive sequence. Transgressions tend to accumulate fine-grained, porous but relatively impermeable rocks (potential petroleum source rocks) on top of coarser-grained, porous and permeable rocks (potential petroleum reservoir rocks): regressions tend to accumulate potential reservoir rocks on top of potential source rocks. We shall refine these generalizations later.

Dominant trends must be distinguished from episodic reversals, for there are usually transgressive episodes in dominantly regressive phases of sedimentary basin development, and vice versa. Consider for a moment such a transgressive episode. It may result from a change in the energy patterns, so that sands accumulate in areas where muds were accumulating, and muds accumulate where sand was accumulating. Such an episode, due perhaps to changing courses of distributary channels in a large delta, may be quite local. Changes on a larger scale could result from a changing geography along a coastline, perhaps with deltas changing their positions, or rivers finding new courses to the sea. Temporary climatic changes could result in a reduction of sediment supply, and episodic accelerations of subsidence could also lead to transgressive episodes. There could also be sea-level changes.

EUSTATIC* SEA-LEVEL CHANGES

There is ample evidence, long recognized, from raised beaches and terraces, and submerged forests, that sealevel has not remained constant relative to the land during the Cenozoic, so the Principle of Uniformitarianism requires us to suppose that changes have taken place throughout the history of the oceans and continents. The evidence of marine borings in the columns of a Roman market in Puzzuoli on the Bay of Naples, noted by Lyell (1875, v. 2, p. 164, and illustrated as the Frontispiece to Vol. 1), indicates that the time scale for relative sea-level changes of some metres can be quite short, and tide-gauge records show significant trends and departures from trends over a few decades (Gutenberg, 1941). We must, however, seek to distinguish between local and regional changes of sealevel, and the world-wide changes that are called eustatic. This is a typical geological problem with many variables and no absolute datum. It is an important problem for petroleum

---

* *Eustatic* was coined by Suess (1888, Vol. 2, p. 680): "Um nun Vorgänge dieser Art näher zu verfolgen, trennen wir von den verschiedenartigen Veränderungen, welchen die Höhe des Strandes unterworfen ist, solche ab, welche annähernd in gleicher Höhe, in positivem oder in negativem Sinne über die ganze Erde sich äussern, und bezeichnen diese Gruppe von Bewegungen als eustatische Bewegungen". In the Sollas' translation (Suess, 1906, Vol. 2, p. 538): ". . . we must commence by separating from the various other changes which affect the level of the strand, those which take place at an approximately equal height, whether in a positive or negative direction, over the whole globe; this group we will distinguish as *eustatic movements.*" (Sollas' italics).

Subsequent definitions usually include a cause, which is undesirable. When a cause can be identified, it can be indicated. There are almost certainly several causes of eustatic changes of sea level.

geology because rising relative sea level can lead to increased rates of sediment accumulation and, perhaps, increases in the biomass of the seas over sedimentary basins.

It is important to realise that the definition of present-day sealevel is far from simple, and that the mean ocean surface (part of the geoid) departs significantly from the surface of an ellipsoid. Figure 1-4 shows the departure of the geoid from its best-fitting ellipsoid, the contours being the elevation of the geoid in metres above or below (−) the ellipsoid. Two features of this map are particularly important: first, there is no obvious relationship between the shape of the geoid and the surface features of the Earth; and secondly, migration of the geoid relative to the surface features of the Earth could give rise to sea-level changes of several tens of metres in *both* senses, rises and falls, without changes in the volume of ocean water.

The main causes of sea-level changes, confining ourselves to those that operate on a time and space scale that can be of significance to the stratigraphic record, are: (1) changes in the volume of sea water; (2) changes in the volume and shape of the ocean basins; (3) changes in the axis of rotation of the Earth; and (4) migrations and changes of shape of the geoid.

The first two lead to eustatic changes of sea level and present no great conceptual difficulties. The retention of water on land during ice ages has long been recognized as a cause of eustatic falls of sea level, with a rise when the ice melts. The last two do not lead to eustatic changes according the Suess' definition because the changes will not be in the same sense over the world, let alone the same height. For these to be important processes, the rate of polar wandering and of geoid migration must be such that adjustment of the crust lags behind the instantaneous changes in the surface of the seas. The last cause listed, changes in the shape of the geoid relative to the geographical features, has only been recognized in the last decade as a possible cause of sealevel changes, largely due to the work of Mörner (e.g. 1976) on geoid maps determined from artificial satellite and surface gravity measurements. All four processes may operate at the same time, and the local changes of relative sea level are the net result of the various processes operating and any local changes due to relative subsidence or uplift.

Despite the difficulties of detail, it is axiomatic that a world-wide change of sealevel, whether eustatic or not, will leave its record in all sedimentary basins that were active and developing at that time. This axiom has its corollary: those periods of geological history that show a word-wide tendency for transgression or regression are likely to be periods of eustatic sea-level changes.

It has been known for about a century (at least since 1888 when Suess discussed "the Cenomanian transgression") that the Cretaceous was a period of world-wide transgressive tendency. It is now known that the transgressions were not strictly synchronous around the world. They were earlier in Australia, South Africa and South America; later in north-west Europe and interior

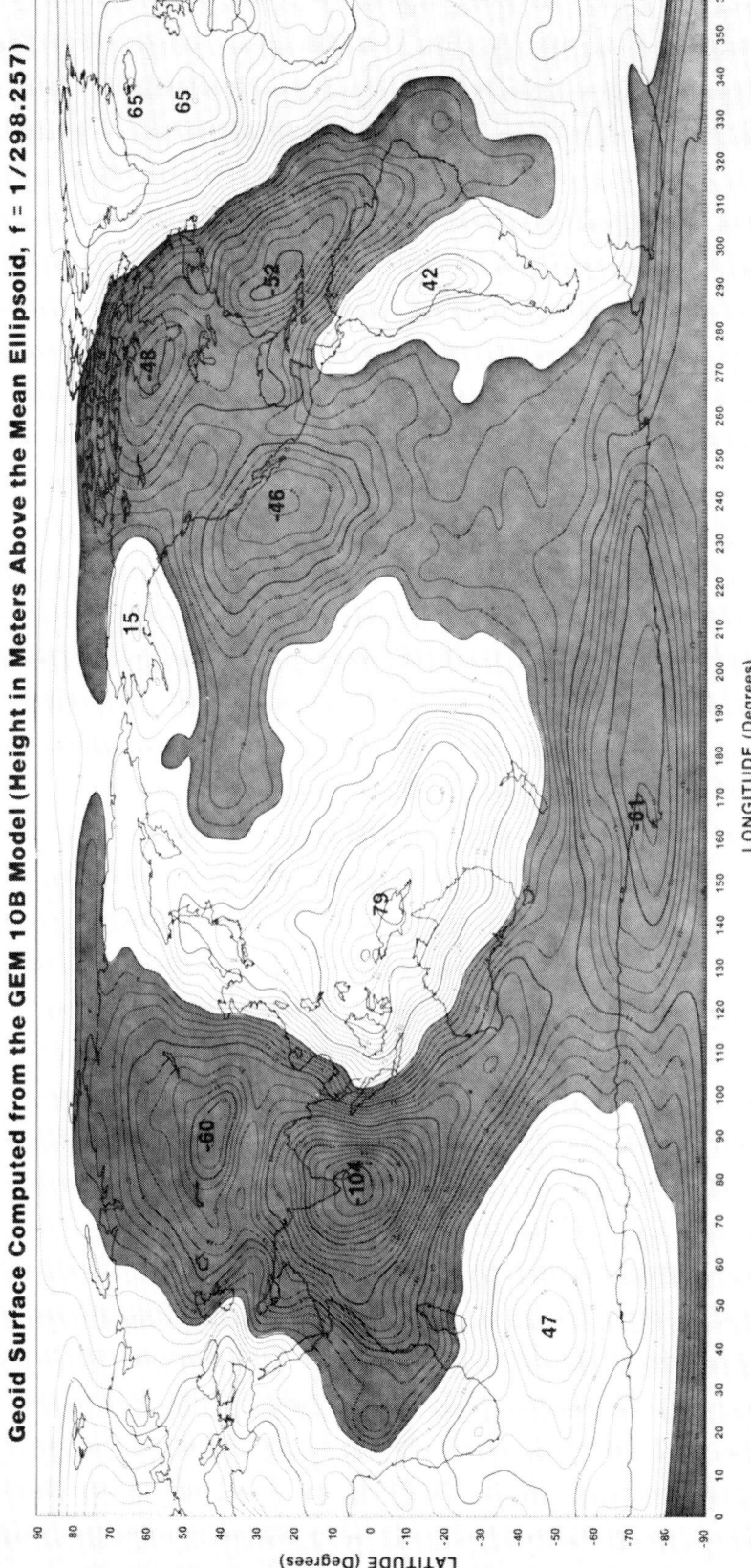

Fig. 1-4. Geoid surface map. Contour interval: 5 m. (Courtesy NASA/Goddard Space Flight Center.)

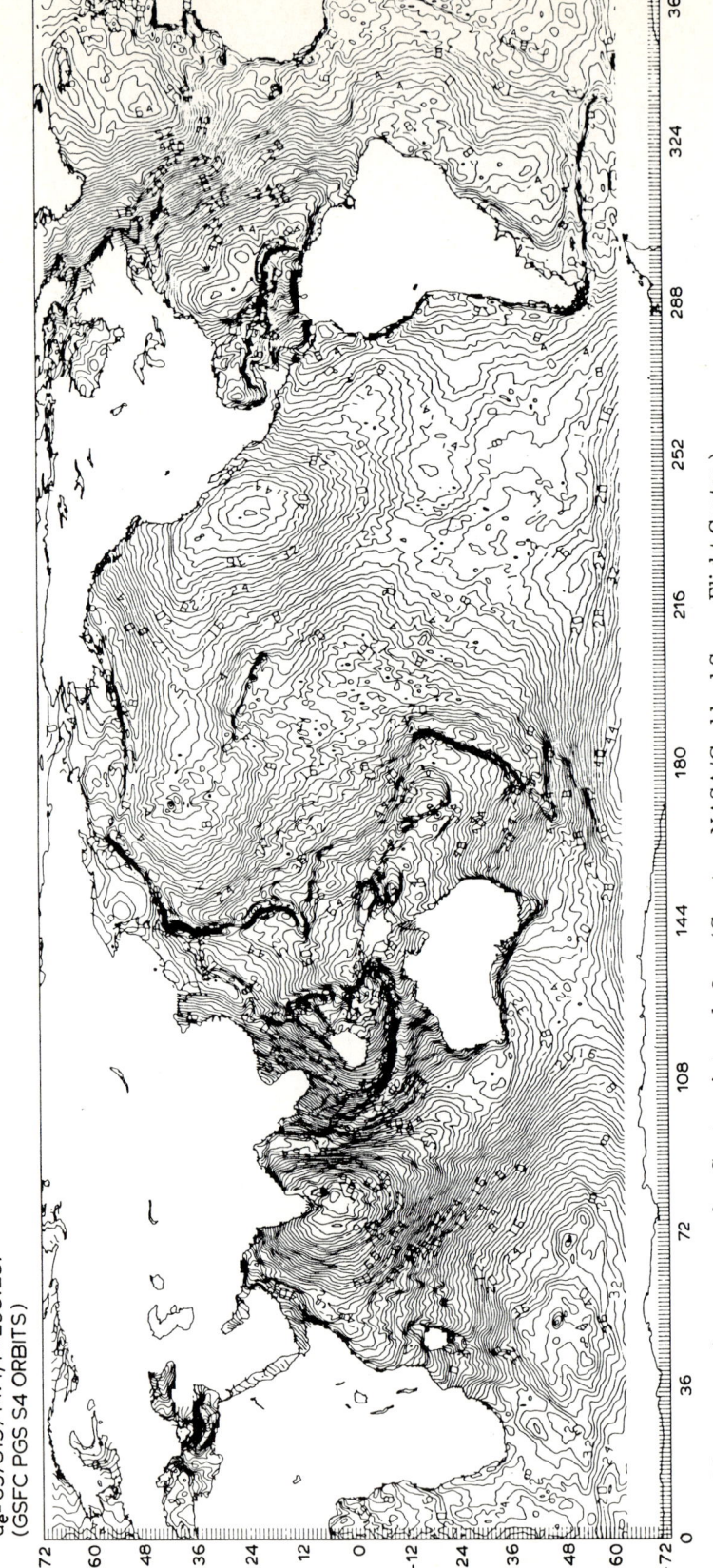

Fig. 1-5. Mean sea-surface topography. Contour interval: 2 m. (Courtesy NASA/Goddard Space Flight Center.)

North America (see Matsumoto, 1980). These events were of great significance for the accumulation of petroleum, and we shall examine them in more detail in a later chapter. We shall see that these transgressions were but an episode in a world-wide Mesozoic tectonic event that may well have involved all four causes listed above.

The fact that transgressions seem to have a world-wide tendency more commonly than regressions is probably due to the amplifying effect of a rising sea level on subsidence of a sedimentary basin compared with falling sealevel on subsidence. Regional variations in subsidence rates may account for the lack of strict contemporaneity because we observe only the net effect of subsidence and sediment accumulation. It is clear that orogeny may generate enough sediment to lead to a regressive sequence in spite of a contemporary rise of sealevel, and so mask it; but it is unlikely that subsidence will so exceed the rate of sea-level fall that a transgressive sequence of any magnitude will accumulate during a period of eustatic fall of sea level. For example, the Catskill delta of the northern Appalachians developed during the middle and late Devonian, as a result of the Acadian orogeny, while transgression was taking place in western Canada, western Australia and north-west Europe.

Vail et al., in a series of papers (see, for example, Vail et al., 1977), concluded from their studies of the seismic stratigraphy of many areas of the world that numerous synchronous, and so global events are recorded in sedimentary basins around the world. These they interpret as eustatic events.

Mörner (1976) believes that true eustatic changes of sea level are unlikely because the geoid is probably unstable even on a short timescale. He pointed out that there is a 180 m difference in geoid level over $50°-60°$ of longitude between the Maldive Islands (-104 m) and New Guinea ($+76$ m on the Smithsonian Standard Earth III geoid map that he used) (Fig. 1-5). An easterly drift of the geoid without crustal adjustment would lead to a fall of sea level of about 180 m around New Guinea (which would reunite it with Australia) and a rise around the Maldives of about 80 m (which would inundate them). Geoid relief may have been greater in the past, but the causes of this relief are not yet well understood. There is some similarity between the geoid surface and the non-dipole magnetic field (see Hide and Roberts, 1961, fig. 2.3b).

Petroleum exploration around the world will doubtless clarify these problems in time as our detailed knowledge and understanding of sedimentary basins is extended.

LITHOLOGICAL ASSOCIATIONS IN SEDIMENTARY BASINS

If one considers the sedimentary rocks broadly as mudstone/shales, sands/sandstones, and carbonates/evaporites, sedimentary basins tend to accumulate either mudstone/shale and sand/sandstone or mudstone/shale and carbonate/evaporite as the dominant association of lithologies. These associations reflect the tendency of physiographic basins to generate, transport and deposit sediments of a similar character over great spans of time. But they do change, and some important sedimentary basins record such changes in associations (for example, the Maracaibo basin in Venezuela, and the East Borneo basin in Indonesia).

More significantly, two associations are evident in transgressive and regressive sequences. There is an association between transgressive sequences and carbonates/evaporites, and between regressive sequences and sands/sandstones. Not all transgressive sequences contain carbonates, but no major regressive sequence contains significant carbonates*.

The upper Devonian organic reefs of the Western Canada basin are well known because many of them are important petroleum reservoirs. The results of much competent research have been published (see, for example, Barss et al., 1970, Hemphill et al., 1970, for accounts of two areas). There is no doubt that the reefs grew during a dominantly transgressive phase of the development of the physiographic basin. In general, the more southerly their position, the younger their age (Fig. 1-6); and many are overlain by mudstones, calcareous mudstones and marls of a deeper-water facies than that implied by the reefs themselves. Are organic reefs always transgressive? What is a transgressive reef?

The matter of transgressive and regressive reefs is complicated and not to be related solely to changes of sea level relative to the land. The reason for the complication lies in the nature of an organic reef. Lowenstam's widely accepted definition of a reef requires that it contain organisms that were frame-builders, the organisms growing to create a wave-resistant structure; and that the organisms were important for retaining sediment and binding it (Lowenstam, 1950). There are thus two important parameters: the biological potential to build and the environmental potential to kill or destroy. These determine the form of the reef, whether biohermal or biostromal, whether isolated as patch or pinnacle reefs, or associated with a back-reef facies that may include evaporites.

A reef is a facies that depends on an environment of a physiographic basin. So the simplest concept is that a reef that migrates landward is transgressive, and a reef that migrates seaward is regressive. Another simple concept, but more important from a petroleum point of view, is that when the true thickness of a reef exceeds the presumed depth tolerance of one or

---

* The thick and extensive carbonate sequences of Iran and Iraq are not clearly transgressive, according to geologists with experience of these areas with whom I have discussed this point; nor are they clearly regressive. Perhaps this is a case of close balance between subsidence and carbonate accumulation without terrestrial clastic supply.

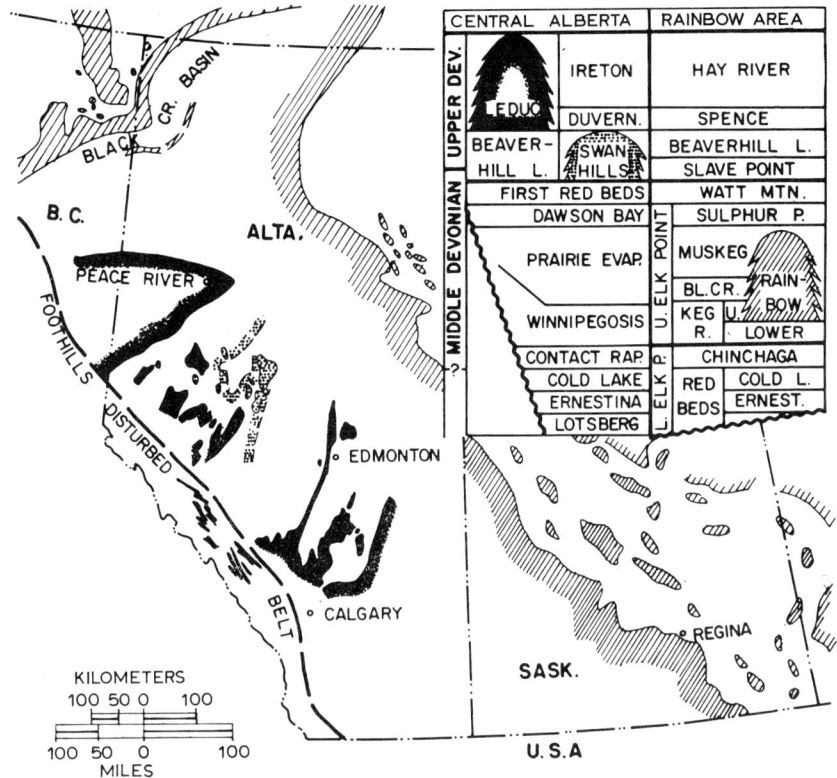

Fig. 1-6. Devonian reefs of western Canada. (Reproduced with permission from Barss et al., 1970.)

more of the reef-building organisms, such a reef is transgressive by inference even though the facies do not migrate.

Modern colonial corals, for example, are sensitive to light, water temperature and salinity; and their depth tolerance is to about 30—45 m (100—150 ft). It cannot be assumed that the same tolerances applied to ancient reef organisms, and as yet there are no reliable means of determining them. It is reasonable, however, to assume that their tendency to form wave-resistant structures, to migrate, and to be exterminated by muddy sediment or hypersalinity, all indicate a rather restricted range of favourable environmental conditions in which light and salinity played an important part.

Ingels (1963) estimated the water depth around the Silurian Thornton reef of north-east Illinois to be about 60 m (200 ft), and Terry and Williams (1969) suggested a similar depth around the Paleocene Intisar "A" bioherm in Libya (this bioherm was known as "Idris" in the literature from discovery in 1967 until 1969, "Intisar" thereafter). But depth of water around a reef is not a reliable indicator of depth of tolerance of the organisms. What is required is the maximum depth at which reef growth can start, for this establishes the maximum thickness of reef that can develop with constant water depth. Whatever the true figure may be for past Periods, it is unlikely to exceed 100 m (300 ft), and may well be half that figure.

This concept places many of the Silurian and Devonian organic reefs of North America (many of them petroleum reservoirs), the Cretaceous reef reservoirs of Mexico, some of the Mesozoic and Tertiary reef reservoirs of the Middle East, and the Paleocene reef reservoirs of Libya, into the transgressive category.

The association of evaporites with carbonates is an important association for petroleum because evaporites can provide seals to the carbonate petroleum reservoirs, as they do in the Middle East, northern U.S.A., some areas of western Canada, and the Permian basin of West Texas and New Mexico. In all these areas evaporites tended to cover large areas eventually. In western Canada, evaporites formed behind the barrier reef that grew in middle and late Devonian times across Alberta into the Northwest Territories (Fig. 1-5), apparently as a direct result of the enclosing of the seas by reefs. We note that evaporites are commonly intercalated with, and eventually terminate the development of extensive carbonate sequences, and that such sequences are significantly associated with petroleum occurrences.

As regards the association between regressive sequences and sands/sandstones, little need be said here. This is due almost certainly not so much to lack of carbonate, but its relegation to insignificance by the quantity of terrigenous clastic material.

These two associations, carbonates and transgressions, and regressions and sands/sandstones, are sufficiently distinct to suggest that the rate of sediment supply is a material factor that determines whether a sedimentary basin preserves the record of a transgressive or a regressive phase in the development of the physiographic basin. Massive sediment supply, such as is observable now on the U.S. Gulf Coast, the Mackenzie Delta in the Arctic, the Niger Delta, the large deltas of the Indian subcontinent, South-East Asia and China, is the principal feature of the present development of these physiographic basins, and so similar features are inferred for the regressive sequences recorded in sedimentary basins.

CLASSIFICATION OF SEDIMENTARY BASINS

Sedimentary basins are of great variety, each being individually unique. Their ages vary, their time spans vary, and the type and distribution of the sediments they accumulated vary. A complete classification would require two separate classifications — one of the physiographic basins, the other of the geometry of sedimentary basins — because, as we have seen, the nature of the sediments that accumulate in a sedimentary basin depends on the position of the sedimentary basin relative to the environments of the physiographic basin. Several classifications have been proposed (some are listed in the references at the end of the chapter) but few are satisfactory because the concepts of physiographic and sedimentary basins are confused. Classifications of continuous variables are always arbitrary and artificial to some extent. It will be sufficient to make but a simple classification here.

Sedimentary-basin geometry

The geometry of sedimentary basins in section, in terms of isochronous surfaces, may be classified as either symmetrical or asymmetrical (Fig. 1-7). These may be further subdivided on the basis of the nature of their margins, that is, faulted or unfaulted (but it must be realized that the nature of the margin of a sedimentary basin is not always determinable because subsequent geological events may have obscured it or it may be beyond the reach of investigation). The asymmetrical basin has received much attention in the literature, and was called a "half-graben" by Weeks (1952) when the asymmetry is due to a fault or fault system.

The term "structural basin" is sometimes used for basins bounded on one or both sides by faults, but no clear meaning has become attached to the term. It is sometimes used for a basin of simple, but large, synclinal form.

It must be emphasized that the symmetry of sedimentary basins in this context relates to the geometrical symmetry of isochronous surfaces, not to the symmetry of facies that accumulated in the basin. It is dangerously fallacious to suppose that deeper-water sediments belong to the deeper parts of the sedimentary basin, and shallower-water sediments belong to the edges.

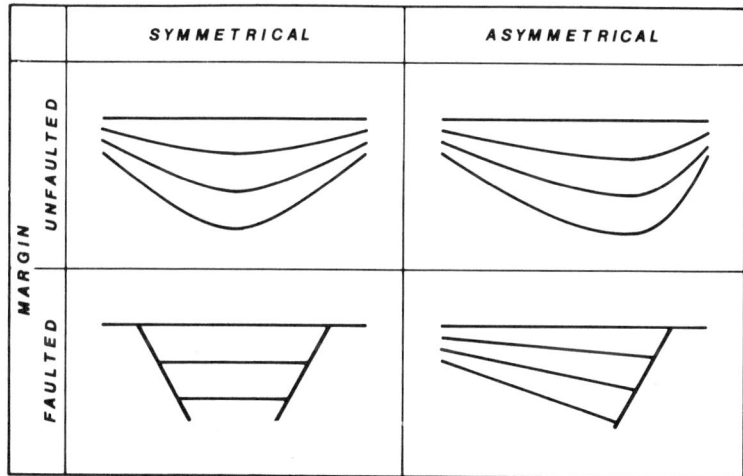

Fig. 1-7. Simple classification of sedimentary basins on basis of geometry of isochronous surfaces.

In plan, sedimentary basins may be circular, oval or almost rectangular (asymmetrical basins usually have one fairly straight side), and they vary in size from a few tens of kilometres to several hundreds in horizontal dimensions. *Geosynclines* may be regarded as particular forms of sedimentary basin of which the length is considerably greater than the width, and in which a great thickness of characteristic sediments and associations of sediments, and volcanic rocks, accumulated (Kay, 1951; Kündig, 1959).

Physiographic basins may be open or closed, silled or unsilled, continental or marine, or a combination of these. The Gulf of Mexico, the northern part of the Bay of Bengal, and the South China Sea, are examples of open basins. The Black Sea, the Baltic and the Mediterranean are examples of closed basins. The area around Lake Eyre in South Australia is an example of a closed continental basin. Lake Eyre is 12 m below sea level, and is evidently subsiding, so it could become a marine basin in time. The environments of physiographic basins are very variable according to the climate (or range of climates) and the general geography and geology of the area. Of particular significance in petroleum geology is the alternation of environments in the course of time between euxinic, or areas of anaerobic, reducing conditions, during the accumulation of fine-grained sediment, and those environments that led to the accumulation of coarser-grained porous and permeable sediments.

The discrete concepts of physiographic and sedimentary basins merge in this respect. Reducing conditions, which are essential for the preservation of organic matter, occur in the aqueous environment near the water/sediment interface, or within the sediment. Hence sediments may either accumulate in a reducing environment of the physiographic basin or be buried to a reducing environment by rapid subsidence.

Generalizing, sedimentary basins usually begin with a dominantly transgressive phase during which the volume created by subsidence exceeds the volume of sediment supplied, and end with a dominantly regressive phase during which the volume of sediment supplied exceeds the volume created by subsidence. Some sedimentary basins record a single, simple dominant *cycle* of sediment accumulation (Fig. 1-3). Others record a transgressive sequence only, and yet others record a long and complicated history with several cycles, doubtless with eustatic changes imposed on the record. The stratigraphic record is nearly everywhere incomplete due to diastems and disconformities, but it records the changes of the physiographic environments on the sedimentary basin with time. An initial carbonate transgression indicates that the physiographic basin had no significant terrestrial topographic relief, whereas regressions, invariably arenaceous, indicate the presence of mountains from which the sediment was derived. Sedimentary basins are commonly intimately associated with orogeny during their development, particularly the later stages of their development.

Sedimentary basins therefore record not only the nature of the physiographic basin where they occur, but also the nature of the margins of the physiographic basin, and the changes of these with time. Stratigraphy and structure, on both small and large scales, are intimately related. Transgressive sequences are typically either not deformed, or deformed only by faulting. Regressive sequences are typically both folded and faulted. The nature of these faults and folds is central to our understanding of petroleum geology, sedimentary basins, and, indeed, geology in general.

REFERENCES

Ager, D.V., 1981. *The nature of the stratigraphic record* (2nd ed.). Macmillan, London, 122 pp.
Aubouin, J., 1965. *Geosynclines.* Elsevier, Amsterdam, 335 pp.
Barrell, J., 1917. Rhythms and the measurement of geologic time. *Bull Geol. Soc. Am.*, 28: 745—904.
Barss, D.L., Copland, A.B. and Ritchie, W.D., 1970. Geology of Middle Devonian reefs, Rainbow area, Alberta, Canada. In: M.T. Halbouty (Editor), Geology of giant petroleum fields. *Mem. Am. Ass. Petrol. Geol.*, 14: 19—49.
Cheney, R.E. and Marsh, J.G., 1981. Oceanic eddy variability measured by GEOS 3 altimeter crossover differences. *Eos, Trans. Am. Geophys. Unions*, 62: 743—752.
Eicher, D.L., 1976. *Geologic time* (2nd ed.). Prentice-Hall, Englewood Cliffs, N.J., 150 pp.
Glaessner, M.F. and Teichert, C., 1947. Geosynclines: a fundamental concept in geology. *Am. J. Sci.*, 245: 465—482; 571—591.
Gretener, P.E., 1967. Significance of the rare event in geology. *Bull. Am. Ass. Petrol. Geol.*, 51: 2197—2206.
Gutenberg, B., 1941. Changes in sea level, postglacial uplift, and mobility of the Earth's interior. *Bull. Geol. Soc. Am.*, 52: 721—772.
Hemphill, C.R., Smith, R.I. and Szabo, F., 1970. Geology of Beaverhill Lake reefs, Swan Hills area, Alberta, In: M.T. Halbouty (Editor), Geology of giant petroleum fields. *Mem. Am. Ass. Petrol. Geol.*, 14: 50—90.
Hide, R. and Roberts, P.H., 1961. The origin of the main geomagnetic field. *Phys. Chem. Earth*, 4: 27—98.
Holmes, A., 1960. A revised geological time-scale. *Trans. Edinburgh Geol. Soc.*, 17 (for 1957—1959): 183—216.
Ingels, J.J.C., 1963. Geometry, paleontology, and petrography of Thornton reef complex, Silurian of northeastern Illinois. *Bull. Am. Ass. Petrol. Geol.*, 47: 405—440.
Jones, O.T., 1938. On the evolution of a geosyncline. *Q.J. Geol. Soc. London*, 94: lx—cx *(Proceedings)*.
Kay, M., 1947. Geosynclinal nomenclature and the craton. *Bull. Am. Ass. Petrol. Geol.*, 31: 1289—1293.
Kay, M., 1951. North American geosynclines. *Mem. Geol. Soc. Am.*, 48: 1—143.
Kidwell, A.L. and Hunt, J.M., 1958. Migration of oil in Recent sediments of Pedernales, Venezuela. In: L.G. Weeks (Editor), *Habitat of oil.* Am. Ass. Petrol. Geol. Tulsa, Okla., pp. 790—817.
Krumbein, W.C. and Sloss, L.L., 1963. *Stratigraphy and sedimentation* (2nd ed.). Freeman, San Francisco, 660 pp.
Kündig, E., 1959. Eugeosynclines as potential oil habitats. *Proc. 5th World Petrol. Con-*

gress, New York, 1959, 1: 461—474.
Lerch, F.J., Putney, B.H., Wagner, C.A. and Klosko, S.M., 1981. Goddard Earth Models for oceanographic applications (GEM 10B and 10C). *Mar. Geodesy*, 5: 145—187.
Lowenstam, H.A., 1950. Niagaran reefs of the Great Lakes area. *J. Geol.*, 58: 430—487.
Lyell, C., 1875. *Principles of geology or the modern changes of the Earth and its inhabitants considered as illustrative of geology* (12th ed.). Murray, London, 2 vols.
Matsumoto, T., 1980. Inter-regional correlation of transgressions and regressions in the Cretaceous Period. *Cretaceous Res.*, 1: 359—373.
Mörner, N.-A., 1976. Eustacy and geoid changes. *J. Geol.*, 84: 123—151.
Sadler, P.M., 1981. Sediment accumulation rates and the completeness of stratigraphic sections. *J. Geol.*, 89: 569—584.
Suess, E., 1885—1909. *Das Antlitz der Erde*. Tempsky, Prague, 3 vols. in 4.
Suess, E., 1904—1924. *The face of the Earth (Das Antlitz der Erde)*. Transl. by H.B.C. Sollas under direction of W.J. Sollas. Clarendon Press, Oxford, 5 vols.
Terry, C.E. and Williams, J.J., 1969. The Idris "A" bioherm and oilfield, Sirte basin, Libya — its commercial development, regional Palaeocene geologic setting and stratigraphy. *In:* P. Hepple (Editor), *The exploration for petroleum in Europe and North Africa*. Institute of Petroleum, London, pp. 31—48.
Vail, P.R., Mitchum, R.M., Todd, R.G., Widmier, J.M., Thompson, S., Sangree, J.B., Bubb, J.W. and Hatlelid, W.G., 1977. Seismic stratigraphy and global changes of sea level. *In:* C.E. Payton (Editor), Seismic stratigraphy — applications to hydrocarbon exploration. *Mem. Am. Ass. Petrol. Geol.*, 26: 49—212.
Weeks, L.G., 1952. Factors of sedimentary basin development that control oil occurrence. *Bull. Am. Ass. Petrol. Geol.*, 36: 2071—2124.
Weeks, L.G., 1958. Habitat of oil and some factors that control it. *In:* L.G. Weeks (Editor), *Habitat of oil*. Am. Ass. Petrol. Geol., Tulsa, Okla., pp. 1—61.

CHAPTER 2

EARLY DEFORMATION OF SEDIMENTARY BASINS: GROWTH STRUCTURES

SUMMARY

(1) Growth structures are folds and faults in which variations of rock-unit thickness are closely related to the structure itself. They result from deformation that took place during sediment accumulation, burial and compaction.

There is an element of interpretation in a short definition of growth structures. The structures that result solely from differential compaction are not strictly growth structures, but some degree of differential compaction exists in growth structures.

(2) Growth faults (there are many synonyms, most common of which are *contemporaneous* and *depositional*) are faults in which the thickness of rock units in the downthrown block is greater than that of the correlative unit in the upthrown block.

(3) Both blocks of a growth fault, in general, had a capacity to accumulate sediment, both were subsiding relative to baselevel; but the downthrowing block subsided faster than the "upthrowing" block, and so had a capacity to accumulate a greater thickness of sediment. The throw of a growth fault tends to increase with depth on account of the thickness contrast across it, but antithetic faults reduce the throw.

(4) The fault plane in transgressive sequences is commonly steep to vertical. In regressive sequences, it is usually curved in plan and in section, concave to the direction of regression and concave up.

(5) Growth faults occur in two major associations: (a) in basins formed by rifting, where basement faulting continued and caused growth faults in the overlying (initially transgressive) sequence. These faults tend to die out upwards, commonly against an unconformity or disconformity; and (b) in regressive sequences, where growth faults occur in the upper, sandy part of the sequence. These die out downwards, and commonly also upwards.

(6) Growth anticlines are anticlines in which rock units thicken from crest to flanks. They grew while the sediment was accumulating and compacting during burial. The whole area of the anticline was subsiding, but the flanks subsided faster than the crest, and so accumulated a greater thickness of sediment. Growth synclines can be formed.

(7) These processes can and do lead to local stratigraphic hiatus, the recognition of which can be extremely difficult — but essential, of course, for the correct interpretation of such areas.

(8) Sedimentary basins are growth structures on a large scale.

INTRODUCTION

A significant geological feature of sedimentary basins revealed first by coal mining, then much more widely by drilling for petroleum, is the common occurrence of faults that have a thicker sedimentary sequence in the downthrown block than the correlative sequence in the upthrown block. Such faults have been interpreted from the earliest days of their recognition as faults that moved while the sediment was accumulating. They are not only evidence of deformation of sedimentary basins early in their development, but also interesting for the light they throw on the processes of sediment accumulation. They are an integral part of sedimentary basin development.

There is similar evidence of contemporaneous folding of strata into anticlines in which the rock units are thinnest over the crests. These are but variations on the theme discussed in the first chapter, that subsidence determines the thickness of rock units.

The reason why such faults and folds are rarely identified at the surface by surface geologists is that their recognition depends on consistent, but relatively slight changes in thickness. Such changes are readily detected in coal mines and boreholes — and, since about 1970, in seismic record sections — but rarely detectable at the surface because thicknesses measured at the surface are composite over an area.

Considerable misunderstanding of these structures has resulted from the application of the old axiom "Folds and faults are younger than the rocks folded or faulted", because this has usually been interpreted on a time scale that is much too long (e.g., Miocene rocks are faulted, therefore the faulting is post-Miocene). The axiom is better rephrased "Folds and faults are not older than the rocks folded or faulted" to allow consideration of contemporaneity.

DEFINITIONS AND TERMINOLOGY

Growth structures are structures in which the variations in sedimentary thickness show a close relationship to the structure itself. They are interpreted as being the result of deformation that was contemporaneous with sediment accumulation. Growth structures are normally thought of as faults or anticlines, but they may also be monoclines and synclines. They are local variations in the development of a sedimentary basin, but the sedimentary basin itself is a growth structure in the broadest sense.

It is difficult to define growth structures without including the interpretation of them in the definition. *Drape*, or *supratenuous* folds are not growth structures in spite of the fact that the sedimentary thicknesses show a close relationship to the underlying structure. Likewise, *contemporaneous deformation*, as used by Billings (1954) and subsequent writers, refers more to the deformation of unconsolidated sediments by slumping and sliding than to folding and faulting on the scale of petroleum accumulations.

The terminology of growth faults is confused. There is little doubt that *concurrent* fault (Tiddeman, 1890) has priority, when applied to a fault that has different *facies* of correlative units across it and, by analogy, to a fault that has a thicker sedimentary sequence on the downthrown side than the upthrown side. Currie (1956) used both *concurrent* and *contemporaneous;* Liechti et al. (1960) used *depositional,* following common U.S. Gulf Coast usage; Ocamb (1961) used *growth,* but Hardin and Hardin (1961) stated that *contemporaneous* had "some claim to priority". In addition to these, *synsedimentary* has been and is still widely used. Less desirable synonyms include *progressive* and *Gulf Coast type* (!). Dennis (1967) accepted *growth fault* for the *International Tectonic Dictionary,* and recommended that all synonyms be dropped.

Because *concurrent* cannot be revived with any hope of acceptance, and the multiplicity of synonyms can serve no useful purpose, we bow to the International Tectonic Dictionary here, and accept *growth fault* as the term to be applied to a fault that separates correlative sequences of different thicknesses, with the thicker sequence on the downthrown side.

The terminology for growth structures other than faults has not received much attention. Growth anticlines are sometimes called growth structures, but "structure" is a wide term that is not synonymous with "anticline". It is desirable to use the same adjective for analogous geological features, so we use the term *growth structure* to embrace all structures that affected the accumulation of sediment in them; and for specific structures, we use the specific terms *growth fault, growth anticline* and *growth syncline.*

GROWTH FAULTS

A typical, but idealized, growth fault is shown in Fig. 2-1. It is interpreted as a fault that was moving during the accumulation of those rock units that show a thickness contrast across the fault. This contrast is not to be expected over the whole fault, and will not exist in those units that had accumulated before the fault was created. The throw of a growth fault generally increases with depth over the interval or intervals of thickness contrast, because of the thickness contrast; but any antithetic faults reduce the throw with depth. The diagnostic feature is therefore the contrast in bed thicknesses: the increase of throw with depth is a consequence of this and is not in itself diagnostic.

Growth faults occur in many — perhaps most, possibly all — sedimentary basins of the world, and in rocks of all ages. Tiddeman's brief description of the Craven fault in England is apparently the first description of a growth fault (Tiddeman, 1890). This fault affected Lower Carboniferous (Mississip-

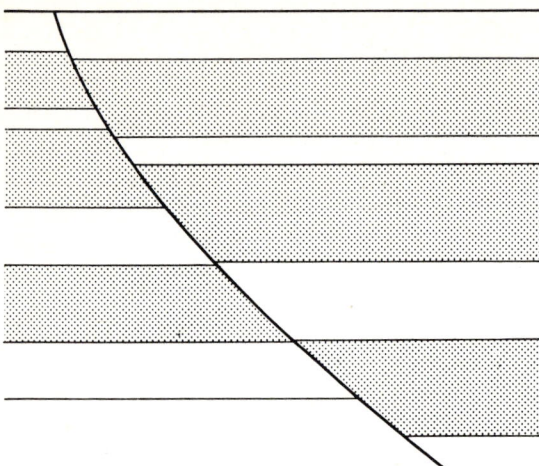

Fig. 2-1. Diagrammatic cross-section through simple growth fault.

pian) sedimentation so that there is a contrast of facies across the fault, and Tiddeman clearly recognized its significance. Dron (1900) later recognized a similar Lower Carboniferous fault in Scotland, involving coal measures. In the coalfields of western Germany, the existence and interpretation of complicated growth structures became well accepted during the 1920s (Böttcher, 1925, 1927), and these concepts were later extended to oil fields in southeast Europe by Stutzer (1930, unfortunately in Abstract only). By the late 1940s, growth faults were recognized in several countries, mainly as a result of coal mining or petroleum development.

Growth faults are important in the Gulf Coast province of North America, and many of the "down-to-the-basin"* faults there are growth faults. They are commonly extensive, with a tendency to be curved in plan and to be generally parallel to the basin margins and the depositional strike. They may form en echelon. Locally they may be conjugate, forming graben. They are commonly associated with flexures ("roll-over anticlines" in the jargon) and with antithetic faults — both on the downthrown side (Fig. 2-2). In section, growth faults commonly flatten with depth, but may also be sinuous. They have been reported more commonly in post-Eocene beds in the Gulf Coast province, but they also affect Mesozoic and Palaeozoic sedimentary rocks.

Much of what we know about growth faults has come from the Gulf Coast province because of the enormous and sustained drilling effort there, but emphasis on this province in the literature must not be taken to mean

---

* This is an ill-conceived but widely used term. They are "down-to-the-basin" in a physiographic sense, perhaps, but may be anywhere in a sedimentary basin.

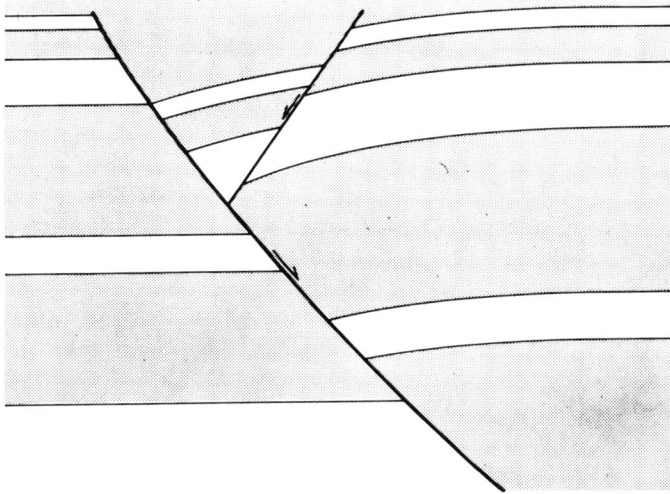

Fig. 2-2. Diagrammatic cross-section through growth fault with "roll-over" anticline and antithetic fault.

that they are more common there than elsewhere. They occur in Europe in the North Sea (Woodland, 1975), in the Vienna basin (Janoschek, 1958), in the Po Basin of Italy (Rocco and Jaboli, 1958); in Nigeria in the Niger Delta (Short and Stäuble, 1967), in Bengal they have been detected on seismic records (Sengupta, 1966). Further east, growth faults have been recognized extensively in north-west Borneo, both on the surface and in the subsurface, in sedimentary rocks of Late Cretaceous and Tertiary ages (Liechti et al., 1960). They also occur in and off Western Australia and the southern margin of Australia (Hosemann, 1971; James and Evans, 1971; Cope, 1972) in rocks of mainly Mesozoic ages. In South America, they occur in the Eastern Venezuela basin (Hedberg et al., 1947; Renz et al., 1958). This list is not exhaustive, but merely a geographic sampling. We shall examine some of these and other areas in context later in the book. Growth faults are also associated with growth anticlines, and these occur as widely as growth faults.

In the Los Angeles basin, post-Cretaceous growth faults have been interpreted as reverse faults (McCulloh, 1960; Shelton, 1968), and here the upthrown blocks were sometimes eroded while the downthrown blocks accumulated sediment. By extension of these concepts, the San Andreas and associated faults of California, and the Alpine fault of New Zealand, are growth transcurrent faults that have affected sedimentation and sediment accumulation (and are still affecting them). However, a transcurrent fault with significant lateral displacement, while clearly having a potential to influence sedimentation and sediment accumulation over a considerable span of time, is a phenomenon that is peripheral to this discussion.

In the Federal Republic of Germany there are some faults that appear to have begun as normal growth faults, to have their sense of movement reversed. There is evidence of thickening of rock units on the overthrust block.

It is evident that growth faults have various expressions and appear in various geological contexts, but there are two main classes:

(a) Those in sedimentary basins formed by rifting, which affect the transgressive and the regressive sequence that usually follows closely in time. These are basement faults that continued to move during the early stages of the development of the basin.

(b) Those in the regressive sequence of a major sedimentary cycle. These are not upward extensions of basement faults, but faults confined to the regressive sequence that die out downwards and many also upwards.

We shall examine here the nature of growth structures as such, leaving discussion of their geological contexts to later chapters.

The nature of growth faults

The quality of correlation across a growth fault varies from poor to excellent. This must be seen as a measure of the degree to which the movement of the fault affected the sediments and their accumulation. Where rock-unit correlation across a growth fault is good, the process was clearly one of transport of sediment over a subsiding sedimentary column, with greater potential for accumulation on the more rapidly subsiding downthrowing block. The continuity of lithologies implied in good correlation suggests that the supply of sediment generally exceeded the capacity of the area to accumulate it. This is a characteristic of regressions. It is instructive to examine the nature of a growth fault in some detail.

Clearly, the deposition of sediment from *suspension* in water cannot lead to the observed results by itself because there is no reason why the rate of deposition from suspension should be different across a subsurface feature, or even a surface feature of modest relief.

These difficulties are removed if the sediment accumulates from the traction load on the sea floor. The clue to this lies with Barrell's concept of baselevel, discussed at some length in Chapter 1. Sedimentary particles are moved by the water until they reach a position in which the energy is insufficient to move them further. There they accumulate. Some particles reach that position near a growth fault, but of these, more reach it on the downthrowing side than on the upthrowing side because of the differing rates of subsidence relative to baselevel. Note that this applies to *muds* as well as sands because mudstones also show thickness contrasts across growth faults.

The accumulation of sediment on the upthrowing side of a growth fault is also evidence that it too was subsiding relative to baselevel, so we are led to the conclusion that *both* blocks of a growth fault were subsiding during fault movement, but one side was subsiding faster than the other, and the thickness difference of correlative units is a measure of the difference in subsidence rates. The term "upthrown" block gives a false impression: it is "up" only relative to the downthrown block.

These conclusions are supported also by field observations because the downthrown block of one growth fault is commonly the upthrown block of another (Fig. 2-3). We can go further.

The quantity of sediment on the sea floor in the area of a growth fault was clearly approximately the same on each side of the fault, but more accumulated on one side than the other. Therefore the supply of sediment exceeded the capacity of the upthrown block to accumulate it (because more accumulated in the downthrown block). Not all the sediment supplied to the upthrowing block accumulated there: the surplus was removed to accumulate elsewhere. Evidence of thickening of a single rock unit across each of several growth faults indicates that there was at least enough sediment to accumulate to form the greatest thickness. Again, the surplus of sediment supply over sediment accumulation is a regressive character.

Fault scarps will form only when the rate of subsidence of the downthrowing block exceeds the capacity of the sediment supply. Evidence of the formation of fault scarps is scanty, but it may be assumed that when correlation across a growth fault is so poor that sediment of different compositions accumulates, a fault scarp may have existed. A growth fault does not require a fault scarp, but growth faults cutting transgressive sequences may well have formed scarps because, as we have seen, the space created by subsidence relative to baselevel during transgression is greater than the volume of sediment supplied. There may still be a thickness contrast because the fluctuations of baselevel affect the upthrowing block more than the downthrowing block. It is difficult to predict the effect of a fault scarp itself on the stratigraphy be-

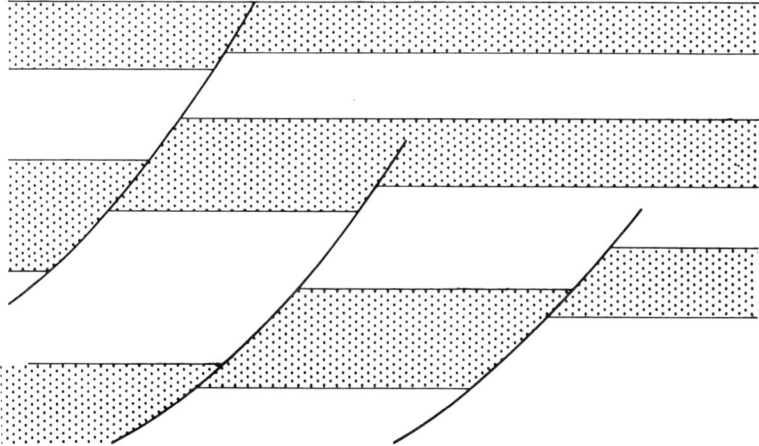
Fig. 2-3. Diagrammatic cross-section through sequence of growth faults.

side a growth fault. It would not be a prolongation of the fault plane at 60° to the horizontal because this is unstable in unconsolidated sediments. Perhaps a sequence of foreset beds inclined at about 30° to the horizontal, bisecting the bedding and the fault plane, would accumulate on the upthrowing side.

The history of movement of a growth fault is recorded in the thicknesses of each rock unit across the fault. This typically shows a gradual acceleration in the rate of movement to reach a peak, and then a gradual deceleration (Fig. 2-4). When there is no movement on the fault, there is no difference of thickness of the rock units that accumulated across the fault during that time; and when all movement ceases, rock units are continuous across the top of the fault.

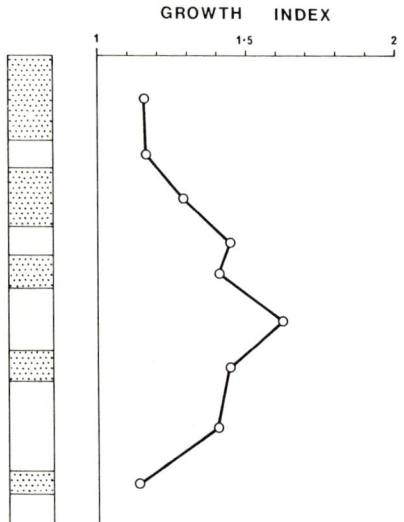

Fig. 2-4. Plot of growth, or expansion, index. The ratio of downthrown to upthrown thickness of each rock unit across a growth fault is plotted at the centre of each unit, regarded as a time series.

In major regressive sequences it is found that the growth fault on the downthrown side of another is further away from the orogeny that caused the regression, and is younger in the sense that it started to move later, reached its peak later, and died later (Fig. 2-3; see also Fig. 15-13 of Seria field, Brunei, and note how the tops of the faults move progressively up the section towards the north-west). Thorsen (1963), in a nice study of growth faults in southeast Louisiana, showed that the maximum rate of movement occurred progressively later in faults towards the south, towards the Gulf of Mexico. He also noted that the maximum rate of growth anticline movement occurred at about the same time as that of the maximum fault growth in nearby faults. South, towards the Gulf of Mexico, is the direction of regression.

The pattern of growth-fault movement in growth faults caused by basement faults is related to the sequence of basement faults. This too can be determined from thickness contrasts, but no detailed study has been published on these.

Growth faults in terminal regressive sequences are characteristically curved in section and in plan, concave to the direction of regression; and they commonly occur en echelon. In section, the dip of the fault plane reduces from a maximum of perhaps 70° at the top, to 40° or even less at the bottom. The simplest, but not the only explanation of this curvature lies in compaction. The growth fault cuts unconsolidated sediments that recently accumulated into the stratigraphic record, so its attitude is determined by the mechanical properties of such material, and its dip will be about 60°. Subsequent compaction reduces the dip of the fault: the greater the compaction, the greater the flattening. There are two influences here. First, the greater compaction with increasing depth, increasing overburden load, tends to flatten the fault with depth. Secondly, such faults in regressive sequences have a decreasing proportion of compactible mudstone with time, so an increasing proportion of mudstone with depth, so accentuating the flattening.

But compaction cannot be the only cause because reduction of dip from 60° to 40° implies about 50% compaction, which would be achieved by reduction of mudstone porosity from 50% to zero. This is much more than is observed: compaction of mudstone from 50 to 20% porosity would reduce the dip from 60° to about 47°. We shall see in Part 3 on regressive sequences that mass flow of relatively imcompetent mudstones at depth could cause the extra flattening of the fault plane. The formation of "roll-over" anticlines in the downthrown block of some growth faults (Fig. 2-2) is probably due to both movement on the curved fault plane and any mass flow at depth.

Once a growth fault has developed and existed during the accumulation of a considerable thickness of sedimentary rocks, differential compaction processes alone must tend to extend its life to some extent. Such processes may also lead to some irregularity in the fault plane because sandstone is less compactible than mudstone. Any irregularities that develop during fault movement will be sheared out by the movement, and a zone of disturbance is to be expected.

It will be appreciated that, by their very nature, growth faults are tilted and folded with the sedimentary sequence that they cut, so that the original dip is measured relative to the dip of the strata (see Fig. 15-15 of Miri field, Sarawak). Any sinuosity introduced into the fault plane during movement may lead to satellite faults that will not have growth characters except where they influenced the rate of accumulation of sediment. If there is sufficient information, much of the history of growth faults and their associated structures can be elucidated.

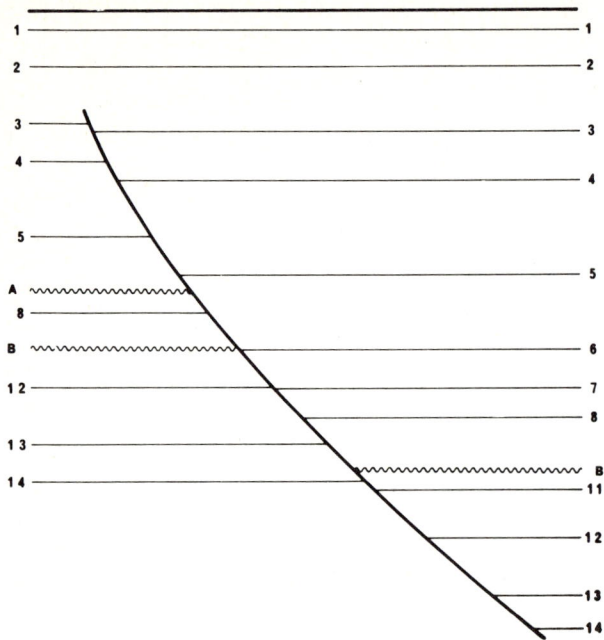

Fig. 2-5. Diagrammatic cross-section through complex growth fault. Fault movement began after accumulation of Marker 12, and ceased after accumulation of Marker 3.

There are several points of detail in the nature of growth faults that are worth examination. Reservoir sandstones not only show volumetric differences due to the different thicknesses but may also have different porosities and permeabilities across a growth fault. The tendency is for the thinner sandstone in the upthrown block to have smaller porosity but larger permeability than the same sandstone in the downthrown block. This is also consistent with the conclusions drawn earlier concerning the accumulation of sediment, because smaller porosity in cleaner sands is to be expected where the accumulation rate was slower.

The stratigraphic-structural relationships about a growth fault may be very complex. In contrast with the simple case in which correlative sequences on either side of the fault differ only in thickness, there are growth faults of which the upthrown, or both, blocks may contain stratigraphic hiatus (Fig. 2-5). These are but variations on the theme of contrasting capacity to accumulate sediment. An eustatic fall of sea level, causing a fall of baselevel that equals the rate of subsidence of the upthrowing block leads to non-accumulation of sediment in the upthrowing block, but continued accumulation in the downthrowing block at a reduced rate. Such is the hiatus A in Fig. 2-5. Similarly, a fall in baselevel that equals the rate of subsidence of the downthrowing block leads to erosion of the upthrowing block and non-accumulation in the downthrowing block. In both cases, the stratigraphic sequence of the downthrowing block is more complete than the other. But in an area with two or more active growth faults, each with different rates of subsidence, the same lowering of baselevel may have different stratigraphic expressions. The difficulty is not so much in the concept as in the recognition of such hiatus.

A stratigraphic hiatus associated with a growth fault may easily be mistaken for a fault in a borehole — a misinterpretation that will be plausible because of the association with known faults. Spurious complexity will be introduced into the geological interpretation that will not only confuse the geologist but also, in a producing field, the reservoir engineer. It will affect reserve estimates and production planning because reservoirs thought to exist in the upthrown block may not exist at all.

The association of growth faults with antithetic and synthetic faults, and stratigraphic hiatus, may therefore present formidable problems of interpretation. There is no certain way of distinguishing hiatus and faults in a borehole, but awareness of the possibilities is a good beginning. It is essentially a geological problem, the solution of which lies in part in the degree of association between "gaps" in the sequence in the borehole and their stratigraphic positions. It is in the geological nature of the phenomena that there is a causal association between hiatus and stratigraphic position, but a chance relationship between faults and their stratigraphic position in a borehole. A hint may also be found in those areas where rock units in the sequence are of comparable thickness in the same block, because reduction in the rate of sediment accumulation may show as a hiatus in the upthrown block but as thinner beds than normal in the downthrown block (see Chapman, 1973, pp. 233—254).

Reverse growth faults would seem to be unlikely, but most of the difficulties that spring to mind at first are found not to be real difficulties. These apparent difficulties stem from the common misconception (aided by the terminology) that a fault separates a block that has been raised from one that has been lowered. As we have seen, the movement is relative, and both may have been downwards relative to baselevel. Topographic expression of growth faults is not a necessary consequence of fault movement provided there is an adequate supply of sediment.

Compaction processes would deform the fault plane of a reverse growth fault in the same way that they deform the fault plane of a normal growth fault. The dip of the fault plane would decrease with depth; but whereas the opposing blocks of a normal fault tend to be separated at the surface and at shallow depth by the movement of the fault, the movement of a reverse growth fault would tend to separate them at depth. This component of horizontal compressive stress induced in the relatively shallow incompetent beds would presumably lead to a zone of disturbance adjacent to the fault.

In general, it is clear that all faulting that takes place at the surface has a potential to influence the deposition of sediment and its accumulation.

When related to the development and evolution of a sedimentary basin, such faults may move — intermittently, perhaps — over significant spans of geological time. Fault movement can rarely (if ever) be regarded as an instantaneous event. A fault is one form of adjustment to the forces acting on a rock mass. Instantaneous forces can only be visualized in catastrophic situations. Whether a fault moved during the accumulation of sediment in a sedimentary basin, or subsequently, the movement must have occupied a span of time, and brought different rock units together across the fault during that time.

It matters to the petroleum geologist when a fault moved, because it may be significant for the migration of fluids and the accumulation of petroleum, as well as important for the understanding of the geology of an area. We shall take this topic into Chapter 9, on fluid migration.

The rate of growth-fault movement is a teasing problem of some interest. Taking an extreme example, assume that while 1000 m of sedimentary rocks accumulated in the upthrowing block, 1500 m accumulated in the downthrowing block at the rate of 500 yrs/m. The fault threw 500 m in 750,000 yrs, or about 7×10^{-4} m/yr — rather less than 1 mm/yr. This is very slow movement, and many growth faults would grow at a much slower rate. Maintenance of a fracture in a thick plastic mudstone is unlikely.

GROWTH ANTICLINES

Growth anticlines are anticlines in which the thicknesses of rock units are greater on the flanks than on the crest (Fig. 2-6). These thickness changes

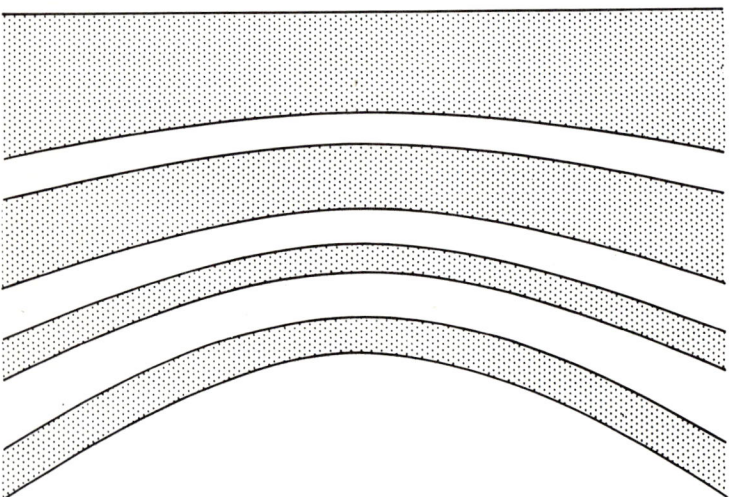

Fig. 2-6. Diagrammatic cross-section through growth anticline.

result in a steepening of the flanks with depth — or, more properly, they result from a steepening of the flanks with depth, but the causes will be considered later in the book. Isopach maps over growth anticlines may be viewed as structural contour maps on the bottom of the isopached interval below the top of the interval as datum. More than one rock unit normally shows crestal thinning, and the development of the growth anticline against the time scale of accumulated sediment will be revealed both by isopach maps of the various rock units, and cumulative isopach maps and cross-sections drawn from them (Figs. 2-7—2-9).

Growth anticlines are commonly closed, oval or dome-like. "Drape structure" is not synonymous with "growth anticline", but, as with growth faults, there is an element of differential compaction in them that will contribute to

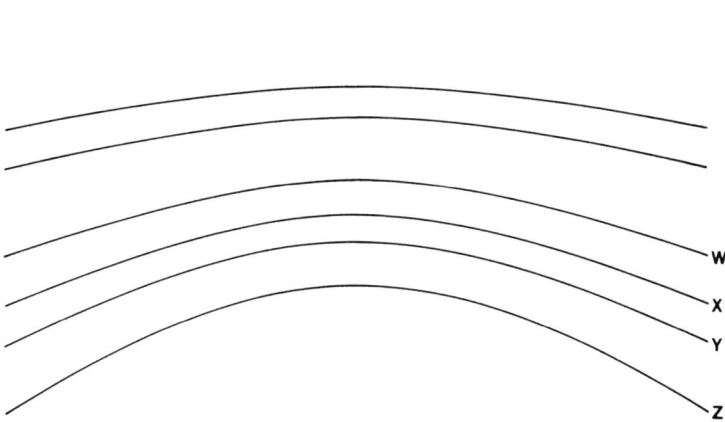

Fig. 2-7. Cumulative isopach maps reveal the history of growth of a growth anticline, shown here in cross-section.

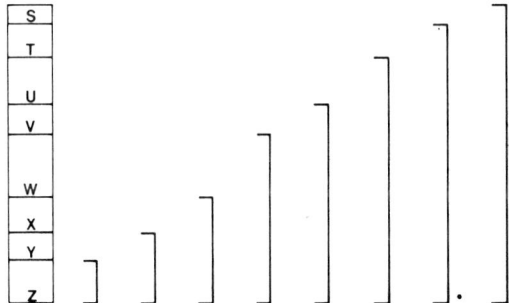

Fig. 2-8. Cumulative isopach intervals are taken from the deepest marker that is sufficiently known.

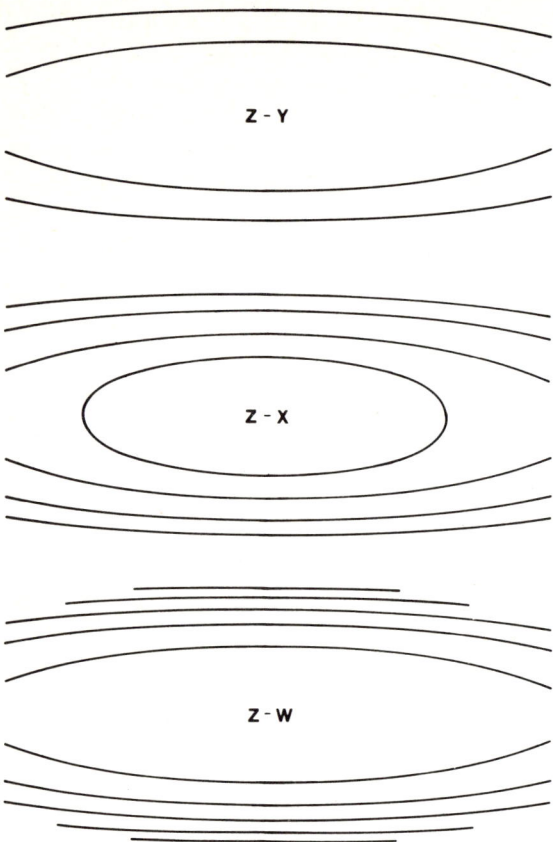

Fig. 2-9. Cumulative isopach maps are drawn from these intervals. They reveal a steepening of the anticline with time. Cross-sections of these maps can also be drawn.

their growth by the greater ultimate compaction of the thicker compactible units on the flanks. Compaction will also tend to be greater on the flanks because of the greater sedimentary load. So isopach maps may tend to overestimate the growth of the anticline as a whole. This growth is perhaps better indicated by the relatively incompactible units.

Growth synclines are analogous, but they are not normal drilling targets, so little is known of them. They are seen in seismic record sections. Growth monoclines are the essence of a sedimentary basin.

Growth anticlines occur, perhaps, even more widely than growth faults, but this would not be surprising because they are just variations in the accumulation history of a basin. They occur extensively in the Gulf Coast province and most of the larger basins of North America; in Venezuela and other areas of South America; in Europe; in Nigeria; in the Middle East and Russia; in the younger basins of South-East Asia, including Borneo and New Guinea; and they occur in Australia. They have received little attention in the literature, yet they are commonly figured. Experience suggests that many of the petroleum-bearing anticlines have a growth component, and that many were initiated as growth anticlines.

As with growth faults, growth anticlines (and other growth structures) are best understood in terms of differential subsidence relative to baselevel. The whole area of the growth anticline was subsiding, but the area of the crest was subsiding more slowly than the flank areas. Hence the flank areas had a capacity to accumulate a greater thickness of sediment. Again, one must see the sediment as being in transport across the area, not deposited from suspension, because one cannot postulate such patterns of deposition from suspension.

Growth anticlines range from those that show continuity of rock units across it, with differences confined to thickness (and perhaps porosity and permeability), through those that influenced sedimentation to such a degree that different facies occurred in the crestal area from the flank areas, to those in which rock units are discontinuous across the crestal area. Reservoir rocks may therefore vary not only volumetrically, but also in reservoir properties. As with growth faults, these are but variations on a theme of differential subsidence.

It is important to bear in mind that the growth of an anticline only affects those sediments that were accumulating during the growth period. This is analogous to the intermittent movement of a growth fault. All sediments that accumulated in the area prior to the beginning of growth of the anticline, and all sediments that accumulated during periods of quiescence and after growth had ceased, will show either no thickness variations, or variations that are not related to the structure. This statement requires two qualifications: it applies to sediments above the agent or cause of growth; and the processes of compaction tend to prolong the growth, as they do of growth faults. One must therefore distinguish between the periods of growth on the time scale of accumulated sediment, and the folding of sedimentary rocks into an anticline subsequently to the accumulation of the sediment. The time or times of deformation may be revealed by the younger sediments.

Growth anticlines may be faulted by growth faults. The very nature of these requires, of course, that *these are contemporaneous deformations.* The simultaneous deformation of a structure by folding and faulting requires careful analysis. Qualitatively, the deformation of the curved fault plane of a growth fault by contemporaneous growth of the anticline may be important in that it sets up stresses that may be relieved by further faulting. Quantitatively, such deformation of the fault surface may be insignificant because it will usually be far finer than the rather coarse control afforded by boreholes.

The deformation of a growth anticline by contemporaneous growth faulting is, of course, significant in terms of the geometry of the anticline and its reservoirs. The growth component due to faulting may obscure the component due to anticlinal growth, so that the recognition of the growth anticline depends on the recognition of convergence of stratigraphic units towards the crest within fault blocks.

The proper use of isopach maps and sections is essential for the elucidation of even simple growth structures. For instance, the axis of minimum sediment accumulation may change position during the development of a growth anticline. Such shifts of position can rarely be seen in structural maps or sections.

We cannot leave the subject of growth anticlines without mention of an alternative hypothesis for the thinning of stratigraphic units: that is, the mechanical attenuation of the beds due to the enlargement of their area, or stretching, over a growing anticline. Confidence in the sediment-accumulation hypothesis rests largely on the analogy with growth faults, the evidence of which is not consistent with mechanical attenuation of the sedimentary rocks because of the abrupt changes of thickness across the fault. Mechanical attenuation is not mutually incompatible with sedimentary thinning, and model studies have shown attenuation of layers over a growing diapir. The key to the problem may lie in the thinning of individual beds of similar lithology. If mudstones, for example, show consistent and greater thinning than sandstones or limestones, then attenuation may be suspected. On the other hand, if the patterns of thinning are variable and not associated with lithology, sedimentary thinning may be the dominant cause. The problem is complex and the evidence ambiguous. However, when growth faults cut an anticline in which the rock units tend to thin towards the crest, the ambiguity for all practical purposes is removed and at least some of the crestal thinning is due to thinner accumulation of sediment there.

Growth anticlines have hardly ever been reported from surface surveys. It is possible that they hardly ever occur at the surface. In active sedimentary basins they will, of course, have no surface expression. In older sedimentary basins where they have been exposed by uplift and erosion, they will be characterized by steepening of dips towards the crest. Such anticlines when mapped may be attributed to diapirism (and we shall see that diapirism is a common cause of growth anticlines), but the feature of crestal steepening of dips is a feature of growth anticlines.

REFERENCES

Billings, M.P., 1954. *Structural geology* (2nd ed.). Prentice Hall, Englewood Cliffs, N.J., 514 pp.
Böttcher, H., 1925. Die Tektonik der Bochumer Mulde zwischen Dortmund und Bochum und das Problem der westfälischen Karbonfaltung. *Glückauf*, 61: 1145—1153 and 61: 1189—1194.
Böttcher, H., 1927. Faltungsformen und primäre Diskordanzen im niederrheinisch—west-

fälisch Steinkohlgebirge. *Glückauf*, 63: 113—121.

Chapman, R.E., 1973. *Petroleum geology: a concise study*. Elsevier, Amsterdam, 304 pp. Paperback edition, 1976.

Cloos, E., 1968. Experimental analysis of Gulf Coast fracture patterns. *Bull. Am. Ass. Petrol. Geol.*, 52: 420—444.

Cope, R.N., 1972. Tectonic style in the southern Perth basin. *Geol. Surv. West. Aust. Annu. Rep. 1971*, pp. 46—50.

Currie, J.B., 1956. Role of concurrent deposition and deformation of sediments in development of salt-dome graben structures. *Bull. Am. Ass. Petrol. Geol.*, 40: 1—16.

Dennis, J.G., 1967. International tectonic dictionary English terminology. *Mem. Am. Ass. Petrol. Geol.*, 7, 196 pp.

Dron, R.W., 1900. The probable duration of the Scottish coalfields. *Trans. Instn. Min. Engrs.*, 18: 194—212.

Hamblin, W.K., 1965. Origin of "reverse drag" on the downthrown side of normal faults. *Geol. Soc. Am. Bull.*, 76: 1145—1164.

Hardin, F.R. and Hardin, G.C., 1961. Contemporaneous normal faults of Gulf Coast and their relation to flexures. *Bull. Am. Ass. Petrol. Geol.*, 45: 238—248.

Hedberg, H.D., Sass, L.C. and Funkhouser, H.J., 1947. Oil fields of Greater Oficina area central Anzoategui, Venezuela. *Bull. Am. Ass. Petrol. Geol.*, 31: 2089—2169.

Hosemann, P., 1971. The stratigraphy of the basal Triassic sandstone, north Perth basin, Western Australia. *J. Aust. Petrol. Explor. Ass.*, 11: 59—63.

James, E.A. and Evans, P.R., 1971. The stratigraphy of the offshore Gippsland basin. *J. Aust. Petrol. Explor. Ass.*, 11: 71—79.

Janoschek, R., 1958. The inner-alpine Vienna basin. In: L.G. Weeks (Editor), *Habitat of oil*. Am. Ass. Petrol. Geol., Tulsa, Okla., pp. 1134—1152.

Koinm, D.N. and Dickey, P.A., 1967. Growth faulting in McAlester basin of Oklahoma. *Bull. Am. Ass. Petrol. Geol.*, 51: 710—718.

Liechti, P., Roe, F.W. and Haile, N.S., 1960. The geology of Sarawak, Brunei and the western part of North Borneo. *Bull. Geol. Surv. Dept. Br. Terr. Borneo*, 3: 1—360.

McCulloh, T.D., 1960. Gravity variations and the geology of the Los Angeles basin of California. *U.S. Geol. Surv. Prof. Pap.*, 400-B: B320—B325.

Ocamb, R.D., 1961. Growth faults of south Louisiana. *Trans. Gulf-Coast Ass. Geol. Socs.*, 11: 139—175.

Quarles, M., 1953. Salt-ridge hypothesis on the origin of the Texas Gulf-Coast type of faulting. *Bull. Am. Ass. Petrol. Geol.*, 37: 489—508.

Renz, H.H., Alberding, H., Dallmus, K.F., Patterson, J.M., Robie, R.H., Weisbord, N.E. and MasVall, J., 1958. The Eastern Venezuela basin. In: L.G. Weeks (Editor), *Habitat of oil*. Am. Ass. Petrol. Geol., Tulsa, Okla., pp. 551—600.

Rocco, T. and Jaboli, D., 1958. Geology and hydrocarbons of the Po basin. In: L.G. Weeks (Editor), *Habitat of oil*. Am. Ass. Petrol. Geol., Tulsa, Okla., pp. 1153—1167.

Sengupta, S., 1966. Geological and geophysical studies in western part of the Bengal basin, India. *Bull. Am. Ass. Petrol. Geol.*, 50: 1001—1017.

Shelton, J.W., 1968. Role of contemporaneous faulting during basinal subsidence. *Bull. Am. Ass. Petrol. Geol.*, 52: 399—413.

Short, K.C. and Stäuble, A.J., 1967. Outline of geology of Niger delta. *Bull. Am. Ass. Petrol. Geol.*, 51: 761—779.

Stanley, T.B., 1970. Vicksburg fault zone, Texas. In: M.T. Halbouty (Editor), *Geology of giant petroleum fields*. Mem. Am. Ass. Petrol. Geol., 14: 301—308.

Stutzer, O., 1930. Absinken, Sedimentation und Faltung — gleichzeitige Vorgänge in manchen Erdölgebieten. *Geol. Rundsch.*, 21: 141 (abstract).

Thorsen, C.E., 1963. Age of growth faulting in south-east Louisiana. *Trans. Gulf-Coast Ass. Geol. Socs.*, 13: 103—110.

Tiddeman, R.H., 1890. On concurrent faulting and deposit in Carboniferous times in

Craven, Yorkshire, with a note on Carboniferous reefs. *Rep. Brit. Ass. Adv. Sci.*, 1889: 600—603.

Walters, J.E., 1959. Effect of structural movement on sedimentation in the Pheasant-Francitas area, Matagorda and Jackson counties, Texas. *Trans. Gulf-Coast Ass. Geol. Socs.*, 9: 51—58.

Weber, K.J., 1971. Sedimentological aspects of oil fields in the Niger delta. *Geol. Mijnbouw*, 50: 559—576.

Woodland, A.W., 1975. *Petroleum and the continental shelf of north-west Europe. 1, Geology.* Applied Science Publishers, London, 501 pp.

CHAPTER 3

COMPACTION OF SEDIMENT AND SEDIMENTARY ROCKS, AND ITS CONSEQUENCES

SUMMARY

(1) Compaction is a diagenetic process that begins on burial and may continue during burial to depths of 9 km (30,000 ft) or more. Compaction increases the bulk density of a rock, increases its competence, and reduces porosity.

(2) Sands compact with relatively little loss of porosity or permeability, but other diagenetic processes may considerably reduce porosity and permeability with authigenic minerals in the pore spaces.

Mudstones compact with serious and permanent loss of porosity and permeability.

Carbonates compact to varying degrees depending on the proportion of plastic material. Most appear to compact by solution processes rather than by mechanical compaction.

(3) Compaction can proceed only with the compression and expulsion of a commensurate proportion of the pore fluids, which move to positions of smaller energy. If mudstones alternate with more permeable beds, the more compactible mudstones expel fluids both upwards and downwards to the more permeable beds. The downward energy gradient in the lower part of the mudstone makes the mudstone a perfect barrier to the upward migration of fluids. Lateral migration takes place in the relatively permeable, intercalated beds.

(4) Pore-fluid pressures in mudstones cannot be measured, but they are inferred to be greater than normal hydrostatic in a compacting mudstone because this is a necessary condition for flow. In a thick mudstone loaded faster than the corresponding rate at which the fluids can be expelled, compaction is retarded and the mudstone retains the mechanical properties it had at shallower depth; and the pore pressures are correspondingly elevated above the normal hydrostatic. The pore fluids are bearing part of the overburden load.

COMPACTION OF SEDIMENT AND SEDIMENTARY ROCKS

Compaction under the gravity load of overlying sediments is a fundamental geological process that is necessarily an important topic of petroleum geology. It affects stratigraphic relationships and thicknesses, pore-water movement and salinities, migration of petroleum and probably also of base metals. All sediments and sedimentary rocks compact to some extent, and the processes lead to changes in some of their properties. Most obviously, they become more dense, less porous and less permeable; and the rock unit becomes thinner. The mechanical strength of a rock increases with compaction, and it becomes less drillable. Compaction normally increases with depth due to the increase in load, the increase in the duration of loading, and the increase in the temperature of the material loaded.

Compaction is a diagenetic process that begins with, or very soon after, sediment accumulation. In early stages, the material has virtually no cohesive strength, and the grains can be rearranged mechanically by shocks due to earthquakes and environmental shocks due to heavy surf and thunderstorms. This rearrangement changes the grain packing from unstable towards stable packing — in a well-sorted sand, for example, from about 45% porosity towards about 30% porosity. Subsequent burial and diagenesis of a chemical nature may reduce the porosity further and give to the material a considerable cohesive strength. These trends are observed qualitatively when drilling boreholes. At shallow depth, the hole is drilled very quickly, but is suffers wash-outs and has a very irregular geometry usually much larger than the diameter of the bit that drilled the hole. At greater depths, the amount varying from province to province, the sediments acquire a cohesive strength: the penetration rate decreases and the hole takes on a regular shape not much larger than the bit that drilled it. Cores can be recovered intact.

Porosity and bulk density of a sedimentary rock are related:

$$\rho_{bw} = f\rho_w + (1-f)\rho_s = \rho_s - f(\rho_s - \rho_w) \tag{3.1a}$$

where ρ_{bw} is the bulk wet mass density of the rock that contains pore fluids of mass density ρ_w and solids of mass density ρ_s, with fractional porosity f. There are practical advantages in expressing this in terms of weight densities because the mean weight density multiplied by depth gives the pressure in that material at that depth:

$$\gamma_{bw} = \gamma_s - f(\gamma_s - \gamma_w). \tag{3.1b}$$

Equations 3.1 show that as the porosity changes from 1 (no solids) to 0 (no fluids) the bulk wet densities change linearly from that of the fluid to that of the solids (Table 3-1). The practical advantage of thinking of these parameters in this way is that as the grain size decreases, porosity becomes increasingly difficult to measure, but bulk weight density easier.

Porosity of a sedimentary rock cannot be reduced without commensurate compression or expulsion of pore water. Pore water can only be expelled if it is free to move away. The rate of compaction is therefore related not only to the rate of subsidence and loading, but also to the permeabilities of the compacting rocks and the permeabilities of all the rocks affected by the consequent water movement. Lithologies therefore also play a part.

TABLE 3-1

Variation of bulk wet mass density and bulk wet weight density of quartz sand with porosity*

Porosity	ρ_{bw} (kg/m^3)	γ_{bw} (kPa/m)	γ_{bw} (kgf cm^{-2} m^{-1})	γ_{bw} (psi/ft)
0.35	2090	20.5	0.209	0.91
0.30	2170	21.3	0.217	0.94
0.25	2250	22.1	0.225	0.98
0.20	2330	22.9	0.233	1.01
0.15	2410	23.6	0.241	1.04
0.10	2490	24.4	0.249	1.08

* Interstitial water mass density, 1050 kg/m^3; grain density, 2650 kg/m^3.

Porosity and permeability are related, as we shall see in Chapter 8 on the nature of petroleum reservoirs. Effective porosity determines the amount of movable water and, with pore size, the resistance of the porous material to movement of the water. Tortuosity of the pore passages increases, in general, with compaction and this too can be related empirically to porosity. Kozeny (1927) found by analysis and experiment that the permeability of unconsolidated sands varies approximately as $f^3/(1-f)^2$. Chapman (1981, p. 60) preferred $f^x/(1-f)^2$ where x varies from 3.5 in unconsolidated sands to about 5.3 in indurated sandstones, because this takes tortuosity better into account. Little is known about the permeability of mudstones, but we may safely assume that similar principles apply. Rearrangement of the grains of an unconsolidated sand that reduces the porosity from 40 to 30% reduces its permeability to about 25—30% of its former value. While this is a considerable relative loss of permeability, the final permeability is usually still quite large because the original permeability may well have been two or three darcies. A sand with 500 millidarcies (md) permeability is still very permeable.

Sandstone compaction is not as simple as that, however, and the effects tend to be obscured by environmental influences. Maxwell (1964) and Selley (1978) found linear trends, and Stephenson (1977) examined the effect of temperature on sandstone compaction.

Sandstone compaction is elastic on a short time-scale, but the removal of load does not lead to restoration of the original properties (no core would ever be extracted from a core barrel if that were not true). Under severe loading, grains may be fractured, resulting in greatly reduced porosity and permeability. Diagenesis also leads to cementation and to authigenic minerals in the pore space. Pendular cement may have very little effect on permeability (Füchtbauer, 1967, p. 359), but authigenic minerals, particularly clays, may have a very adverse effect (see Plates 8-1 and 8-2 on pp. 162—163 and 166—

167). Nevertheless, the permeability of sandstones usually remains high enough not to impede the expulsion of pore water significantly.

Calcarenites may well share some of the compaction features of sands and sandstones, but solution processes appear to be more important than mechanical compaction in carbonate rocks. McCrossan (1961) found that the dry density of some Devonian mudstones in Canada increased with $CaCO_3$ content as well as with depth; and the higher the carbonate content, the smaller was the relative compaction. If stylolites are but a special case of compaction of carbonates under load, the loss of both fluid and solids (in solution) may occur.

Compaction of mudstones ("shale" in petroleum jargon), which may contain 50% silt, is a complex process involving irreversible deformation of the ductile grains and also chemical diagenesis. When mudstone first accumulates into the stratigraphic record, it has a porosity of about 50% (it is misleading to think of larger initial porosities in muds because the commonly quoted figure of 80% is almost certainly not applicable to mudstone in the stratigraphic record — and muds grade up to dirty water). The early stages of mudstone compaction probably include significant grain rearrangement. Hedberg (1936), in a careful study of Tertiary mudstone in some wells in Venezuela, concluded that there were four overlapping stages of mudstone compaction: first, mechanical rearrangement; then dewatering; then mechanical deformation, and finally recrystallization. These processes can expel vast quantities of pore water.

Consider the compaction of a unit cube of water-saturated mudstone from 30% porosity $12\frac{1}{2}\%$ porosity under gravity. Porosity is not as convenient a parameter as *void ratio* (ϵ) in these computations: $\epsilon = f/(1-f)$, the ratio of pore volume to solids volume. The initial cube with 30% porosity has a void ratio of 0.43 (Fig. 3-1). When compacted to $12\frac{1}{2}\%$ porosity, the void ratio becomes 0.143. If we make the assumption that the solids volume remains constant at 0.7 units (an assumption that may prove untenable when the diagenetic history of mudstone is better understood) the volume of liquid retained is $0.7 \times 0.143 = 0.1$ units. The bulk volume of the initial cube is therefore reduced to 0.8 units by the expulsion of 0.2 units of water. These figures are

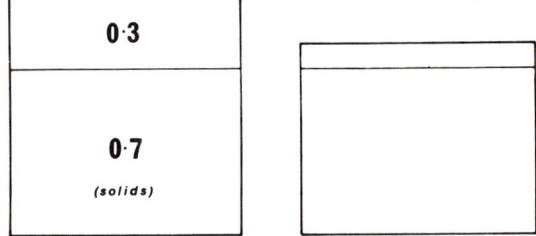

Fig. 3-1. Compaction of unit volume of mudstone from 30% to 12.5% porosity results in the explusion of 0.2 units of pore water.

proportional to thickness when compaction is due to gravity without lateral spreading. The immense volumes of water expelled will be appreciated when it is considered that 1 km³ of water-saturated mudstone compacted from 30% porosity to 12.5% porosity expels 0.2 km³ of water. This is equal to the *total* pore volume of 1 km³ of sandstone with 20% porosity. This is why there are difficulties with the concept of connate water (pp. 77—78). The chemistry of mudstone pore water may differ considerably from that of the displaced sandstone pore water. By the same token, if this water cannot escape from the mudstone (or, indeed, from an intercalated sandstone) considerable quantities are retained and buried to greater depths and higher temperatures for a longer time.

Mudstone compaction, like other sediment compaction, involves chemical and physicochemical processes as well as the physical, but these are not sufficiently well understood yet for general rules to be formulated. In confining ourselves to the physical processes, however, we believe that these are sufficient to explain the observed and deduced effects and that it is a valuable — even essential — simplification.

The matter of mudstone compaction as a function of depth, overburden load and bulk densities, has been the subject of several studies (Hedberg, 1926, 1936; Athy, 1930a; Dickinson, 1951, 1953). Hedberg's (1936) and Athy's curves (Fig. 3-2) differ from each other in some interesting respects, but we cannot claim to understand these differences fully.

First, it is essential to understand clearly the nature of such curves. They are, of course, a plot of present porosity (or wet bulk density, as the case

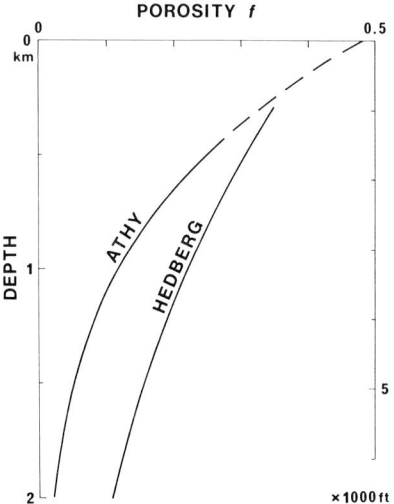

Fig. 3-2. Mudstone compaction curves of Athy (1930) and Hedberg (1936), the latter generalized.

may be) against present depth. Time and temperature are not included explicitly: depth embraces time, temperature, intergranular pressures and pore-water pressures. Temperature and pressures may be taken as linearly related to depth from a practical point of view, but it is by no means certain that time is. We cannot interpret such curves as describing the *history* of compaction — that mudstone now 2 km deep followed the curve during its burial to 2 km. We cannot assume that a mudstone unit would have one history-of-burial curve; it is much more likely that the very top and bottom would have a history much like these curves, but that the centre's would be very different because pore-water expulsion from the centre would have been retarded.

Compaction curves may be determined by direct measurement, and by indirect measurement through geophysical properties. Hedberg (1936, p. 254, table 1) determined the bulk density, grain density and porosity of mudstones from a well in Venezuela by direct measurement, and from these he estimated the overburden pressures and recorded the data against depth. The deepest sample was from 6175 ft (1882 m), the shallowest from 291 ft (89 m). He regarded porosity as a better indicator of compaction than bulk density, and pointed out that the relationship is not between depth and porosity (although this has practical value) but between pressure and porosity. He found that the depth in feet was *approximately* numerically equal to the overburden pressure in pounds per square inch (psi), that is, an overburden pressure gradient of 22.6 kPa/m. He recognized three distinct compaction trends:

— From 0—800 psi (to 5.5 MPa) he found the exponential relationship:

$$P = 67.214\, G^{-0.1047}$$

where P is the porosity in percent, and G is the overburden pressure in psi, which was estimated from the measurements.

— From 800—6000 psi (5.5—41 MPa) he found the *linear* relationship:

$$P = 34.86 - 0.00421\, G$$

— From 6000—10000 psi (41—69 MPa) he found the linear relationship:

$$P = 13.93 - 0.0006935\, G$$

While we accept Hedberg's insistence that pressure, not depth, is the cause of compaction, we also accept (as Hedberg did) that there is practical value in porosity-depth relationship. Since the depths and pressures listed by Hedberg (1936, p. 254, table 1) are very closely related by $G = 0.70\, z^{1.04}$ (where z is the depth in feet) we may substitute this into his formulae above, obtaining:

$$\begin{aligned}P &= 69.82\, z^{-0.1086}\\ P &= 34.86 - 2.93 \times 10^{-3}\, z^{1.04}\\ P &= 13.93 - 4.82 \times 10^{-4}\, z^{1.04}\end{aligned} \quad (3.2)$$

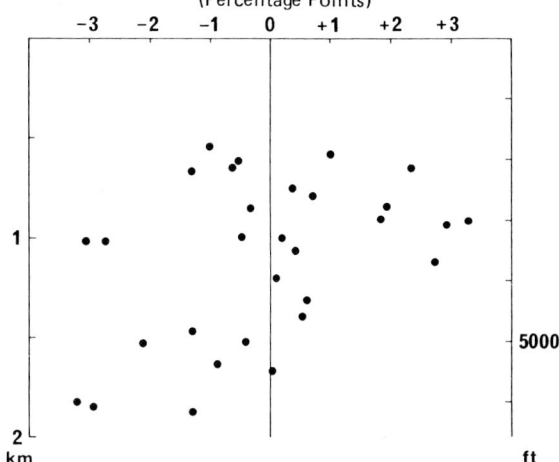

Fig. 3-3. Discrepancies between Hedberg's measured porosities and those given by eq. 3.3. Data of Hedberg, 1936, p. 254, table 1.

He also accepted that there is practical value in a single formula. His general formula, $P = 40.22\,(0.9998)^G$, can similarly be converted to a depth relationship, and it can be shown to correspond closely with

$$f = 0.41\,e^{-6.5 \times 10^{-4} z} \tag{3.3}$$

where z is in metres, and the exponential factor 6.5×10^{-4} has the dimension of inverse length (L^{-1}). Figure 3-3 shows the discrepancies between the measured porosities of his well AB (Oficina 1 in eastern Venezuela) and those predicted by this formula. The errors are small and of no practical significance, but there appears to be a systemic error that suggests that the true formula may not be of this form.

Athy (1930a) made direct measurements on Palaeozoic mudstones in Oklahoma. His curve, which has been widely used, has two disadvantages: the area has suffered some tectonic disturbance, and he had to extrapolate the top 1,400 ft (430 m). He found the relationship:

$$f = 0.48\,e^{-1.42 \times 10^{-3} z} \tag{3.4}$$

and proposed the general form:

$$f = f_0\,e^{-az} \tag{3.5}$$

where f_0 is the fractional porosity when $z = 0$.

There are advantages in writing this:

$$f = f_0\,e^{-z/b} \tag{3.5a}$$

where b is a scale length in the same units as z, and z/b can be regarded as a dimensionless depth.

The differences between Hedberg's and Athy's curves are not to be explained by compromise. Almost certainly they are due to the role of time and temperature, neither of which is explicitly included in the formula. It is apparent that there is no such thing as a simple depth-porosity formula that has general validity. The local constants, f_0 and b, are not universal and their local value takes local conditions into account.

Mudstone compaction curves determined by direct measurement are too time-consuming to be in general use, and we must resort to indirect measurement through geophysical measurements made in the borehole of the rocks close to the wall of the borehole. Satisfactory mudstone compaction curves can be constructed from the sonic log, which measures the inverse of the sonic velocity in the rocks close to the wall of the borehole (see p. 130) and records this in units of microseconds per foot or per metre against depth. If the logarithm of the sonic transit time in mudstone (Δt_{sh}) is plotted against depth, a linear trend (at least at depths to about 2 km) is usually found, suggesting a relationship of the form:

$$\Delta t_{sh} = \Delta t_0 \, e^{-cz} \tag{3.6}$$

as seen in Fig. 3-4. Δt_0 is the extrapolated value of the transit time at $z = 0$, and c is the slope (ln $\Delta t/z$; we shall omit the suffix sh where it is clear that mudstone transit times are intended). It is found that Δt_0 is commonly about 165 µs/ft (540 µs/m) corresponding to a sonic velocity of about 1850 m/s (6000 ft/s). The slope, however, varies from area to area.

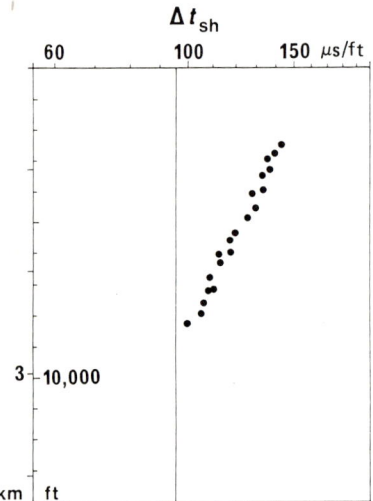

Fig. 3-4. The logarithm of shale transit time plotted against depth is normally approximately linear.

It must be noted at once that while this formula 3.6 is satisfactory for shallow depths, to perhaps 2 or 3 km, it cannot be correct because as porosity is eliminated with depth, the transit time approaches the matrix transit time (~ 55 μs/ft, ~ 180 μs/m. We shall therefore seek a better one.

There is little reliable experimental data on mudstone sonic transit times and the corresponding porosities, but Magara (1968, p. 2474, table II) found a linear relationship. Accepting this, we proceed as follows:

Define Δt_0 as the value of Δt when the porosity is f_0;
Δt^* as the dimensionless transit time $\Delta t / \Delta t_0$;
Δt^*_{ma} as the dimensionless transit time at zero porosity $\Delta t_{ma}/\Delta t_0$;
and
f^* as the relative porosity f/f_0.

Then $f^* = 1$ when $\Delta t^* = 1$, and $f^* = 0$ when $\Delta t^* = \Delta t^*_{ma}$, and we can write:

$$f^* = (\Delta t^* - \Delta t^*_{ma})/(1 - \Delta t^*_{ma}) \qquad (3.7)$$

or:

$$f/f_0 = (\Delta t - \Delta t_{ma})/(\Delta t_0 - \Delta t_{ma}). \qquad (3.7a)$$

The value of Δt_{ma} is close to 55 μs/ft (180 μs/m) in mudstones. The values of Δt_0 and f_0 are about 165 μs/ft (540 μs/m) and 0.5 (to one significant figure) and these figures should be used when there is no reliable data suggesting others. Again, note carefully that we are concerned with the values when the mudstone has accumulated into the stratigraphic record, not the superficial values of 200 μs/ft and porosities of 60% and more.

Magara's data is plotted on Fig. 3-5, using the above values of the material constants.

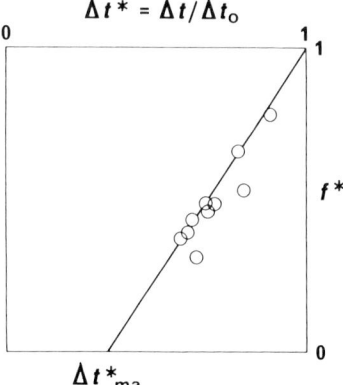

Fig. 3-5. Plot of dimensionless time against relative porosity for mudstones. Points are computed from data of Magara (1968, p. 2474, table II) using $\Delta t_0 = 165$ μs/ft, $\Delta t_{ma} = 55$ μs/ft and $f_0 = 0.5$.

We take the relationship between porosity and sonic transit time in mudstones to be:

$$f = f_0(\Delta t - 55)/(\Delta t_0 - 55) \simeq (\Delta t - 55)/220. \tag{3.7b}$$

Equation 3.7b gives meaning to the empirical constants of Magara's equations in the form $f = m\Delta t + n$.

It must be pointed out here that error in the determination of Δt is not symmetrical about the mean. In a perfect hole, its true value will be closely determined by the sonic log, but any geometrical irregularities in the wall of the borehole lengthen the travel path and so lead to a larger value of Δt than the true one. Also, in shallow and in thick mudstones there may be a degree of undercompaction near the middle, which also increases the transit time. So we are concerned with the smallest values of Δt, and with values in thin mudstones and near the top and bottom of thick mudstones for the determination of true normal compaction curves. Likewise, we are concerned with the smaller values of porosity, which represent maximum compaction.

Combining eqs. 3.5a and 3.7a, we obtain:

$$f/f_0 = e^{-z/b} = (\Delta t - \Delta t_{ma})/(\Delta t_0 - \Delta t_{ma})$$

from which we write:

$$\Delta t = (\Delta t_0 - \Delta t_{ma}) e^{-z/b} + \Delta t_{ma} \tag{3.8}$$
$$\simeq 110\, e^{-z/b} + 55\, \mu s/ft.$$

This equation is an improvement on eq. 3.6 in that it satisfies the boundary conditions of Δt_0 at $z = 0$, and Δt_{ma} when porosity is eliminated. It remains to determine the value of the scale length b.

The scale length can be determined by two methods, the choice of which depends on the nature of the data. First, by setting the dimensionless depth, z/b, equal to unity and solving eq. 3.8, b is equal to the depth z at which Δt is 95—96 μs/ft *in normally compacted mudstones* (z being measured from ground level or seafloor). If there is no reliable data around 95—96 μs/ft, take the deepest level at which Δt is thought to represent normal compaction and solve eq. 3.8 for b:

$$b = -z/\ln\{(\Delta t - 55)/(\Delta t_0 - 55)\} \tag{3.9}$$

the units being the same as those of depth, and the transit time being in μs/ft.

We test this result using the data of Hottmann and Johnson (1965, p. 719, fig. 2), which has hitherto been accepted as satisfactory evidence that a plot of the logarithm of the transit time against depth is linear to depths of about 4 km in the U.S. Gulf Coast region. Inserting into eq. 3.8 the values $\Delta t_0 = 165$ μs/ft, $\Delta t_{ma} = 55$ μs/ft, and $z/b = 1$, gives $\Delta t = 95.5$ μs/ft when $b = z$. This value is reached at a depth of about 12,500 ft (3800 m) using the trend of shorter transit times, indicating this value for the scale length. Alternatively,

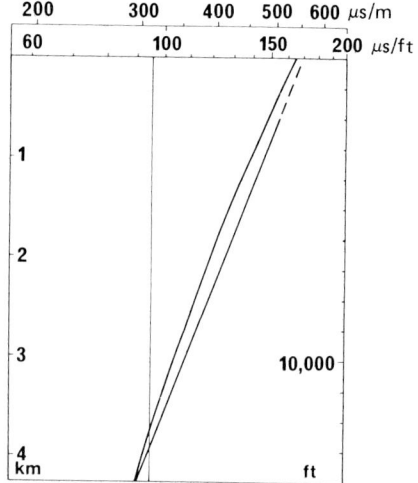
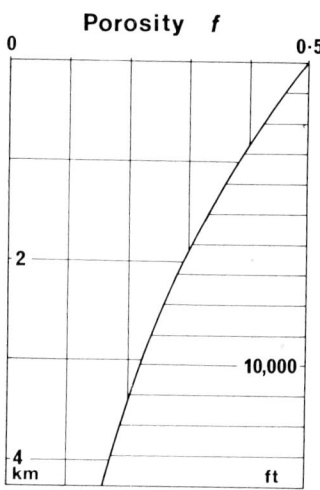

Fig. 3-6. Hottmann and Johnson's regression line (straight) and the curve of eq. 3.8 using a scale length of 12,200 ft (3700 m).

Fig. 3-7. Porosity-depth curve corresponding to Fig. 3-6.

taking Δt to be 90 μs/ft at 14,000 ft, we find from eq. 3.9 the value of 12,200 ft (3700 m) for b. This difference is insignificant. Figure 3-6 shows Hottmann and Johnson's regression line and the curve of eq. 3.8 using a scale length of 12,200 ft. The curve bounds the data points closely on the side of shorter transit time, or greater compaction, as it should. The corresponding porosity-depth curve obtained by inserting this value of b into eq. 3.5a and taking $f_0 = 0.5$ is shown in Fig. 3-7.

As a further check on this method of deriving porosity-depth curves, we take Margara's (1968, p. 2474, table II) data on some Miocene mudstones in Japan, here tabulated in simplified form in Table 3-2. The value of the scale length lies where the value of Δt is 95—96 μs/ft. This is between 3063 and 3205 m, nearer 3205 m. We take b = 3150 m. The values of Δt predicted from eq. 3.8 for the depths of the data are tabulated beside those measured. Using the same value of the scale length in eq. 3.5a, and $f_0 = 0.5$, the porosities predicted are tabulated beside those measured in cores. Note that the porosities are better predicted from eq. 3.5a than directly from the measured transit times using eq. 3.7b, again suggesting that the transit times may err on the long side.

These results leave little doubt that eqs. 3.5a and 3.8 satisfactorily describe the porosity-depth and transit time-depth relationships below depths of about 2 km (6000 ft) in normally compacted mudstones. The underestimation of porosity and transit times at shallower depths could well be due to undercompaction of the mudstone.

TABLE 3-2

Comparison of data of Magara (1968, p. 2474, table II) with estimated quantities (^) from eqs. 3.5a and 3.8 using the value of b determined from the sonic data

Depth (m)	Δt (μs/ft)	$\hat{\Delta t}$ (μs/ft)	Porosity f	$\hat{f}$
1029	145	134	0.390	0.36
1610	127	121	0.332	0.30
1809	130	117	0.266	0.28
2151	109	111	0.246	0.25
2296	114	108	0.243	0.24
2444	110	106	0.231	0.23
2607	102	103	0.218	0.22
3063	99	97	0.196	0.19
3205	95	95	0.188	0.18
3505	104	91	0.159	0.16
3701	∼92	89	0.146	0.15

The values of the scale length b determined from the sonic log in Tertiary mudstones seem to lie commonly between 2500 and 4000 m (8000 and 13,000 ft). The scale length of Hedberg's (1936) data — 1550 m or 5000 ft — is unusually short. Older sequences tend to have shorter scale lengths. Athy's Palaeozoic mudstones have a scale length of about 700 m (2300 ft). Cretaceous mudstones of western Canada have one of about 520 m (1700 ft) (Magara, 1973, p. 11, fig. 6). But, as can be seen, the value of the scale length is not solely determined by age.

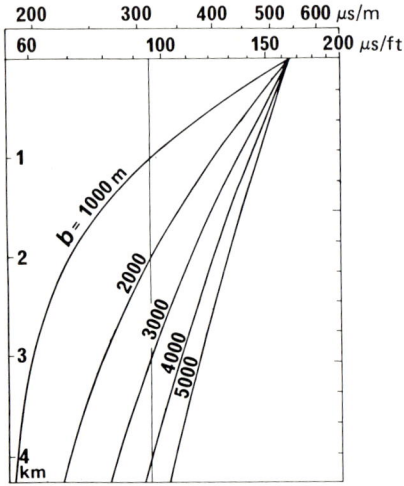

Fig. 3-8. Family of transit-time — depth curves for various values of the scale length.

Figure 3-8 shows the family of curves of Δt for various values of scale length. It will be noted that departure from linearity (log Δt with depth) is slight at shallow depths and with larger values of b. The scale length is therefore interpreted as a parameter that takes time, temperature, and the compressibility of the mineral grains and the bulk material into account.

It is possible that the compaction of sandstones can be described by similar equations with a scale length so large that the porosity-depth relationship is almost linear.

PORE-WATER EXPULSION

Sediments and sedimentary rocks can only compact if a commensurate volume of pore water can be expelled. Compaction leads to a reduction of bed thickness as well as a reduction of porosity. Earlier in this chapter we noted that very large volumes of water must be expelled during compaction of mudstones, and it is becoming increasingly important for quantitative geological and geochemical studies of source rocks and primary migration to be able to assess the quantities of pore fluids expelled, and to reconstruct the stratigraphic thicknesses and depths at earlier stages of the accumulation of a sequence of sedimentary rocks.

The volume of pore water expelled during compaction from fractional porosity f_1 to f_2, expressed as a proportion of the bulk volume at porosity f_2, is approximated by:

$$q = (f_2 - f_1)/(1 - f_1) \tag{3.10}$$

assuming constant volume of solids and incompressibility of water. (This equation is quite general: if f_2 is larger than f_1, q is positive and represents the additional proportional volume required for porosity f_1 to be increased to f_2.)

To estimate the thickness at porosity f_1 when the present thickness at f_2 is known, the known thickness is multiplied by the *compaction factor*, $1 - q$.

Equations 3.7 permit us to estimate these quantities from the sonic log data. For example, compaction from Δt_1 = 120 μs/ft to Δt_2 = 100 μs/ft corresponds with compaction from f_1 = 0.295 to f_2 = 0.205. The change of volume, as a proportion of the volume at Δt_2 and f_2, is —0.13. This amounts to 130,000 m^3 km^{-2} per metre of thickness at 20.5% porosity. The original thickness was greater by a factor of 1.13.

Conversely, when mudstones are undercompacted for their depth, there is a volume of water that has not been expelled. The resulting porosity is greater than normal for that depth, and the transit time longer than normal. For example, the sonic log shows a transit time of 115 μs/ft in mudstone at a depth at which the normal transit time (from eq. 3.8) would have been 90 μs/ft. So f_1 = 0.159 and f_2 = 0.273. The proportion of the present volume at

f_2 retained is +0.14, and one metre thickness will be reduced to 0.86 m when the porosity of 15.9% is reached.

The retention of this relatively large volume of water in the mudstone gives to the mudstone the properties of bulk density, effective viscosity or competence, and all the related properties that a normally compacted mudstone would have had at a much shallower depth — the depth at which a normally compacted mudstone would have a porosity of 27.3%.

INTERSTITIAL WATER PRESSURES

It is a matter of common observation, amply confirmed by water wells, and boreholes drilled for petroleum, that the porous and permeable sediments and sedimentary rocks at shallow depths in the continents and the continental shelves are completely saturated with water (locally, also oil or gas), and that the pressures in this water are closely approximated by:

$$p = \gamma_w z \qquad (3.11)$$

where γ_w is the weight density of the water, and z is the depth of measurement below the surface (consistent units being used throughout)*.

Equation 3.11 implies that if the water is free to rise in the borehole, it will reach the surface. It is, however, a generalization because the weight density varies with the salinity of the water, its temperature, and its depth (pressure); and because the level $z = 0$ should strictly be the water table, not the ground surface. The weight density of water, which determines the pressure gradient, varies from the superficial value of 9.8 kPa/m (0.1 kgf cm^{-2} m^{-1}; 0.433 psi/ft) when it is fresh near the surface, to about 10.6 kPa/m (0.11 kgf cm^{-2} m^{-1}; 0.47 psi/ft) when it has 130,000 ppm total dissolved solids.

Pressures that conform to weight densities within these limits in eq. 3.11 are known as *normal hydrostatic pressures;* and the *gradients* that conform to this range of weight densities, but not necessarily to eq. 3.11, are known as *hydrostatic gradients* (Fig. 3-9).

The existence of normal hydrostatic pressures indicates that the pore spaces containing the water are connected to the surface, however tortuously. The grains of sediment may be regarded as contained within a water reservoir with as much validity as the more usual statement that the rocks contain water. The sediments of many environments, of course, accumulated in water.

It is also a matter of observation (though not so common) that water pressures are found in the subsurface that do not correspond with normal hydro-

---

* It is conventional to measure z negative downwards below a datum such as sea level, but positive downwards from the surface of the ground or rig floor.

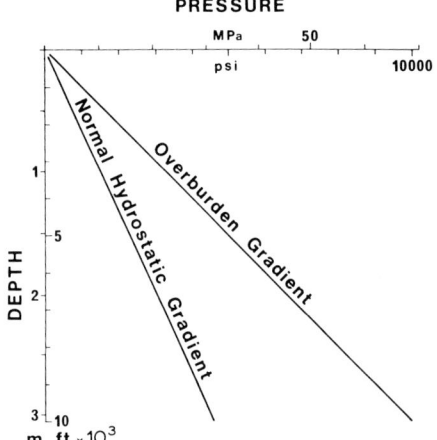

Fig. 3-9. Normal hydrostatic gradient, and overburden (or geostatic) pressure gradient, generalized.

static pressures. In some areas they are significantly lower than normal hydrostatic; more commonly, they are higher. Pressures below normal hydrostatic, called *subnormal* pressures, are usually due to a water table or aquifer outlet that is considerably below ground level of the well in question. Similarly, some modest excess pressure may be due to artesian conditions in which the intake area of the aquifer is elevated above the ground level at the well. However, we are concerned more with those areas in which measured pressures far exceed the normal hydrostatic, and cannot be explained in artesian terms (for example, when such pressures are encountered in places such as the Niger delta and the U.S. Gulf Coast). These high pressures, known as *abnormal pressures*, have a limiting value close to the pressure exerted by the overburden, solids and fluids:

$$p_{max} \rightarrow \overline{\gamma}_{bw} \, z \qquad (3.12)$$

where $\overline{\gamma}_{bw}$ is the mean bulk wet density of the overburden above depth z (Fig. 3-9). The limiting gradient and the pressures on it are called *overburden*, *geostatic*, or *lithostatic* gradients or pressures (in that order of priority). Pressures lying between the overburden value and the normal hydrostatic value are called *abnormal pressures* or *geopressures* — or *superpressures* when they are very high. Abnormal pressures are only abnormal in the sense that they do not conform to the normal hydrostatic.

The overburden pressure gradient is also generalized as a straight line, but of course it depends on lithologies and the state of compaction. Its value is commonly taken as 22.6 kPa/m (0.23 kgf cm^{-2} m^{-1}; 1 psi/ft), but this is usually on the high side. Because porosities tend to decrease with depth, the overburden pressure gradient tends to increase with depth.

Abnormal pressures imply that if the water is free to rise in a borehole it would flow at the surface or, if we could insert a manometer into the formation, the water would rise in it to a level well above the surface of the ground (Fig. 3-10). If we take sea level as the datum surface, the total head (see p. 165, Fig. 8-10) may well be of the same order of magnitude as the depth — that is, the pressure head may well be about twice the elevation head. Because abnormal pressures in all but a few exceptional areas lie deeper than normal hydrostatic pressures (shown in schematic but typical form in Fig. 3-11) we infer that low permeability has retarded water flow to the normally pressured part of the sequence.

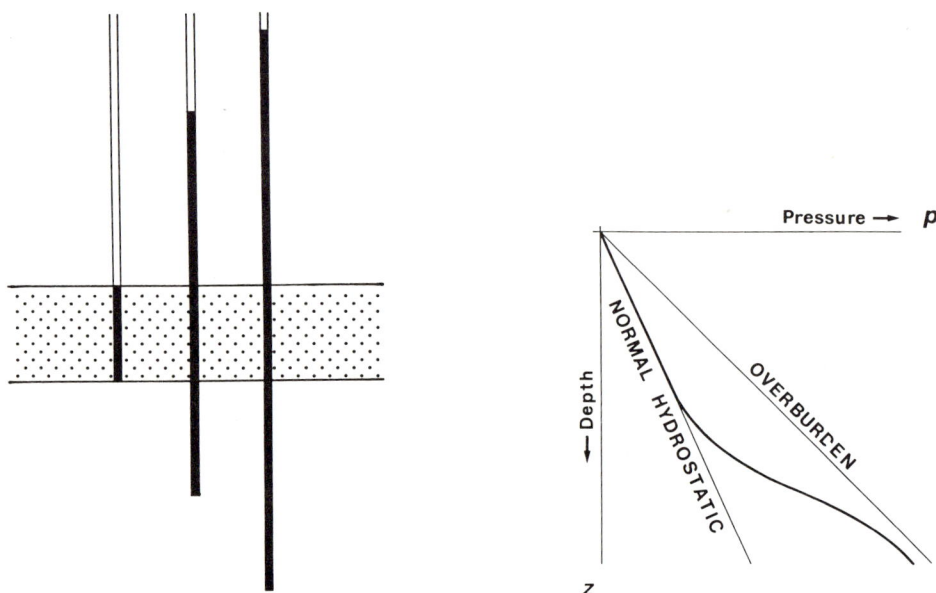

Fig. 3-10. If manometers could be inserted into a sedimentary sequence, the pore pressures would be indicated by the elevation of the water level in the manometer.

Fig. 3-11. Diagrammatic, but typical pressure—depth plot for pore fluids in a regressive sequence.

Geologically, the great majority of abnormally pressured sequences have the tops of abnormal pressures (as Dickinson, 1951 and 1953, noted for the Louisiana Gulf Coast) near the top of the thick mudstone in a dominantly regressive sequence of sedimentary rocks, and the normal hydrostatic pressures occur in the sandier part of the sequence. It appears, therefore, that abnormal pressures do not occur where the permeability is sufficiently large to drain the water of compaction. Harkins and Baugher (1969) found 5 to 10% sand to be the regional indicator of impending high pressures when drilling in the U.S. Gulf Coast region.

MECHANICAL ASPECTS OF PORE WATER IN SEDIMENTARY ROCKS

As a permeable sedimentary rock compacts during burial, the reduction of porosity is accompanied by the expulsion of a commensurate part of its pore water. As the load is applied, it is borne first by the water which, by flowing away, transfers it to the grain-to-grain contacts. A less permeable rock compacting under the same conditions takes longer to compact to mechanical equilibrium because it takes longer for the commensurate amount of pore water to escape. The loading creates a potential (energy) gradient in the water, which will flow, if it can, to positions of smaller potential; the incremental load borne by the water increases the pressure in the water.

Geologists may therefore be in intuitive agreement with Terzaghi (1936) who postulated that the total load on sediment is borne partly by the solid framework and partly by the pore fluid:

$$S = \sigma + p \tag{3.13}$$

where S ($= \overline{\gamma}_{bw} z$) is the total vertical component of overburden pressure, σ is the *effective stress* transmitted through the solid matrix, and p is the pore-fluid pressure (which Terzaghi called the *neutral stress*). He found that it is the effective stress, σ, that compacts a sedimentary rock, and that this quantity is the difference between the total stress and the pore pressure (Fig. 3.12). Equation 3.13 is called Terzaghi's relationship.

Hubbert and Rubey (1959, p. 142) introduced a useful parameter λ, which is the ratio of pore-fluid pressure to total overburden pressure:

$$\lambda = p/S. \tag{3.14}$$

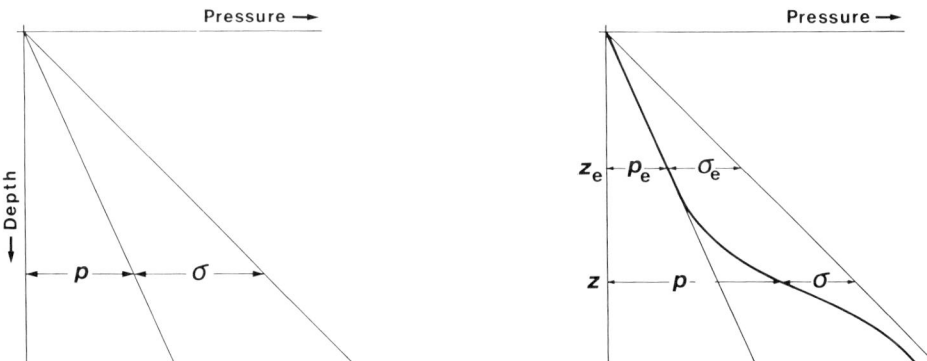

Fig. 3-12. The effective (intergranular) stress is the difference between the total stress due to the overburden and the pore-fluid pressures.

Fig. 3-13. The effective stress, σ, in a mudstone at depth z is equal to that of a normally compacted mudstone at depth z_e.

This is a dimensionless parameter representing the proportion of the total overburden supported by fluid pressure. Its value ranges from about 0.45 under normal hydrostatic conditions to near 1 when pore pressures are severely abnormal. Rearranging eq. 3.13 and substituting eq. 3.14, we obtain:

$$\sigma = S - p = (1 - \lambda) \overline{\gamma}_{bw} \, z. \tag{3.15}$$

From this we see that as the pore pressure approaches the overburden pressure at depth z, $\lambda \to 1$ and $\sigma \to 0$; that is, the effective stress that tends to compact a sedimentary rock decreases as the pore pressure increases relative to the overburden pressure, and the rock's compaction is retarded.

This value of effective stress, however, corresponds with that of a normally compacted but otherwise identical rock at a shallower depth, z_e (Fig. 3.13). At this shallower depth (using the suffix e to denote equilibrium compaction conditions*):

$$\sigma = (1 - \lambda_e) \overline{\gamma}_{bw} \, z_e. \tag{3.16}$$

Equating eq. 3.15 and 3.16 and accepting that for practical purposes the value of $\overline{\gamma}_{bw}$ above depth z_e does not differ significantly from that above depth z, we obtain:

$$z_e/z = (1 - \lambda)/(1 - \lambda_e). \tag{3.17}$$

This quantity has been assigned the symbol δ (Chapman, 1972), so:

$$z_e = \delta \, z. \tag{3.18}$$

The parameter δ is a dimensionless quantity that takes pore-fluid pressures into account. Its value varies from 1 (when $\lambda = \lambda_e$) to 0 (when $\lambda = 1$) and it may be regarded as a non-linear measure of the extent to which mechanical compaction equilibrium has been achieved by the expulsion of pore water.

For example, the water pressure measured in a thin sandstone lens within a thick mudstone was found to be 62 MPa (632 kgf cm^{-2}; 8992 psi) at 3250 m (10,660 ft). Assuming an overburden pressure gradient of 22 kPa/m (0.24 kgf cm^{-2} m^{-1}; 0,97 psi/ft) and a normal hydrostatic pressure gradient of 10 kPa/m (0.102 kgf cm^{-2} m^{-1}; 0.442 psi/ft) we compute:

$S \;\; = 22 \times 10^3 \times 3250 = 71.5$ MPa
$p_e \;= 10 \times 10^3 \times 3250 = 32.5$ MPa
$\lambda_e \;= 0.45$
$\lambda \;\; = 62 \times 10^6/71.5 \times 10^6 = 0.87$
$\delta \;\; = (1 - 0.87)/(1 - 0.45) = 0.24$.
So, $z_e = 0.24z = 768$ m (2520 ft).

---

* Strictly, λ_e is the proportion of the overburden supported by the *ambient* fluid — air, if subaerial; water, if submarine — but almost all, if not all abnormal pressures are below sea level, so we may take normal hydrostatic pressures to define λ_e (see Chapman, 1979).

We would expect, therefore, that the state of compaction at 3250 m under these conditions will be comparable to an identical, but normally compacted rock at about 768 m. The common occurrence of drilling breaks at the top of abnormal pressures lends support to this approach.

There is a real geological meaning to be attached to the equilibrium compaction depth, z_e. It is an estimate of the *maximum* thickness of overburden at the time compaction equilibrium was lost in the sedimentary rock now at depth z, and this overburden was approximately the thickness z_e on top of z. We shall return to this in Chapter 14 when we consider the time of generation of abnormal pressures.

These results can now be incorporated into our various compaction curve formulae, eqs. 3.5a, 3.8 and 3.9. These equations relate strictly to equilibrium compaction depths, z_e in the notation now developed, so we write the dimensionless depth more generally as $\delta z/b$. The formulae so modified to take pore pressure into account, with others used with them, are shown in the box on p. 63; and the values of the dimensionless depth for the range of Δt from 56 to 165 μs/ft are tabulated in Table 3-3. If the value of Δt and any two components of dimensionless depth are known, the third can be simply calculated. Note also that the effective compaction depth, z_e, is a function of the sonic transit time, Δt, and can therefore be estimated directly from the sonic log. Practical hints on the construction of Δt plots and the use of these formulae will be found in Appendix I.

The formulae enable us to estimate pore-fluid pressures in *mudstones* from the sonic log through the parameter δ. Given a plot of mudstone or

TABLE 3-3

Values of dimensionless depth, $\delta z/b$, for Δt_{sh}. Read tens in lines and units in columns (e.g., the value of $\delta z/b$ for $\Delta t_{sh} = 117$ μs/ft is 0.573)

Δt_{sh} (μs/ft)										
	0	1	2	3	4	5	6	7	8	9
5							4.700	4.007	3.602	3.314
6	3.091	2.909	2.755	2.621	2.503	2.398	2.303	2.216	2.136	2.061
7	1.992	1.928	1.867	1.810	1.756	1.705	1.656	1.609	1.565	1.522
8	1.482	1.442	1.405	1.368	1.333	1.299	1.266	1.235	1.204	1.174
9	1.145	1.117	1.090	1.063	1.037	1.012	0.987	0.963	0.939	0.916
10	0.894	0.872	0.850	0.829	0.809	0.788	0.769	0.749	0.730	0.711
11	0.693	0.675	0.657	0.640	0.623	0.606	0.590	0.573	0.557	0.542
12	0.526	0.511	0.496	0.481	0.466	0.452	0.438	0.424	0.410	0.396
13	0.383	0.370	0.357	0.344	0.331	0.318	0.306	0.294	0.282	0.270
14	0.258	0.246	0.235	0.223	0.212	0.201	0.190	0.179	0.168	0.157
15	0.147	0.136	0.126	0.116	0.105	0.095	0.085	0.076	0.066	0.056
16	0.047	0.037	0.028	0.018	0.009	(Δt_0)				

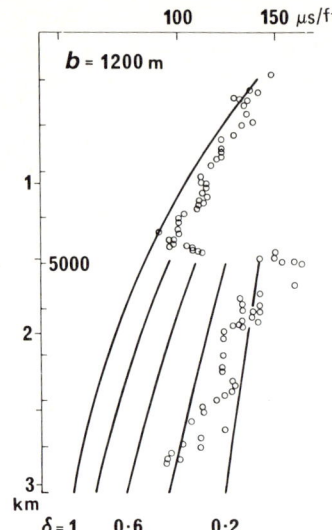
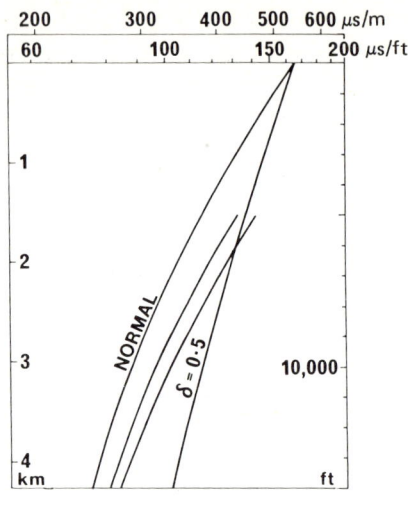

Fig. 3-14. Family of curves of constant δ superimposed on transit-time — depth plot of Tertiary mudstones in a well in Borneo. (Data courtesy the Royal Dutch/Shell Group.)

Fig. 3-15. Mudstone transit-time — depth trends for hydrostatic gradients in abnormally pressured, permeable beds are parallel to the normal curve but displaced vertically downwards.

shale transit time versus depth, a family of curves can be superimposed using eq. 3.8 with the modified dimensionless depth term, $\delta z/b$, each line representing a constant value of δ (Fig. 3-14)*. The greater the departure of the transit time from that corresponding to normal compaction for its depth, the greater the pore pressure relative to normal hydrostatic. The trends of transit time with depth can therefore be interpreted as trends of pressure with depth, and so trends of energy or potential with depth. Normal hydrostatic pressures are indicated by the normal compaction line, δ = 1, and hydrostatic gradients at abnormally low or high pressures by the slope of the normal compaction line projected vertically upwards or downwards (respectively) to the depth of interest (Fig. 3-15). Any trend of transit time that decreases at a greater rate than the normal hydrostatic indicates a downward energy or potential gradient and downward water flow. The hydrostatic trend indicates constant energy and no flow. Other trends indicate upward flow (as in the transition zone at 1.5 km depth in Fig. 3-14).

---

* These can be converted to mud weights and used as a check from drilling data (see Appendix I).

FLUID MIGRATION FROM ABNORMALLY HIGH PORE PRESSURES

Water tends to flow from positions of higher energy or potential to positions of lower energy, and the directions of flow are normal to the surfaces of equal potential (called *equipotential* surfaces) through the water. For water to flow through a sedimentary rock, there must be a potential gradient within the pore water and the rock must have permeability to water. The hydraulic conductivity (K) of a rock, or its coefficient of permeability, has the dimensions of a velocity, LT^{-1}. The dimension of time means that it is unlikely that any natural sedimentary rock is totally impermeable. Indeed, we can and have measured mudstone permeabilities (see, for example, Magara, 1971, p. 241, fig. 9), so they are not normally impermeable to water even on a short time scale.

In a mudstone compacting under the force of gravity, so that the vertical component of total stress is a function of the overburden thickness, surfaces of equal total stress will be horizontal or nearly horizontal. The vertical component of effective stress (σ) is a function of both total stress and pore-fluid pressure, and surfaces of equal effective stress will also be horizontal or nearly horizontal. Lack of mechanical equilibrium between solids and liquids therefore induces equipotential surfaces in the liquids that are also horizontal, or nearly so; and pore-water movement tends to be vertical (upwards *or* downwards) when induced by gravitational compaction. Pore water tends to migrate from the high-energy or high-potential zones within the mudstone to positions of lower energy above *or below* it. Commonly, an extensive permeable

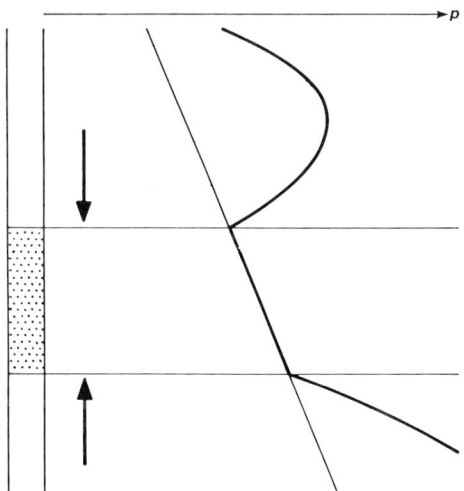

Fig. 3-16. Pressure-depth plot for sandstone interbedded with abnormally pressured mudstones. Mudstone pore-fluid migration is downwards to the upper sandstone interface, upwards to the lower interface.

sandstone at normal hydrostatic pressures acts as a drain to the water expelled by compaction of the adjacent mudstones (Fig. 3-16). In general, therefore, the pore water moves vertically across the bedding of the mudstones, then laterally within the sandstone*.

The geological context of abnormally pressured mudstones is dominantly regressive sequences in which the sandstones tend to thicken towards the land of the time, and wedge out in the seaward direction (Fig. 3-17). The geometry of the drains therefore imposes a landward direction of migration within the sandstones. The top of abnormal pressures tends, as Dickinson (1951, 1953) found for the Louisiana Gulf Coast, to become younger in the direction of the regression. We shall examine these features in more detail in Chapter 14.

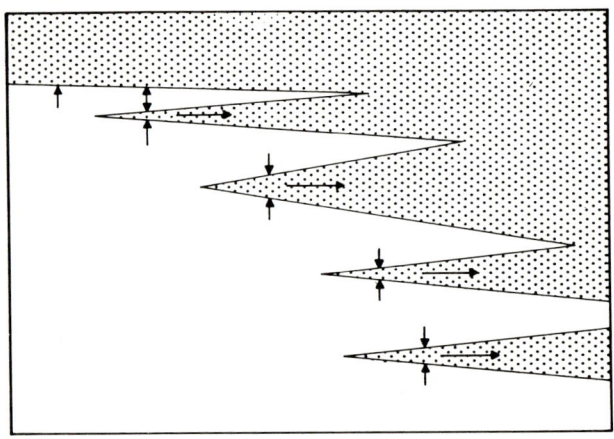

Fig. 3-17. Regional pore-water migration in sandstones of regressive sequences tends to be towards the land from which the regression came.

---

* It is sometimes argued that mudstone pore water migrates laterally "because that is the direction of greatest permeability". We have seen that permeability is but one of the components of fluid flow, and flow direction is determined by the direction of the fluid potential gradient. The quantity of water flowing across a surface in unit time is proportional to the area normal to flow and to the potential or hydraulic gradient (the loss of total head divided by the length of the porous material in the direction of flow; $\Delta h/l$). To get a feel for the quantities and directions of compaction flow, consider a rectangular mass of mudstone, $10 \times 10 \times 0.5$ km, entirely surrounded by permeable sands at normal hydrostatic pressures. During compaction, the hydraulic gradient from a point in the centre of the mudstone is $x/0.25$ vertically and $x/5$ horizontally — a factor of 20 in favour of vertical flow. The two horizontal surface areas total 200 km²; the four vertical surfaces total 20 km² — a factor of 10 in favour of vertical flow. So horizontal permeability would have to be of the order of 200 times the vertical permeability for *equal* quantities to be lost from the sides as from the top and bottom. In reality, mudstones tend to have lenticular interfaces with sandstones (except at faults), and the ratio of thickness to lateral extent is commonly much greater than that used in this simple model. There is no evidence that anisotropy in mudstone permeability is anywhere near sufficient to lead to significant quantities of lateral flow in mudstones.

Formulae for obtaining porosity and pore pressure in mudstone from sonic transit time (Δt in $\mu s/ft$)

1. $f \quad = f_0 (\Delta t - \Delta t_{ma})/(\Delta t_0 - \Delta t_{ma})$
$\simeq (\Delta t - 55)/220$ (3.7)

2. $\Delta t \quad = (\Delta t_0 - \Delta t_{ma}) e^{-\delta z/b} + \Delta t_{ma}$
$\simeq 110 \, e^{-z/b} + 55 \, \mu s/ft$ (3.8)

3. $b \quad = -\delta z/\ln\{(\Delta t - 55)/(\Delta t_0 - 55)\}$ (units as z) (3.9)

4. $\delta \quad = -(b/z) \ln\{(\Delta t - 55)/(\Delta t_0 - 55)\}$

5. $z_e \quad = \delta z = -b \ln\{(\Delta t - 55)/(\Delta t_0 - 55)\} = bz^*$

6. $z^* \quad = \delta z/b = -\ln\{(\Delta t - 55)/(\Delta t_0 - 55)\}$

7. Estimated maximum pore pressure in mudstone:

$\hat{p}_{max} = \lambda S = (1 - 0.55\delta) \, \overline{\gamma}_{bw} \, z$
$\simeq (1 - 0.55\delta) \, 22.63 \times 10^{-3} \, z \quad MPa$

$h_{max} = (\hat{p}/\overline{\gamma}_w) - z \simeq (\hat{p}/10.18 \times 10^3) - z \quad m$

8. Equivalent Δt for mudstone to balance mud specific gravity:

$\Delta t_{mud \, eq.} = (\Delta t - 55) e^x + 55 \, \mu s/ft$

where $x = (z/b)\{(s.g. - 2.3)/1.3\}$

REFERENCES

Athy, L.F., 1930a. Density, porosity, and compaction of sedimentary rocks. *Bull. Am. Ass. Petrol. Geol.*, 14: 1—24.
Athy, L.F., 1930b. Compaction and oil migration. *Bull. Am. Ass. Petrol. Geol.*, 14: 25—35.
Boatman, W.A., 1967. Measuring and using shale density to aid in drilling wells in high-pressure areas. *J. Petrol. Technol.*, 19: 1423—1429.
Bredehoeft, J.D. and Hanshaw, B.B., 1968. On the maintenance of anomalous fluid pressures, I. Thick sedimentary sequences. *Geol. Soc. Am. Bull.*, 79: 1097—1106.
Burst, J.F., 1969. Diagenesis of Gulf Coast clayey sediments and its possible relation to petroleum migration. *Bull. Am. Ass. Petrol. Geol.*, 53: 73—93.
Chapman, R.E., 1972. Clays with abnormal interstitial fluid pressures. *Bull. Am. Ass. Petrol. Geol.*, 56: 790—795.
Chapman, R.E., 1979. Mechanics of unlubricated sliding. *Geol. Soc. Am. Bull.*, 90: 19—28.
Chapman, R.E., 1981. *Geology and water: an introduction to fluid mechanics for geologists*. Nijhoff/Junk, The Hague, 228 pp.
Dickinson, G., 1951. Geological aspects of abnormal reservoir pressures in Gulf Coast region of Louisiana, U.S.A. *Proc. 3rd World Petrol. Congress, The Hague, 1951*, sect. 1: 1—16; discussion: 1: 16—17.
Dickinson, G., 1953. Geological aspects of abnormal reservoir pressures in Gulf Coast Louisiana. *Bull. Am. Ass. Petrol. Geol.*, 37: 410—432.
Fertl, W.H., 1976. *Abnormal formation pressures*. Elsevier, Amsterdam, 382 pp.

Füchtbauer, H., 1967. Influence of different types of diagenesis on sandstone porosity. *Proc. 7th World Petrol. Congress*, 2: 353—370.
Gretener, P.E., 1969. Fluid pressure in porous media — its importance in geology. *Bull. Can. Petrol. Geol.*, 17: 255—295.
Hanshaw, B.B. and Bredehoeft, J.D., 1968. On the maintenance of anomalous fluid pressures, II. Source of layer at depth. *Geol. Soc. Am. Bull.*, 79: 1107—1122.
Hanshaw, B.B. and Zen, E-an, 1965. Osmotic equilibrium and overthrust faulting. *Geol. Soc. Am. Bull.*, 76: 1379—1386.
Harkins, K.L. and Baugher, J.W., 1969. Geological significance of abnormal formation pressures. *J. Petrol. Technol.*, 21: 961—966.
Hedberg, H.D., 1926. The effect of gravitational compaction on the structure of sedimentary rocks. *Bull. Am. Ass. Petrol. Geol.*, 10: 1035—1072.
Hedberg, H.D., 1936. Gravitational compaction of clays and shales. *Am. J. Sci.*, 31: 241—287.
Hottmann, C.E. and Johnson, R.K., 1965. Estimation of formation pressures from log-derived shale properties. *J. Petrol. Technol.*, 17: 717—722.
Hubbert, M.K. and Rubey, W.W., 1959. Role of fluid pressure in mechanics of overthrust faulting, I. Mechanics of fluid-filled porous solids and its application to overthrust faulting. *Bull. Geol. Soc. Am.*, 70: 115—166.
Keep, C.E. and Ward, H.L., 1934. Drilling against high rock pressures with particular reference to operations conducted in the Khaur field, Punjab. *J. Instn. Petrol. Technol.*, 20: 990—1026.
Kidwell, A.L. and Hunt, J.M., 1958. Migration of oil in Recent sediments of Pedernales, Venezuela. In: L.G. Weeks (Editor), *Habitat of oil*. Am. Ass. Petrol. Geol., Tulsa, Okla., pp. 790—817.
Kok, P.C. and Thomeer, J.H.M.A., 1955. Abnormal pressures in oil- and gas reservoirs. *Geol. Mijnbouw*, 17: 207—216.
Kozeny, J., 1927. Über kapillare Leitung des Wassers im Boden, (Aufstieg, Versicherung und Anwendung auf die Bewässerung). *Sitzungsber. Akad. Wiss. Wien*, Math. Naturwiss., Abt. IIa, 136: 271—306.
Magara, K., 1968. Compaction and migration of fluids in Miocene mudstone, Nagaoka Plain, Japan. *Bull. Am. Ass. Petrol. Geol.*, 52: 2466—2501.
Magara, K., 1971. Permeability considerations in generation of abnormal pressures. *J. Soc. Petrol. Engrs.*, 11: 236—242.
Magara, K., 1973. Compaction and fluid migration in Cretaceous shales of western Canada. *Geol. Surv. Can. Pap.*, 72—18: 81 pp.
Maxwell, J.C., 1964. Influence of depth, temperature, and geologic age on porosity of quartzose sandstone. *Bull. Am. Ass. Petrol. Geol.*, 48: 697—709.
McCrossan, R.G., 1961. Resistivity mapping and petrophysical study of Upper Devonian inter-reef calcareous shales of central Alberta, Canada. *Bull. Am. Ass. Petrol. Geol.*, 45: 441—470.
Perry, E. and Hower, J., 1970. Burial diagenesis in Gulf Coast pelitic sediments. *Clays Clay Miner.*, 18: 165—177.
Powers, M.C., 1967. Fluid-release mechanisms in compacting marine mudrocks and their importance in oil exploration. *Bull. Am. Ass. Petrol. Geol.*, 51: 1240—1254.
Rieke, H.H. and Chilingarian, G.V., 1974. *Compaction of argillaceous sediments*. Elsevier, Amsterdam, 424 pp.
Rittenhouse, G., 1971a. Pore-space reduction by solution and cementation. *Bull. Am. Ass. Petrol. Geol.*, 55: 80—91.
Rittenhouse, G., 1971b. Mechanical compaction of sands containing percentages of ductile grains: a theoretical approach. *Bull. Am. Ass. Petrol. Geol.*, 55: 92—96.
Rubey, W.W. and Hubbert, M.K., 1959. Role of fluid pressure in mechanics of overthrust

faulting, II. Overthrust belt in geosynclinal area of western Wyoming in light of fluid-pressure hypothesis. *Bull. Geol. Soc. Am.*, 70: 167—206.

Selley, R.C., 1978. Porosity gradients in North Sea oil-bearing sandstones. *J. Geol. Soc. London*, 135: 119—132.

Skempton, A.W., 1970. The consolidation of clays by gravitational compaction. *Q. J. Geol. Soc. London*, 125 (for 1969): 373—411.

Smith, J.E., 1971. The dynamics of shale compaction and evolution of pore-fluid pressures. *J. Int. Ass. Math. Geol.*, 3: 239—263.

Stephenson, L.P., 1977. Porosity dependence on temperature: limits on maximum possible effect. *Bull. Am. Ass. Petrol. Geol.*, 61: 407—415.

Terzaghi, K., 1936. Simple tests determine hydrostatic uplift. *Eng. News Rec.*, 116 (June 18): 872—875.

Thomeer, J.H.M.A. and Bottema, J.A., 1961. Increasing occurrence of abnormally high reservoir pressures in boreholes, and drilling problems resulting therefrom. *Bull. Am. Ass. Petrol. Geol.*, 45: 1721—1730.

Weaver, C.E. and Beck, K.C., 1971. Clay water diagenesis during burial: how mud becomes gneiss. *Geol. Soc. Am. Spec. Pap.*, 134: 96 pp.

CHAPTER 4

A SYNOPSIS OF PETROLEUM GEOLOGY

INTRODUCTION

Geology is concerned with observing facts about the Earth, studying the relationships between these facts, and interpreting them in terms of natural processess in space and time. It is a descriptive science, in the first place, because the diverse data of geology cannot be interpreted until they have been accurately described and represented on a scale that the human mind can embrace. It is an electric science because it must borrow from many disciplines, notably biology, chemistry and physics. It is a speculative science in part (like archaeology and astronomy) because hypotheses based on geological data are rarely verifiable; but speculation is a valid scientific activity provided it is based on reliable observations. So the geologist's work falls into two main activities: the description of present geology and the interpretation of that geology in terms of past geology and geography. Geologists tend to specialize in such topics as palaeontology, sedimentology, geochemistry, structural geology and geophysics, and seek to advance our knowledge by concentrating on a special facet of the science.

Petroleum geology is general geology with a specific aim, and all these things apply to petroleum geology. The petroleum geologist's work also has its descriptive and interpretive aspects, but the emphasis tends to linger on the descriptive because the goal of petroleum geology is a deterministic model of the area under study — ultimately, the oil or gas field. To achieve this goal, the specialists of petroleum geology tend to work in teams (which also broadens their minds).

Petroleum geologists, whatever their speciality, tend to become either exploration geologists or development geologists. The difference is not only a matter of scale, but also of outlook that can be so different that there is danger of the one not understanding the other properly. Petroleum exploration, in its simplest terms, consists of studying large regions that do or could contain petroleum, identifying progressively smaller areas of progressively greater interest in these until a prospect worth drilling has been identified, and discovering oil or gas in one or more of these. The development geologist starts with the discovery well and a detailed seismic survey, and locates appraisal wells to assess the size and nature of the accumulation or accumulations. If petroleum is found to be in commercially viable quantities, the development geologist seeks to obtain an accurate model of the accumulation on maps and cross-sections that can be used for estimating the recoverable reserves and the siting of development wells that will produce these reserves as efficiently as possible.

The exploration geologist is concerned with regional geology deduced from surface outcrop, geophysical surveys, and the results of any boreholes drilled in the area. In spite of the enormous advances in geophysical techniques, the stratigraphy of the area may only be determinable in a rather general sense. The development geologist, on the other hand, is concerned with the detailed stratigraphic sequence, and its structure, over a relatively small area. Yet this detailed stratigraphic sequence is not obtained from a study of the rocks themselves, but rather from the electrical and other geophysical responses obtained in the boreholes.

It is commonly said that the petroleum geologist has the advantage of working without financial restraint. This is rarely, if ever, true. It is true that many areas are investigated using several disciplines, some of which (geophysics, drilling) are very expensive to apply, but all will be operating under some financial restraint. It is not so often said, but is nevertheless true, that most petroleum geological work has a *time* constraint (which is also a financial constraint) and so the conclusions may be based on inadequate data. Much of this work is carried out in parts of the world that would otherwise have waited decades for investigation, and it is carried out with time limitations. The danger of false inference is always present. When such work finds its way into print, the conclusions are likely to be accepted by geologists with no local knowledge. There are also competitive restraints. It would be invidious to mention specific published articles, but we have all experienced the frustration of papers that give formation names but no lithologies, the stratigraphy of the petroleum-bearing part of the sequence but not of those overlying and underlying it.

The great advances in geological knowledge, thought and understanding during the 19th Century are attributable largely to the construction of coal mines, railways and canals. A study of petroleum geology suggests that the construction of boreholes in the 20th Century has not contributed as much to modern geological thought as it could, and that some of the difficulties we encounter are due to the extension of concepts developed from surface and near-surface geology to the subsurface, and a failure to revise these concepts in the light of the borehole data. We shall return to this topic in later chapters.

GEOLOGICAL INFERENCE

Geological inference is a process of argument based on observations and inductions. The nature of the argument may be logical, as when determining the relative ages of superimposed strata, or by analogy with areas where the evidence is more conclusive. Conclusive geological arguments are those in which several independent lines of argument lead to the same conclusions. They are unfortunately rare. We are also guided by principles that have come to be accepted by common experience and consent, but even the Principle of Superposition applies strictly only in a vertical sequence in one place; if applied carelessly to diachronous units, it leads to erroneous conclusions.

It has been noted by previous writers that there is more disagreement than agreement amongst geologists, and that consequently many of us are in error much of the time. The major cause of disagreement is lack of understanding of all the disciplines involved. One ordinary person cannot know enough, and ignorance of one facet of a problem means that judgment is incomplete. Conflicting evidence is also a cause of disagreement, but that too is the result of ignorance.

Take for example the movement of oil through permeable rock. Gravity segregation of gas, oil and water, in that order of increasing density with depth in a single reservoir, is demonstrable in a great many petroleum reservoirs. When a well is put onto production, oil moves through the rock to the well. There is, therefore, no difficulty at all in accepting the migration of oil through permeable rocks. Yet those familiar with capillary forces insist that oil *disseminated* as a separate phase in water cannot easily be moved through the pore spaces. When we produce oil from a reservoir, we can produce only about 1/3 of the oil in place, and significant quantities of oil are left in the pore spaces of a reservoir that is being depleted. We know that it is very difficult to displace this residual oil, and impossible to displace all of it artificially with water. Evidently, physical continuity of the oil phase is necessary for oil movement; but even then, not all the oil moves. Most newly discovered oil reservoirs show not the slightest trace of oil below the oil/water interface (and those that do can usually be explained by production from

	Depth	Temperature	Water pressure	Water salinity	Water density	Oil density	Clay density
Depth	1						
Temperature	+	1					
Water pressure	+	+	1				
Water salinity	(+)	(+)	(+)	1			
Water density	(+)	(+)	(+)	(+)	1		
Oil density	(−)	(−)	(−)	(−)	(−)	1	
Clay density	(+)	(+)	(+)	(+)	(+)	(−)	1

Fig. 4-1. Qualitative correlation half-matrix of common associations with depth. Brackets indicate a tendency.

another part of the reservoir rock). We are therefore forced to conclude that oil migrates into a reservoir as a "stream" in which physical continuity of the oil is maintained. Such streams have not been identified with confidence. There is much conflicting evidence in this fundamental topic of petroleum geology, and there is no general agreement amongst geologists on the nature of oil migration.

Many geological and petroleum geological inferences are based on associations. The association of lithologies in a sequence, and the association of the contained fossils, is the basis for correlating rock units and for the interpretation of the environment in which the sediment accumulated. Consistent association of a particular fauna with a lithology leads to the conclusion that the fauna is "facies-bound". The association of petroleum with anticlines rather than synclines indicates accumulation under the influence of gravity, with an upward component of movement of the less dense petroleum through the denser formation* water.

Associations are notoriously difficult to use in geological (and other) inference because a causal relationship does not necessarily exist. Associations such as those between the consumption of alcohol and teachers' salaries, or between Denmark's birthrate and her stork population, are amusing. Perhaps we would find some geological associations amusing if we knew all the facts. Consider the following common associations with depth:
— Temperature increases.
— Fluid pressure increases.
— Water salinity tends to increase.
— Water density, as measured at the surface, tends to increase.
— Oil density tends to decrease in successive reservoirs in one field.
— Mudstone bulk density tends to increase.

These six associations with depth, seven items in all, lead to 21 consequent associations that can be shown in a correlation half-matrix (Fig. 4-1). Some of these are clearly not causal relationships. For example, the association of increasing water density with increasing temperature is demonstrably false (indirect) because water expands when heated more than the pressure compresses it, and its density decreases (Fig. 4-2). Is the increase in water salinity with depth due to increasing temperature? . . . or pressure? . . . or both? Or is it perhaps related to the increase in mudstone bulk density, that is, with the expulsion of pore water on compaction? Is the commonly observed decrease of oil density in successive reservoirs with depth to be attributed to temperature, or the salinity of the water, or to none of these?

The increase in temperature is *caused* by none of the listed factors, nor is the increase in mudstone density. It could be that the only causal relation-

---

* "Formation" does not have a strict connotation in petroleum geology outside its purely geological context.

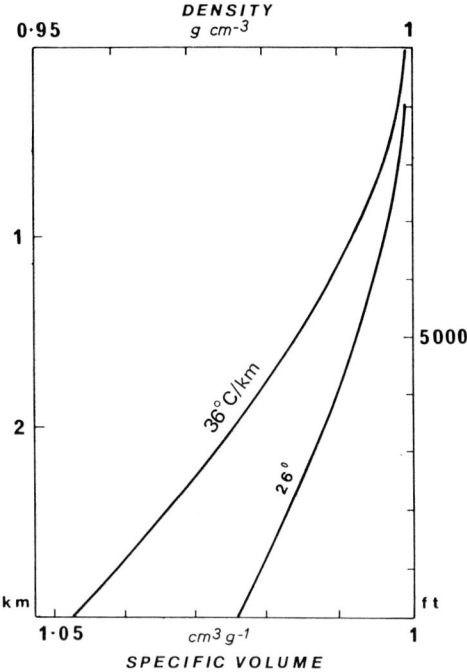

Fig. 4-2. Density and specific volume of fresh water in the subsurface at normal hydrostatic pressures with geothermal gradients of 26 and 36°C/km (14.3°F and 19.9°F/1000 ft).

ship in the table of Fig. 4-1 is that between water salinity and water density.

Two conclusions are evident: associations must be searched and researched for causal relationships, and the significant departures from normal associations must be studied. Abnormality must not be dismissed as a freak, for in abnormality (as we shall see) may lie the clues to the true causal relationships.

For example, in some oil fields of eastern Venezuela and Indonesia the density of the oil in successive reservoirs increases with depth, or a trend of decreasing density is reversed. It is also observed in some of these fields that the salinity of the associated interstitial water decreases as the density of the oil increases. Water salinity and oil density are apparently closely related — but again, closeness of relationship is not necessarily to be interpreted as a causal relationship, and causal relationships are necessary for understanding. These reversals indicate that temperature is not the dominant influence. The real association is probably environmental, affecting both the source material of the oil and the salinity of the water; these anomalously heavy oils may be generated from organic matter associated with a terrestrial environment, with fresher water. Even in the U.S. Gulf Coast, where the progressive decrease of mean density with depth has long been recognized (Barton, 1934) and used as an example, detailed study led Haeberle (1951) to conclude that variations of facies were intimately associated with the variations of oil density. This association in its turn suggests that the source of the oil is stratigraphically close to the reservoirs. However, the environmental influence is unlikely to be the only influence because temperature demonstrably affects the diagenesis of organic matter to crude oil.

The goal of petroleum geology and the industry cannot be to drill every anticline (one is reminded of the definition of an engineer as one who can do for $1 what any fool can do for $2). The goal of petroleum geology must be to understand the processes of oil and gas generation, migration, and entrapment so well that the speculative element in the industry is significantly reduced and finally eliminated. Even if this goal is unattainable, it is worth seeking; and some measure of success must be achieved if the remaining undiscovered reserves are to be discovered economically. Petroleum geology is not just a technically based variety of geology. It contibutes significant ideas as well as important data to the geological understanding of the Earth, and will certainly continue to do so.

PETROLEUM GEOLOGY

The study of petroleum geology centres logically around the three main processes — petroleum generation, migration, and accumulation. However, the emphasis is placed in the reverse order. Entrapment is the heart of the industry. It is observable, definable and measurable.

The experience of the industry is that petroleum occurs more commonly and in larger quantities in *sedimentary basins* than in areas of thin and incomplete sedimentary sequences; more commonly and in rather larger quantities in rocks of Mesozoic and Tertiary ages than in older or younger rocks; and more commonly in anticlinal traps than in other types of trap. Experience also indicates that petroleum is found more cheaply in sedimentary basins where petroleum has been found before, and in areas where petroleum has been found before.

Petroleum

The chemistry of petroleum is a subject in itself, and some knowledge of it is essential for petroleum geologists because of the importance of geochemistry in petroleum exploration and the urgent need to solve the problems of petroleum genesis and alteration in nature. The books of Tissot and Welte (1978) and Hunt (1979) are recommended for serious study.

Petroleum is a natural substance that occurs in the Earth as semi-solids, liquids, or gases — or mutual solutions of these. There are two types of petroleum liquids: crude oil, which leaves a residue of compounds of high molecular weight on distillation, and gas condensate, which leaves no residue on distillation. Petroleums are mixtures of hydrocarbons (compounds of hydrogen and carbon only) usually with some contaminant non-hydrocarbon compounds of nitrogen, sulphur and oxygen in small amounts; also of vanadium and nickel. The hydrocarbons form a great range of compounds, and each accumulation of oil and most accumulations of gas are unique in that they differ in the proportions of the constituent hydrocarbons and the contaminants. They differ also in that each hydrocarbon compound may have the atoms of its molecule arranged in two or more different spatial arrangements, called *isomers*. The isomers of a compound differ slightly in physical and chemical properties while sharing the general properties of the *Series*.

The Alkane (Paraffin) Series

A single carbon atom can attach itself to other atoms through four valency bonds, and the simplest hydrocarbon molecule is therefore one carbon atom to which are attached four hydrogen atoms. This is *methane* (CH_4), the commonest petroleum gas. A molecule with two carbon atoms may have six hydrogen atoms attached.

Methane (CH_4) Ethane (C_2H_6)

The attachment of a third carbon atom leads to yet another compound with eight hydrogen atoms (propane, C_3H_8), and continuing along these lines, we have a hydrocarbon series with the general formula C_nH_{2n+2} — the Alkane Series. The next member of the series, butane (C_4H_{10}), is also a gas; but those with *carbon numbers* from 5 to 16–20 are liquids, after which they are waxy solids. Members of a series are said to be *homologous*.

When each carbon atom is linked to the next by a single bond and the remaining bonds are to hydrogen atoms, the hydrocarbon is said to be *saturated*. Most hydrocarbons in nature are saturated.

The Alkane Series is divided into two: the straight-chain *normal-alkanes* (*n*-alkane) and the isomeric branch-chain *iso-alkanes* (*i*-alkanes), for example (leaving out the hydrogen atoms).

n-nonane (C_9H_{20})   i-nonane (C_9H_{20})

The Cyclo-Alkane Series

There are two cyclo-alkane series, one consisting of rings of five carbon atoms (cyclo-pentanes), the other of rings of six carbon atoms (cyclo-hexanes). Both have the general formula C_nH_{2n} with carbon numbers greater than five or six, as the case may be, having carbon chains attached to the ring shown below.

Cyclo-pentane (C_5H_{10})   Methyl-cyclo-pentane (C_6H_{12})

Cyclo-hexane (C_6H_{12})   Ethyl-cyclo-hexane (C_8H_{16})

The Aromatic Series

The simplest hydrocarbon of this series is benzene. It is an unsaturated cyclic hydrocarbon series, with carbon atoms forming double bonds to other carbon atoms. Two series are formed, one by substitution of hydrogen atoms by alkanes (the alkyl benzene series), the other by addition of benzene rings.

Benzene (C_6H_6)   Toluene ($C_6H_5 \cdot CH_3$)   Naphthalene ($C_{10}H_8$)

Tissot and Welte (1978, pp. 370—377) proposed a classification of crude oils based on the relative proportions of alkanes, cyclo-alkanes, and aromatics plus N, S, O compounds. These are best seen in a ternary diagram (Fig. 4-3).

While the composition of crude oil varies widely from area to area and field to field, even reservoir to reservoir in one field, its elemental composition varies very little. Carbon content ranges from about 83 to 87%, hydrogen from about 12 to 15% by weight, with nitrogen, sulphur and oxygen (commonly abbreviated to NSO or NOS) in amounts usually less than 5%. Some

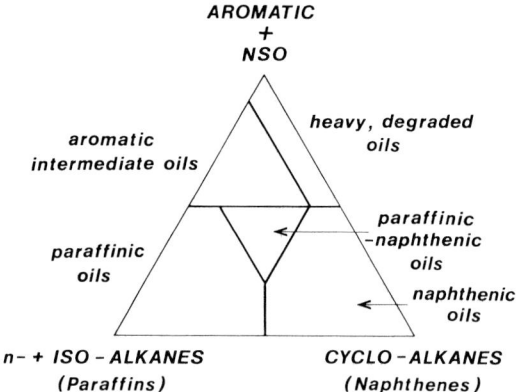

Fig. 4-3. Ternary diagram illustrating Tissot and Welte's proposed classification of crude oils on the basis of relative proportions of alkanes, cyclo-alkanes, and aromatics plus NSO compounds. (After Tissot and Welte, 1978, p. 373, fig. IV.2.1.)

crude oils contain paraffin wax in solution — up to about 15%, rarely more — that may pose problems during production and transport when cooling results in the precipitation of the wax. Wax content is geologically important, so the geologist should encourage its determination for all reservoirs.

Crude oil is almost invariably a rather dark, viscous liquid at the surface. But when we see it, it is "dead", cool, and without gas in solution. In the ground, its colour is irrelevant; but its viscosity is highly relevant. For oil of any one series, its viscosity decreases with increasing temperature, and with increasing gas in solution. Its viscosity also increases with increasing density ("gravity" in the jargon), and with increasing numbers of carbon atoms in the molecule ("carbon number" in the jargon). The density of crude oil varies, but almost all have mass densities less than 1000 kg m^{-3} (1 g cm^{-3}), and most fall in the range 750—900 kg m^{-3}. The density of crude oil is reported in *degrees A.P.I.*:

$^{\circ}$API = (141.5/s.g.) — 131.5
s.g. = 141.5/($^{\circ}$API + 131.5).

Although specific gravity is not the same thing as mass density, it is sufficiently close numerically for practical purposes.

Hydrocarbons are variably soluble in water. In the alkane series, solubility is inversely proportional to carbon number. Hydrocarbons are soluble in other hydrocarbons, and very soluble in chloroform, carbon tetrachloride, and carbon disulphide. This latter property is important to the geologist for testing rock samples for traces of petroleum. Crude oil discolours the solvent.

Crude oil and solid hydrocarbons are also fluorescent, as is the crude oil in solution in one of the solvents. When an oil-stained rock sample or cutting is examined under ultra-violet light, it fluoresces in greens or blues.

Crude oil is said to be *saturated* when no more gas can be taken into solution — or *unsaturated*. It may also be saturated, with an associated gas cap.

Petroleum gas is usually a mixture of several hydrocarbons and exists in the subsurface as separate accumulations, in association with oil accumulations, in solution in oil, and in solution in formation water. Gas is compressible, and under certain conditions of temperature and high pressure, some of the components may be liquid in the subsurface (their physical properties then being similar to those of oil). It is important as a substance in its own right as a fuel and as a feedstock for industrial chemical plants. It is also the prime source of energy in many oil reservoirs. Methane is the dominant constituent of natural gas, usually with smaller amounts of alkanes with higher carbon numbers. A gas may be *wet* and contain liquid oil vapours, or *dry*. Carbon dioxide is usually present as a contaminant; nitrogen is found in some areas, notably the German North Sea. Sulphur is a common contaminant that must be removed before the gas (or oil) is used as a fuel. Most of the world's commercial sulphur comes from this source.

Solid hydrocarbons are relatively rare, and are perhaps best known from Pitch Lake in south-west Trinidad, where the semi-solid bitumen is mined at the surface. There are other surface occurrences, such as that of the Bermúdez Pitch Lake in eastern Venezuela. They also occur as bitumen dykes. Bitumen is also valuable to the geologist in its elastico-viscous properties, for it demonstrably flows yet can be broken by a hammer.

Varieties of solid bitumen include albertite, elaterite, gilsonite, grahamite and wurtzillite. *Kerogen* is a solid bituminous substance disseminated in sedimentary rocks, and in "oil shales" and coals. There are several varieties, which will be considered in some detail in Chapter 10. It consists of about 80% carbon, with oxygen, hydrogen, sulphur, and some nitrogen. It is a *pyrobitumen*, yielding hydrocarbons on heating (in the laboratory, to temperatures much higher than those in petroleum reservoirs). Some varieties appear to be primary source material for oil, others for gas; but kerogen itself is the insoluble residue of diagenesis of organic matter, and so may be the residue of early petroleum genesis.

Water

Water is the most common fluid in pore spaces in sedimentary rocks in the subsurface. It is found in many parts of the world within a few metres of the surface; and in most parts of the world within a few tens of metres of the surface. Water has been found in the deepest wells drilled, and is probably only eliminated as free water during metamorphism.

Water is important, of course, as a natural resource when it is relatively free of dissolved solids. For the geologist, though, the role of water in the rocks is of fundamental importance, equal to that of the solid constituents. For the petroleum geologist, it could almost be said to be more important than the solids because of its association with petroleum. Geology that ignores the water in the rocks is largely meaningless: one can describe rocks without considering their water, but one cannot understand them.

All water in the rocks contains some dissolved solids. These solids, in terms of their dissociated ions, are:

calcium	Ca^{2+}		carbonate	CO_3^{2-}	
magnesium	Mg^{2+}	cations	bicarbonate	HCO_3^-	anions
sodium	Na^+		sulphate	SO_4^{2-}	
potassium	K^+		chloride	Cl^-	

There is a long list of minor constituents that includes iron, aluminium, boron, fluoride, copper, silver, tin and vanadium (the last also occurs in oil shales and in association with petroleum).

The chemistry of ground water, like that of petroleum, is a subject in itself. For general purposes, however, the geologist needs to know the *salinity* of the water in a particular rock unit, usually expressed in *parts per million* (p.p.m.) *total dissolved solids*. For electrical log interpretation, this may be converted to *NaCl equivalent* — the solution of sodium chloride that would have the same electrical conductivity as the more complex solution.

The salinity of surface water varies from almost nil in many rivers and lakes, to that of sea water, which is about 35,000 p.p.m. total dissolved solids. In the subsurface, however, salinities vary from very low in freshwater aquifers to well over that of sea water. Not only does the salinity commonly differ significantly from that of sea water, but also the ionic composition. The relative proportions of the ions vary greatly. Hence it is an unjustifiable over-simplification to regard formation water as sea water buried with the rocks. Formation waters commonly contain 80,000—100,000 p.p.m. total dissolved solids; and salinities around 300,000 p.p.m. are known.

Variations in ionic composition (as we have noted) are considerable, and rarely reflect the composition of sea water. Sulphate is significantly low in most oil-field brines, but commonly high near base metal sulphide deposits. Carbonate and bicarbonate are unimportant in sea water and usually unimportant in oil-field brines; but they are relatively important in many fresh waters. However, it must be remembered that the great bulk of formation water analyses comes from water associated with petroleum in the subsurface, and may not therefore be representative.

It is pointless to speculate on the origin of the salinity of formation water at this stage. Fresh ground water near the surface may be regarded as *meteoric* (rain water that has infiltrated and percolated into the rocks), and much of the water of artesian aquifers (when fresh or brackish) may also be regarded as meteoric. Interstitial water that is regarded as original, in the sense that it was the medium in which the sediment accumulated, is called *connate* —

but there are difficulties with this concept. The compaction of sediments, as we saw in Chapter 3, results in important redistribution of the original interstitial water. Changes in salinity may be related to this redistribution. There are therefore not only the well-known difficulties of reconciling the composition of sea water with the compositions of the surface waters discharging into the sea, but also those of reconciling the compositions of formation waters with those of sea water and fresh surface water. Reservoir engineers call the water associated with petroleum in reservoirs "connate", and the composition of this water may differ from that of the water below the petroleum in the same rock unit.

The salinity of formation water is also of interest to geologists because it affects its density, and so also the pressure exerted by a column of water.

Water also exists in sedimentary rocks as part of the molecular structure of some minerals — such as gypsum, $CaSO_4 \cdot 2H_2O$, and smectite, $Al_4Si_8O_{20}(OH)_4 \cdot nH_2O$. Under certain conditions of temperature and pressure, this water can be released to the free interstitial water. The role of molecular or lattice water is not yet fully understood, but it may have significance in both petroleum geology and structural geology through the influence its release could have on interstitial fluid pressures.

Generation of petroleum

The occurrences of petroleum in the world strongly suggest that petroleum rarely, if ever, originates in the reservoir in which it is found, but rather in other rocks, known as source rocks, from which the petroleum migrates to the trap.

The study of petroleum generation is hampered by the fact that we cannot scale the possible processes in the laboratory with confidence, so all hypotheses are based on interpretations of geochemical observations. It is also hampered by the fact that we do not know with confidence what the source rocks are for most known petroleum accumulations. If source rocks cannot be identified with certainty, they cannot be studied with understanding.

There is general agreement that the main source of petroleum is organic matter buried with a fine-grained sediment, usually a mudstone; and that diagenesis of this organic matter leads to a "proto-petroleum" which, before or during migration, becomes modified by the physical and chemical environment — particularly by increasing temperature during burial — until it sooner or later becomes the petroleum we find in the accumulation.

There is general agreement that the conditions at the depositional surface are critical for the preservation of organic matter, that aerated, oxidizing conditions are inimicable to the preservation of organic matter for subsequent alteration to petroleum. This must not be interpreted too qualitatively, because the amount of organic matter supplied to an environment is an important factor. One should view it rather as the difference between the supply of organic matter and the ability of the environment to destroy it. Large amounts of organic matter may lead to reducing conditions in an environment that would otherwise be oxidizing.

There is no obvious correlation between the composition of crude oils of different ages and the record of life on Earth, except possibly the occurrence of waxy oils in rocks of Mesozoic age and younger, after the establishment of vegetation on the continents. There is as much variation between crude oils in rocks of the same age as between those in rocks of different ages. From this it is inferred that the precursors of petroleum are the fundamental biological molecules or compounds, such as protein, fats, waxes, humus and the like.

The doubts and uncertainties that surround the processes of petroleum generation (and migration), despite the great advances in petroleum geochemistry, arise from one main factor: many scientific disciplines are clearly required for the elucidation of the processes, but we have only a limited capacity to master the knowledge we need. Judgements therefore tend to be based on inadequate knowledge of geology, biology, chemistry and physics.

Much of the evidence is apparently contradictory. Research into Holocene sediments of the Orinoco delta at Pedernales indicated that petroleum of a sort is being generated at depths shallower than 60 m (200 ft), some of it accumulating in a sand at a depth of about 35 m (120 ft) that is probably less than 10,000 yrs old from ^{14}C dating (Kidwell and Hunt, 1958). But there seems to be a consensus amongst petroleum geochemists that the ceiling of major oil generation is at depths around 1500 m (5000 ft). Not all oil fields can be reconciled with such a deep ceiling, and Hunt (1979, p. 355, table 8-2) gave the richest depth interval for the 236 largest fields known in 1956 as only little deeper than 1000 m.

Research into the petroleum content of mudstones in the Los Angeles basin, California, tends to support a conclusion of late generation, because petroleum hydrocarbons were not apparently generated at shallow depth, and the composition of the hydrocarbons in the fine-grained rocks does not approach that of the accumulated petroleum until depths of about 4000 m (12,000—13,000 ft; Philippi, 1965).

We therefore return to the problems of geological inference. Does the absence of petroleum at shallow depths in an area indicate that it was only generated deeper? . . . or it is that there are no source rocks at shallow depth? Should a petroleum source rock contain petroleum of the same composition as that expelled? . . . or do some components migrate more readily than others? . . . or does petroleum in the reservoir alter during burial? Does the presence of hydrocarbons at shallow depth in the Orinoco delta mean that it will become a significant accumulation, given the right conditions, in a few million years time? . . . or is this part of the immense quantities of petroleum that are considered to be lost through lack of a trap?

One conclusion is clear: we cannot reasonably postulate that petroleum generation is not going on today, for that would make this period of time unique in the last 600 m.y. or so of world history. Until the present-day source rocks can be identified with certainty and studied, the processes will necessarily remain rather speculative. Nevertheless, geological arguments can narrow the areas of doubt and unify to some extent the conflicting evidence (as we shall see in Chapters 9—11). Above all, it must be remembered that the problems of petroleum in the rocks are fundamentally geological problems. No chemical or physical explanation can be satisfactory unless the geological explanation is also satisfactory. If a geologist feels in need of encouragement in this point of view, he should read some of the geological pronouncements of the great 19th-Century physicist, Sir William Thomson (Lord Kelvin)!

Migration of petroleum

Migration of petroleum into accumulations is thought of largely in terms of permeability paths. It must also be thought of in terms of fluid potential gradients or energy gradients — that is, deviations from hydrostatic equilibrium in the fluids in the pore spaces of the rocks. Permeability, as we have seen, is but one component necessary for flow; it is a measure of the transmissibility of fluids through a rock. An energy gradient in the fluid is necessary for it to flow, and migration paths are paths of continuously decreasing energy, such that the energy of each particle of the fluid migrating decreases continuously during its migration.

Migration paths may be long or short. Hypotheses vary from simple migration from the source rock into the adjacent *carrier bed*, and through the carrier bed to the trap, on the one hand, to multiple-stage migration from a distant source, some with intermediate trapping from which the petroleum is spilled by subsequent geological events, on the other. Much of the uncertainty is because of the uncertainty about the true source rocks and their positions.

Migration is divided into two stages: *primary migration* from the source rock to the permeable carrier bed; *secondary migration* from there to the trap, through one or more carrier beds (Illing, 1933, p. 232).

Opinions differ widely on the state of petroleum during migration, whether in solution, in colloidal solution, or as a separate phase in water. Petroleum occurs in solution in formation waters; it occurs as emulsions in the production processes; and it occurs as a separate phase in the trap. Each possibility has its merits and its problems.

Migration in aqueous solution has the merit of requiring least work for its transport through the rocks. The problems relate to the quantitative sufficiency of this process in view of the relatively low solubility of petroleum in water, and the need for some physical or chemical change to be imposed on the solution for the release of petroleum to a separate phase in (or on the way to) the accumulation.

At the other extreme, migration as a separate phase in water has the merit that this is the state found in the accumulation. The problems relate to the mechanical difficulty of transporting petroleum as a separate phase when it is disseminated through the pore spaces with water (and disseminated it must be at some stage between generation and accumulation, because the source material itself is disseminated). This difficulty is particularly great if primary migration through a fine-grained source rock as a separate phase is postulated.

Transport in a colloidal state has the merit that it reduces the difficulties of the alternatives — but it also has a problem relating to the need for an agent to emulsify the petroleum and an agent to de-emulsify it.

Combinations of these have also been proposed, of which perhaps the most attractive is transport as a separate phase, with residual petroleum being dissolved in the formation water.

From the great diversity of processes proposed, the only conclusions that would receive general support from petroleum geologists would be the following:

(1) Mudstones are the principal source rocks of petroleum, so primary migration takes place in mudstones along energy gradients generated by compaction. In some areas, notably the Middle East, fine-grained carbonate rocks appear to have been petroleum source rocks.

(2) Petroleum, probably as a separate phase, is expelled from the source mudstone during compaction, and migrates through permeable carrier beds under the influence of gravity (and the hydrodynamic field) to the trap — or the surface.

Entrapment of petroleum

The accumulation of petroleum in a trap is accomplished when the physical properties and geometry of the rocks prevent further migration. Since petroleum is less dense than water, the barrier to further migration is, in general, such that it prevents upward, or an upward component of, migration.

The main forms of petroleum traps are well known: they consist of the *anticlinal* trap, the *fault* trap, and the *stratigraphic* trap (with which *unconformity* traps are classified in spite of the fact that this is a structural term). The entrapment of petroleum in anticlines was one of the first principles of petroleum geology, and it has dominated petroleum-geological thinking. Its essential features are (Fig. 4-4): (a) a geometrical closure, that is, the structural contours on top of the reservoir form closed rings; (b) a reservoir rock that has permeability (which implies porosity); and (c) a fine-grained, relatively impermeable cap rock that overlies the reservoir and seals it.

One must be specific and state that the closure must be on top of the

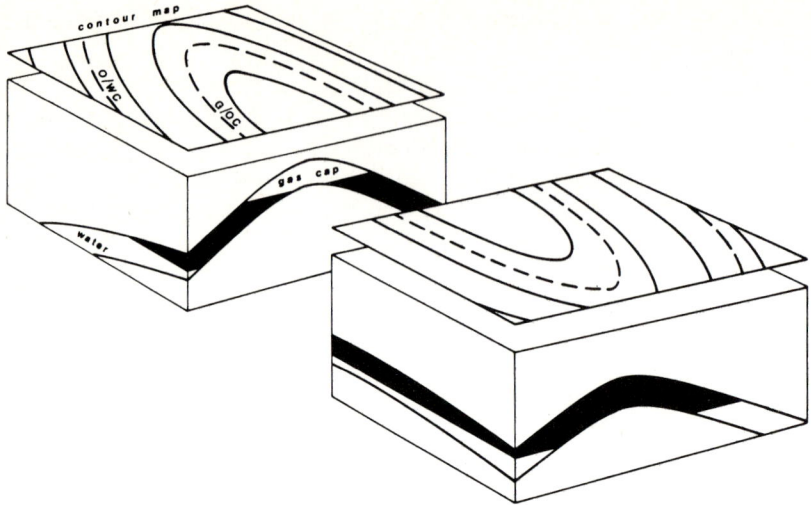

Fig. 4-4. Block diagram of petroleum in an anticlinal trap.

reservoir rock and under the cap rock because there are many anticlines in which the geometry of the individual rock units changes with depth.

Entrapment may be in a single reservoir, or in multiple reservoirs in the same anticline. A single reservoir accumulation is sometimes referred to as a "pool", but the terminology is not standard over the world. The general term for an accumulation from which production has started is a *field* — oil field or gas field, as the case may be.

The nature of the petroleum and its composition vary from one field to another; and they may vary from one reservoir to another in the same field. Variation between petroleums of different reservoirs in a single field is commonly from heavy oil with little or no gas near the surface, through intermediate oils, to light oils with gas, gas and condensates at depth.

Within a reservoir, oil lies on water; and the interface — the *oil/water contact* (O/WC or OWC) — is a horizontal or nearly horizontal surface. If free gas is present when the reservoir is discovered, it lies on the oil and the *gas/oil contact* (G/OC or GOC) will be horizontal. Gas that segregates as a result of changes in the physical state of the reservoir during production will not necessarily have a horizontal interface with the oil because the accumulation will depend on the *effective permeability* of the reservoir to gas, and this may not be uniform.

The oil/water and gas/oil contacts of different reservoirs in the same field are usually at different depths below datum. Occasionally the oil/water contact of two or more reservoirs is found at the same depth below datum. When this happens, it is argued that these reservoirs are connected in some way, leading to a common plane of physical equilibrium. Gas/oil contacts may also coincide, but they are more difficult to assess unless they can be shown to be original gas/oil contacts that have not been affected by production.

The quantity of petroleum, in terms of the volume of oil or gas *in place* in the reservoir also varies from reservoir to reservoir. Reservoirs are rarely full, in the sense that the petroleum column rarely extends down to the point at which it would spill out of the trap (the *spill points*). They are never full in the sense of occupying all the available pore space, because petroleum does not displace all the water in the pore spaces. *Irreducible water saturation* of 20 to 40% of the pore space is quite common (this is the reservoir engineer's "connate water"). However, if the reservoir exists by virtue of cracks, joints and fissures, the total volume of these may be very nearly filled with petroleum due to their large volume in relation to the enclosing surface area and the consequent relative reduction in the water adsorbed to solid surfaces and trapped in the interstices.

The volume of petroleum in a field, and its distribution in different reservoirs, is clearly important. The volume of oil or gas in place in a reservoir is estimated by estimating the total volume of rock that contains oil or gas, multiplying this by the best estimate of the mean porosity of the reservoir, and then multiplying the product by the best estimate of the proportion of oil in the pore spaces (the oil saturation). The volume of *recoverable oil* — oil that is recoverable by present technology at present prices — is obtained by multiplying the volume of oil in place by a *recovery factor* (usually 25—30%). The volumes of gas are given at standard temperature and pressure, such as 15°C and 760 mm of mercury (the standard always being stated). During production, these estimates are revised by reservoir engineers on the basis of performance, new data, trends and theory.

The sizes of anticlines and anticlinal accumulations vary greatly. The Middle East is famous for anticlines that extend for several hundreds of kilometres and accumulations 100 km (60 miles) long, such as Kirkuk in Iraq. We shall examine the size-distribution of oil fields in Chapter 7, but it is worth noting here that 2% of the world's petroleum fields contain about 80% of the known recoverable reserves.

The *area* occupied by a petroleum accumulation is important mainly for its influence on the chances of finding it, but it also influences the economics of producing it. A thin extensive reservoir is more expensive to develop than a thicker and less extensive reservoir with the same porosity and permeability, and the same volume of original oil in place.

The entrapment of petroleum by a *fault* (Fig. 4-5) involves: (a) an inclined reservoir; (b) a cap rock; (c) a fault that forms an up-dip barrier across the reservoir either by virtue of the fine-grained, relatively impermeable material in the fault plane itself, or by virtue of such material being in juxtaposition across the fault; and (d) some barrier in the reservoir along the fault, to prevent lateral migration.

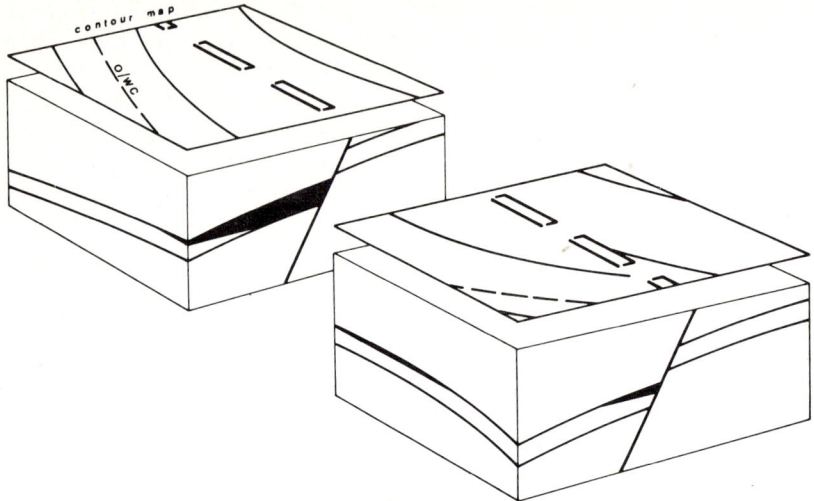

Fig. 4-5. Block diagram of petroleum in a fault trap.

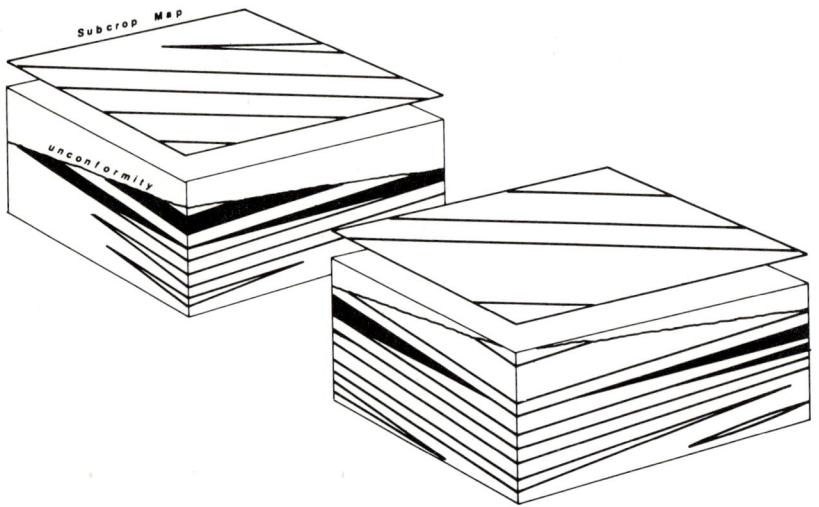

Fig. 4-6. Block diagram of petroleum in a stratigraphic (mainly unconformity) trap.

As with anticlinal traps, fault traps may include single or multiple reservoirs, and the oil/water contacts of multiple reservoirs are usually at different levels, but may be the same. Displacement of a fluid contact in the same reservoir (lithological) unit across a fault is taken as evidence of fault trapping; and conversely, observation of different fluid contacts in the same lithological unit is regarded as evidence of an intervening sealing fault.

The accumulation of petroleum in *stratigraphic* traps (Fig. 4-6) exists essentially because of a restraint on further migration that is due to stratigraphic causes. Of the many possibilities, the two most important are organic reefs enclosed above by mudstones, and unconformities where the subcrop of the permeable formations is sealed by mudstone on the unconformity. Some very large fields are palaeogeomorphic traps, where transgression has led to porous and permeable beds against a contemporary topographic feature that was covered by mudstone as the transgression proceeded. Wedging, lensing, and convergence of rock units also cause stratigraphic traps.

In practice it is not always possible to assign an accumulation exclusively to a single class of trap. Many anticlinal traps are faulted, and are classed as *structural traps*. Most unconformity traps are also faulted, some also folded. Organic reefs and palaeogeomorphic traps are usually not folded, but may be faulted, and they are usually purely stratigraphic traps.

GEOPHYSICS

It is not my purpose here to review the role of geophysics in petroleum geology, but rather to pick out some salient points to illustrate the importance of the topic in petroleum geology.

Geophysics is the study of the physical properties of the Earth and the study of the Earth through its physical properties. It is concerned with the distribution of matter in and around the earth, particularly that within the Earth to depths to 10 km or so. It is concerned with the force of gravity and its variation from place to place; the strength and direction of the magnetic field and their variations from place to place; the occurrence of earthquakes and the transmission of their energy, and the propagation of artificially generated elastic (seismic) waves through the earth.

As applied to economic geology, geophysics consists of measuring the physical properties of the rocks beneath an area (gravity, magnetism, radioactivity) and measuring the response of the rocks to artificially generated energy fields (seismic, electric or electro-magnetic). The scale of such investigations varies from the global to the very local (such as the measurement of the velocity of sound in the wall of a borehole). The application of geophysical techniques to the search for petroleum has not only been of immense importance to the petroleum industry, it has also provided an important stimulus to the broader development of the science of geophysics. It is the practical application, largely, that has kept geophysics closely related to geology.

Although many of the early regional exploratory surveys were based on magnetic and gravity methods, and gravity surveys were used to detect salt domes and other structures, their importance in petroleum exploration has declined. The enormous improvements in seismic technology since 1960, largely due to improvements in computer technology, in turn, often stimulated by the needs of the exploration industry, have made seismic surveys the most important geophysical tool in petroleum geology.

Just as we can get an echo from across a steep valley when we shout or fire a gun, so can we get an echo from rock surfaces beneath us when we cause an elastic wave or pulse to pass down through the rocks. Just as we can estimate the distance to the cliff echoing our voice by measuring the time taken for sound to cross the valley and return at a speed of about 330 m/s, so can we estimate the depth to a subsurface reflector by measuring the *two-way time*, and estimating or measuring the velocity of the energy wave through the porous rocks (typically 2—4 km/s). A seismic reflection survey consists of generating energy waves from *shot points* and recording the reflections by means of *geophones* in known positions. At each point, a reflection trace is recorded, and these are assembled side-by-side into a *record section* (Fig. 4-7). A record section is therefore a graphic representation of the geology along the line of survey, with the vertical scale in units of *time* — almost invariably, two-way time — and the horizontal scale in units of shotpoints. This basic data, which is also recorded in forms suitable for com-

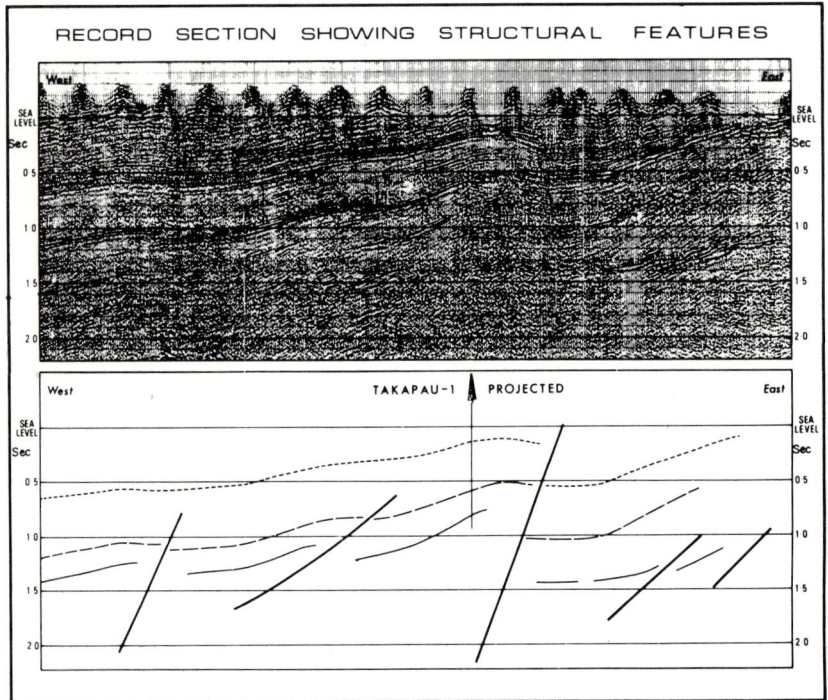

Fig. 4-7. Seismic record section (reflection profile), with line interpretation. (Courtesy of Beaver Exploration Australia NL, from Leslie and Hollingworth, 1972, *J. Aust. Petrol. Explor. Ass.*, 12, p. 42.)

puter storage and analysis, can be processed to give depth rather than time, and to correct for geometrical effects (when they are called *migrated sections*).

It is important to realize that record sections are distorted, and can be seriously distorted, representations of the geology. In the first place, the elastic energy wave travels radially outwards from its source, as an expanding hemispherical shell; but since the velocity of propagation is a function of rock density, it increases with depth, in general, and the hemisphere is soon distorted by the greater vertical velocity. We can therefore ignore the expanding wave and concentrate on the path of that part that will be reflected to a geophone from a surface.

At each reflecting surface, part of the energy is returned as a reflected wave, part passes on. If this reflecting surface is horizontal, the echo will come from vertically below the line of shot points. But if, as is usually the case, the attitude of the beds is not horizontal and the dip not strictly parallel to the line of shot points, reflections in general come from outside the line of the record section. Because record sections are constructed by placing the record vertically beneath the shot point, the reflection may appear too shallow, and displaced laterally.

The second distortion is due to the fact that different lithologies propagate the energy pulse at different velocities, so distorting the hemisphere even more. This distortion is not usually serious in uniform beds that are nearly flat because it affects all reflections in much the same way. But if there is a concentrated mound of material that propagates the energy wave at a velocity different from the surrounding material, such as a fossil coral

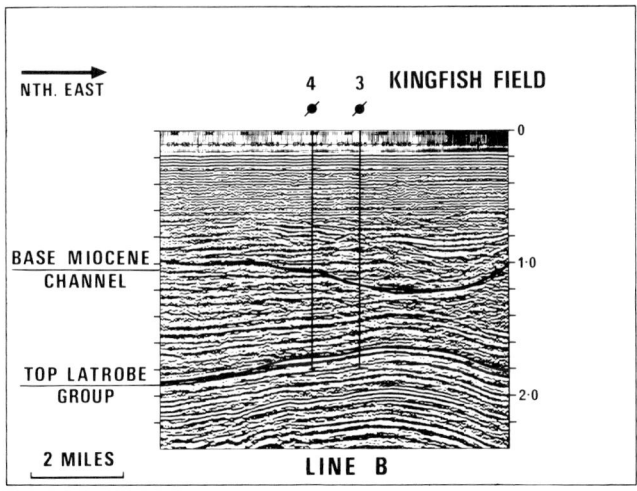

Fig. 4-8. Seismic record section across Kingfish field, Gippsland basin, S.E. Australia. (Courtesy of Esso Australia Ltd.)

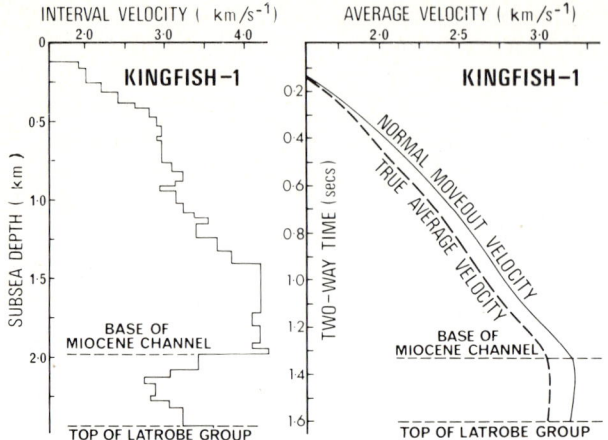

Fig. 4-9. Velocity profiles, Kingfish field. The true average vertical velocity is less than that indicated in a seismic record section because the reflections do not come from vertically beneath the shot point. (Courtesy of Esso Australia Ltd.)

reef or a volcanic cone, the echo from the base returns sooner than that from a reflector outside the mound, and so appears higher. This is called *seismic pull-up*. A fault may have a similar effect by causing a lateral velocity discontinuity in the section, so distorting the record section. A gas accumulation may depress deeper reflectors (*pull-down*) by virtue of reduced velocity in the gas-bearing sand (but not all gas sands cause pull-down).

A good example of pull-up occurs in the Gippsland basin of south-east Australia. The seismic record section across the Kingfish field (Fig. 4-8) shows an anticline underlying a channel. This Miocene channel is filled in places with a dense micritic limestone, and borehole data revealed that it has a velocity up to about 4.5 km/s, whereas the mudstones above and below have velocities of 2.4—2.8 km/s (Fig. 4-9). The high-velocity channel fill has distorted the crest of the anticline, "pull-up" causing it to appear immediately beneath the channel. The true crest on this line is near the well Kingfish 4. The same distortion appears areally, the "time" crest being aligned with the channel*.

Not all reflections on a seismic record section are necessarily simple reflections; some are spurious, and called *multiples*. There are various causes of multiples, but all relate to reflections between layers, from the bottom of one back to the top of a deeper reflector, and then back to the geophone, and so appear on record sections as parallel events due to the constant time

---

* From a talk by D.I. McEvoy, "Velocity analysis from seismic data in the Gippsland basin, Australia", presented at the 44th Annual Meeting of the Society of Exploration Geophysicists, Dallas, 1974.

delay. Record sections can be processed to remove multiples when they occur.

The seismic trace at a given point not only records reflections from layers at depth, but also a considerable amount of "noise", geological and other. Random noise is eliminated by making repeated records in the same position so that when the records are combined, random noise tends to cancel out while the real events are enhanced.

Record sections are worked into seismic maps by correlating reflections within record sections and between record sections, and plotting them either in units of two-way time, or depth. These are contour maps, and they are read in the same way as other contour maps (with the reservations mentioned above for time contours); and isochron or isopach maps can be constructed. These are usually the first maps of subsurface data in an area, and from them the most prospective areas are chosen. Some idea of the rock types and their distribution is also required.

The interpretation of seismic record sections in terms of probable lithologies is based largely on the apparent velocity of propagation. The length of modern geophone spreads, commonly over 3 km, is such that the difference in time taken for the reflected wave to travel to different geophones can be used to compute velocities. The interval velocity between two points is the difference between the average velocity above the bottom point, and the average velocity above the top point, of the interval. Interval velocity analysis is carried out by computer and shown in a plot like that in Fig. 4-9, except that the time intervals are rather coarser when there is no well control. Record sections can be processed and replotted in terms of interval velocities, and from these the distribution of lithologies can be inferred.

Seismic stratigraphy

The spectacular improvements in seismic technology that resulted in record sections of great clarity with little noise, such as in the accompanying figures, also led to the development of what naturally came to be called "seismic stratigraphy" (see Payton, 1977). Individual reflectors could be followed over considerable distances, and their relationships with other reflectors, above and below, studied and mapped. As studies extended over wider and wider areas, certain characteristic patterns emerged, analogous to those known in classical stratigraphy, and with them, a new terminology (Vail et al., 1977, pp. 205—212).

It will be recalled from Chapter 1 that Barrell's (1917) diastems, which result from minor fluctuations of baselevel, can probably be equated to bedding planes and interpreted as isochronous surfaces through a formation; and if the fluctuations are due to global sea-level changes, and the diastem or disconformity to a global fall of sealevel, then such surfaces will be synchronous in all active marine sedimentary basins of the world. Such surfaces

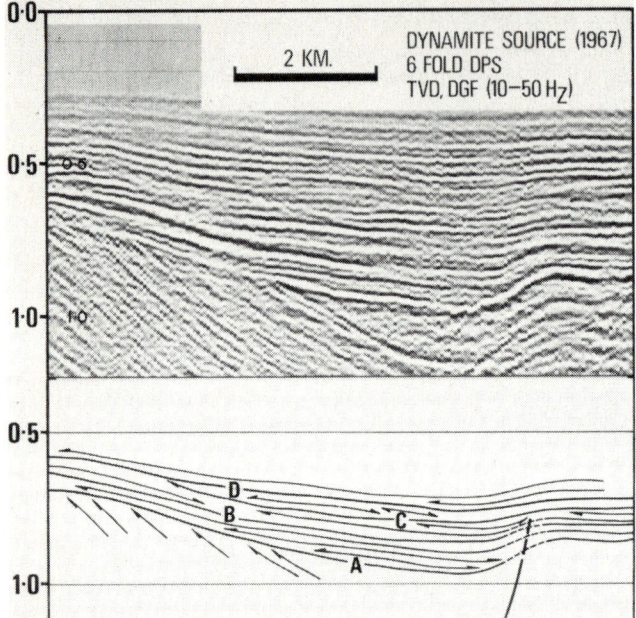

Fig. 4-10. Seismic record section showing stratigraphic detail. (Courtesy of Esso Australia Ltd.)

will not necessarily extend over the whole area of a sedimentary basin because the sediment that cannot accumulate in the area of a diastem will move on until it reaches a position in which it can accumulate, and that position will normally lie within the sedimentary basin.

From a study of many seismic record sections in many areas of the world, drawing on well control, Vail and his colleagues concluded that most reflectors are isochronous surfaces (and so, in effect, diastems and disconformities)*; and that many of the mappable hiatuses were indeed correlatable around the world, and were therefore due to eustatic sealevel falls. It can perhaps be argued that the frequency of diastems and disconformities in Barrell's concept is such that a degree of coincidence is inevitable, and precise correlation difficult to prove. Nevertheless, the value of seismic stratigraphy in elucidating the stratigraphy and structure of an area is well demonstrated.

---

* Some years ago I saw a record section from the U.S. Gulf Coast that had many good reflectors, suggesting a sequence of sandstones alternating with mudstones. The interval velocities were rather low, suggesting more mudstone than sandstone. The well drilled an almost continuous sequence of mudstones more than 3 km thick. This illustrates the point that not all reflectors are lithological boundaries in the accepted sense, despite the fact that reflection is a function of density contrasts.

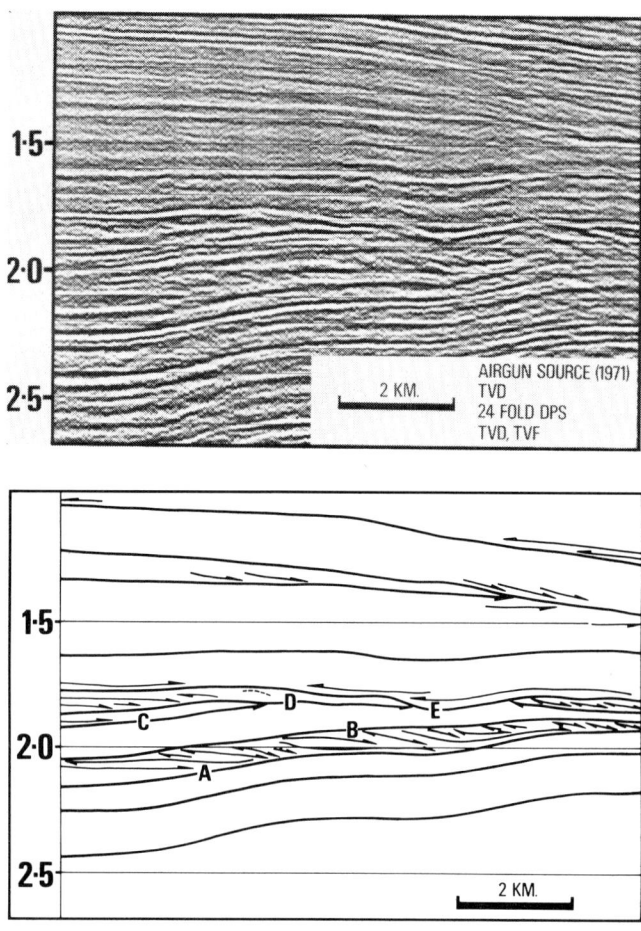

Fig. 4-11. Seismic record section showing stratigraphic detail. (Courtesy of Esso Australia Ltd.)

The first step is to identify *sequences* of conformable reflections and the *sequence boundaries*, which are 'unconformities" (an unfortunate choice of term, but perhaps its usage has become so imprecise that another matters little), and their correlative disconformities and conformities. In Fig. 4-10 there is a clear unconformable boundary (a true unconformity, also) marked A, which truncates the reflectors below. In Fig. 4-11, two clear offlapping sequences can be seen, $A-B$ and above and to the left of the letter D, where reflectors terminate on the lower boundary. The boundary B is an unconformable boundary in seismic stratigraphic terminology, but here it is more in the nature of truncated foreset beds.

Sequences and groups of sequences can be correlated, and their geological

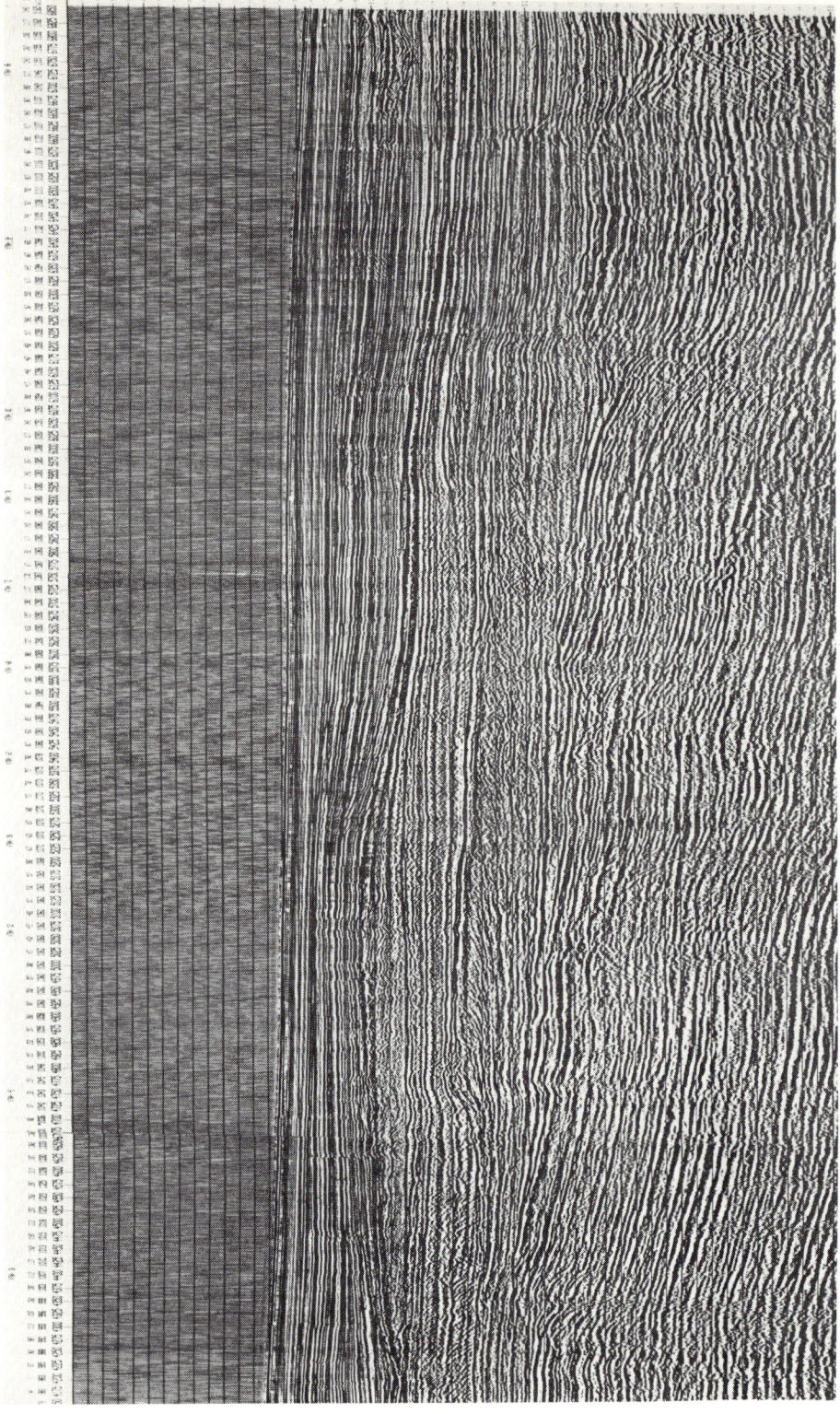

Fig. 4-12. Seismic record section across part of the Exmouth Plateau, western Australia, revealing a delta. Water depths are nearly 1 km. (Courtesy of Esso Australia Ltd.)

nature inferred from the bed geometry, so building up a detailed, three-dimensional picture of an area. Obviously, the orientation of the seismic line in relation to the sequence boundaries is important because a section parallel to the depositional strike will show spuriously conformable sequences. Figure 4-12 is part of a seismic record section across the Exmouth Plateau off Western Australia, in a water depth of about 1 km. It reveals a delta that could be mapped.

The detail obtainable in modern seismic record sections extends to gas/liquid contacts on some. These "bright spots" or "flat spots" result from reflection from the interface, but they do not always indicate commercial accumulations and their absence is certainly never to be construed as an absence of petroleum. The final test of whether or not there is a commercial accumulation of petroleum can only be made by drilling.

REFERENCES

Barrell, J., 1917. Rhythms and the measurement of geologic time. *Bull. Geol. Soc. Am.*, 28: 725—904.
Barton, D.C., 1934. Natural history of the Gulf Coast crude oil. *In:* W.E. Wrather and F.H. Lahee (Editors), *Problems of petroleum geology*. Am. Ass. Petrol. Geol., Tulsa, Okla., pp. 109—155.
British Petroleum Company Limited, 1977. *Our industry petroleum* (5th ed.). The British Petroleum Company Limited, London, 600 pp.
Dalton, L.V., 1909. On the origin of petroleum. *Econ. Geol.*, 4: 603—631.
Dobrin, M.B., 1976. *Introduction to geophysical prospecting* (3rd ed.). McGraw-Hill, New York, N.Y., 630 pp.
Dohr, G., 1981. *Applied geophysics: introduction to geophysical prospecting* (2nd ed.). (Transl. by A. Franc de Ferrière and R.A. Dawe.) Enke, Stuttgart, 231 pp.
Dott, R.H. and Reynolds, M.J., 1969. Sourcebook for petroleum geology. *Mem. Am. Ass. Petrol. Geol.*, 5, 471 pp.
Engler, C., 1888. Zur Bildung des Erdöles. *Ber. Deutsch. Chem. Ges.*, 21: 1816—1827.
Engler, C., 1889. Die Zersetzung der Fettstoffe beim Erhitzen unter Druck. *Ber. Deutsch. Chem. Ges.*, 22: 592—597.
Haeberle, F.R., 1951. Relationship of hydrocarbon gravities to facies in Gulf Coast. *Bull. Am. Ass. Petrol. Geol.*, 35: 2238—2248.
Hedberg, H.D., 1964. Geologic aspects of origin of petroleum. *Bull. Am. Ass. Petrol. Geol.*, 48: 1755—1803.
Hedberg, H.D., 1967. Geologic controls on petroleum genesis. *Proc. 7th World Petrol. Congress*, 2: 3—11.
Hobson, G.D. and Tiratsoo, E.N., 1981. *Introduction to petroleum geology* (2nd ed.). Gulf, Houston, 352 pp.
Höfer, H., 1888. *Das Erdöl (Petroleum) und seine Verwandten*. Vieweg, Braunschweig, 179 pp.
Hunt, J.M., 1979. *Petroleum geochemistry and geology*. Freeman, San Francisco, Calif., 617 pp.
Illing, V.C., 1933. The migration of oil and natural gas. *J. Instn. Petrol. Technol.* (now *J. Instn. Petrol.*), 19: 229—260; Discussion 19: 260—274.
Illing, V.C., 1938. The origin of pressure in oil pools. *In:* A.E. Dunstan (Editor), *The science of petroleum*, vol. 1. Oxford University Press, London, pp. 224—229.

Kidwell, A.L. and Hunt, J.M., 1958. Migration of oil in Recent sediments of Pedernales, Venezuela. *In:* L.G. Weeks (Editor), *Habitat of oil.* Am. Ass. Petrol. Geol., Tulsa, Okla., pp. 790—817.

Levorsen, A.I., 1967. *Geology of petroleum* (2nd ed.). Freeman, San Francisco, Calif., 724 pp.

McQuillin, R., Bacon, M. and Barclay, W., 1979. *An introduction to seismic interpretation.* Graham and Trotman, London, 199 pp.

Munn, M.J., 1909a. Studies in the application of the anticlinal theory of oil and gas accumulation. *Econ. Geol.,* 4: 141—157.

Munn, M.J., 1909b. The anticlinal and hydraulic theories of oil and gas accumulation. *Econ. Geol.,* 4: 509—529.

Perrodon, A., 1980. *Géodynamique pétrolière: genèse et répartition des gisements d'hydrocarbures.* Masson/Elf Aquitaine, Paris, 381 pp.

Philippi, G.T., 1965. On the depth, time and mechanism of petroleum generation. *Geochim. Cosmochim. Acta,* 29: 1021—1049.

Steele, R.J., 1976. Some concepts of seismic stratigraphy with application to the Gippsland basin. *J. Aust. Petrol. Explor. Ass.,* 16: 67—71.

Stuart, M., 1910. The sedimentary deposition of oil. *Rec. Geol. Surv. India,* 40: 320—333.

Thomson, W., 1868. On geological time. *Trans. Geol. Soc. Glasgow,* 3: 1—28.

Thomson, W., 1877. Presidential Address, Mathematics and Physics. *Notices Abstr., Brit. Ass. Adv. Sci.,* (Glasgow, 1876), pp. 1—12.

Tissot, B.P. and Welte, D.H., 1978. *Petroleum formation and occurrence.* Springer, Berlin, 538 pp.

Vail, P.R. and Todd, R.G., 1981. Northern North Sea Jurassic unconformities, chronostratigraphy and sealevel changes from seismic stratigraphy. *In:* L.V. Illing and G.D. Hobson (Editors), *Petroleum geology of the continental shelf of north-west Europe.* Heyden/Institute of Petroleum, London, pp. 216—235.

Vail, P.R., Mitchum, R.M., Todd, R.G., Widmier, J.M., Thompson, S., Sangree, J.B., Bubb, J.W. and Hatlelid, W.G., 1977. Seismic stratigraphy and global changes of sea level. *In:* C.E. Payton (Editor), Seismic stratigraphy — applications to hydrocarbon exploration. *Mem. Am. Ass. Petrol. Geol.,* 26: 49—212.

CHAPTER 5

DRILLING

INTRODUCTION

Those who use and interpret data must know how the data were acquired. This is particularly true for petroleum geologists because drilling operations can affect the nature of the data acquired from them. We shall begin with a short review of cable-tool drilling because it illustrates some of the principles, but it is unlikely that the reader will ever be concerned with data from wells drilled by cable tool.

CABLE-TOOL DRILLING

The young oil industry did not have to invent drilling: it simply adopted and adapted the equipment and techniques of the water-well drillers. The petroleum industry owes even more than that to the water-well drillers because it was the accidental occurrence of oil and gas in water and brine wells that encouraged the early attempts at drilling specifically for oil. Oil had been found in Pennsylvania, U.S.A., in a borehole drilled for salt, 40 years before Drake's famous well of 1859.

Holes were drilled in the ground by the cable tool, or percussion, method. The essential mechanism is shown in Fig. 5-1. The bottom of the hole is struck by a bit suspended on a cable. The cyclic motion at the surface is converted to vertical reciprocal motion, and the repeated blows cut and break the rock. By feeding the cable from the drum by means of the brake, the hole is deepened; and the drum can be connected to the motor for pulling the bit for dressing (sharpening, and restoring it to its proper diameter).

To drill a straight and round hole, it is essential that the bit rotates. This is accomplished by the lay of the rope, and a rope socket connection to the tools that allows the rope to rotate freely when slack. When lifting the tools off bottom, the weight stretches the rope, imparting a torque through the lay. This turns the bit slightly. On impact with the bottom of the hole, the rope is momentarily slack, and twists back to its natural lay, removing the torque. Lifting again imparts a slight rotation to the bit, so that on impact it is not quite in the same orientation as before — and so on. Modern ropes are of flexible steel wire with a left-hand lay, which imparts a clockwise rotation to the bit when viewed from above. Drilling not only dulls the bit, but also wears the shoulders so that it becomes under-gauge and drills a slightly

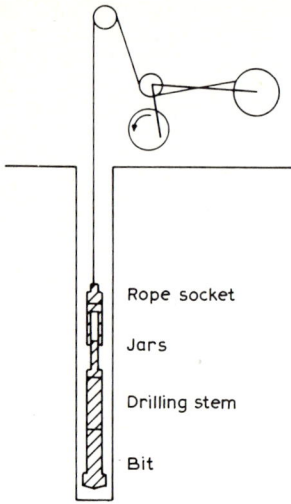

Fig. 5-1. Cable-tool mechanism, and drilling assembly.

smaller hole. A dulled bit is sharpened by heating it in a fire, restoring the cutting edge and the gauge by hammering, and then re-tempering it.

The straightness of a borehole is also important, mainly because crooked holes lead to drilling difficulties. Straightness and verticality is achieved with cable tools by drilling with a "tight" line. If the bit is hung about 5 cm off bottom before setting the rig in motion, the bit will begin to strike the bottom because of stretch in the rope and because of movement of the crown block against a spring in a slide. The rope is thus in tension when the bit strikes. This tends to keep the hole vertical and straight.

The rock fragments, or *cuttings*, that are broken off by the bit tend to accumulate at the bottom of the hole, and eventually impede drilling and wear the shoulders of the bit, making it undergauge. They are removed by bailing. The bailer is simply a length of pipe, open at the top, but with a flap-valve at the bottom that allows entry from below (Fig. 5-2). After pulling the bit, the bailer is run on the *sand line*. It is then raised and lowered in a pumping action for a few strokes. The turbulence below the bailer on the upstroke tends to suspend cuttings in the water or mud, which pass into the bailer on the downstroke. These cuttings are the fragments of the rock that have been drilled since the last bailing, contaminated to some extent by cuttings that were not bailed previously and by cavings from higher up the hole, dislodged by the rope or by driving casing. When they are viewed against drilling performance and the "feel" of the rope, a very complete and accurate log of the rock types drilled, and their depths, can be drawn up.

The role of water in the borehole is also important. It provides suspension for the cuttings, during both drilling and bailing, and it cools the bit. In most

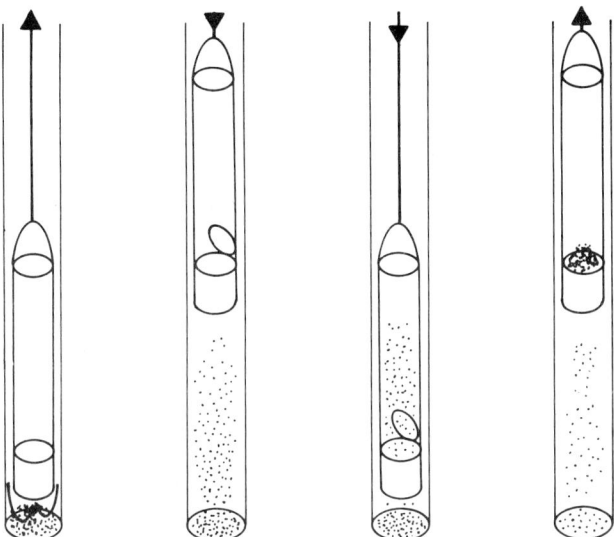

Fig. 5-2. Bailer operation.

parts of the world, any hole drilled in the ground will be found to contain some water within about 10 m of the surface. If there is no ground water, then water must be put into the hole for it cannot be drilled far without.

It was part of the traditional knowledge of cable tool drilling that if the hole contained mud rather than water, drilling performance and bailing were improved because of improved suspension of the cuttings in the mud. Often the sedimentary rocks drilled contained sufficient clay to turn the ground water in the hole into mud. If not, then clay could be added with advantage, so increasing the viscosity.

The *static water level* is always well known to cable tool drillers, and this level is watched and noted. The new bit can be heard splashing into the water when running in, and the water level can be felt on the sand line with the bailer. If the hole is straight and vertical, daylight (or sunlight from a mirror) is reflected back from the surface of the water. This is a standard check on straightness and verticality. Any change in the static water level during drilling is significant, for a sudden rise or fall indicates that an aquifer has been penetrated that has different hydraulic properties from those already drilled. If normal bailing results in a lowering of the level in the borehole, then clearly the rocks penetrated so far have little capacity to yield water. Rates up to about 1 l/s (800 gallons/h) can be achieved with a bailer at shallow depth. The *drawdown* (Fig. 5-3) is thus a measure of the permeability of the rocks to water.

In hard rocks, the bit drills a hole only marginally larger than the bit, and the margin can be so slight that a bit pulled undergauge may have been drill-

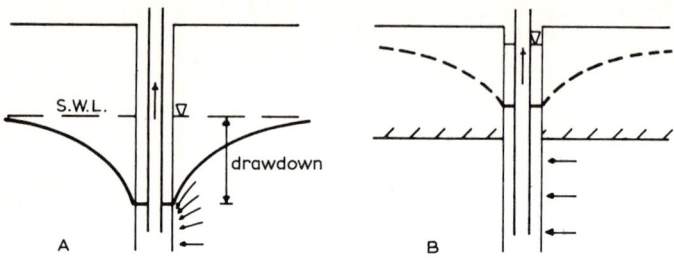

Fig. 5-3. Drawdown in producing water wells. *A:* unconfined aquifer. *B:* confined aquifer. Static water level indicated by triangle. The drawdown forms a cone of depression.

ing a hole too small to take a full-gauge dressed bit. However, in soft rocks the hole drilled is larger — sometimes substantially larger — than the bit, due to the surge of mud past the bit at each stroke.

The hole must be protected from cavings, for these can cause a bit to stick (and a stuck bit may be freed by an upward blow through the jars). This protection is given by lining the hole with steel pipe, or *casing*. At the surface, a few joints of large diameter conductor pipe will normally be cemented in the hole through the soil layer. Below this, when drilling in soft rock, it is often possible to drive a string of casing with an O.D. (outside diameter) less than the hole, and an I.D. greater than the bit. This provides almost continuous protection while drilling in soft rock. Once the casing "freezes", the hole has to be drilled ahead without this protection — but the freezing usually indicates that the borehole has reached more competent rocks. As each string of casing becomes necessary and is run, so the hole diameter has to be reduced. In general, the deeper the planned depth of the hole, the larger the diameter of the bit used to start the hole.

This method of casing a hole is satisfactory for water wells in general, but it does lead to some waste when artesian water is struck. Because only the surface conductor pipe is usually cemented (unless artesian water is expected) artesian water, when confined by well-head installations, can pass outside the casing and enter porous and permeable beds in which the water is at normal hydrostatic pressures. Clearly these were very unsatisfactory features of cable tool drilling when used for drilling for petroleum.

Although the water-well drillers' techniques could be borrowed by the petroleum industry with little or no modification, there was another disadvantage that was both wasteful and dangerous. That is that the hole during drilling was necessarily open to the atmosphere. When oil or gas was struck, there was nothing to stop it flowing out at the surface other than gravity. Oil would flow at the surface even under normal hydrostatic conditions because the column of oil in the borehole would not normally balance the column of water outside the borehole. One trip with the bailer could swab oil into the hole. Gas coming out of solution in the oil because of the reduction in pressure could accelerate the flow by reducing the mean density of the fluids in the borehole. Gas without oil, unless contained by a column of mud, would tend to pass through the mud and, by expansion of the bubbles, tend to empty the hole of liquids. "Gushers" may have been the delight of drillers, but they were wasteful both of natural resources and of reservoir energy.

Cable-tool rigs are still used extensively in the world for water drilling; and they remained competitive in some areas for oil drilling, notably in Pennsylvania (where it all started) until the 1950s. But for petroleum drilling, they have been replaced by the rotary rigs.

ROTARY DRILLING

With the introduction of rotary drilling early in this century, much of the terminology and jargon of cable-tool drilling — such as crown block, drilling line, sand line, cuttings, fishing, striking oil or gas — were passed on, and many of the principles were consciously or unconsciously adopted. While the overall performance of rotary drilling exceeded that of cable tool, the advantages were not in every respect. The quality of direct geological data tended to diminish, and the petroleum geologist tended to lose touch with the rocks themselves. These losses were to be compensated for to an important extent by the introduction and development of electrical logging.

The process of rotary drilling is too well known to need much description here. The drilling string is made up of an *assembly* of a bit, drill collars (with reamers or stabilizers), and then drill pipe to the surface. Into the drill pipe below the drill floor is screwed the kelly — a pipe of square or hexagonal cross-section — by which the rotary motion is applied to the string from the rotary table. Mud is circulated from the suction tank through the pumps to the swivel on top of the kelly; from there down the inside of the drill pipe to the bit; then up the annulus to the shale-shaker at the surface, and so back to the suction tank after passing through settling tanks in which much of the solids settles out.

At the bottom of the hole, the mud serves various purposes: it cools the bit, it assists the drilling process not only by removing the cuttings and keeping the bit clean, but also by actively scouring the bottom of the hole by virtue of nozzles in the bit that accelerate the mud into jets. The hydraulic energy of mud accelerated to a velocity of 100 m/s or more makes a significant contribution to the penetration rate and general bit performance.

The drilling mud has another essential function. Its density ("weight" in the jargon) can be adjusted so that the pressure it exerts at the bottom of the hole is greater than that exerted by the formation fluids. It thus excludes formation fluids from the borehole. Indeed, excessive mud weight can flush petroleum from the immediate vicinity of the borehole in permeable formations so that it is not detected in subsequent logging operations.

While rotary drilling was a great advance in the engineering aspects of the petroleum industry, it did not benefit the geologist in his study of the rocks penetrated by the drill. Cuttings there are, but they have been so abused by the time they reach the surface that they may be almost useless. When a more substantial sample of the subsurface rock is required, cores are obtained either by conventional coring, or by taking samples from the wall of the borehole. Conventional coring is done with a special bit and core barrel. The bit drills an annulus, leaving a core of rock to pass into the core barrel. When the length of the core barrel (less a little) has been drilled, the core is broken off by fast rotation of the bit. The broken-off core should then be retained in the barrel by spring retainers. The core assembly is pulled slowly and carefully. Each *stand* of pipe (3 × 30 ft, *joints* of drill pipe about 27 m in all) is unscrewed not by spinning the string in the hole with the rotary table as normal, but by unscrewing the stand above the slips. The expense of a core in terms of rig time alone requires justification for the core sample; and its expense demands careful treatment of it. The core is extracted from the barrel not by hanging it over the core boxes and hammering, but by taking the barrel off the drill floor and extracting it with a hydraulic pump in a horizontal position. Government regulations may require a fixed or minimum core programme.

The penetration rate while drilling is of general interest. The action of the bit on the rock at the bottom of the hole is a matter of mechanical engineering, rock mechanics and geological engineering. In the final analysis, what matters is not so much the length of hole drilled per bit, or the rate of drilling per bit, but the overall performance of the operation with its associated down-time for round trips, reaming an under-gauge hole, and other delays. From the geologists' point of view, optimization of the penetration rate, which is a parameter in the overall economy, is concerned largely with the optimization of the destruction of the rock and the removal of the cuttings.

The main parameters of the penetration rate are: (1) bit tooth geometry; (2) weight on bit; (3) rotation rate; (4) hydraulic energy and properties of the mud in circulation; (5) the fluid potential gradient across the bottom of the borehole; and (6) the "drillability" or competence of the rock at the bottom of the hole under the stresses existing in it and imposed upon it. If the first four are kept constant, the penetration rate reflects changes in the last two parameters. Geological use can be made of a penetration-rate log.

The elimination of formation fluids from the borehole is one of the essential features of rotary drilling; but it also denies the geologist the insight he used to receive from the variations in the static water table. He is no longer aware of the hydraulic properties of permeable formations, nor of their capacity to yield water. Formation fluids only enter the borehole if their energy is sufficient to displace the mud column. Although the petroleum geologist is no longer in touch, as it were, with the static water table, the geology of interstitial fluids in the subsurface is one of his major concerns.

Fluid inflow into the borehole, or loss of mud to the formations, is detected by observing the level in the suction tank. Modern rigs record this level automatically from a float, and an alarm bell is rung on the rig floor if the level reaches pre-set limits. The physical protection against a blowout (which may be defined as an uncontrolled flow of fluid through the borehole) is provided by the casing and the surface equipment, consisting of blowout preventers (BOPs). The BOP stack includes one that closes the open hole (BOP with blind rams), one that closes around the drill pipe (pipe rams), and one that combines these needs with an expandable rubber compound (Hydril).

The casing in a borehole drilled by rotary has a dual purpose: the first is protection against caving (as with cable tool holes) and the second is for the isolation of fluids by sealing the casing to the wall of the borehole.

The first string of casing is the *conductor* (Fig. 5-4), which is a large diameter pipe that may be driven into the surface materials. The purpose of this casing is to conduct the bit into the hole from below the rotary table, to prevent the unconsolidated surface materials from collapsing into the hole, and to return the mud through the shale-shaker to the tanks.

The *surface* casing is cemented into drilled hole, and also serves several functions. It is chiefly a protection against caving of relatively unconsolidated sediments at shallow depth, but also serves the important function of preventing contamination of fresh-water aquifers by the drilling mud. The depth at which this casing is set, or landed, varies with local conditions but is usually at least 100 m, and may be set at about 10% of the planned total depth. Its diameter depends on the depths scheduled for the well, and the size of the anticipated production string (also called the oil string).

The *protective* casing is the casing on which the safety of the drilling operations will depend. It is therefore set in competent rocks, but not so deep that any risk is run before it is set. It may be necessary to run and cement two protective strings before the oil or production string is run. In areas where abnormal pressures occur at depth, it will be common to set the first at a relatively shallow depth, followed by the second at the first kick or drilling break (when the penetration rate increases significantly). Protective casing is usually cemented to the surface. This may involve dual, or multiple stage cementing in which the lower part is cemented as usual by displacing cement from the casing to the annulus, and the upper part is cemented through a sleeve in the casing that is opened by a plug pumped down inside the casing. The geologist should note that while this procedure reduces the pressure necessary to displace the cement, it does not reduce the pressure on bottom due to the total column of liquid cement in the annulus. This may approach the overburden pressure.

The *production* string serves the dual purpose of isolating the reservoir or reservoirs to be produced from other reservoirs (petroleum or water), and of providing easy access for the tubing, packers, etc., that will be necessary during its producing life. The sealing function is of paramount importance. As

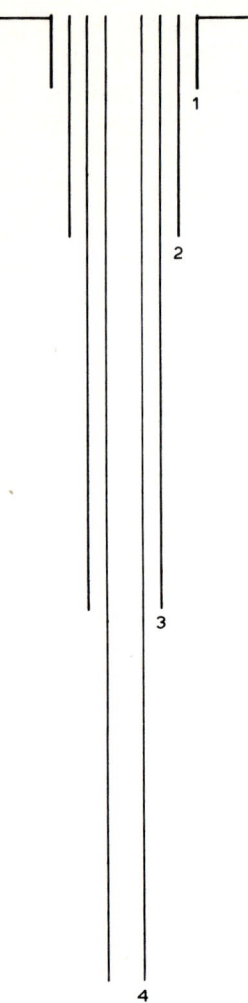

Fig. 5-4. Casing strings. *1* = conductor; *2* = surface string; *3* = protective string; *4* = production or oil string.

with the uncemented casing in a cable-tool hole that enters an artesian aquifer, so petroleum reservoirs of different energies can communicate with each other behind casing if it is not sealed. The production string is not necessarily cemented to the surface, but to a level comfortably within the protective casing. The reservoirs will be produced through perforations in the casing made by bullets or shaped charges fired from a gun lowered into the hole on a cable.

Geological considerations enter into the casing programmes of all wells. Such considerations are particularly important in exploratory holes drilled in areas without experience for guidance. The protective string must be set well into consolidated sediments, preferably landed with the shoe in an impermeable bed, such as a mudstone. Boreholes are logged before running each string of casing (except the surface string, usually), and the precise depth for the casing shoe is chosen after examination of the log. Geological considerations are not the only ones. Drilling engineering aspects, such as the amount of open hole (uncased) that it is considered safe to carry, are also important. And in many areas, government regulations determine the depths of strings other than the production string.

Depth measurements are made from the level of the rotary table, the height of which above the surface of the ground and the survey datum level is determined. All depths for geological use are converted to depths below datum. On completion of the well, the elevation of the rotary table (or derrick floor, D.F., or kelly bushing, K.B.) is recorded with reference to the top of a casing flange, so that depth measurements in service and work-over operations can be related to the driller's depths, recorded in the drilling reports.

Deviated or directional drilling

The efficient development of a petroleum reservoir sometimes requires drainage points that are vertically beneath sites that are impossible to drill from, or unpractical due to expense. Such points are reached by drilling boreholes that are intentionally deviated from the vertical below a practical drilling site (Fig. 5-5). These skills, which have grown with experience since the 1930s, are now such that it is normal practice to develop offshore fields by

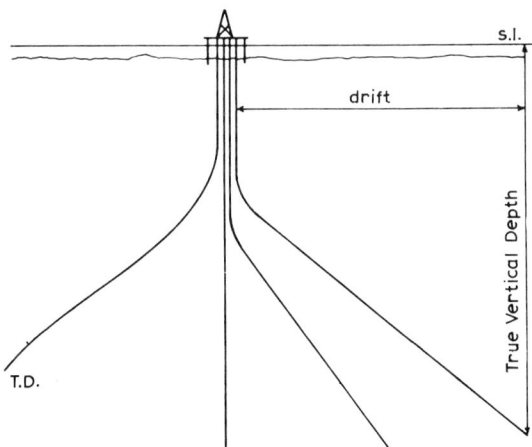

Fig. 5-5. Deviated boreholes drilled from a marine platform. *T.D.* = total depth, measured along the borehole from the rotary table or derrick floor. More than 30 wells can be drilled from a platform.

drilling 10, 20 or more wells from a single platform. The attainable precision of deviated drilling is illustrated by its other important application: the control of a well that has blown out and destroyed the rig or platform, or required the rig to be removed. A deviated hole is then drilled from a safe position, to penetrate the troublesome formation as close to the original well as possible. Water, mud and cement are then pumped from the one to the other.

The principles of deviated drilling are best understood by comparison with those of drilling vertical holes. To drill a vertical hole, the number of drill collars made up in the assembly is such that their weight in mud is greater than the weight to be put on the bit. Thus, when drilling normally, the lower part of the drill collar assembly is in compression, for which the collars are designed, while the upper part and the drill pipe are in tension. The neutral point is within the drill collars, and it moves down when the weight on the bit is reduced, and up when it is increased. A crooked hole is straightened by reducing the weight on the bit, gravity then helping as with the taut line in cable tool drilling.

A deviated hole is started vertically, and drilled to the "kick-off" point*. The bit is then pulled, and a whipstock run. The whipstock is essentially a wedge that forces the bit to deviate about $\frac{1}{2}°$ from the line of the hole. This initial deflexion is oriented in the desired direction. In areas of soft sediments, the whipstock has been replaced by a bit with one eccentric nozzle through which the mud jet blasts a hole slightly to one side, giving the same effect.

The whipstock is now pulled, and an ordinary drilling bit is run on a short assembly, or even on drill pipe without drill collars. Applying a light weight, the assembly bends, and the hole is deviated beyond the $\frac{1}{2}°$ of initial deflexion. The rate of deviation, or build-up, is controlled by controlling the weight on the bit, and so controlling the bending. The assembly is progressively lengthened to the normal length by the addition of drill collars (one of which will be of non-magnetic monel metal for deviation surveys using a compass) and, at intervals between the collars, stabilizers or reamers are inserted to hold the assembly centrally in the hole. The angle of deviation is then controlled by the stabilizer spacing at the bottom of the assembly, the length of the short drill collar inserted between the lowest stabilizer and the bit (sometimes called a "stinger"), and the weight on the bit. The course is controlled to some extent by the speed of rotation, but may also be controlled if necessary by a primary deflexion tool — the whipstock, or the bit with an eccentric nozzle.

---

* Belco Petroleum Corporation designed a "tilt rig" for developing shallow offshore reservoirs that could not otherwise be developed because the conventional method would have a kick-off point within or below the reservoirs. Grames and Reyner (1967) report on the use of these rigs for developing a shallow Peruvian offshore reservoir that lies at depths between 150 and 400 m.

Deviated drilling complicates the geologists' work by distorting the logs, compared with those of vertical holes, and by requiring correction of depths measured along the borehole to true vertical depths. In the geological planning of deviated boreholes, the execution of the plan should be borne in mind. The tolerances of position — the "tunnel" within which the borehole should be confined — should be as large as is compatible with the objectives of the borehole. Course corrections are expensive, and an unnecessarily restricted deviated hole will take longer to drill and cost more than a less restricted one.

The measurement of deviation in a vertical hole is not normally a matter that concerns the geologist greatly during drilling. The course of the well (for it is most unlikely to be strictly vertical because that too costs money through being too restrictive) will be known with sufficient precision on completion from the directional survey data acquired during drilling and logging (from the dipmeter, for instance). But in a deviated hole, the positions at different depths along the borehole vary considerably, and may vary considerably from those planned. Ultimately, the positions of significant parts of the borehole must be known as precisely as the data allows.

It is important for the petroleum geologist to be familiar with the drilling operations from which his basic data are acquired, and to learn the jargon of those who drill the boreholes. He must understand the problems of getting the data because he must make his own assessment of the value of the information he seeks and the risks involved in getting it. He must also clearly understand the limitations imposed on the data by the operations themselves. A microfossil found in a sample bag labelled 2175 m, for example, is most unlikely to have come from that depth because sample bags are usually labelled with the drilling depth at the time the sample was taken from the shale shaker. If it is the first occurrence of a species in that well, its probable depth depends on drilling rate and mud-circulation rate in the annulus.

Above all, it must be remembered that drilling and the acquisition of geological data from the boreholes are but a part of a larger operation, and there are limits on the time and money that can be spent on them.

REFERENCES

British Petroleum Company Limited, 1977. *Our industry petroleum* (5th ed.). The British Petroleum Company Ltd., London, 600 pp.
Grames, L.R. and Reyner, R.R., 1967. Unique tilt rig drills shallow offshore wells. *World Oil*, 165 (Nov.): 86—93.

CHAPTER 6

THE LOGGING OF BOREHOLES

SUMMARY

(1) Wire-line logging devices provide borehole logs that are essential documents for the description and interpretation of the subsurface geology of petroleum accumulations. They are run after drilling a section of the hole, before running casing, and they are used both qualitatively and quantitatively. While drilling, a plot of penetration rate against depth gives a valuable indication of the rock units penetrated, and it is commonly more reliable than the cutting samples obtained.

(2) The basic log is an electrical log that records resistivity and spontaneous potential (S.P.) against depth. The resistivity of a porous rock is a function of its porosity, the pore-space geometry, the resistivity of the interstitial fluids and their salinity, and their temperature. The S.P. log distinguishes mudstones and shales from other lithologies. The deflexions on the log are due to natural electrical currents circulating around the intersection of the borehole with a lithological boundary. These currents are partly due to the contrast in salinity between the drilling mud and the formation fluids, partly due to mudstone or shale acting as a semi-permeable membrane, and, to a small extent, to the movement of fluid from the borehole to the formation.

(3) Radioactivity logs record either spontaneous or induced radioactivity of rocks around the borehole. They are used for determining lithology, porosity, fluid content, and bulk density of the rocks penetrated. The gamma-ray log, which records the natural gamma radiation, can be run in cased holes.

(4) The sonic log provides a record of the velocity of sound through the rocks adjacent to the borehole by timing a signal over a fixed distance and recording the transit time. It is used for integrating borehole and seismic data, and for estimation of porosity, and also pore-fluid pressures in mudstones.

(5) The dipmeter is an electrical device that measures the dip of the strata in the borehole, the attitude of the borehole, and computes the true dip. It is used principally for the determination of structure, but can also help the elucidation of the environment of deposition of a sequence of sedimentary rocks.

GENERAL

The desire and need to know the succession of rock types penetrated by a borehole dates from the earliest days of drilling; and as the geological basis for drilling increased, so did the need for accurate and informative borehole logs. The primary need was probably for geological information from which the shape of the structure could be determined; but with increasing knowledge of the technological problems of petroleum production, a parallel need for physical data on the rocks also grew. It became necessary not only to know at what depths the various rock types were encountered, but also to know the nature of the rock, its porosity and permeability, its temperature, the nature of the contained fluids (known in this context as "formation fluids") and the depths of any gas/oil and oil/water contacts.

The process of drilling holes in the ground consists necessarily of breaking the rock and removing the cuttings from the hole. When drilling by cable tool, the cuttings were bailed from the bottom of the hole at intervals, and these were reliable samples of the rock penetrated since the last bailing. Contamination was largely confined to caving from higher parts of the hole (from driving casing, or from the lash of the cable), and the only loss was a tendency for mudstones to form mud with the water in the borehole. Fluid samples were also obtained when bailing, but these tended to be contaminated. Once oil or gas was encountered, this was usually quite obvious because it flowed into the borehole, and sometimes at the surface until it could be controlled. The compilation of all these data was incorporated into the driller's log.

The introduction of rotary drilling significantly altered the nature of the problem of logging the borehole. The mud column confined the fluids within the formations, and samples of these could only be obtained (intentionally or unintentionally) by producing them into the borehole.

The rotary drilling process is so destructive of the cuttings that it has become much more difficult to determine the nature of the rock from which they came. As holes are drilled deeper, and ways are found of improving drilling performance, so the destructive forces applied to the bottom of the hole increase. The time taken by the cuttings in transit to the surface increases as the depth of the borehole increases, so the sample taken at the shaleshaker belongs not to the present depth of drilling, but some shallower depth. From the time the cutting is broken by the bit tooth and blasted by mud that has been accelerated through nozzles in the bit, to the time it comes onto the shaleshaker, it has been thoroughly abused. It has been accelerated up the annulus between the drill collars and the wall of the borehole; decelerated at the top of the drill collars, accelerated past each tool joint at about 10 m (30 ft) intervals, all in a spiral motion, and no doubt hammered from time to time by the drill pipe. Clay fractions become part of the mud. The larger fragments travel more slowly than the smaller, the round more slowly than the angular, the denser more slowly than the less dense. The cuttings collected at the shaleshaker can hardly be thought of as reliable samples of the rocks drilled; and there are occasions when the well-site geologist should not be required to follow tradition and examine and describe all samples, because the reward cannot recompense the labour.

The sample must be taken, however, because it may contain fossils, and these form an important part of the geological record of the borehole. It must be remembered, though, that if Nature has been very selective in the matter of which organisms were preserved as fossils, the drilling process is further selective in destroying all but the micro-fossils, and even then perhaps in destroying some of the larger, or more delicate, forms. The biostratigraphic record of a borehole also suffers from uncertainties.

An essential aid to the logging of boreholes is to have installed (and properly functioning!) a penetration-rate recorder that enables one to make a plot of penetration rate against depth. If the normal operational drilling parameters are kept constant (that is, weight on bit, rate of rotation, hydraulic energy and properties of the mud) the rate of penetration is a function of bit wear, fluid potential gradient across the bottom of the hole, and the mechanical properties of the rock being drilled. In practice, when drilling normally pressured rocks, the rate of penetration varies significantly with different lithologies, and only towards the end of a bit's run does the dulling of the bit mask these changes. If a plot is kept of "minutes to drill 1 m" against depth, the resulting penetration log will commonly be found to correlate closely with the Spontaneous Potential or Gamma-ray log when it has been run. The penetration log is therefore an indispensable part of the drilling record. It is a valuable aid to making sense of the cuttings while drilling, and to choosing the logging depth when wishing to case off particular sands.

Contamination of the drilling mud is also an important part of the drilling record. Part of the fluid content of the rock drilled is retained in the cutting, the rest is in the mud. *Gas-cutting*, dilution with formation water, and any other significant changes in mud properties must be recorded, and these changes compared with the penetration log and, eventually, with the electrical logs. Modern rigs run a continuous sampling of the mud returns for hydrocarbons.

There is always a need to know as fully as possible the sequence of rocks, their properties and their fluids, as they are drilled. This need can be urgent when drilling towards a sequence with abnormal pore-fluid pressures, or when wishing to land casing at a particular stratigraphic horizon, or below a particular reservoir. The nature of the record is necessarily incomplete while drilling is in progress because so much of the evidence is destroyed.

Coring is generally too slow, and so too expensive, for general use. At the best, a core will provide a few metres of fairly reliable information, but there is a danger here of regarding it as more reliable than it really is. The core has been cut by drilling around it, broken off by spinning the pipe, removed from the subsurface conditions of temperature and pressure to the surface, and extracted from the core barrel. Samples are taken, cleaned and flushed with solvents. Parameters such as porosity and permeability can be measured with great precision from these samples, and this precision beguiles us into accepting them as valid in situ. Nevertheless, they are very valuable samples to the geologist because they reveal the lithology and environment of the sedimentary rocks cored, and provide a valuable check on porosity and permeability data obtained by other means.

PRINCIPLES OF ELECTRICAL LOGGING

Electrical logging of boreholes was developed in the late 1920s by the French engineer, Schlumberger, and the techniques grew with the demands until it is now an essential service using a wide range of sophisticated tools for subsurface logging and interpretation. Techniques have developed so fast that no attempt will be made here to discuss the details of the tools available. These details, and the description of interpretive methods, can be obtained from the handbooks of the companies performing these services. The petroleum geologist — indeed, any geologist today — must acquire a knowledge of the principles of borehole logging with wire-line devices if he is to read his logs intelligently.

The basic log used by the petroleum geologist consists of a recording of the resistivity of the rocks and the spontaneous potential in the borehole against depth (Fig. 6-1). This log is obtained in a single run in the borehole. The device is contained in a *sonde* that is lowered down the hole on a cable within which electric cables pass. Most logs are run from the bottom of the hole to the top because this direction, with a taut cable, gives more positive depth control on the log, which is measured from the cable movement at the surface. Temperature logs are run while running in; and other logs should be run when running in if the condition of the hole is such that there is danger of losing the device.

The readings from the sonde are converted to a signal at the surface that is recorded on photographic film that is wound past the signal at a speed scaled to the speed of the sonde in the borehole. The data can also be stored on tape for subsequent computer processing, but a visible record is usually required at the well site at the time of logging.

Resistivity

Nearly all the common rock-forming minerals are non-conductors of electricity. Dry porous rocks are non-conductors because the fluid in the pore spaces (air) is also a non-conductor. In the ground, however, porous rocks contain water, and the water is usually saline. Saline water, an electrolyte, conducts electricity by the movement of charged ions that result from the dissociation of salts in solution in water.

The conductivity, or capacity of ground water to conduct electricity, is

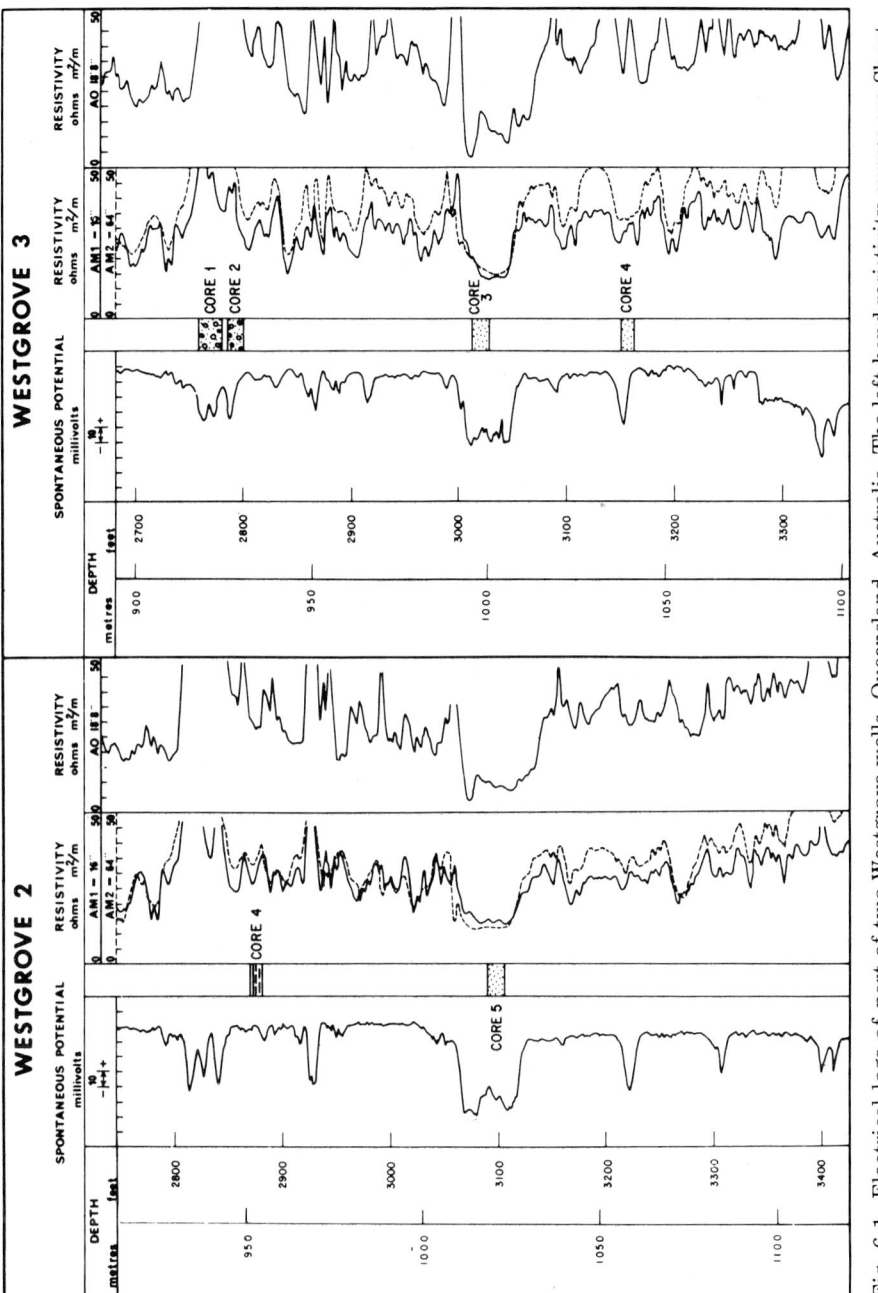

Fig. 6-1. Electrical logs of part of two Westgrove wells, Queensland, Australia. The left-hand resistivity curves are Short and Long Normals; the right-hand curves are the Inverse. Electrode arrangements are as in Fig. 6-7, but the nomenclature for the Inverse, AO, refers to the Lateral arrangement. (Courtesy of AAR Ltd, Brisbane.)

proportional to the number of ions, each of which can conduct a specific amount of electrical charge. Ground water contains many different ions, the common ones being Na^+, Ca^{2+}, Mg^{2+} (cations); Cl^-, SO_4^{2-}, CO_3^{2-} and HCO_3^- (anions). Each of these ions has a different conduction capacity, so a valuable simplification is to consider an *equivalent* NaCl solution that would have the same electrical properties as the more complex reality. In general, the more saline the solution, the more conductive and less resistant it is.

The resistivity of a material is a measure of the difficulty with which an electrical current flows through the material. It is the inverse of conductivity. Some electrical logging devices measure conductivity, but most measure (directly or indirectly) resistivity.

Ohm's law states (with certain limits on the magnitude of the quantities) that for a given conductive material, the ratio of the voltage (potential) across the ends and the current in the conductor is constant*:

$$\frac{V \text{ (volts)}}{I \text{ (amps)}} = \text{constant} = \text{resistance } r \text{ (ohms)}. \tag{6.1}$$

If the cross-sectional area A of the conductor is uniform, the resistance is proportional to the length l and inversely proportional to A. Thus, the *resistivity* R is related to resistance by:

$$r = Rl/A. \tag{6.2}$$

If the resistance is measured in ohms (Ω) and the unit of length is the metre, then the unit of resistivity is ohm $\cdot$ m^2 m^{-1}, or ohm-metre (ohm-m). Resistivity is numerically equal to the resistance of a metre cube of the conductive material.

Consider a cube of non-conductive material, $1 \times 1 \times 1$ m. The resistivity of this cube is infinite. If a straight hole of 1 cm^2 cross-sectional area is drilled through the cube normal to the two faces between which the current will be passed, and this hole is filled with water of resistivity (R_w) one ohm-metre, the *resistance* of this water in the hole will be:

$$\frac{R_w \cdot \text{length}}{\text{area}} = \frac{1 \cdot 1}{0.0001} = 10{,}000 \ \Omega$$

and the *resistivity* of the cube will be 10,000 ohm-m. If a second hole of the same size is drilled parallel to the first, and both holes are filled with water of resistivity $R_w = 1$ ohm-m, the resistivity of the cube will be 5000 ohm-m. If there are 2000 such holes, the resistivity of the cube will be 5 ohm-m, and the "porosity" will be 20%. The resistivity of the cube is thus inversely propor-

---

* The notation R for resistance and ρ for resistivity is not usually followed in well-log analysis.

tional to the "porosity" when the resistivity of the contained water is constant.

If the holes are now filled with water of a different resistivity, the resulting resistivity of the cube will be different. Other things being equal, the larger the resistivity of the water, the larger the resistivity of the cube.

In nature, the pore passages through rocks are neither straight nor of constant diameter; and almost all are interconnected in a permeable rock. (We are concerned only with *effective* porosity because pores that are completely enclosed in non-conductive material do not contribute to the flow of electricity — or, of course, the flow of fluid.) Electrical current passes only through the electrolyte saturating the pore spaces, so the paths are tortuous and longer than the direct path. The mean path length between opposing faces of a metre cube is greater than one metre. The ratio of the mean effective length (l_t) and the macroscopic length (l) is known as *tortuosity** and is a dimensionless geometrical property of the rock. On account of tortuosity, a metre cube of rock with 20% porosity, the pores being saturated with water of one ohm-m resistivity, would have a resistivity greater than 5 ohm-m.

It will be convenient to consider at this point a fundamental dimensionless material constant that is used in electrical-log interpretation: the ratio of the resistivity of a porous rock that is saturated with an electrolyte and the resistivity of the electrolyte. This is called the *Formation Resistivity Factor (F)* or *Formation Factor*, which was defined by Archie (1942):

$$R_0 = F R_w \tag{6.3}$$

where R_0 is the resistivity of a rock saturated with an electrolyte of resistivity R_w. Experiment showed that the Formation Resistivity Factor is a measure of porosity; but its significance is perhaps clearer when considered in the following manner.

Consider a rectangular block of porous and permeable sandstone through which an electrical current will be passed between opposing faces of area A separated by the length l. The fractional porosity, f, of this block is the volume of the pores divided by the volume of the block, and the volume of the pores is given equally by fAl and $A_t l_t$, where A_t is the true area normal to the tortuous electrical flow paths and l_t is the true mean length of these paths. So:

$$fAl = A_t l_t$$

and:

$$A_t = fAl/l_t. \tag{6.4}$$

---

* Some authors define tortuosity as the square of this ratio, but the simple ratio is preferred here.

If we now pass an electrical current between the two faces, the *resistance* of the block is $R_w l_t/A_t$ or, substituting eq. 6.4, $R_w l_t^2/fAl$. The *resistivity* of the block is therefore:

$$R_0 = \frac{R_w l_t^2}{fAl} \frac{A}{l} = \frac{R_w l_t^2}{fl^2}$$

Substituting this result into the definition of F (eq. 6.3), we obtain:

$$F = \frac{R_0}{R_w} = \frac{(l_t/l)^2}{f} = T^2 f^{-1}. \tag{6.5}$$

The Formation Resistivity Factor is a dimensionless material constant that is proportional to the square of the tortuosity and inversely proportional to the porosity of the material. *It is independent of the resistivity, and so also of the salinity of the pore fluid.* Not all rocks with, say, 20% porosity will have the same formation factor because their tortuosities will normally differ. Note that the tortuosity of a rock, which is also important for fluid flow, can be obtained from $\sqrt{Ff}$ (but see Winsauer et al., 1952: they measured the tortuosity by an independent electrical method and concluded that $Ff = T^{1.67}$).

Archie (1942) also inferred from experimental data that:

$$F = af^{-m} \tag{6.6}$$

where the factor a is close to unity, and the exponent m varies between 1.4 for unconsolidated sands and about 2.3 for indurated sandstones (it is unity for straight pores). This empirical result seems to bear little relationship to eq. 6.5, but it is intuitively reasonable to suppose that tortuosity is an inverse function of porosity — the greater the porosity, the smaller the tortuosity. Winsauer et al.'s tabulation (1952, p. 266, table II) may be used to obtain an empirical relationship between porosity and tortuosity. If we take their measured tortuosity, it is closely approximated by $f^{-2/3}$: if we take $\sqrt{Ff}$ as the measure of tortuosity, it is closely approximated by $0.9 f^{-\frac{1}{2}}$. Substituting these results into eq. 6.5, the first suggests that $F = f^{-2.3}$, while the second suggests $F = 0.8 f^{-2}$. If the aberrant (third) data point is omitted, the second suggests $F = 0.6 f^{-2.2}$. These data are plotted on Fig. 6-2.

In addition to Archie's formula, as eq. 6.6 is called, the following are also in common use, and there is no significant practical difference between them:

$F = 0.81 f^{-2}$ for sand

$F = f^{-2}$ for hard sandstone

$F = 0.62 f^{-2.15}$ "Humble Formula"

These results, amply justified by decades of use, suggest that $\sqrt{Ff}$ is a better measure of tortuosity than the direct measurement used by Winsauer and his

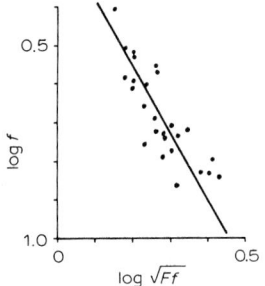 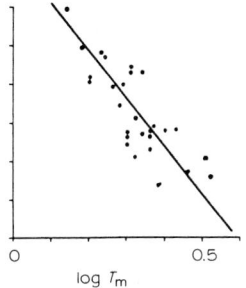

Fig. 6-2. Two measures of tortuosity plotted against porosity (log-log). Data of Winsauer et al., 1952, p. 266, table II. $\sqrt{Ff}$ is the tortuosity term derived from eq. 6.5. T_m is the tortuosity measured electrically. The lines are linear regression lines.

colleagues (who, it is interesting to note, worked for the Humble Oil and Refining Company). However, it must be clearly understood that none of these is a precise relationship, as the scatter about the regression lines in Fig. 6-2 shows.

Thus, if the value of the Formation Factor can be measured by saturating a representative rock sample with an electrolyte of known resistivity and measuring its resistivity, the resistivity of that rock saturated with a fluid of another resistivity can be calculated if the resistivity of that fluid is known. Alternatively, knowing or estimating the Formation Factor and the true resistivity of the brine-saturated rock, the resistivity (and hence the salinity) of the pore fluids can be calculated or estimated. An estimate of the Formation Factor leads to an estimate of the porosity, and vice versa, but there are better tools for measuring porosity in situ and from these the Formation Resistivity Factor can be better estimated.

So far we have considered only clean porous rock saturated with an electrolyte. These relationships do not hold for "dirty" sands, that is, sands with an appreciable clay content, because wet clay contributes to the electrical conductivity of a rock. The evaluation of dirty sands presents problems that are best referred to a specialist well-log analyst, or petrophysicist.

Contamination of the formation fluid by petroleum is very much the petroleum geologist's business. Oil and gas are non-conductors of electricity. Petroleum in a permeable rock with intergranular porosity does not displace all the water that was originally present in the pores. The oil or gas occupies the central parts of the pores in a water-wet reservoir rock (Fig. 6-3). The electrical effect is one of reduced porosity and therefore higher resistivity, but there are two components to this. First, the true area available for current flow (A_t) is small: secondly, the tortuosity is large because the current flow paths are confined to the peripheral parts of the pore spaces, where the water is.

There are great theoretical difficulties with the concept of true resistivity

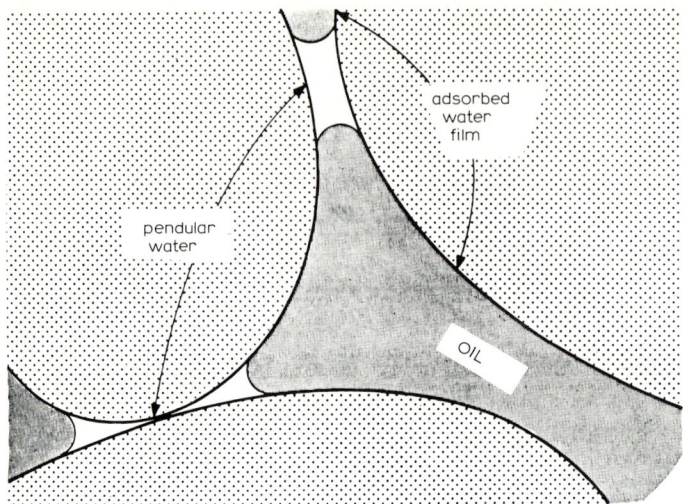

Fig. 6-3. Oil or gas occupies the central parts of the pores in a water-wet reservoir rock. The "connate" water is concentrated in pendular rings around the grain contacts.

of an oil or gas reservoir rock because there are cogent physical arguments that the water is largely confined to what are called *pendular rings* (*cf.* pendular cement) around the grain contacts, and that these pendular rings are separated from each other by a thin film of adsorbed water no thicker than 1 nm (3 or 4 water molecules). We shall consider this in more detail in Chapter 8, and must accept meanwhile that electrical current can flow through pores that are largely saturated with petroleum, and that there are reasonably satisfactory methods of estimating the water saturation (and so the petroleum saturation) of reservoirs (see Morrow, 1971a, b, for a discussion of these points). This is the principal quantitative use to which resistivity logs are put.

Saturation is expressed as a fraction or percentage of the *pore* space occupied by water, oil or gas. It is of central importance to determine the oil or gas saturation of a reservoir both for reserve estimation and for producibility assessments, the latter because water saturation above some critical value means that water only will be produced.

By definition:

$s_w + s_o = 1.$

The resistivity of an oil-bearing reservoir rock is a function of the water saturation s_w and it has been found experimentally that:

$$s_w \simeq (R_0/R_t)^{1/n} \qquad (6.7)$$

where R_t is the true resistivity of the rock containing some oil, and R_0 is the resistivity it would have if the oil saturation were zero (that is, if the pores were saturated with the water that occurs in the reservoir), and n is a saturation exponent with a value about 2. Inserting this value of n and substituting eq. 6.3 into 6.7, we get:

$$s_w \simeq (FR_w/R_t)^{1/2} \tag{6.8a}$$

or:

$$R_t \simeq FR_w/s_w^2. \tag{6.8b}$$

As mentioned before, the theoretical validity of these formulae is dubious. Resistance of adsorbed water films to lateral electrical flow is almost certainly involved. The saturation exponent n is probably also a function of saturation because the true area available for flow, and the tortuosity through the water, are also involved. With these reservations, the formulae can be put to practical use.

The true resistivity of a petroleum-bearing rock is only relatively higher than that of the same rock saturated with the same water. The true resistivity may be quite low when the Formation Factor is small (large porosity) and the resistivity of the connate water is low (large salinity, high temperature). The *resistivity index*, R_t/R_0, is used as a *guide* to possible production (it is approximately equal to the inverse square of the water saturation). Local experience will be the guide, but many reservoirs with 30% water saturation produce oil with no appreciable water, so resistivity ratios > 10 may well produce clean oil.

The determination of the true resistivity of formations (R_t) is one of the main goals of electrical logging. It is not simple. Its determination is in the realm of the specialist well-log analyst or petrophysicist.

Resistivity in the subsurface.

Resistance and, by calibration or measurement, resistivity are measured by systems of four electrodes. A known current is passed between two of them (Fig. 6-4) and a potential is developed as a consequence between the other two. This can be measured. For example, a current is passed between electrodes 1 and 4 in Fig. 6-4, and a voltmeter reads the potential between electrodes 2 and 3.

It has also been found that if the role of the electrodes is interchanged, and the same current passed between electrodes 2 and 3, the potential between electrodes 1 and 4 is identical to that originally found between 2 and 3. This interchangeability of electrodes, known as reciprocity, is used in subsurface devices so that one electrode can fulfil more than one function.

If a current is passed between two electrodes in a very large volume of homogeneous conductive material, as in Fig. 6-5, a potential field is generated throughout the material and current flows down paths from high potential to low potential. These current paths form a three-dimensional pattern such

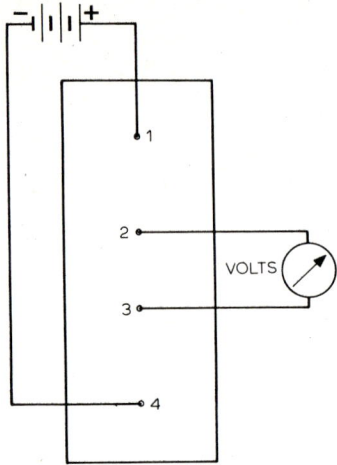

Fig. 6-4. The electrical resistivity of a block of material can be measured by passing a known current between one pair of electrodes and measuring the voltage (potential drop) between another pair.

as would be generated by the rotation of Fig. 6-5 about the axis through the electrodes. Since each path begins at the same potential and each path ends at the same potential, surfaces exist in the conductor on which all paths have the same potential. These, as with fluid flow, are called *equipotential surfaces*. Flow paths are everywhere normal to the equipotential surfaces because this is the direction of maximum potential gradient. The pattern in the field depends on the resistivity of the conductor.

In the immediate vicinity of the lower electrode, the current flows radially inwards from all directions, and the equipotential surfaces approximate spheres. If we now insert another electrode in an arbitrary fixed position, the potential between this electrode and another on any point of a sphere relatively close to the lower current electrode A would be the same (very nearly) whether to one side or vertically above or below. Any of these electrode devices could be calibrated to give the resistivity of the conductor directly.

When, therefore, we put a current electrode down a borehole filled with a resistive mud (the volume of the borehole being negligible relative to the volume of rock around it) and pass a current to it from an electrode near the surface, a potential field is set up in the rocks around the borehole (Fig. 6-6). If a pair of measuring electrodes is now inserted into the borehole, one being close to the current electrode A, the potential between the two measuring electrodes is related to the resistivity of the rocks in the neighbourhood of electrode A. This potential will be very similar to the potential that would have been obtained if it had been possible to put M into the rocks a lateral distance AM from A. The *spacing AM* is therefore regarded as a measure of

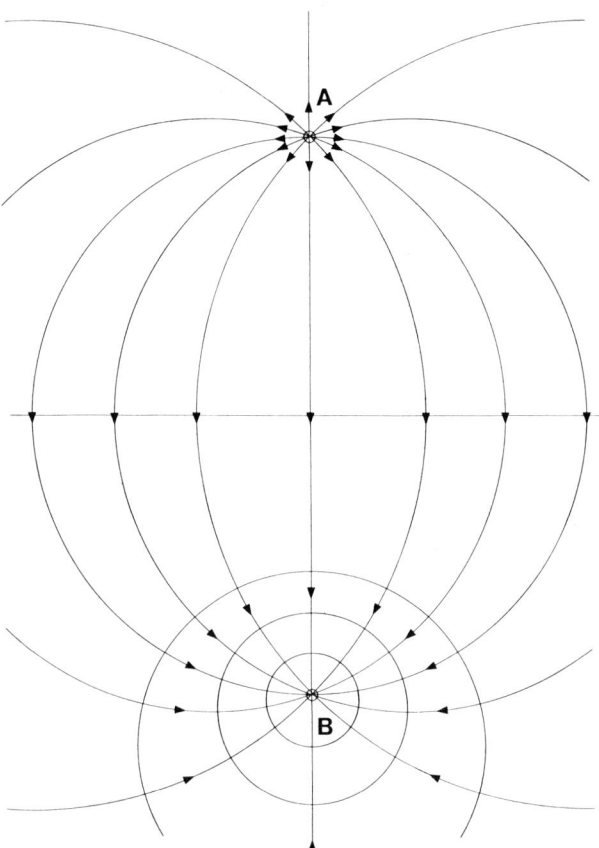

Fig. 6-5. Potential field due to electrical current passing from electrode A to electrode B in a very large volume of homogeneous material.

the lateral *depth of investigation*. The greater the spacing, the greater the "depth" of investigation.

Rock sequences are never electrically homogeneous: they consist of layers of different resistivities, and it is the purpose of electrical resistivity logging to measure the resistivities of the various layers by measuring the potential at a fixed distance close to the current electrode as they are moved up the hole. However, it will be evident that the resistivity of a bed cannot be measured if the bed is thinner than the spacing. In fact, the apparent resistivity of a bed must be corrected for thicknesses up to a few times the spacing because beds of different resistivities above and below affect the potential field. Other corrections concern the resistivity of the mud and the diameter of the borehole. If a large diameter hole is filled with a saline, conductive mud, most of the current passing between the electrodes will pass down the hole.

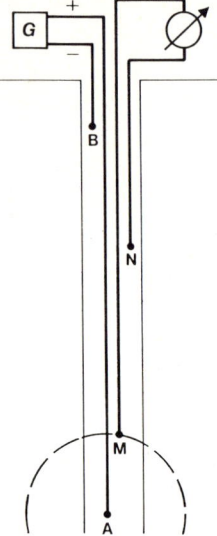

Fig. 6-6. A current passed between electrode A, in the borehole, to electrode B near the surface sets up a potential field in the rocks around the borehole. Measuring electrodes M and N, both in the borehole, will measure a voltage similar to that that would have been measured had it been possible to place M a lateral distance AM into the formations.

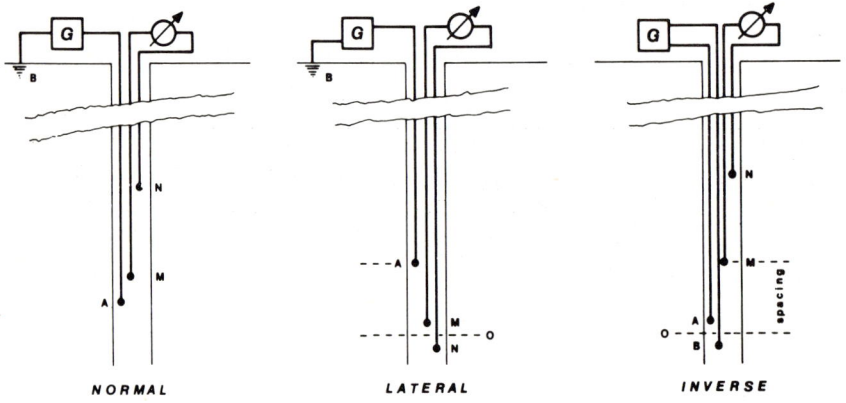

Fig. 6-7. Electrode arrangements of Normal, Lateral, and Inverse resistivity devices. A and B are current electrodes; M and N are measuring electrodes. The Inverse is equivalent to the Lateral because interchanging the electrode functions does not alter the measurement. The Inverse allows duplication of electrodes in the sonde for running Short Normal, Long Normal and Inverse simultaneously. (Courtesy of Schlumberger Seaco Inc., Sydney.)

It is worth noting before passing on that Ohm's law is mathematically and physically analogous to Darcy's law for the flow of liquids through porous media, and that if the current electrode B were replaced by a point-source of water of constant density and viscosity, and electrode A by a point sink, the water would flow in a similar pattern to the electrical flow*.

In each well of a field or district, a standard electrical log is run that will be used for comparative and correlative purposes between wells, as well as for specific detail in each well. The choice of devices in the standard log will depend on local conditions and preferences, but will usually consist of a Spontaneous Potential (SP; which will be discussed shortly), two Normal resistivity and one Lateral or Inverse resistivity devices (Fig. 6-7). These are all run in one sonde and the logs obtained simultaneously. The logging programme of a well will include many other types of logs, and the whole logging operation may well take 10—20 hours of rig time.

It is common practice to run two Normal devices with spacings (AM) of 16 and 64 inches (0.41 and 1.63 m). The first is called the Short Normal, the second, the Long Normal. The Lateral or Inverse spacing is usually 15 ft (4.6 m) or more; the Schlumberger Inverse spacing is 16 ft 8 in. (5.69 m). The reason for the different spacings is that depth of penetration can only be achieved at the expense of stratigraphic or bed detail. We need this detail, and we need to know the effects of the borehole on the formations close to the borehole, but we also need to determine the true resistivity of the formation beyond the influence of the borehole.

The Short Normal therefore gives the best bed detail, but its depth of penetration is small. Thus, if the resistivity curves across a particular water-bearing sand show that the Short Normal measures a resistivity larger than that measured by the Long Normal, it is inferred that the natural formation water salinity is greater than that of the mud filtrate (which is known). Likewise, if the sand is oil or gas bearing, the Long Normal may indicate much greater resistivity than the Short Normal, reflecting displacement of oil or gas by mud filtrate close to the borehole. A sandstone of very little porosity and permeability shows large resistivities on both Normals because there is little effect on the Short Normal from invasion by mud filtrate. The true resistivity of an oil or gas sand may differ little from that of a tight sand (of small porosity), but the physical effect of drilling through such sands creates differences that are detectable with the two Normals.

Bed thickness, borehole diameter, sonde diameter, and mud resistivity all

---

* Hubbert (1969, p. 11) records that his paper on the theory of ground-water motion (Hubbert, 1940) came to be written because, while conducting an electrical resistivity survey, he wondered what the flow pattern would be if water wells (one injecting, the other producing at the same rate) replaced the electrodes driven into the ground. Look at Fig. 6-5 as a map rather than a section. See also De Wiest (1965, p. 248, fig. 6-5).

affect the electrical logs obtained, and so must be taken into account for quantitative evaluation. This treatment is beyond the scope of this book, and reference should be make to the manuals prepared by the electrical logging companies, and to specialized books (see References). Particular care must be taken with beds thinner than the Long Normal but thicker than the Short Normal spacing because these affect the two Normals differently.

The third resistivity device is usually a Lateral or Inverse of large spacing (the Inverse is essentially the Lateral with the roles of the electrodes interchanged, on the principal of reciprocity). These differ from the Normals in measuring the potential drop between two measuring electrodes, M and N, that are close together relative to their distance from A. The effective measuring point is the mid-point between M and N (or A and B in the Inverse), usually labelled O. Thus the spacing is AO in the Lateral. All spacings are recorded on the log heading.

The Lateral and Inverse logs belong almost exclusively to the realm of the specialist well-log analyst because there is little that can be deduced from them qualitatively, and their fluctuations usually bear little obvious relationship to the beds penetrated by the borehole.

The standard electrical log will also include the SP or a gamma-ray log, which will be discussed shortly.

The contamination of formation fluids by drilling fluids is both a help and a hindrance: in any case, it is unavoidable. Drilling mud in the borehole is given a larger pressure gradient than that in the formation fluids in order to contain them, so at any given depth there is a pressure gradient, or fluid potential gradient, outwards from the borehole into the rocks. The resulting flow of fluid is insignificant into the relatively impermeable beds such as mudstones, marls, silts, and some evaporites, during the time interval between drilling through them and logging them. But it is significant into the more permeable beds, such as sands, sandstones, limestones, and dolomites.

The wall of the borehole in a granular permeable bed acts as a filter to the mud, with the result that a *filter cake* (also called a *mud cake*) forms on the wall of the borehole. Through this filter cake, mud filtrate passes to the pore spaces of the permeable rock. This contamination ranges from nearly 100% in the water-bearing sands in the immediate vicinity of the borehole (virtually complete flushing of the original fluids) to nil at some distance from it. In oil-bearing sandstones, the flushing is not complete and residual oil remains in the pore spaces. The proximal zone is known as the "flushed zone" (and parameters relating to it are given the suffix xo, as in R_{xo}) and the contaminated zone beyond this and including the flushed zone is known as the "invaded zone" (suffix i; Fig. 6-8). The dimensions of these zones are not only variable from bed to bed, but they are also likely to be variable at different depths within the same bed. In representing them as circular in plan, one makes this assumption from ignorance of their true geometry. There is also the dimension of *time* involved in the invasion of the rock unit by mud filtrate.

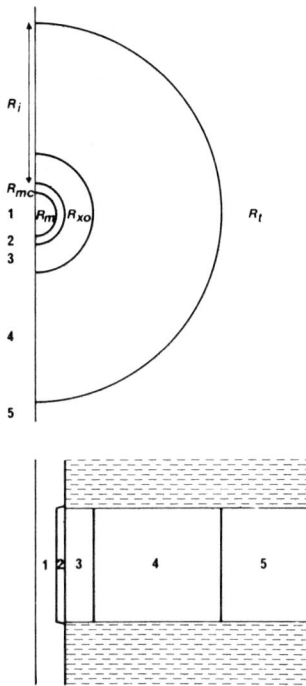

Fig. 6-8. Invasion of homogeneous, permeable, water-saturated bed by mud filtrate (plan and section). 1 = borehole filled with mud; 2 = mud cake; 3 = flushed zone; 3 and 4 = invaded zone; 5 = uncontaminated zone. Invasion of petroleum-bearing permeable beds differs mainly in that there is a greater tendency for gravity segregation, invasion tending to be deeper at the bottom of the bed.

In spite of these difficulties, it will be clear that within the flushed zone there will be a fluid the resistivity of which can be measured. If the temperature of the filtrate in the rock is known, then the Formation Factor can be obtained, in theory, by applying eq. 6.8b:

$$R_{xo} = FR_{mf}/s_{xo}^2 \tag{6.9}$$

where R_{mf} is the resistivity of the mud filtrate at the temperature of the flushed zone, R_{xo} and s_{xo} the resistivity and water saturation of the flushed zone. In a water sand, $s_{xo} = 1$. It must be remembered, however, that these measurements are in the zone most affected mechanically by the drilling, so the value of the Formation Factor obtained may not be representative of the bed.

The *Microlog*, which is the most important resistivity device for the geologist after the Normals, makes use of these disturbances close to the wall of the borehole with electrodes of very short spacing. It consists of a two-inch Normal (5 cm) and a one-inch by one-inch Inverse in an insulating pad

(Fig. 6-9). This is applied to the wall of the borehole by a spring-loaded arm that acts also as a Caliper (the hole diameter is thus also logged and recorded). When lowered down the hole with the caliper closed, the true mud resistivity in the hole is measured and logged. Into the relatively impermeable beds there is virtually no invasion by mud filtrate, and the two devices measure essentially the same resistivity. Opposite permeable beds, where invasion has led to the formation of a filter cake on the wall of the borehole, the one-inch Inverse measures the resistivity of the mud cake (mostly) while the two-inch Normal measures the resistivity of the mud cake and part of the invaded zone. These two resistivities normally differ, so the two curves on the log are separated. The caliper log opposite permeable beds may show a hole diameter slightly less than the diameter opposite the relatively impermeable beds because of the mud cake, and it may even be less than the nominal diameter of the hole.

The Microlog can therefore be used qualitatively to identify and measure the thickness of the permeable parts of the sequence penetrated by the borehole. Quantitative analysis leads to estimation of the Formation Factor, through eq. 6.9, with all the reservations of that discussion and the discussion of eq. 6.8 from which it was derived.

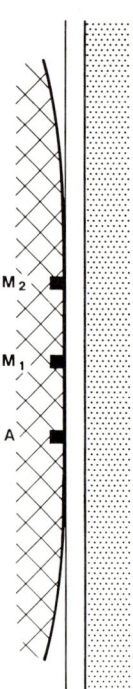

Fig. 6-9. Electrode arrangement of the Microlog.

The Lateral and the Microlog both have analogous devices in which the current flow patterns are focussed between two current electrodes, the second of which is maintained automatically at the same potential as the normal current electrode. The potential difference between the two measuring electrodes depends then largely on the resistivity of the material in the narrowest part of the beam.

Resistivity is the inverse of conductivity, and there is one electrical logging device that measures conductivity — the Induction Log. Eddy currents coaxial with the borehole are generated in the rocks around the borehole by a high-frequency alternating current passing through coaxial coils in the sonde. These Eddy currents induce a signal in receiver coils that are coaxial with the transmitter coils, and the signal is proportional to the conductivity of the formation. The unit of conductivity used is the *millimho* (= 1000/ohm), the reciprocal of which is usually also recorded, in ohms.

Spontaneous potential (SP) log

The SP log is run concurrently with one or more resistivity devices and is an important part of the basic log used by the petroleum geologist, or the geologist working on any area in which borehole logs are available. It records the potential difference between a fixed electrode at the surface and a movable electrode in the borehole. The unit of measurement is the millivolt (mV). The base line for the measurement is not a line on the scale of the log, but the line recorded opposite mudstones or shales. It is known as the *shale base line* or *shale line*. Deflexions from this are measured to the left of this line (negative) or to the right (positive). The deflexions result from natural electric currents in the borehole that are caused by electromotive forces of electrochemical and electrokinetic origins. The position of the base line is arbitrary with respect to the scale, and it is normally placed in a position that allows deflexions (observed by the operator when running in) to fall conveniently within the scale area.

An *electrochemical potential* (E_c or E_{ch}) results from the introduction of a conductive borehole fluid (the drilling mud) across porous rocks with fluids of a salinity different from that of the mud and its filtrate. This potential, which contributes most (if not all) of the deflexion on the log, consists of two components in a sequence of alternating lithologies with mudstones or shales.

Consider a porous, permeable bed between two thick, porous, but relatively impermeable mudstones in a borehole (Fig. 6-10). Let us assume that the electrolytes — the formation water, the mud and its filtrate — are NaCl solutions, and that the mud and its filtrate are less saline than the formation water. The composition of mudstones and shales is extremely complicated and variable, but the clay minerals can be considered as grains in which there are layers of Al, Si, and O atoms with some layers of water that are bound to

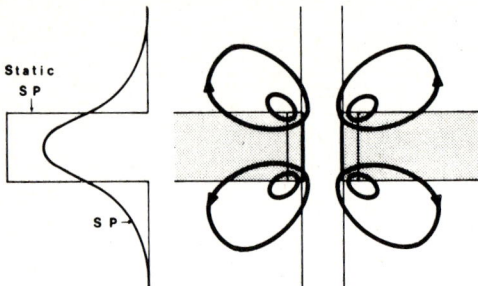

Fig. 6-10. Spontaneous potential diagram.

the lattice (not interstitial) by residual charges due to the substitution of one positive atom in a clay mineral by another of lower valency (e.g., Mg^{2+} replacing Al^{3+}). O^- tends to occur on the outer margins of the layers, with the result that mudstones are more permeable to Na^+ ions than to Cl^- ions. Na^+ tends to pass through the clay from the more saline solution (assumed for discussion to be the formation water) to the less saline solution in the borehole, and this is the direction of the electrical current (+). This is known as the *membrane potential*.

Within the permeable bed, however, invasion of mud filtrate gives rise to contrasting salinities in the pore spaces, across the interface between the mud filtrate and the formation water. Cl^- are more mobile than Na^+ ions, so there is a net flow of Cl^- from the more saline formation water to the less saline mud filtrate, with a resulting current flow in the opposite direction. This is known as the *liquid junction potential*.

These two components give rise to an electric current that circulates around the interface between the mudstone and the permeable unit near the borehole, and the potential of this current is measured in the borehole. The intensity of the electrochemical potential is greatest at the level of the interface in the borehole. If the static SP could be measured, that is, the SP that is not reduced by the resistance encountered by the current flowing through the rocks and their fluids, it would be rectilinear and mark the changes of lithology at the borehole precisely. Nevertheless, the slope of the SP as recorded is proportional to the electromotive force in the borehole, and the lithological boundary is indicated by the *inflexion point* of the curve. This may well not coincide with the mean value of the deflexion. The SP log is therefore of fundamental geological value in locating the boundaries of rock units penetrated by the borehole.

When the salinity of the mud and its filtrate is greater than that of the formation water, the direction of the currents is reversed, and the deflexion opposite the permeable beds will be positive with respect to that opposite a mudstone or shale. Fresh-water aquifers, commonly in the higher part of the borehole, are so revealed (see Fig. 6-1). Holes drilled with sea water or a mud based on sea water also lead to SP reversal over at least the upper part of the hole, and where the salinity of the mud and its filtrate are equal to that of the formation water, no deflexion of the SP occurs.

The total electromotive force corresponding to the electrochemical potential can be expressed:

$$E_{ch} = -K \log \frac{a_w}{a_{mf}} \qquad (6.10)$$

where a_w and a_{mf} are the chemical activities of the formation water and the mud filtrate, and K is a coefficient proportional to the absolute temperature. However, if the concentrations of the two NaCl solutions are less than about 80,000 ppm, their chemical activities are approximately inversely proportional to their resistivities, so that eq. 6.10 can be approximated by:

$$E_{ch} \simeq -K \log \frac{R_{mf}}{R_w}. \qquad (6.10a)$$

Of these parameters, E_{ch} can be estimated and R_{mf} measured.

The electrokinetic (or streaming) potential (E_k) in a borehole is considered by some to be very small, and by others to be non-existent from a practical point of view. This potential is attributable to the flow of mud filtrate across the mud cake into the permeable rock unit. It is probably large, but approximately equal to that opposite the mudstones. The contribution of E_k is the difference between the e.m.f. across the mud cake and that into the mudstone (for few mudstones are so impermeable that no flow takes place into them). There is no argument that whatever its value, the electrokinetic potential makes at most a small contribution to the total deflexion of the SP on most logs.

In theory, many parameters influence the shape and amplitude of the deflexion of the SP opposite the more permeable beds. They include the thickness and true resistivity of the permeable bed, and the resistivity of the adjacent beds; the resistivity of the mud and its filtrate; the resistivity of the invaded zone and its diameter; and the diameter of the borehole. It is not strictly a permeability log because a small fraction of a millidarcy (mD) is sufficient for an e.m.f. to be generated. Nevertheless, it is an invaluable log from which mudstones and shales can be distinguished from more permeable lithologies, fresh water can be distinguished from brine (by its reversed SP and high resistivity). Quantitative analysis leads to the determination of formation water salinities without the need of a sample at the surface. It is a log from which lithological boundaries and their depths are recorded, and from which the thicknesses of rock units are obtained.

Radioactivity logs

Some rock types, particularly mudstones, emit gamma rays spontaneously because of the presence of ^{40}K in clay minerals. These gamma rays are absorbed by the surrounding rocks, the rate of absorption being roughly proportional to the rock's density. Gamma rays emitted from the zone within about 0.3 m of the wall of the borehole are detectable in the borehole. The *Gamma Ray log* is a log of the natural gamma radiation in the borehole as measured by a scintillation counter in the sonde. It is thus essentially a log of clay content, in the petroleum context, in the immediate vicinity of the borehole. This is independent of the salinity of the mud of the borehole, so the Gamma Ray log can be run instead of the SP when the salinities are unfavourable for the SP.

The Gamma Ray log is recorded on the left-hand side of the log in absolute units of microgram radium equivalent per ton (µg Ra eq./ton) or API units (16.5 API units = 1 µg Ra eq./ton), increasing to the right. Clean sands (with but few exceptions due to heavy minerals) have no radioactivity, but mudstones may have 200 API units (12 µg Ra eq./ton) or more. A Gamma Ray log therefore records a sandstone/shale sequence in the same sense as would be recorded by a SP log if $R_{mf} > R_w$.

Because gamma rays will also penetrate a small thickness of cement and steel, logs can be obtained from cased holes and used for geological purposes (if the casing was run, for one reason or another, without electrical logging) and for production purposes, to discover the exact depth at which casing should be perforated to produce a sandstone or carbonate reservoir. When run in casing, the Gamma Ray device is run with a Casing Collar Locator, which is similar to the Induction device and detects the extra thickness of steel in the casing couplings.

All radioactivity measurement has a statistical component, and all radioactivity logs should record the statistical variations opposite a bed of low radioactivity and opposite one of high radioactivity, so that real changes of radioactivity level can be distinguished from statistical fluctuations. This is achieved by holding the sonde opposite a sand, for example, while winding film through for a minute or so.

The Gamma Ray log sometimes reveals beds of unusually low radioactivity (e.g., coal seams) or unusually high radioactivity (e.g., the Kimmeridge Clay in the North Sea) which may be useful for correlation — even when the cause of the abnormality cannot be identified with confidence.

The *Neutron log* is obtained with a device that measures the response of the rock around the borehole to bombardment with neutrons. The sonde includes a source that emits fast neutrons, which are slowed down mainly by collision with light nuclei, hydrogen and to some extent chlorine, until they become "thermal" neutrons. These do not lose energy as they move from one collision to another, but eventually they are absorbed by hydrogen atoms, mostly, and gamma rays of capture are emitted. Those gamma rays of capture that are not absorbed in the rock are detected in the sonde. The Neutron log therefore reflects the proportion of hydrogen in the rock around the borehole. Since virtually all the hydrogen in a sedimentary rock is confined to the pore fluids, the Neutron log will detect porosity in general, and distinguish between gas and liquids in particular. The differences of hydrogen concentration between oil and water are too small to be detected (but these can be distinguished by electrical logs). Deep invasion of a gas reservoir by mud filtrate may mask the effect of gas.

The Neutron log deflexion is inversely proportional to the logarithm of porosity, to a close approximation, because in large porosities, neutrons are slowed down and captured over a shorter distance than in small porosities, so fewer gamma rays of capture reach the detector. However, a neutron "does not know" if the hydrogen it has collided with is in free water or in the crystal lattice of a clay mineral such as smectite (montmorillonite), or of gypsum, so the effective porosity may well be less than that indicated by the Neutron log, and care must be taken with dirty sands and lithologies in which such minerals may occur. The depth of penetration is usually less than 0.5 m, and this rock may not be representative of the natural material due to drilling effects.

The Neutron log can be run in cased holes, but it cannot be used reliably to determine porosity quantitatively behind casing. It usually indicates changes in porosity, so it may be useful qualitatively.

Formation Density log (Fig. 6-11). Gamma rays emitted from a sonde that

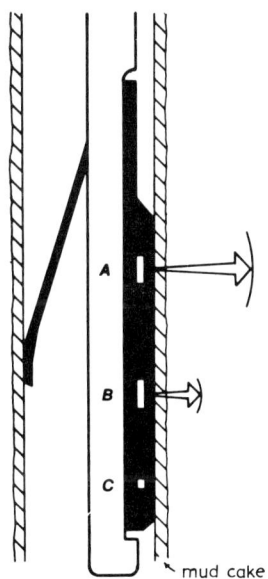

Fig. 6-11. Formation density logging device (Schlumberger F.D.C.). A = long spacing detector; B = short spacing detector; C = source. (Courtesy of Schlumberger Seaco Inc., Sydney.)

is held to the wall of the borehole, ploughing through the mud cake, lose energy by collision with electrons in the formation. Thus, the greater the absorption in the formation, the lower the reading from the detectors. Because the absorption rate is approximately proportional to the rock density (the proportion of electrons is proportional to density) the bulk density can be determined. This log is affected by the thickness of mud cake that may intervene between the sonde and the wall of the borehole, so there is a correction to be applied to the raw log, and considerable care in its use is required. Porosity can be estimated from eq. 3:

$$\rho_{bw} = f\rho_w + (1-f)\rho_g$$

into which can be inserted the appropriate values of ρ_g according to the lithology and, in this context, ρ_w is the density of the mud filtrate because the depth of penetration of the device is very small.

Sonic log

The velocity of sound through sedimentary rocks is a parameter of considerable geophysical and geological interest, and this log provides a profile of the *transit* or *travel time* (μs/ft or μs/m, the inverse of velocity) — usually on the right hand side of the log, with the SP or Gamma Ray on the left from which lithologies may be distinguished. The old device (Fig. 6-12), with one signal generator and three detectors, illustrates the principles. The

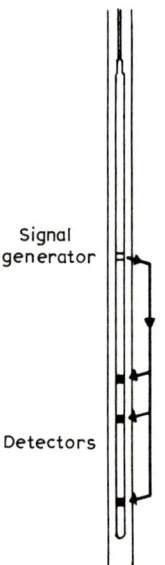

Fig. 6-12. Sonic logging device. (Courtesy of Schlumberger Seaco Inc., Sydney.)

velocity of sound through liquids is very much slower than that through solids, so the detectors pick up the *first arrival* of a signal that has passed down the wall of the borehole. The device measures the time of arrival at the 1st and 2nd (or first and third, depending on which spacing is being used) and the difference is recorded as the transit time. If the sonde is symmetrical in the borehole, the passages through mud to each detector will be equal, and the recorded transit time is the true transit time over the distance separating the detectors (1 or 3 ft: 0.30 or 0.91 m). The newer, compensated type is so designed that the average of an upward and a downward signal is correct even if the sonde is not symmetrical with the borehole.

The geometry of the wall of the borehole is important, and if it is irregular due to washouts (see the Caliper log) the results may be erratic with the travel times too long. A signal not received at a detector (due perhaps to attenuation) leads to wild fluctuations in the trace on the log. This is called *cycle skipping*, and such sections must be ignored.

Together with the transit time is recorded the integrated transit time, each pip on this track indicating an increase of one *milli*second in the total travel time (except opposite cycle skipping). Interval velocities can therefore be computed by counting the pips between the limits of interest. Hence the sonic log can be used to help seismic interpretation not only by identifying lithologies with velocities, but also by the generation of *synthetic seismograms* for the identification of reflectors.

The Sonic log, as we saw in Chapter 3, can also be used to estimate porosity. This is a complex matter that still has no good theoretical basis. Work by Wyllie and his co-workers (Wyllie et al., 1956, 1958) led to the so-called *time-average formula:*

$$f = (\Delta t - \Delta t_{matrix})/(\Delta t_{fluid} - \Delta t_{matrix}). \tag{6.11}$$

The transit time through solids (Δt_{matrix}; i.e., at zero porosity) is about 52—55 μs/ft in quartz, 45—50 μs/ft in carbonates, and about 190 μs/ft in liquids, so the porosity can be estimated once the transit time, Δt, has been measured. This time-average formula has two weaknesses, long recognized. First, it is a linear equation in porosity and transit time, although there is a great deal of empirical evidence suggesting that the true relationship is not linear. Secondly, correction factors are needed for all but small porosities. It seems therefore that these correction factors are required because the time average formula is not correct.

Chapman (1981, p. 223) assumed that the sonic path lies in solids only, by virtue of the fact that the transit time in liquids is about four times that in solids, and that porosity affects the length of this path (a sort of tortuosity in solids). He suggested the formula (abbreviating the suffix):

$$f = 1 - (\Delta t_{ma}/\Delta t)^x \simeq 1 - (\Delta t_{ma}/\Delta t)^{1/2}. \tag{6.12}$$

This is a non-linear equation that gives satisfactory results in several areas

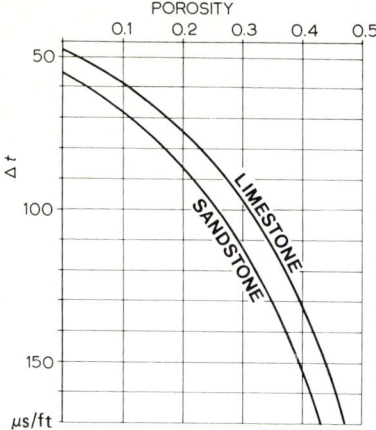

Fig. 6-13. Porosity-transit time relationship in granular beds from eq. 6.12.

when Δt_{ma} is taken as 55 µs/ft (180 µs/m) in sands and 47 or 48 µs/ft (155 µs/m) in carbonates. Local experience may indicate other values, and for the exponent x; and it has been used successfully to obtain porosity profiles over incompletely-cored reservoirs, and those where core recovery was poor (the parts lost being mainly those with larger porosity).

Since any sonic-log derived porosity is an average over the detector spacing, detailed correlations with porosity determined from plugs taken from cores is not to be expected if there is much vertical variation in porosity. There is an undoubted gas effect in some reservoirs, but by no means all, the transit times being abnormally long.

The relationship between transit time in mudstones (Δt_{sh}) and porosity, already discussed at length in Chapter 3, does appear to be represented best by a time-average, linear formula. This is probably due to the platy nature of the grains and the consequent flattened geometry of the pores.

Dipmeter

The dipmeter, as its name suggests, provides a profile of dips and strikes of the strata in the borehole. The principles are best illustrated by the old tool, which was a centralized sonde with three electrodes (resistivity) at 120° to each other in a plane normal to the axis of the sonde, held in pads to the wall of the hole. If the borehole is perfectly perpendicular, any dipping layer with characteristic resistivity will be detected by each electrode at slightly different depths, the amount of the difference being a function of the borehole diameter and the orientation of the sonde relative to the direction and amount of dip. By means of a small gyrocompass, the orientation of a particular electrode is recorded; and this allows computation of the apparent dip and strike of the bed in the borehole. By means of a similar device that records the angle of the borehole from the vertical, and the azimuth of such deviations from the vertical, the true dip of the bed can be computed.

The modern tool works on the same principle, but has four arms so that dips can be obtained when the hole geometry is irregular. The data are recorded for computer processing.

This is the geologists' rather than the engineers' device, but it must be remembered that it is a bulky tool with greater than normal risk of sticking and loss in the hole. While this risk is normally quite acceptable, there may be occasions when it is not. For example, a 4 km hole that penetrated mudstone for the last kilometre might well present risks that are out of all proportion to the value of the data that might be obtained if the dipmeter were to be run from bottom. In such a case, if the engineer in charge requested it, it would be reasonable to run the dipmeter from 500 m or so above bottom.

To the student who has been encouraged to draw sections with beds of regular geometry and equal thickness, it may come as a shock to find evidence that both dip and strike may change significantly in a borehole without unconformities. Some changes are due to faults, others to unconformities, and to internal features such as cross-bedding. There may be parts where the vertical sequence of dips suggests the environment of deposition of the sediments with some clarity.

TEMPERATURE

Ideally we should know the true temperature distribution in the subsurface both at the time of logging and when thermal equilibrium has been reestablished. We need to know the temperatures for quantitative interpretation of resistivity logs, and for the proper evaluation of reservoir performance (because temperature affects fluid viscosities and volumes). Geologists and geochemists also need to know temperatures so that they can assess the thermal role in petroleum generation and the local prospects of petroleum accumulation.

First, it must be emphasized that the *Temperature log* is *not* the log to use. The commonest use of this log is to locate the top of cement behind casing. The setting of cement is an exothermic reaction, so the location of setting cement behind casing can be detected by its elevated temperature*.

The sonde can take a maximum-reading thermometer, and the maximum reading on each run is recorded on the log heading. This bottom-hole tem-

---

* At least one published temperature profile was a temperature survey to locate top of cement in the well.

perature (BHT) is lower than the true bottom-hole temperature because of the circulation of cooler mud. Once mud circulation stops, heat flow tends to restore thermal equilibrium. Because this cooling effect is negligible ahead of the bit while drilling, the significant period for a *bottom-hole* temperature determination is the period of circulation after drilling and before pulling out of the hole for logging.

Dowdle and Cobb (1975) showed that a Horner plot (see p. 174) will lead to satisfactory approximations to true "formation" temperature for short circulation times (which were not specified, but may be taken as less than five hours or so — but any intelligent estimate may well be better than none). They assumed that the relationship between true bottom-hole temperature and the measured, perturbed temperature was of the form:

$$t_m = t_{bh} - C \log \frac{T + \Delta T}{\Delta T} \qquad (6.13)$$

where t_m is the temperature measured at the bottom of the hole at time ΔT hours after stopping circulation that had been going on for T hours; C is a constant, and t_{bh} is the true bottom-hole temperature.

A Horner plot makes use of the fact that $(T + \Delta T)/\Delta T$ approaches unity, and its logarithm zero, as ΔT increases indefinitely. Thus a linear relationship was assumed between the measured temperature and the logarithm of the dimensionless time, and extrapolation of the dimensionless time to zero gives an estimate of the true bottom-hole temperature if the hole were allowed to approach thermal equilibrium indefinitely.

What is required is:

— The duration of circulation after drilling, before pulling out (T hours) from properly logged drilling operations.

— The elapsed time (ΔT hours) from stopping circulation until the sonde was *on the bottom, for each run* (i.e., a calibrated maximum-reading thermometer read after *each* run and recorded on the log heading).

For example, at a depth of 2284 m (below kelly bushing, KB) drilling is stopped and mud circulated for 1.5 h. Six hours and 5 min later, the electrical log sonde was on bottom and a temperature of 111°C was later re-

TABLE 6-1

Circulation data

T	ΔT	$(T + \Delta T)/\Delta T$	$\log (T + \Delta T)/\Delta T$	t_m (°C)
1.5	6.08	1.25	0.0957	111
	9.38	1.16	0.0644	116
	12.42	1.12	0.0492	119.5

corded. Nine hours and 23 min after stopping circulation, the Sonic log was on bottom, and a temperature of 116°C was recorded. Twelve hours and 25 min after stopping circulation, a bottom-hole temperature of 119.5°C was recorded. The data are shown in Table 6-1.

A graphical plot (Fig. 6-14) or a pocket calculator with statistical functions indicates that the extrapolated bottom-hole temperature is 128°C. This can be designated BHT* to distinguish it from the actual measurements.

It is important to understand clearly what this means and what it does not mean. It means that the temperature in the rock unit at the bottom of the hole at a depth of 2284 m below kelly bushing was about 128°C, as nearly as we can determine from the data available, before it was disturbed by drilling. So, if the elevation of the kelly bushing is about 8 m above ground level (it makes no sense to take any other datum) the geothermal gradient can be computed if the mean ambient surface temperature is known. Assume it to be 15°C. Then the estimate of geothermal gradient is:

$$\frac{128 - 15}{2284 - 8} = 0.05°C/m$$

or 50°C/km.

None of the figures tabulated gives, or can be used to give the temperature at any other depth in the borehole at the time of logging. The formations above the bottom of the hole have had much longer perturbation times while the deeper part of the hole was being drilled, and there is a lateral temperature gradient increasing away from the borehole to the true formation temperature at some distance from the borehole. Short Normal devices may therefore be affected more by borehole temperature perturbation than the devices with long spacing.

Boreholes usually have complete logging programmes, taking several hours, before setting intermediate casing and before completion or abandonment.

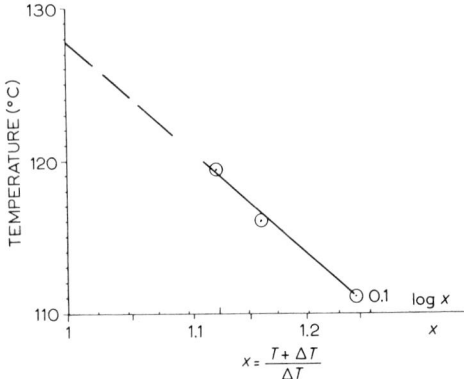

Fig. 6-14. Horner plot for estimating true formation temperature.

Deep holes may have more than two. For each logging depth the BHT* can be obtained, and so geothermal gradients determined (including interval geothermal gradients between two such determinations). These will be of considerable interest to geologists and geochemists, and the interval geothermal gradient between two logging depths may well be more interesting than those from the surface for extrapolating to greater depths.

The necessary data are acquired at virtually no extra cost: they should be acquired at every opportunity, and steps should be taken to ensure their reliability. Subsequent reservoir tests may yield a better temperature, but these estimates can be obtained before the logging programme is complete, and so may be used to assist in the decision whether to drill deeper or not. But, as always, extrapolations are made at one's peril. There are numerous examples in Kenyon and Beddoes (1977) of reservoir temperatures measured during tests that show a deeper reservoir to be cooler than a shallower, or at much the same temperature.

REFERENCES

Archie, G.E., 1942. The electrical resistivity log as an aid in determining some reservoir characteristics. *Trans. Am. Inst. Min. Metall. Engrs.* (Petroleum Division), 146: 54—62.
Archie, G.E., 1950. Introduction to petrophysics of reservoir rocks. *Bull. Am. Ass. Petrol. Geol.*, 34: 943—961.
Barsukov, O.A., Blinova, N.M., Vybornykh, S.F., Gulin, Yu.A., Dakhnov, V.N., Larionov, V.V. and Kholin, A.I., 1965. *Radioactive investigations of oil and gas wells.* (Transl. from Russian by J.O.H. Muhlhaus, edited by N. Rast.) Pergamon, Oxford, 299 pp.
Chapman, R.E., 1981. *Geology and water: an introduction to fluid mechanics for geologists.* Nijhoff/Junk, The Hague, 228 pp.
Cole, D.I., 1976. Velocity/porosity relationships in limestones from the Portland Group in southern England. *Geoexploration*, 14: 37—50.
De Wiest, R.J.M., 1965. *Geohydrology.* Wiley, New York, N.Y., 366 pp.
Dowdle, W.L. and Cobb, W.M., 1975. Static formation temperature from well logs — an empirical approach. *J. Petrol. Technol.*, 27: 1326—1330.
Gardner, G.H.F., Gardner, L.W. and Gregory, A.R., 1974. Formation velocity and density — the diagnostic basics for stratigraphic traps. *Geophysics*, 39: 770—780.
Hubbert, M.K., 1940. Theory of ground-water motion. *J. Geol.*, 48: 785—944.
Hubbert, M.K., 1969. *The theory of ground-water motion and related papers.* Hafner, New York, N.Y., 310 pp.
Kenyon, C.S. and Beddoes, L.R., 1977. *Geothermal gradient map of southeast Asia.* Southeast Asia Petrol. Explor. Soc., Singapore and Indonesian Petrol. Assoc., Jakarta.
Moore, C.A., 1963. *Handbook of subsurface geology.* Harper and Row, New York, N.Y., 235 pp.
Morrow, N.R., 1971a. The retention of connate water in hydrocarbon reservoirs, I. A review of basic principles. *J. Can. Petrol. Technol.*, 10: 38—46.

Morrow, N.R., 1971b. The retention of connate water in hydrocarbon reservoirs. II. Environment and properties of connate water. *J. Can. Petrol. Technol.*, 10: 47—55.

Owen, J.E., 1952. The resistivity of a fluid-filled porous body. *Trans. Am. Inst. Min. Metall. Engrs.* (Petroleum Branch), 195: 169—174.

Sarmiento, R., 1961. Geological factors influencing porosity estimates from velocity logs. *Bull. Am. Ass. Petrol. Geol.*, 45: 633—644.

Winsauer, W.O., Shearin, H.M., Masson, P.H. and Williams, M., 1952. Resistivity of brine-saturated sands in relation to pore geometry. *Bull. Am. Ass. Petrol. Geol.*, 36: 253—277.

Worthington, P.F., 1975. Quantitative geophysical investigations of granular aquifers. *Geophys. Surv.*, 2: 313—366.

Wyllie, M.R.J., 1963. *The fundamentals of well log interpretation* (3rd ed.). Academic Press, New York, N.Y., 238 pp.

Wyllie, M.R.J., Gregory, A.R. and Gardner, L.W., 1956. Elastic wave velocities in heterogeneous and porous media. *Geophysics*, 21: 41—70.

Wyllie, M.R.J., Gregory, A.R. and Gardner, G.H.F., 1958. An experimental investigation of factors affecting elastic wave velocities in porous media. *Geophysics*, 23: 459—493.

Wyman, R.E., 1977. How should we measure residual-oil saturation? *Bull. Can. Petrol. Geol.*, 25: 233—270.

CHAPTER 7

THE NATURE OF OIL AND GAS FIELDS

SUMMARY

(1) Oil and gas fields are not purely geological entities, because they involve engineering, economics and finance, and the normal range of goods and services for the people who operate them.

(2) Current known reserves recoverable by present technology at present prices are dominated by relatively few very large fields. At the end of 1977, about 85% was in only 288 fields.

(3) The volumetric size distribution of *accumulations*, rather than fields, appears to follow Zipf's law in which the relative sizes when ranked are approximated by the successive terms of the harmonic series, 1, 1/2, 1/3, 1/4, ..., $1/n$, so that the product of rank number and size is approximately constant. For the world, this constant appears to be at least 120×10^9 barrels of oil (19×10^9 m^3).

(4) The sum of this harmonic series suggests that the total ultimate recoverable reserves of oil, by present technology at present prices, will be about $1,800 \times 10^9$ bbl (286×10^9 m^3), which is comparable with some other estimates.

(5) The as-yet-undiscovered recoverable reserves are probably in relatively few very large fields, but the proportion is probably less than that of the discovered fields because the very large fields are easier to find.

(6) Increasing geological, geophysical, geochemical, engineering and financial skills will be required to find the remaining reserves.

Our purpose here is to seek the wider context of petroleum geology as a sort of intermezzo and, after reviewing the present distribution of oil fields by size, venture into the speculative domain of the total world resource of oil. It must be realized that reliable statistics of the world's oil fields are never up-to-date because it takes some years for new discoveries to be accurately assessed, and more before the figures are published. Perhaps Hedberg was right when he wrote that ultimate reserve estimation was a topic that belonged more to the after-dinner speech than serious writing!

An oil or gas field is a petroleum accumulation that has been discovered and found to contain enough petroleum of sufficient quality to be worth more in the market place than the total cost of getting it there. It remains an oil or gas field until the value of the production in the market place no longer exceeds adequately the cost of getting it there. It is then abandoned. It is not therefore a purely geological entity, but one that involves engineering, transport, marketing, finance and economics. During its life it involves people with a great range of skills in the various activities required. These people require the same varieties of goods and services as people in other activities, and so oil and gas fields include houses, shops, schools, recreation, and the power to run these. The only difference between an oil-field community and others is, perhaps, that the average age is rather younger.

Oil and gas fields, like people, come in all shapes and sizes, some onshore, some offshore. In plan, they may be long and narrow, or nearly circular; in section, they can be thick or thin, deep or shallow. In size, they can be very large or rather small — but they must all contain some minimum quantity of recoverable oil or gas of marketable quality. This quantity may be quite small near markets or near other fields, or rather large if offshore at some distance from markets. There are very many accumulations that are too small to become fields. Oil and gas fields are commonly in trends (that are geological trends) or groups or groups of trends, and these collectively form petroleum provinces.

An oil or gas field begins as an anomaly on a map, revealed by regional geological and geophysical surveys. The anomaly is investigated with more detailed mapping, and if this shows features that could trap petroleum, the company must decide whether it is sufficiently promising to drill and if they can obtain the finance to drill it (or seek partners). The drilling site is chosen on the basis of a detailed seismic reflection survey. Such is the precision of seismic surveys that there will be great confidence in the geometry of the anomaly and in the *general* nature of the sedimentary rocks in it, and, if there is some local knowledge, also the ages of the rocks. There may even be direct indications of gas ("bright spots", "flat spots"), but the survey will not generally provide any information on whether oil was actually generated and came to accumulate there. That can only be found out by drilling.

The first well drilled to a prospect is an *exploration* well, sometimes called a *wildcat*. It is designed to acquire information, not necessarily to produce what it finds on a permanent basis. This exploration borehole will be designed and drilled to the greatest depth of practical interest, and the geological results — the stratigraphy, ages of the rocks, dips, sonic velocity characteristics, maturation levels of organic matter in the rocks, and fossils recovered in cores and samples — all contribute to a refinement of the original model that was based on the seismic surveys. More seismic work may be done. If petroleum is not found, the information gained may lead to the drilling of further exploratory wells in the area. If petroleum is found, it will be tested and analysed, and the pressures and flow rates will be measured.

On the basis of the refined model, *appraisal* wells will be sited and drilled to establish as quickly and as economically as possible the minimum size of the accumulation. This is not just a matter of drilling elsewhere on the anomaly but rather of deciding how big such an accumulation must be if it is to be commercial, and drilling near the limits so indicated. The appraisal wells may therefore be some kilometres from the discovery well.

With the drilling of the appraisal wells, the nature of the accumulation begins to emerge. With each well, the information gained is used to refine the model of the accumulation until the point is reached when it can be stated with some confidence that the accumulation is sufficiently large and contains sufficient recoverable oil or gas to make the development of the field commercially sensible.

In a simple onshore structure, these appraisal wells may be designed so that they can produce petroleum eventually; but their real purpose is to obtain more information. Offshore, the appraisal wells will not be capable of more than limited testing of reservoirs because the economics of offshore production requires production platforms that will be used first to drill a number of deviated wells to develop part of the field, then to produce them to a central facility for preliminary processing (such as gas separation) and transport through pipelines to the shore facilities. More than 20 wells are usually drilled and produced to one platform, so the appraisal programme must be sufficiently comprehensive offshore to permit planning of these wells and the platforms.

It is in the nature of petroleum fields that the decision to develop is made without full knowledge of the accumulation. Full information is available only when it is no longer needed. The appraisal phase ends with the decision either to abandon the prospect as non-commercial, or to develop it as an oil or gas field.

After the decision to develop is made, *development* wells are drilled for the production of oil (or gas, as the case may be); and their siting is such that each reservoir will be developed as economically as possible (which is not usually the same as as cheaply as possible) on the model that has evolved with the added information, which may include additional seismic data as well as the borehole data. This is an exercise in engineering economics based on geology that must also take transport of the production into account. It will normally take at least five years from discovery to put a new field onto production, and during this time there is no income from the field to offset the expenses incurred in its development.

The rate at which a new field can be produced economically is part of the nature of an oil field. It depends on the number of reservoirs, their size, their productivity and other factors that will be discussed later. The rule of thumb is that the faster the rate of production, the smaller the ultimate yield of the field; but the rate must be fast enough to generate sufficient income from the investment. So the production rate chosen is a balance between these opposing interests such that the development as a whole is as commercially attractive as possible within the constraints imposed by Nature and the government regulating such activities.

In the course of development there will be some surprises, and these will modify the earlier decisions. As the field becomes better known and documented, possible field extensions will become apparent. These will be investigated with *outstep* or *outpost* wells. It is in the nature of such marginal areas that some of these will be failures, either failing to find what was expected and hoped for, or finding that the quantity or quality does not justify development.

The nature of an oil field with a single large pool is quite different from one with many pools and from one with many pools in a faulted structure, and the economics also differ. A single large pool may be developed with relatively few wells spaced several kilometres apart if the field permeability is good, or a few wells if the oil column is thick and the field area relatively small, whereas the same volume of oil in several faulted reservoirs may require wells at close spacing for each reservoir in each fault block. Part of the development planning of such fields with multiple reservoirs involves siting wells for multiple completion on more than one reservoir (dual completions, even triple completions). It is in the nature of oil fields that those in transgressive sequences tend to be large single-pool accumulations and those in regressive sequences tend to be multiple-pool accumulations.

The extraction of oil tends to result in a gradual pressure decline in the reservoirs. This not only impairs the productivity of the field but also leads to surface subsidence through the compaction of the depleted reservoirs and, in some fields, the movement of oil or water from undepeleted reservoirs to depleted reservoirs in juxtaposition across faults. Pressure-maintenance schemes are planned and put into operation early in the life of a field, usually involving the injection of water at the field perimeter, or the substitution of peripheral wells as injection wells when they go wet.

GIANT OIL FIELDS

Very large oil fields are called *giants* when they contain at least 100×10^6 bbl (15.9×10^6 m^3) of *recoverable* oil with present technology and at present prices; and gas fields are giants when they have at least 1 Tcf (trillion standard cubic feet, 10^{12}; 28.3×10^9 m^3) of recoverable gas. Definition of a giant has not been universally accepted, and some would have it as large as 500×10^6 bbl and 10 Tcf, or 10^9 bbl and 10 Tcf.

If we take North America as typical of a continent that has been (and is being) intensely explored, we find that in 1968, according to Moody et al. (1970), there were 26,250 oil fields with recoverable reserves totalling 132×10^9 bbl (21×10^9 m^3) for an average of 5×10^6 bbl (755,000 m^3) each. Of these, the 45 largest, each with more than 500×10^6 bbl (80×10^6 m^3) recoverable oil, had a total of 46×10^9 bbl (7×10^9 m^3) of recoverable oil, for an average of 10^9 bbl (159×10^6 m^3) each. *Less than 0.2% of the oil fields contained 35% of the total known North American recoverable oil.* The largest of these, the East Texas field*, contained 5.6 × 10⁹ bbl (890 × 10⁶ m³), or over 4% of the North American recoverable oil.

The surface area underlain by these giant fields varies from 6 km² (1515 acres) to 1580 km² (390,000 acres), with nearly half of them less than 100 km² (25,000 acres). The thickness of productive zones varies from 10 to 439 m (30—1440 ft), with half having less than 30 m (100 ft). The depth of principal production varies from 305 m to 3475 m (1000—11,400 ft), with virtually no oil production from below 4270 m (14,000 ft). The age of the reservoir rock ranges from Ordovician onwards, with about half the fields younger than mid-Cretaceous (100 m.y.). Two thirds of these giants have sandstone reservoirs, and the rest have carbonate reservoirs. The recoverable reserves are in about the same proportion.

Similar statistics have not yet been compiled for the world, but Fitzgerald (1980) found that of the total recoverable oil, including gas converted to oil equivalent, discovered in the world up to the end of 1977, about 85% was in only 288 fields. The importance of the giants remains, and the worrying aspect of the statistics is that the volumetric rate of discovery of giants has been declining since the 1960s, while the rate of discovery of the smaller fields was still increasing in 1980. Overall, the rate of discovery of reserves appears to have been declining since the 1960s.

Zipf's law

In response to our desire to impose order on the disordered, various attempts have been made at describing the size distribution of oil fields in mathematical terms. Since Folinsbee's stimulating presidential adress to the Geological Society of America (Folinsbee, 1977) the use of Zipf's law (Zipf, 1949, 1965) has become common for petroleum as well as other resources.

Zipf found that a range of social phenomena such as the frequency distribution of words used in books, the distribution of salaries, and the size distribution of cities, follows a simple "law": if they are ranked by size, the product of the rank number and the size is approximately constant. Folinsbee (1977) found that the flow rate of the world's major rivers followed Zipf's law. In the context of petroleum, Zipf's law would postulate that if all the oil *accumulations* could be ranked according to size, the product of rank number and the reserves of that accumulation would be found to be constant, and equal to the reserves of the largest accumulation. The reserves of the largest accumulation would be twice those of the second largest, three times those of the third largest, and so on (Fig. 7-1). In other words, the total reserves would be distributed in accumulations in a harmonic series, $1 + 1/2 + 1/3 +$

---

* Prudhoe Bay in the Alaskan arctic had been discovered by the time these figures had been published. This illustrates the impossibility of accurate, up-to-date figures.

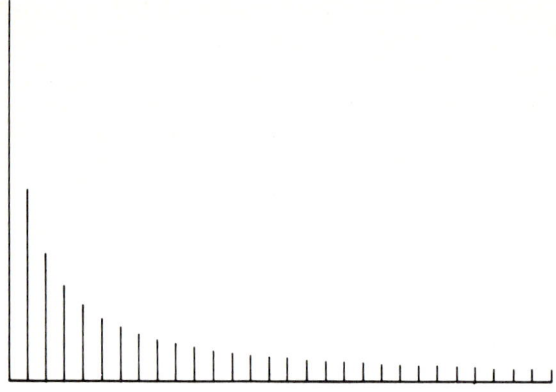

Fig. 7-1. Harmonic size distribution, $1 + 1/2 + 1/3 + \ldots + 1/30$.

TABLE 7-1

Expected distribution of reserves in the eight largest of 400 giant accumulations according to Zipf's law

Rank	%	Cumulative (%)
1	15.2	15.2
2	7.6	22.8
3	5.1	27.9
4	3.8	31.7
5	3.0	34.8
6	2.5	37.3
7	2.2	39.5
8	1.9	41.4

$\ldots + 1/n$, the sum of which for large n (> 50, for example) is very closely approximated by $\ln n + (1/2n) + 0.5772$ (the last number being Euler's Constant). From these it is a simple matter to compute the expected distribution of accumulations by size, as a proportion of the total. For example, if the world actually contains 400 giant oil accumulations, Zipf's law predicts that the proportional distribution of their total reserves will be as in Table 7-1, and 40% of the total reserves of these 400 giants will be in only 8 accumulations.

This is a plausible result, and may one day be found to be accurate; but only about 300 of these accumulations have so far been discovered. We can rank these according to size, but *we do not know the true rank of these accumulations.*

In practice, Zipf's law has come to be applied more to the distribution of *oil fields* ranked according to initial *recoverable* reserves, rather than oil in place. This is very different from the demonstrable validity of Zipf's law to the flow rate of the world's major rivers, all of which are known. The distinction between oil in place and recoverable reserves is probably not significant for oil fields, because great precision is not to be expected, and the variation of recovery factor in giants may not be great. But the very large tar-sand deposits are oil accumulations even if the recovery factor is very small.

Zipf's law has also been applied on different scales — the world, a continent, by countries (which imposes a rather artificial subdivision on the geological distribution) and by oil provinces. The reader may care to try this on the ranking of countries by reserves in Tables 7-7—7-9. The fact that Zipf's law seems to work quite well under these modified conditions means that it must be used and interpreted with caution.

The statistics of Moody et al. (1970) given above can be used as our example. On the information available to them, 26,250 oil fields contained 132×10^9 bbl of recoverable oil; and of these, 45 fields collectively contained 46×10^9 bbl of recoverable oil. Zipf's law predicts the following statistics from these figures:

— The sum of the harmonic series to $n = 45$ is 4.3950 and to $n = 26,250$ is 10.7526, so the accumulations ranked 1—45 should contain nearly 41% of the total reserves, or 54×10^9 bbl recoverable oil for an average of about 1.2×10^6 bbl each.

— Rank 45 will contain about 273×10^6 bbl of oil initially, and only 24 would contain more than 500×10^6 bbl each (which was the criterion for listing the first 45).

—Ranks 1 to 10 of the accumulations would be as given in Table 7-2, first

TABLE 7-2

Comparison of Zipf prediction of size of largest ten oil fields in North America on basis of 132×10^9 bbl ultimate recoverable oil from 26,250 fields with actual ranking and sizes in 1968, and possible future ranking of these fields. Figures in millions of barrels

	Zipf	Actual, 1968		Possible	
1	12,276	5600	E. Texas		
2	6138	2500	Wilmington	5600	E. Texas
3	4092	2334	Eunice		
4	3069	2100	Poza Rica	2500	Wilmington
5	2455	1773	Pembina	2334	Eunice
6	2046	1683	Kelly-Snyder	2100	Poza Rica
7	1754	1647	Panhandle	1773	Pembina
8	1535	1318	Swan Hills	1683	Kelly-Snyder
9	1364	1303	Elk Hills	1647	Panhandle
10	1228	1205	Golden Lane	1318	Swan Hills

TABLE 7-3

North America's ten largest oil fields in 1968 ranked according to ultimate recoverable reserve estimates (from Moody et al., 1970, p. 17), with the estimates of the Zipf constant. Figures in millions of barrels

		Ultimate reserves	Rank × ult. reserves
1	East Texas, Texas	5600	5600
2	Wilmington, California	2500	5000
3	Eunice-southward, New Mexico	2334	7002
4	Poza Rica, Mexico	2100	8400
5	Pembina, Canada	1773	8865
6	Kelly-Snyder, Texas	1683	10,098
7	Panhandle, Texas	1647	11,529
8	Swan Hills, Canada	1318	10,544
9	Elk Hills, California	1303	11,727
10	Golden Lane, Mexico	1205	12,050

and second columns, the Zipf constant being 12.3×10^9 bbl of recoverable oil.

The first result above is in reasonable agreement with the actual figures of 35%, or 46×10^9 bbl, in the 45 largest fields, but the size predicted for Rank 45 is little more than half the minimum size required for inclusion in this group of 45 fields.

Moody et al. (1970, p. 17) list the North American giants of this group, and Table 7-3 shows the top-ranking ten fields from this list with their estimated ultimate recoverable reserves and each estimate of the Zipf constant (reserves × rank number). Tables 7-2 and 7-3 show that the predictions are not very good, but they suggest that the East Texas field is not the largest accumulation in North America, and that Wilmington may only rank 4th or 5th eventually.

Zipf's law is also applied with yet another modification. The largest field is taken as Rank 1; Rank 2 will have half Rank 1's reserves, Rank 3, one third, and so on. Table 7-4 shows the results of manipulations on this basis. The results are not very convincing. This could be because Zipf's law does not apply to North American (or any) oil fields, or because there were more giants to be found that would rank amongst the ten largest.

In the event, Prudhoe Bay field (which, but for an event in recent history, would be classified as a Russian oil field) had been found before the list of Table 7-3 had been published in 1970, and its recoverable reserves were soon to be estimated at 20×10^9 bbl, later to be reduced to 9.6×10^9 bbl — 70% larger than the East Texas field. This was to be followed during the 1970s with about 44 giants in North America (Fitzgerald, 1980), and when the ultimate recoverable reserves of these can be fairly estimated, no doubt a new table would look different again.

TABLE 7-4

Zipf predictions based on East Texas field as Rank 1 (column 1) and Rank 2 (column 2); figures in column 3 assume that Prudhoe Bay was next discovery. Other figures are Rank 1 divided by rank number. Figures in millions of barrels

Rank	Actual	1	2	3
1	5600	**5600**	11,200	9600
2	2500	2800	**5600**	4800
3	2334	1867	3733	3200
4	2100	1400	2800	2400
5	1773	1120	2240	1920
6	1683	933	1867	1600
7	1647	800	1600	1371
8	1318	700	1400	1200
9	1303	622	1244	1067
10	1205	560	1120	960
Total	21,463	16,402	32,804	28,118

It can hardly be said that Zipf's law is yet a law that describes the size distribution of oil *fields* very well, but it is an interesting way of looking at the statistics. Bearing in mind that there is a tendency to rank oil fields too highly, with too small a rank number, and to underestimate the ultimate recoverable reserves in them, we can perhaps improve on the estimate of the Zipf constant. Note how the value of this constant on the 1968 information in Table 7-3 tends to increase with increasing rank, suggesting that in the largest fields, the rank number is too small or the reserve estimate too small, or both. A value of 12×10^9 bbl for this constant in North America could still prove to be a good value.

We can extend the argument, following Folinsbee (1977), to estimate the world's ultimate total recoverable oil. Let us assume that the sum of the harmonic series approaches 12, with about 100,000 accumulations that could be produced economically. These numbers are not very sensitive: for 50,000, it is 11.4, so Ranks 50,000 to 100,000 would only contribute about 5% to the total. The ultimate recoverable reserves are given by the product of this number and the Zipf constant for the world. Taking the Zipf constant to be the largest suggested by the world's top-ranking ten oil fields in 1968 (Table 7-5, from Halbouty et al., 1970, Table I, facing p. 504), as a conservative figure, the indicated ultimate recoverable reserves are probably at least $120 \times 10^9 \times 12 = 1450 \times 10^9$ bbl. If the Zipf constant is really about 150×10^9 bbl, the ultimate recoverable reserves will be about 1800×10^9 bbl.

TABLE 7-5

The world's ten largest oil fields in 1968 ranked according to ultimate recoverable reserve estimates (from Halbouty et al., 1970, table I, facing p. 504). Figures in thousands of millions of barrels

		Reserves	Zipf constant
1	Ghawar	75	75
2	Burgan	66	122
3	Bolívar	30	90
4	Safaniya-Khafji	25	100
5	Prudhoe Bay	20	100
6	Samotlor	15	90
7	Kirkuk	15	105
8	Romashkino	14.3	114
9	Rumaila	13.6	122
10	Abqaiq	12.0	120
100	Quirequire	1.0	100
150	Santa Fe	0.615	92

TABLE 7-6

The sizes of the ten largest oil accumulations in the world predicted by Zipf's law if the total ultimate recoverable reserves are 1800×10^9 bbl. Figures in thousands of millions of barrels

1	2	3	4	5	6	7	8	9	10
150	75	50	38	30	25	21	19	17	15

If the ultimate recoverable reserves are 1450 thousand million barrels, Rank 300 would contain 400 million barrels of recoverable oil and Rank 100,000 would contain 1.2 million. If there are 1800 thousand million barrels, Rank 300 would contain 500 million, and rank 100,000 would contain 1.5 million. We have already found more than 300 fields with more than 400 million barrels — indeed, nearly 300 with more than 500 million — so even the larger figure may be conservative. This figure of 1800×10^9 bbl is comparable with the 2000×10^9 bbl estimated by Moody and Esser (1975) by other means. The largest 15 accumulations would contain 28% (3.32/12) of the total, and the 264 largest accumulations would contain 51% (6.16/12). The 1975 figures (Moody and Esser, 1975, p. 17, table IV) are 35 and 72%, respectively, suggesting that *if* Zipf's law is valid for oil accumulations, there has been a successful bias towards the larger accumulations. The ten largest accumulations predicted by Zipf's law for total ultimate recoverable reserves of 1800×10^9 bbl are shown in Table 7-6. This is a plausible result, and

Ghawar with 65×10^9 bbl may be the second largest, rather than the largest accumulation in the world. Again, one wonders whether oil in place should not be the basis, in which case the Athabasca tar sands with 868.7×10^9 bbl (reported by Hackbarth and Nastasa, 1979) would rank number one.

The account sheet therefore looks like this*:

Ultimate recoverable reserves:	1800×10^9 bbl
Cumulative production to end-1980:	446×10^9 bbl
Known recoverable reserves at end-1980:	649×10^9 bbl
Reserves yet to be discovered, approximately:	700×10^9 bbl

The disadvantage of Zipf's law is its empirical nature, with no obvious relation to geology or geochemistry (but perhaps to communities of organic life?). Its advantage is that it emphasizes the giants where most of the oil is. Whatever the true size distribution of oil accumulations may be, it seems almost certain that there are relatively few super giants left to be discovered, and that most of the undiscovered oil is in large, but more modest accumulations. The trends detected by Fitzgerald (1980) support this conclusion. It will be from better understanding of petroleum geology and geochemistry, and from improved exploration and production technology that the rest will be discovered. If a way could be found to extract almost all of the oil, instead of leaving about 2/3 in the ground, the recoverable reserves would be immediately doubled, and our concern for the future halved.

THE OIL-RICH COUNTRIES

Just as most of the known reserves of crude oil is in relatively few fields, so is most of the known reserves in relatively few countries. Fifteen countries possess about 90% of the known reserves of oil recoverable with present technology and prices.

The measure of oil-wealth of a country has traditionally been her annual oil production, usually expressed as the average daily rate of production in barrels. Ranking of countries on the basis of production was quite satisfactory when oil production bore some fairly constant relationship to reserves, but by the end of the 1970s it had become misleading because political and economic influences began to dominate, and production is now a poor indicator of reserves. Tables 7-7—7-9 show the fifteen countries with the largest known

---

* The figure for cumulative production was obtained by taking Moody and Esser's (1975) figure of 297×10^9 bbl at end 1973, and adding the annual figures from the *Oil and Gas Journal's* World Reports. The figure for known recoverable reserves at end-1980 is also from the *Oil and Gas Journal's* World Report.

TABLE 7-7

Ranking of oil-rich countries — 1970

	By reserves end-1970 (10⁹ bbl)	Cumulative production (10⁹ bbl)	By 1970 production	(10⁶ b/day)
1 Saudi Arabia	(128.5)	11.8	U.S.A.	(9.51)
2 U.S.S.R.	(77)		U.S.S.R.	(∼7.1)
3 Iran	(70.0)	11.6	Iran	(3.75)
4 Kuwait	(67.1)		Venezuela	(3.69)
5 U.S.A.	(37.0)		Saudi Arabia	(3.44)
6 Iraq	(32.0)	12.7	Libya	(3.39)
7 Libya	(29.2)	5.0	Kuwait	(2.74)
8 Neutral Zone	(25.7)	1.4	Iraq	(1.52)
9 China	(20)		Canada	(1.28)
10 Venezuela	(14.0)	25.6	Nigeria	(1.00)
11 Abu Dhabi	(11.8)	1.0	Algeria	(0.98)
12 Canada	(10.8)	4.3	Indonesia	(0.86)
13 Indonesia	(10.0)	4.0	Abu Dhabi	(0.64)
14 Nigeria	(9.3)	0.9	Neutral Zone	(0.49)
15 Algeria	(8*)	2.5	China	(∼0.4)
World	590			44.9

* Reported reserves at end-1970 were grossly exaggerated at 30 × 10⁹ bbl, a figure that would have seriously distorted the ranking. The figure of 8 × 10⁹ bbl is retained from end-1969 reports.

recoverable reserves, ranked according to published* reserves at the end of 1970, 1975 and 1980. They show also the same fifteen countries ranked according to their average daily production. Note how by 1980 a tendency had developed for the countries that ranked high on reserves to rank low on production.

Another common measure is the reserves/production ratio, which is the number of years for which the proven recoverable reserves could be produced at the current annual rate. Table 7-10 shows the ranking of the same 15 countries of 1980 on the basis of this ratio.

The message is clear, because the first seven countries all produce very much more than they consume, while only Algeria, Nigeria and Venezuela in the lower half produce substantially more than they consume.

Reserves/production ratios do not, of course, mean that crude oil will run

---

* The *Oil and Gas Journal* publishes a Worldwide Review in its last issue each year. All figures used in this section come from these reviews.

TABLE 7-8

Ranking of oil-rich countries — 1975

	By reserves end-1975 (10⁹ bbl)	Cumulative production (10⁹ bbl)	By 1975 production	(10⁶ b/day)
1 Saudi Arabia	(148.6)	23.1	U.S.S.R.	(9.74)
2 U.S.S.R.	(80.4)		U.S.A.	(8.37)
3 Kuwait	(68.0)	16.6	Saudi Arabia	(7.00)
4 Iran	(64.5)	21.4	Iran	(5.60)
5 Iraq	(34.3)	10.5	⎧ Iraq	(2.40)
6 U.S.A.	(33.0)		⎩ Venezuela	(2.40)
7 Abu Dhabi	(29.5)	2.5	Kuwait	(1.95)
8 Libya	(26.1)	8.7	Nigeria	(1.85)
9 Nigeria	(20.2)	4.2	Abu Dhabi	(1.50)
10 China	(20)		Libya	(1.40)
11 Venezuela	(17.7)	31.7	⎧ China	(∼1.3)
12 U.K.	(16.0)	0.0	⎩ Indonesia	(1.30)
13 Indonesia	(14.0)	6.0	Algeria	(0.94)
14 Mexico	(9.5)	5.4	Mexico	(0.71)
15 Algeria	(7.4)	3.8	U.K.	(0.04)
World	659			53.9

TABLE 7-9

Ranking of oil-rich countries — 1980

	By reserves end-1980 (10⁹ bbl)	Cumulative production (10⁹ bbl)	By 1980 production	(10⁶ b/day)
1 Saudi Arabia	(165.0)	38.8	U.S.S.R.	(12.05)
2 Kuwait	(64.9)	20.0	Saudi Arabia	(9.62)
3 U.S.S.R.	(63.0)		U.S.A.	(8.65)
4 Iran	(57.5)	29.4	Iraq	(2.60)
5 Mexico	(44.0)	7.5	China	(2.17)
6 Iraq	(30.0)	15.2	Venezuela	(2.15)
7 Abu Dhabi	(29.0)	5.8	Nigeria	(2.10)
8 U.S.A.	(26.4)		Mexico	(1.96)
9 Libya	(23.0)	13.0	Libya	(1.78)
10 China	(20.5)		U.K.	(1.60)
11 Venezuela	(18.0)	35.8	Indonesia	(1.57)
12 Nigeria	(16.7)	8.0	Kuwait	(1.40)
13 U.K.	(14.8)	1.7	Abu Dhabi	(1.38)
14 Indonesia	(9.5)	8.8	Iran	(1.28)
15 Algeria	(8.2)	5.6	Algeria	(1.00)
World	649			59.7

TABLE 7-10

Ranking of oil-rich countries of Table 7-9 on the basis of the reserves/production ratio, 1980

Rank	Country	Reserves/production ratio (years)
1	Kuwait	127
2	Iran	123
3	Mexico	62
4	Abu Dhabi	58
5	Saudi Arabia	47
6	Libya	35
7	Iraq	32
	(world average	30)
8	China	26
9	U.K.	25
10	Venezuela	23
11	Algeria	22
12	Nigeria	22
13	Indonesia	17
14	U.S.S.R.	14
15	U.S.A.	8

out in the years stated. New reserves will be found. But if we focus on the United States of America (as one tends to do in the oil industry), a declining trend is seen through the 1970s. Figure 7-2 shows the comparison with the ratios for the world.

Such statistics should never be extrapolated far into the future because the influences on the figures change. The OPEC price rises that started in 1974 changed not only the consumption figures by making products more expensive for those countries that must import all or part of their requirements, but also the reserves because the economically recoverable reserve figure is influenced by price — and this effect was also felt in those non-OPEC countries that adopted "world parity" pricing policies. At the higher price, exploration is encouraged (unless the Government takes away what it regards as "excess profits"). For each influence there is also an opposite, and the net effect is impossible to determine.

Two conclusions are certain:

(1) The world will not run out of crude oil by the year 2000, or by the year 2050, or (most probably) by the year 2100. But certain countries will, for all practical purposes, run out of oil.

(2) There is a great and rewarding challenge for petroleum geologists to better their understanding of the geology of petroleum and, by using the ever-improving technology to the full, to find the reserves needed.

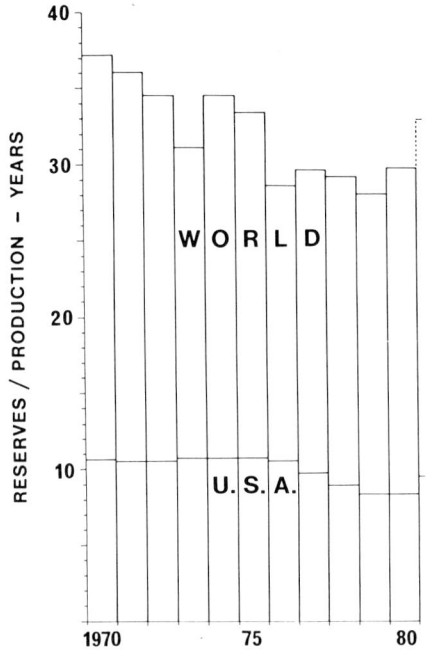

Fig. 7-2. Reserves/production ratios, 1970—1980, for the world and for the U.S.A. (Data from *Oil and Gas Journal's* World Reports.)

REFERENCES

Fitzgerald, T.A., 1980. Giant field discoveries 1968—1978: an overview. In: M.T. Halbouty (Editor), Giant oil and gas fields of the decade 1968—1978. Mem. Am. Ass. Petrol. Geol., 30: 1—5.
Folinsbee, R.E., 1977. World's view — from Alph to Zipf. Geol. Soc. Am. Bull., 88: 897—907.
Hackbarth, D.A. and Nastasa, N., 1979. The hydrogeology of the Athabasca Oil Sands area, Alberta. Bull. Alberta Res. Council, 38, 39 pp.
Halbouty, M.T., Meyerhoff, A.A., King, R.E., Dott, R.H., Klemme, H.D. and Shabad, T., 1970. World's giant oil and gas fields, geologic factors affecting their formation, and basin classification. Part. 1. Giant oil and gas fields. In: M.T. Halbouty (Editor), Geology of giant petroleum fields. Mem. Am. Ass. Petrol. Geol., 14: 502—528.
Haun, J.D., 1981. Future of petroleum exploration in the United States. Bull. Am. Ass. Petrol. Geol., 65: 1720—1727.
Howarth, R.J., White, C.M. and Koch, G.S., 1980. On Zipf's law applied to resource prediction. Trans. Instn. Min. Metall., Sect. B, Appl. Earth Sci., 89 (Nov.): B182—B190.
Meyerhoff, A.A., 1976. Economic impact and geopolitical implications of giant petroleum fields. Am. Sci., 64: 536—541.
Moody, J.D. and Esser, R.W., 1975. An estimate of the world's recoverable crude oil resource. Proc. 9th World Petrol. Congress, 3: 11—20.

Moody, J.D., Mooney, J.W. and Spivak, J., 1970. Giant oil fields of North America. *In:* M.T. Halbouty (Editor), Geology of giant petroleum fields. *Mem. Am. Ass. Petrol. Geol.*, 14: 8—17.

Zipf, G.K., 1949. *Human behavior and the principle of least effort.* Addison-Wesley Press, London, 573 pp.

Zipf, G.K., 1965. *Human behavior and the principle of least effort.* Hafner, New York, N.Y., 573 pp. (Facsimile ed.).

CHAPTER 8

THE NATURE OF PETROLEUM RESERVOIRS

SUMMARY

(1) Petroleum reservoirs contain gas and water, or oil and water, with the petroleum usually in the central parts of the pores and the water in pendular rings around the grain contacts. This water is apparently immobile, and the irreducible water saturation is commonly 20—40% of the pore volume.

(2) Pressures in the petroleum are higher than normal hydrostatic water pressures because of its smaller weight density.

(3) There are two measures of permeability: *intrinsic* permeability (symbol k, dimensions L^2), which is a property of the rock independent of the fluid in it, and *coefficient of permeability* or *hydraulic conductivity* (symbol K, dimensions LT^{-1}), which includes the properties of the fluid. *Relative* permeability (dimensionless) is the ratio of the *effective* permeability to oil or gas to the permeability to water when the rock is fully saturated with water.

(4) Petroleum flows to a well by virtue of the effective permeability of the rock to oil or gas, and an energy gradient induced by the well. The effective and relative permeabilities to oil and gas decrease as the water saturation increases, but the relative permeability to oil and gas may be close to unity when the irreducible water saturation is low. Most of the oil or gas is produced without significant water volumes.

(5) The energy of subsurface fluids is most easily represented by *head* (dimension L). The *total head*, h, is the algebraic sum of the *elevation* head, z, relative to an arbitrary datum (negative downwards), and the *pressure* head, $p/\rho g$ (pressure divided by weight density).

(6) A reservoir is said to have water drive if water expansion is the main driving force replacing the volume of oil produced; gas drive, if gas-cap expansion replaces the oil produced. The energy of a reservoir must usually be maintained by injecting water below the oil/water contact, or by injecting gas into the gas cap.

(7) About two thirds of the oil remains in the reservoir and cannot be produced mechanically by conventional methods. There have been some succesful attempts at recovering some of this oil with solvents.

The reservoir, as we have seen, consists of porous and permeable rock — usually sedimentary rock, but some are of volcanic and other igneous rocks — of which the pores and fissures are filled with water and oil, water and gas, or (rarely) water, oil and gas. The fluids and solids are at elevated temperatures, relative to the surface, depending on the depth and the geothermal gradient.

Geothermal gradients very widely, commonly between 25 and 40°C/km, so the temperature in a typical reservoir at a depth of 2000 m will be between 65° and 95°C. The fluids and the solids are also at elevated pressures, depending generally on the depth and the pressure gradients due to the weight of the overlying materials. We shall consider the pressures in more detail later.

The accumulation is bounded above by a relatively fine-grained material such as a mudstone — or, more precisely, by a material with small pores. The criterion here is that the capillary injection pressure required for the oil or gas to penetrate the cap rock is greater than that existing in the reservoir fluid. The accumulation is usually bounded below by a material of small pores, and, within the reservoir rock unit, by the oil/water or gas/water contact below which the pores are entirely saturated with water that is usually saline. This interface between the petroleum and the water is horizontal or nearly horizontal. The accumulation is sometimes bounded laterally by a fault which may itself be a barrier to further migration on account of the fine-grained fault gouge in the fault plane, or which may bring fine-grained material with small pores into juxtaposition with the reservoir across the fault. Within the reservoir pore spaces, both water and petroleum exist. Reservoir engineers call this "connate" water. We shall examine this first.

Water saturation

When a properly constructed well produces from a virgin oil reservoir, from a zone well away from the oil/water contact or gas/water contact, it typically produces most of its ultimate yield without water. Towards the end of its life, when the oil/water contact approaches the producing zone, the water-cut typically increases and the total liquid production decreases until the well is no longer economic.

If a core is cut in such a reservoir using oil-base mud, so that the connate water is neither displaced nor contaminated, the water saturation (always expressed as a proportion or percentage of the pore space) is found to be between 5 and 60% (rarely, even higher), with values between 15 and 40% common. There is no marked tendency towards lower saturations with increased elevation above the oil/water contact except in a thin zone at the oil/water contact (Fig. 8-1). This paradoxical result is of considerable interest to geologists as well as petroleum engineers.

Bruce and Welge (1947, p. 235, table 1) and Thornton and Marshall (1947, p. 73, table 1) made measurements of water saturation in cores cut in an oil reservoir using oil-base mud, and estimated the water saturation by simulation, using a restored-state, capillary pressure method. In such restored-state experiments it was found that the water saturation reached a figure that was virtually independent of the capillary pressures. This is called the *irreducible water saturation*. The correlation between the two sets of figures is statistically highly significant, and Fig. 8-2 shows the data plotted as correlation lines over

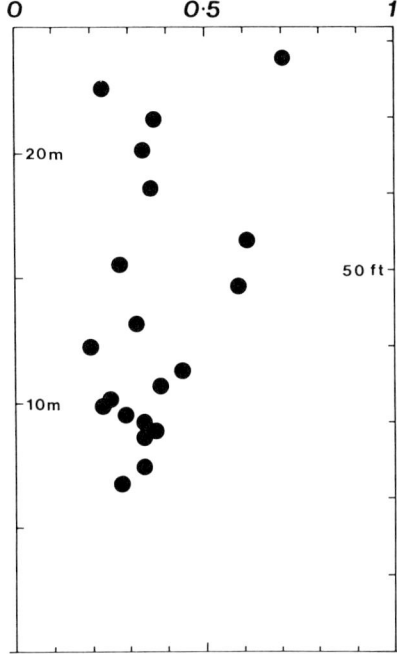
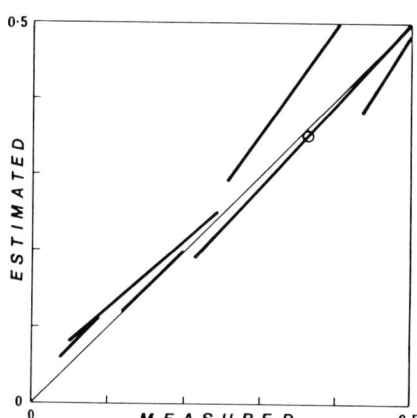

Fig. 8-1. Plot of water saturation against elevation above the oil/water contact. Note that there is no evident tendency for water saturation to decrease with elevation. (Data of Thornton and Marshall, 1947, p. 73, table 1.)

Fig. 8-2. Irreducible water saturation measured from cores cut using oil-base mud plotted against that estimated from capillary pressure measurements. The data is plotted as least-squares correlation lines over approximate range of each set of data. Except for aberrant top line (four measurements), all correlation coefficients were statistically significant at 5% level or less. Data from Bruce and Welge, 1947, p. 235, table 1. The line with a circle is the data of Thornton and Marshall, 1947, p. 73, table 1. Reservoir ages range from Carboniferous to Tertiary.

the approximate range of each set of data. There is little doubt that the restored-state estimates are very close to the measured water saturations, and that estimates from capillary pressure measurements are sufficiently accurate and much cheaper than measurements from cores cut with an oil-base mud.

How is this water distributed in the pore space? It has commonly been assumed that it forms a film around the grains, confining the oil (or gas) to the central parts of the pores. This is what is suggested when we talk of a sandstone reservoir's being water-wet (i.e., preferentially wetted by water rather than oil).

We can make simple estimates of how thick such a film of water would be about the grains. Consider a packing of equal spheres. The total surface area of n spheres, neglecting their effective areas of contact, is $n\pi d^2$; their volume is $n\pi d^3/6$, and they occupy $(1-f)$ of the bulk volume, where f is the fractional porosity. So the ratio of surface area to bulk volume, known as the *specific surface*, is approximated by:

$$S \simeq 6(1-f)/d. \tag{8.1}$$

The saturation due to a film of water of thickness t around the grains is approximated by:

$$s_{wa} \simeq 6t(1-f)/fd. \tag{8.2}$$

If we consider a reservoir rock consisting of spherical grains of 75 μm diameter, corresponding to fine sand, 26% porosity, and an irreducible water saturation of 5%, eq. 8.2 suggests that the film of water, if evenly distributed about the grains, would be about 0.2 μm thick. If the water saturation were 20%, the film thickness would be about 1 μm.

The question now arises, what is the thickness of an adsorbed film of water? We find clues to the answer in the work of Nordberg (1944) and Debye and Cleland (1959) on a porous glass with the brand-name VYCOR. VYCOR has rather uniform pores with radii about 2 nm (Nordberg, 1944) and it was found that the flow of water (and acetone and *n*-decane) followed Darcy's law (Fig. 8-3, see p. 163, and Chapman, 1981, p. 64). From this we infer that the maximum thickness of an adsorbed film on this material at room temperature is about 1 nm. This is two to three orders of magnitude smaller than the thickness of the water film inferred for 5—20% water saturation. An adsorbed film of water 1 nm thick, which is 3 to 4 molecular layers of water, would mean a water saturation (eq. 8.2) of about 0.02%, which is at least an order of magnitude smaller than the precision of our measurements.

We therefore conclude that at the irreducible water saturations we find in reservoirs (> 5%) the water is not evenly distributed about the grains because,

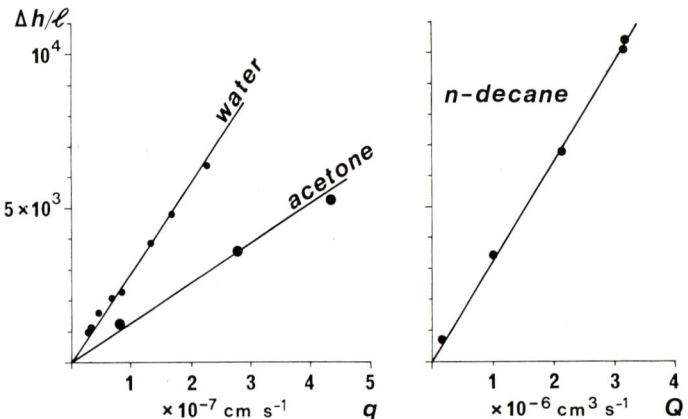

Fig. 8-3. The flow of water, acetone, and *n*-decane through the porous glass VYCOR follows Darcy's law. Data of Nordberg (1944) on left, Debye and Cleland (1959) on right.

if it were, it would be thick enough to flow and would drain to leave virtually zero water saturation. It therefore follows that this water must be largely concentrated around the grain contacts in pendular rings (Fig. 8-4) that are not in hydraulic continuity with each other — as can be seen in drained packings of glass spheres.

This leads to two other questions: how big can such pendular rings be without impinging upon neighbouring pendular rings? and, what is the maximum water saturation due to these rings?

Versluys (1916) approximated the shape of a pendular ring as a solid of revolution of the area bounded by the arcs of three circles (Fig. 8-5), and Rose (1958) showed this to be a close approximation. It can be shown (Chapman, 1982) that the half-volume of a pendular ring so approximated, relative to the volume of a single sphere (again, we must idealize the geometry) is given by:

$$V_{pr}/2 = \frac{3}{4}\left(\frac{1}{\sin\alpha} - 1\right)^2 \left(1 - \frac{\alpha}{\tan\alpha}\right) \qquad (8.3)$$

where $\alpha = (\pi/2) - \beta$. And it can also be shown by an argument similar to that used for the specific surface, that the saturation is then given by:

$$S_{pr} = (V_{pr}/2) N (1-f)/f \qquad (8.4)$$

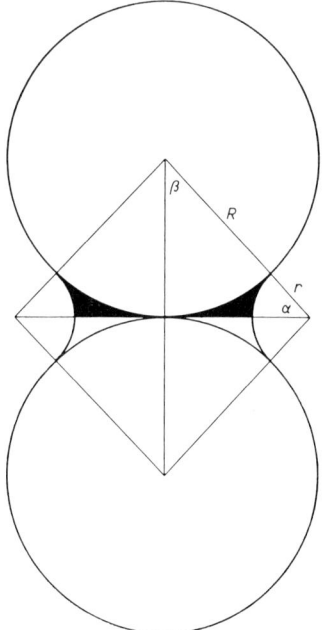

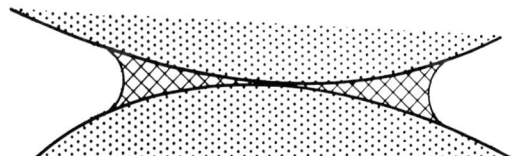

Fig. 8-4. Water is concentrated as pendular rings around the grain-contacts.

Fig. 8-5. Idealized pendular ring geometry.

where N is the average number of point contacts per sphere.

The closest possible packing of equal spheres is close hexagonal, with 25.95% porosity, 12 point contacts per sphere, and a maximum value of β of $30°$. The maximum saturation due to pendular rings in such material is therefore 24%. Cubic packing, with 47.64% porosity, 6 point contacts per sphere, and $\beta = 45°$, can have up to 18% saturation. Figure 8-6 shows the saturations of these two extremes as a function of β. Clearly, saturations well into the observed range are geometrically possible with pendular rings.

Higher saturations can be caused by irregularities of sorting (Morrow, 1971c) and it seems likely that the considerable variation of irreducible saturation found in heterogeneous reservoir sands is due to smaller pores remaining water-filled, the oil preferentially occupying the large pores with smaller injection-pressure requirements.

At the oil/water contact, the evidence of well logs and of capillary pressure experiments suggests that the water saturation increases from irreducible to 100% over a thin transition zone (direct measurement in this zone is not possible because part of this water is mobile).

Real reservoir rocks differ from our ideal packing in two important respects. The sizes and shapes of the grains vary, depending on the sorting, and

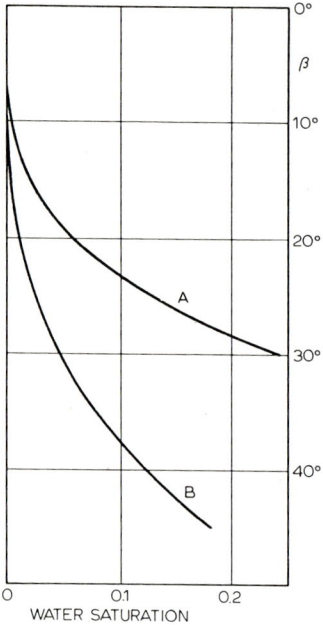

Fig. 8-6. Water saturation in idealized granular material as a function of the angle β (see Fig. 8-5) for close hexagonal packing of equal spheres (A) and cubic packing of equal spheres (B).

authigenic minerals, particularly clays, may form in the pore spaces (see Plate 8-1). These have the effect of reducing porosity and increasing the specific surface, and both tend to reduce the permeability of the rock. Even if the specific surface is increased by two orders of magnitude, most of the water saturation will still be due to free water trapped at grain contacts and in small pores.

Pressures

Within the oil reservoir, above the level at which the oil forms a continuous phase (which we take to be the oil/water contact), the pressure in the oil decreases upwards according to the relationship:

$$\Delta p / \Delta z = \rho_o g \qquad (8.5)$$

where ρ_o is the mean mass density of the oil. This pressure is imposed on the pendular rings of water, which, because they are isolated from each other, does not lead to drainage of the water. Figure 8-7 shows the relationship between pressure, water saturation, and depth in an idealized homogeneous reservoir. Note in passing that a considerable pressure discontinuity may exist at the top of a reservoir with a thick oil or gas column — particularly the latter, because of its smaller weight density.

Permeability and relative permeability

When a well is put onto production, the oil (or gas) flows through the reservoir rock to the well. The rate at which the oil flows (volume divided by time) depends on a number of factors in the reservoir and in the well. We shall concentrate on those in the reservoir. To understand these properly, we must

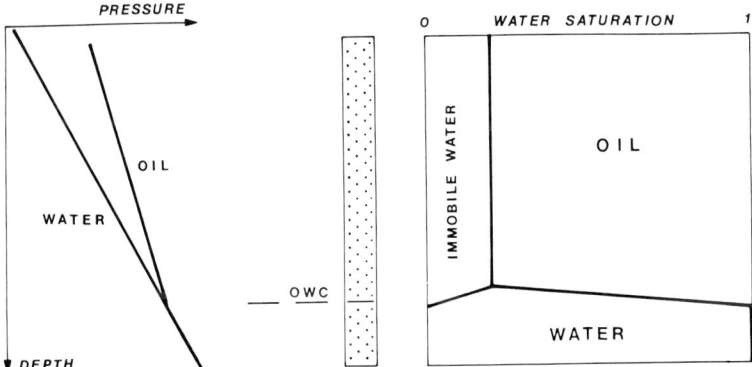

Fig. 8-7. Idealized vertical distribution of pressures and water saturation in sandstone oil reservoir, but the water pressures at left are those that would be found if no oil were present.

PLATE 8-1

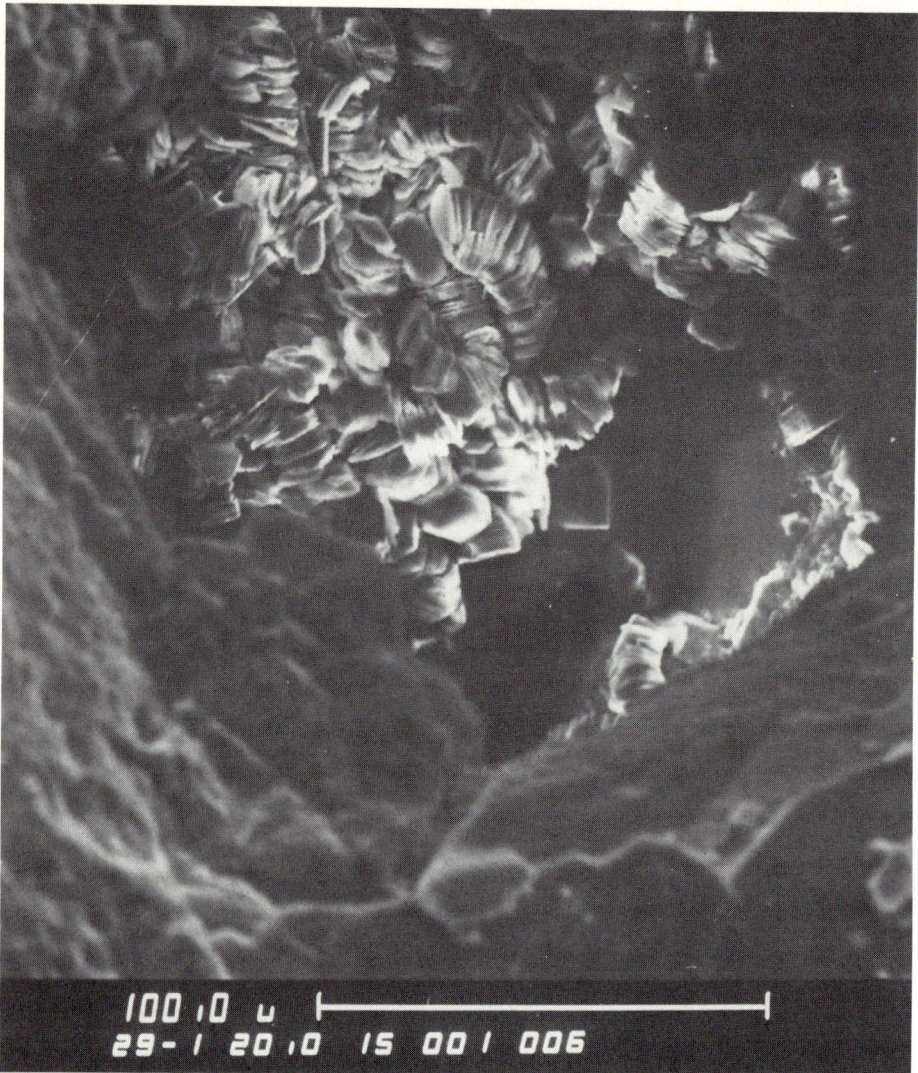

Left. Mungaroo Formation, Gorgon 1, North-West Shelf, Australia. Scanning Electron Microscope images. View looking into a pore space between quartz grains (foreground) showing the large decrease in porosity due to the formation of secondary kaolinite. Depth 3759.7 m.
Right. View of an area dominated by two carbonate minerals, siderite and dolomite. Their authigenesis has been partly responsible for the low porosity and permeability of the rock. Depth 3767.4 m. (Reproduced, with permission, from Campbell and Smith, 1982, p. 108, Plate 2; courtesy of West Australian Petroleum Pty Ltd.)

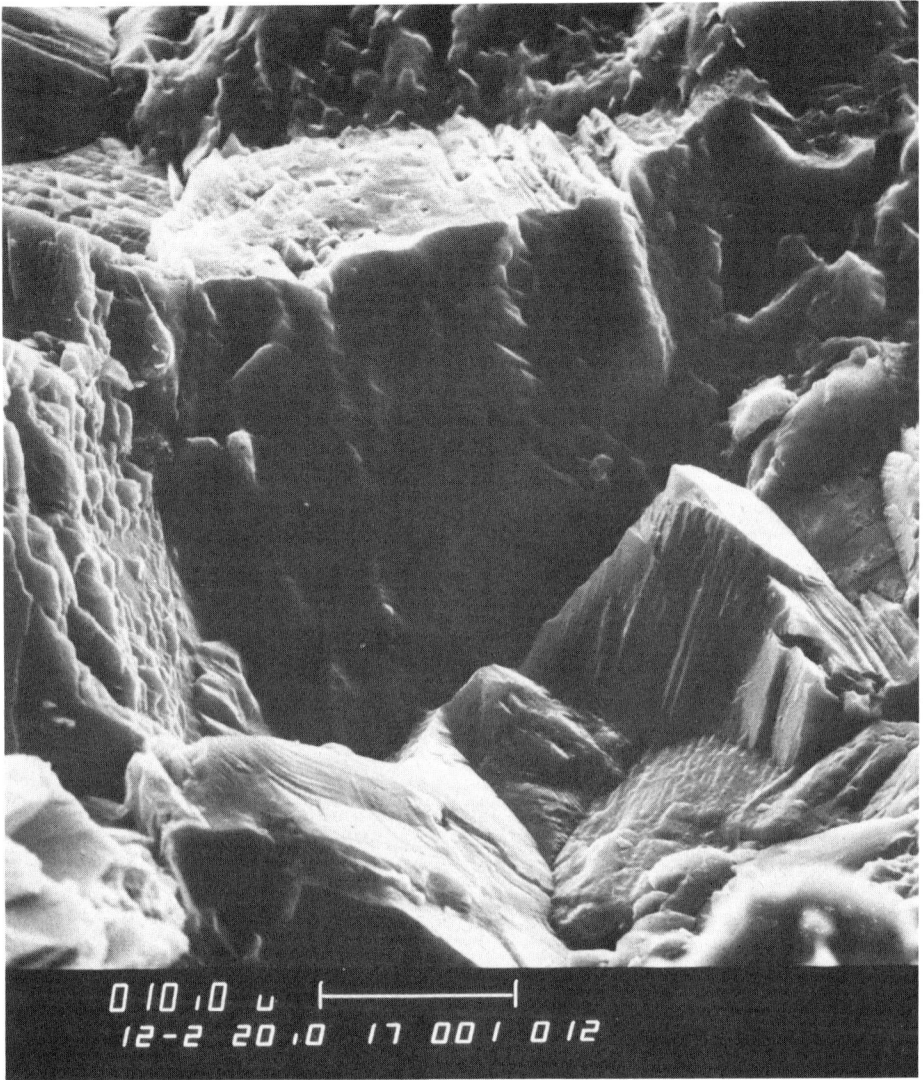

first consider the flow of water through porous rocks: *Darcy's law* (see Chapman, 1981, pp. 49—72, for a fuller account).

Henry Darcy (1856; his name is so spelled) experimented with sand filters in the context of public water supply to the town of Dijon in France. Using apparatus shown diagrammatically in Fig. 8-8, he found that the volumetric rate of discharge was proportional to the difference of levels in the manometers and inversely proportional to the length of the sand filter (Fig. 8-9). The equation he deduced is, in the notation of this book:

$$q = Q/A = K\Delta h/l \tag{8.6}$$

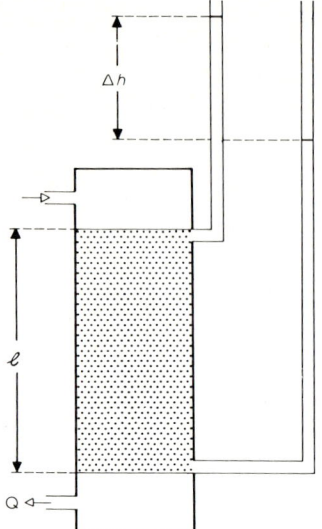
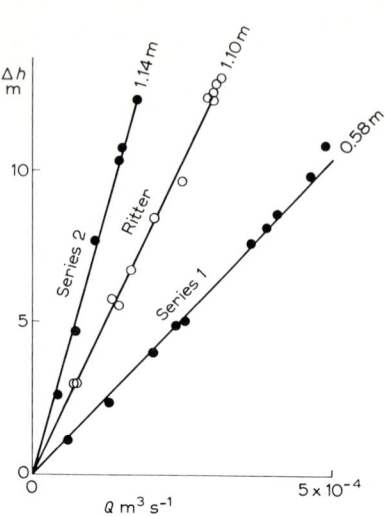

Fig. 8-8. Diagram of Darcy's apparatus.

Fig. 8-9. Darcy's results. The slope increases as the length of sand filter increases.

where K is "a coefficient depending on the permeability of the bed" of sand. K is known as the *coefficient of permeability*, or *hydraulic conductivity*: it has the dimensions of a velocity, LT^{-1}. The quantity $\Delta h/l$ is known as the *hydraulic gradient*: it is dimensionless, being a length divided by a length. Q is the volumetric rate of flow, and A its gross cross-sectional area; so q, which is known as the *specific yield*, has the dimensions of a velocity ($L^3 T^{-1} L^{-2} = LT^{-1}$).

The coefficient of permeability takes several influences into account. If Darcy had used liquids other than water, his results would have been numerically different. He did use different sands, with different numerical results. The main influences are the nature of the sand and the nature of the liquid. These can be separated, and Darcy's law written:

$$q = k \frac{\rho}{\eta} g \frac{\Delta h}{l} \tag{8.7}$$

where k is called the *intrinsic permeability* (dimensions L^2) and it relates solely to the porous material. The ratio of mass density to dynamic or absolute viscosity, ρ/η, is the inverse of the kinematic viscosity of the liquid (the dimensions of kinematic viscosity are those of the product of a length and a velocity, $L^2 T^{-1}$). And the term $g\Delta h/l$ is known as the *potential gradient* and relates to the loss of energy incurred by the flow (see Versluys, 1917; and Hubbert, 1940, pp. 796—803).

Because the acceleration due to gravity is a constant in this context, the energy of the water is indicated for practical purposes by its *total head, h:* this is the algebraic sum of the elevation head z relative to an arbitrary datum (measured negative downwards) and the pressure head, $p/\rho g$ (Fig. 8-10). Note most carefully that the pressure head is measured from the point at which the pressure is measured, and the elevation head from an arbitrary datum, commonly sea level. All heads have the dimension of *length.*

Intrinsic permeability, k, with the dimensions of an area, is a material constant that is a function of porosity, the tortuosity of the flow paths through the pores, and the *harmonic* mean diameter of the grains. Tortuosity can be approximated by a porosity term, and the combined expansion is:

$$k = Cf^x d^2/(1-f)^2 \tag{8.8}$$

where x varies from about 3.5 for unconsolidated sands to about 5 for well-consolidated sands. The dimensionless coefficient C is not a true constant because it takes some undeterminable properties of the material into account.

Before proceeding, we must pause to note that eq. 8.7 is commonly and erroneously written:

$$q = (k/\eta)\Delta p/l.$$

That this is an incomplete expression of Darcy's law is evident from the fact that when the fluid is at rest, $q = 0$, and this implies that $\Delta p = 0$. This is only true on a horizontal plane in the water-saturated porous material, so that the equation is only valid for horizontal flow. It is on the basis of this equa-

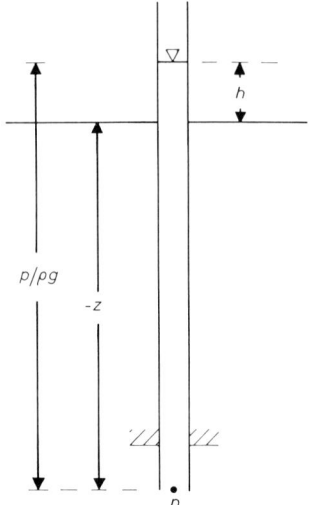

Fig. 8-10. The total head, h, is the algebraic sum of the pressure head, $p/\rho g$, and the elevation head, z.

PLATE 8-2

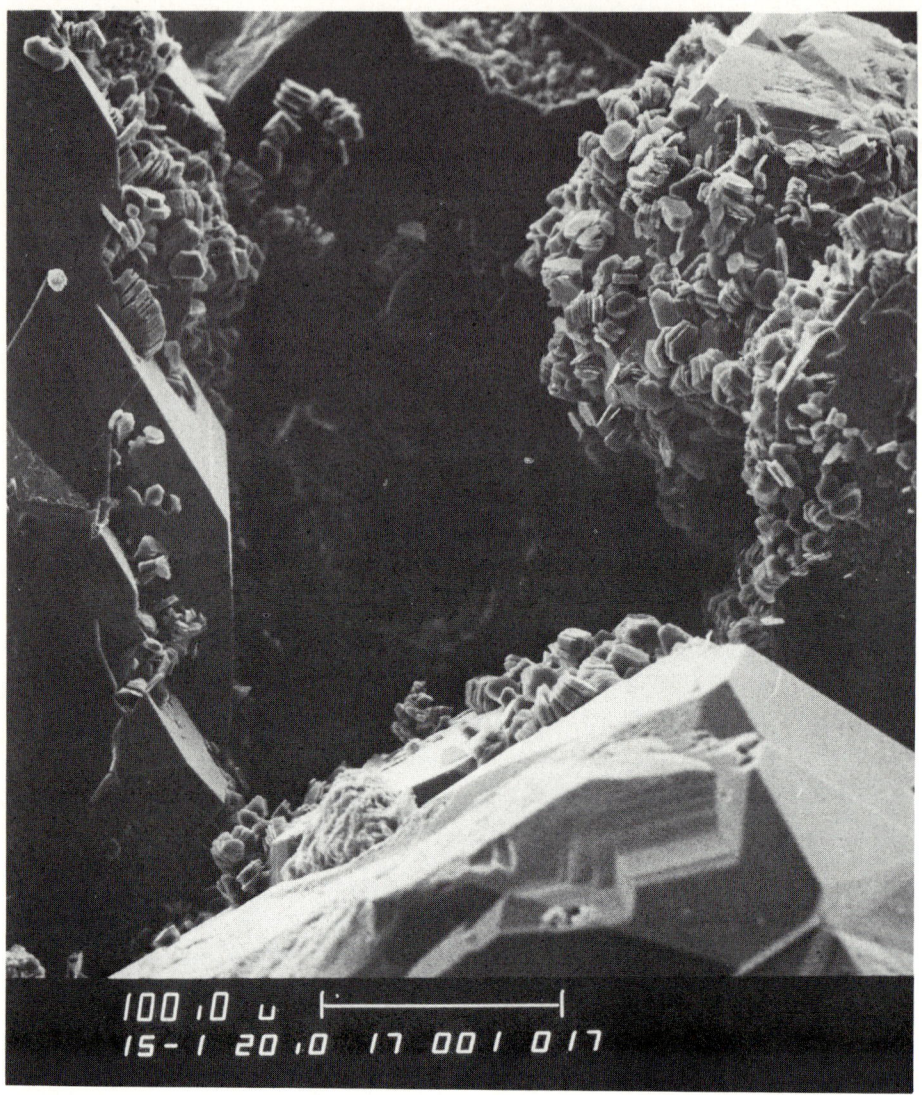

Left. Mungaroo Formation, Gorgon 1; SEM images. Kaolinite lining a pore, and several quartz grains exhibiting secondary overgrowths. Depth 3973.6 m.
Right. High-magnification view of well-crystallized kaolinite on a quartz grain, seen in the left hand image at lower magnification. Depth 3973.6 m. (Reproduced, with permission, from Campbell and Smith, 1982, p. 108, Plate 2; courtesy of West Australian Petroleum Pty Ltd.)

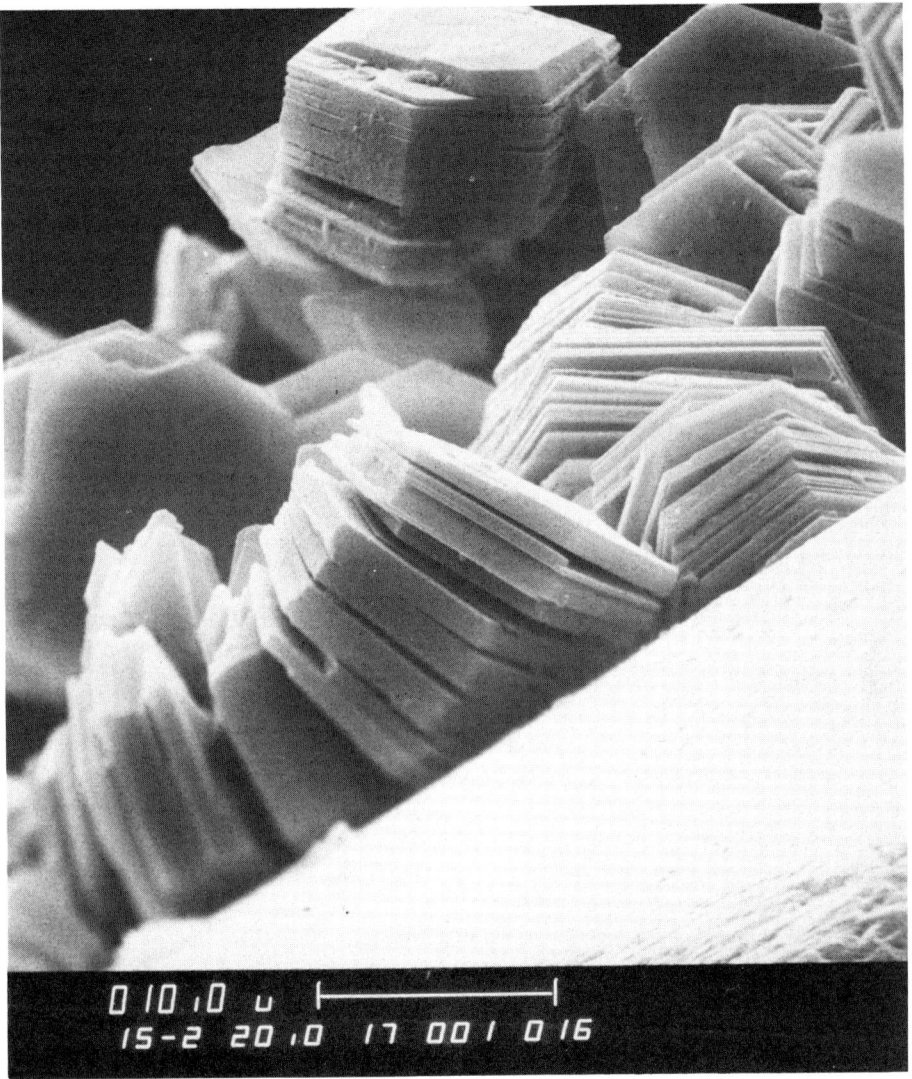

tion that the unit of intrinsic permeability, the *darcy*, was defined (Wyckoff et al., 1933, 1934).

Equation 8.7 is the general form of Darcy's law for liquids, and applies to the flow of any liquid. The intrinsic permeability is independent of the liquid flowing. As with all such equations, care must be taken with units. Since the unit of intrinsic permeability is traditionally cm^2, mass density should be in $g\ cm^{-3}$, dynamic viscosity (η) in *poise* ($g\ cm^{-1}\ s^{-1}$), and $g = 980\ cm\ s^{-2}$. For practical purposes, the darcy is $10^{-8}\ cm^2$; so the unit of intrinsic permeability in SI units is likely to be the μm^2.

Darcy's law has an upper limit, but this is rarely (if ever) exceeded in nature. No lower limit has been demonstrated. The intrinsic permeability of VYCOR, mentioned above, is of the order of 10^{-16} cm^2, or 10^{-8} darcies.

For any porous and permeable material, the intrinsic permeability can be determined experimentally using any liquid — for example, oil or water:

$$k = q_w \eta_w l / \rho_w g \Delta h_w = q_o \eta_o l / \rho_o g \Delta h_o. \tag{8.9}$$

When both liquids are present as two immiscible phases in the pore spaces, we can determine the permeability of the material to oil and the permeability to water. These are known as *effective permeabilities* — the effective permeability to oil (k_o) or gas (k_g) and the effective permeability to water (k_w). Clearly these quantities are functions of water saturation: the greater the water saturation, the greater the effective permeability to water and the smaller the effective permeability to oil (or gas). But at 100% and zero water saturations, the effective permeability is equal to the intrinsic permeability, so it is convenient to express these as *relative permeability*: $k_{rw} = k_w/k$, $k_{ro} = k_o/k$, and $k_{rg} = k_g/k$. Effective permeability is in the nature of an intrinsic permeability because each fluid reduces the effective porosity for the other and each affects the tortuosity of the flow path of the other.

In the experimental determination of relative permeabilities, it has been found that there is hysteresis. If one starts with 100% water saturation and determines the relative permeabilities for decreasing water saturations and increasing oil saturations, the figures are different from those determined for increasing water saturations and decreasing oil saturations (Fig. 8-11). The curves obtained for decreasing water saturations are called *drainage* curves;

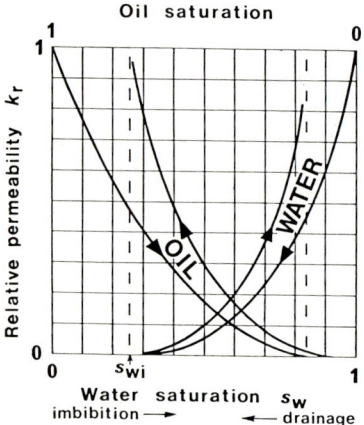
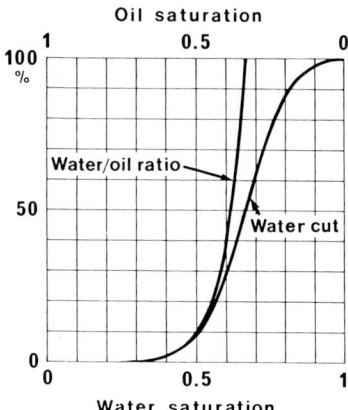

Fig. 8-11. Experimental relative permeability curves obtained with increasing water saturation (imbibition) differ from those obtained with decreasing water saturation (drainage).

Fig. 8-12. Diagrammatic relationship between water saturation, water/oil ratio and water cut (ratio of water produced to total liquid produced).

those for increasing water saturations, *imbibition* curves. From the previous discussion of irreducible water saturation, it will be evident that a drainage experiment is terminated at irreducible water saturation (as it must be), and an imbibition experiment is terminated at the irreducible oil saturation. This is reached when the oil saturation is insufficient to maintain a continuous oil phase through the pore space, and the capillary pressure that exists in the oil droplets so formed is insufficient to drive them from one pore to the next, through the constrictions or "throats".

There are several points of interest in relative permeability curves. First, the sum of the relative permeabilities is always less than unity. Secondly, the drainage relative permeability to water may become negligibly small *before* the irreducible water saturation is reached. Thirdly, the drainage relative permeability to oil is close to unity at irreducible water saturation (it has even been claimed to be greater than unity). From a practical reservoir engineering point of view, such curves can be used to predict production rates, water/oil ratios, and water cut (proportion of water in total production or yield) at various saturations.

The reason for the high drainage relative permeability to oil at irreducible water saturation is that the oil in a water-wet sand (for example) is excluded from the pendular spaces, which contribute little to the flow of a single liquid. The reason for the very low relative permeability to water at low saturations above the irreducible is that the water is denied access to the central pore space and must flow in thin streams or films around the grains — paths of much greater tortuosity. The reason for the sum of the relative permeabilities being less than unity is that each liquid interferes with the other, and at saturations between the two irreducible limits, the tortuosity of each component is greater than the tortuosity of a single-phase flow. That these two tortuosities are unequal is shown by the feature that the relative permeabilities are equal at water saturations greater than 50%, commonly 60—65%.

For production purposes we are more interested in imbibition curves that will predict well behaviour as water saturations increase. In migration studies, we shall be more interested in drainage curves. Figure 8-12 shows the predicted water/oil ratio and water cut derived from the imbibition relative permeability curves of Fig. 8-12.

Mechanics of production

The energy of a petroleum reservoir comes from the expansion of the water and/or the gas to replace the volume of petroleum produced. If water expansion below the oil/water contact is the main driving force, the reservoir is said to have *water drive;* if gas expansion above the gas/oil contact is the main driving force, the reservoir is said to have *gas drive*. Because it is necessary, as we shall see, to conserve the natural reservoir energy, gas is never produced intentionally when it exists as a gas cap to an oil reservoir that is economically producible.

We shall consider only the simplest aspects of oil-well production that will enable us to reach the point of understanding the nature of the process. We shall first consider an oil well producing without gas or water from an inclined reservoir over a small interval at a depth of 2000 m, with the oil/water contact at 2050 m (Fig. 8.13).

At the oil/water contact, the pressure in the oil is equal to that in the water; and, if it is normal hydrostatic, it will be:

$$p = \rho_w gz = 1020 \times 9.8 \times 2050 = 20.5 \text{ MPa}$$
$$= 2973 \text{ psi.} \tag{8.10}$$

The total head will be nominally zero, and pressures in the water outside the reservoir will decrease upwards according to the relationship, $\Delta p_w/\Delta z = \rho_w g$. This is a stable pressure gradient in the water, and the water will be at rest.

Within the oil reservoir, the pressure in the oil will also be 20.5 MPa (2973 psi) at the oil/water contact, and the pressures in the oil will decrease upwards according to the relationship, $\Delta p_o/\Delta z = \rho_o g$. This is a stable pressure gradient in the oil, and the oil will be at rest. If the mass density of the oil is 780 kg m^{-3}, the pressure at the bottom of the interval to be produced at 2000 m

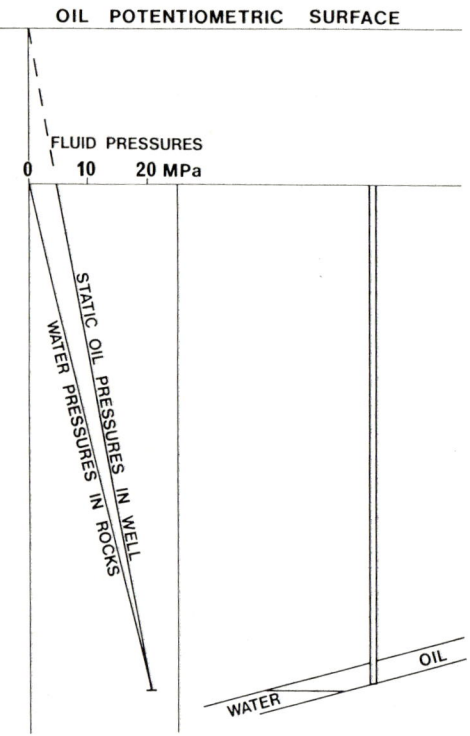

Fig. 8-13. Static pressures in an oil well that is closed in (diagrammatic).

will be 20.1 MPa (2915 psi) before the well is put onto production. If the tubing is now swabbed to oil, the static pressure at the well head will be:

$$p = (\rho_w - \rho_o) gz = 4.7 \text{ MPa} = 682 \text{ psi}. \tag{8.11}$$

This pressure is unbalanced, so the well will flow if the valve is opened. The rate at which the well will flow depends on the energy losses in the entire system — energy lost by flow to the well, and energy lost by flow up the tubing and through any surface pipes. In the reservoir, the energy losses depend on the intrinsic permeability, the relative permeability to oil, and the kinematic viscosity of the oil. In the well, the energy loss depends on the internal diameter and length of the tubing, and also on the kinematic viscosity of the oil.

Within the reservoir, there is now a potential gradient towards the well from *all* directions, and the oil flows radially into the well — downdip from above the well, updip from below the well, and horizontally from along strike. Intuition is not always reliable in these matters, so we must examine the flow more carefully.

When a liquid is at rest, the entire body of liquid is at constant potential, that is:

$$\Phi = gh = g \left(\frac{p}{\rho g} + z\right) = \text{constant} \tag{8.12}$$

so that the change of pressure head to different levels is exactly equal to the change of elevation. When the well is put onto production, the potential in the well is made less than that in the reservoir, and a potential gradient is created down which the oil flows to the well.

The potential is proportional to the total head of the oil in the reservoir:

$$h_o = \frac{p}{\rho_o g} + z = 2681 - 2050 = 631 \text{ m}. \tag{8.13}$$

So the energy of the oil reservoir can be represented by a conceptual surface known as a *potentiometric surface*, and while the oil is static, this surface is horizontal and elevated (in our example) 631 m above the datum surface, taken here to be the level of the well head. When a well is put onto production, a cone of depression is imposed on this surface (Fig. 8-14) and the *equipotential lines* or contours on this surface form circles around the well (Fig. 8-15). Flow in the reservoir is normal to the equipotential surfaces — normal to the equipotential lines in plan — that is, radial to the well. The better the effective permeability to oil, the shallower the cone of depression.

The oil flows parallel to the bed surfaces, as it must, and normal to the equipotential surfaces that form concentric cylinders about the well. Ignoring compressibility, equal volumes of oil cross any of these concentric surfaces in unit time, so the oil is accelerating towards the well. Regarding the well as normal to the reservoir for simplicity, the area of any concentric surface is

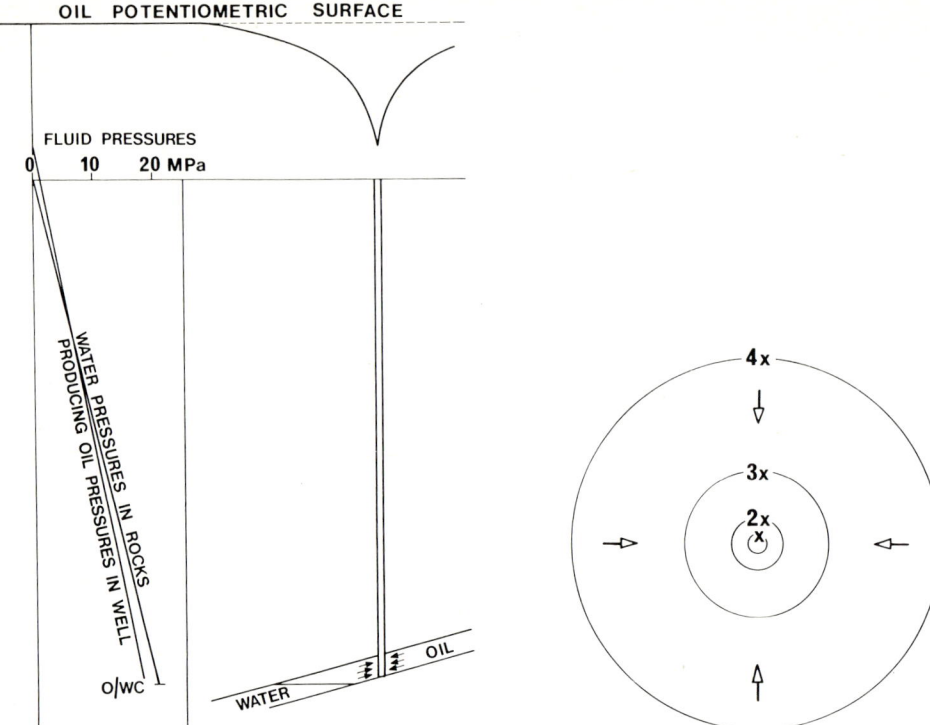

Fig. 8-14. When put onto production, the pressures in the well of Fig. 8-13 are reduced by friction. Oil flows radially in the reservoir to the well.

Fig. 8-15. Diagrammatic map of potentiometric surface near a producing oil well (contours at interval x above lowest point on surface). Elevation of the potentiometric surface increases as the natural logarithm of the distance from the well.

$2\pi rt$, where t is the thickness of the reservoir and r is the radius to the concentric surface. In the well, this area is:

$$A_b = 2\pi r_b t \qquad (8.14)$$

where r_b is the radius of the borehole. Let q_b be the specific discharge, Q/A_b, into the well, and q the specific discharge across a concentric surface at radius r, then:

$$q_b \times 2\pi r_b t = q \times 2\pi rt$$

and:

$$q = q_b r_b/r. \qquad (8.15)$$

Thus the specific discharge varies inversely with distance from the well.
From Darcy's law, assuming a nearly horizontal reservoir:

$$q = k_o \frac{\rho_o}{\eta_o} g \frac{\Delta h}{\Delta r} = q_b r_b / r \tag{8.16}$$

from which the hydraulic gradient is seen to be:

$$(\Delta h / \Delta r) = q_b r_b \eta_o / r \rho_o g k_o. \tag{8.17}$$

If the total head in the borehole is h_b, the total head at radius r is given by

$$\begin{aligned} h_r &= h_b + (q_b r_b \eta_o / \rho_o g k_o) \int_{r_b}^{r} 1/r \, dr \\ &= h_b + (q_b r_b \eta_o / \rho_o g k_o) \ln (r/r_b). \end{aligned} \tag{8.18}$$

Because $q_b = Q/2\pi r_b t$, we may substitute $Q/2\pi t$ for $q_b r_b$ in eq. 8.18:

$$h_r - h_b = \frac{Q\eta}{2\pi t \rho_o g k_o} \ln \frac{r}{r_b}. \tag{8.19}$$

This equation indicates that the elevation of the potentiometric surface, when steady production has been achieved at rate Q, increases as the natural logarithm of the distance from the well. It also indicates that h_r increases indefinitely as the radius increases, so that steady flow is theoretically impossible in a finite reservoir. Nevertheless, it is evident that h_r approaches the original total head at some radius r at any steady production rate. This is the radius of influence of the well. When the optimum rate Q has been determined, the optimum well spacing can also be determined.

Looking at this development of a potential gradient in the context of time, we see that it takes time to develop steady production and a stable potentiometric surface profile to the well. By the same token, if the well is closed in after steady production has been achieved, it will take time for the pressure in the well to recover its original value.

The modification of eq. 8.19 to take account of semi-steady flow, and the development of the pressure build-up equation, are beyond the scope of this book (see, for example, Dake, 1978, chapters 4—7), but it can be shown that for small values of time since closing-in, the build-up equation is similar in form to eq. 8.19:

$$\frac{4\pi t \rho_o g k_o}{Q \eta_o} (h - h_b) = \ln (T + \Delta T)/\Delta T \tag{8.20}$$

where T is the time during which the well has been produced at rate Q (for irregular production, the cumulative production divided by time of production is taken as Q), and ΔT is the closed-in time. This equation is usually written for horizontal flow:

$$\frac{4\pi t k_o}{Q \eta_o} (p^* - p) = \ln (T + \Delta T)/\Delta T \tag{8.20a}$$

where p^* is the reservoir pressure estimated from infinite closed-in time, and p is the pressure measurement made at time ΔT after closing-in. When p is plotted against the logarithm of the dimensionless time term, a straight line is found. This method is due to Horner (1951) and such plots are called Horner plots.

If the well being tested is a new discovery well, the Horner plot gives two pieces of valuable information (Fig. 8-16). The linear part of the plot extrapolates to infinite time where the logarithm of the dimensionless time term equals zero; and this gives the initial reservoir pressure at the level at which the pressures were measured. The slope of this line is equal to $Q\eta_o/4\pi k_o t$. After the properties of the oil have been found from tests, only the effective permeability to oil is unknown, so it can be determined. This value is the mean value for the reservoir in the drainage area of the well.

There is another point of interest to geologists. If the reservoir is bounded by a fault or a marked change of permeability near the well within its radius of influence, the plot shows two straight lines (Fig. 8-17), the second, at larger

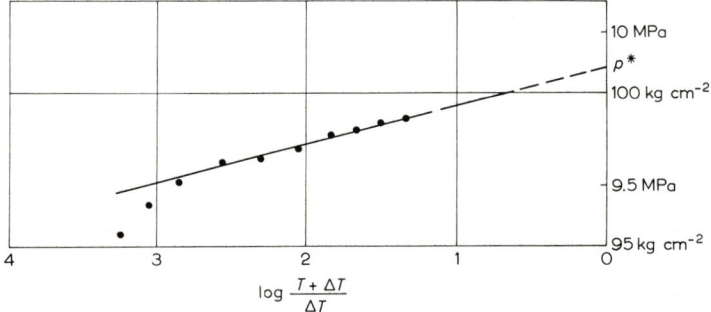

Fig. 8-16. Horner plot of pressure build-up in a well. The linear part extrapolates to p^*, which is the original pressure at depth of measurement; reservoir parameters can be determined from slope.

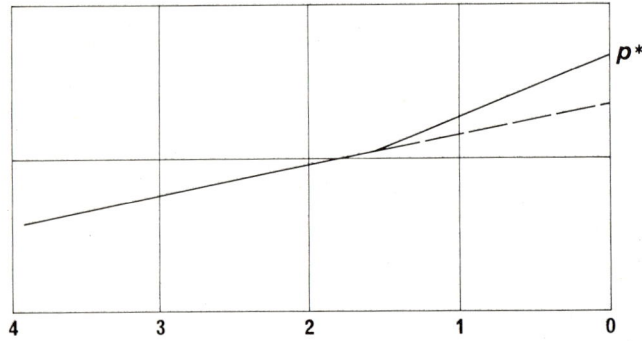

Fig. 8-17. If there is a permeability barrier (such as a fault) near the well, the Horner plot shows two linear trends — the second theoretically with twice the slope of the first.

ΔT, having theoretically twice the slope of the first. It is theoretically possible to estimate the distance of the fault from the well. (The slope of the first part extrapolates to a p^* that is too small, but gives the correct value of effective permeability; the second line extrapolates approximately to the true value of p^*.)

If the energy of the reservoir is, or becomes, insufficient to generate flow, or insufficient to generate optimum flow rate, it can be artificially increased by *gas-lift* or by *pumping*. Gas-lifting involves the injection of gas through valves into the tubing so that the effective density of the oil column in the tubing is decreased. This has the effect of increasing the potential gradient. Natural gas coming out of solution in the oil as it rises in the tubing to lower pressures commonly assists production in the same way.

As oil is abstracted from the reservoir by production, its place must be taken by water expansion or gas expansion (or both), or the pressure loss will result in some mechanical compaction of the reservoir rock. It will be evident from the discussion of pressure build-up in a well that water drive and gas drive mechanisms require time, because the volume created by expansion must flow through the reservoir rock, displacing the oil/water contact upwards and the gas/oil contact downwards.

Compaction drive is therefore also a production mechanism, but an undesirable one in most oil fields because it leads to surface subsidence. Its main interest to geologists is that it demonstrates mechanical compaction and the qualitative validity (at least) of Terzaghi's relationship, which we studied in Chapter 3. The seriousness of compaction in an oil field with multiple reservoirs is well illustrated by the history of the Wilmington field in California (Mayuga, 1970). This oil field, discovered in 1932 and found to be one of the largest in North America, lies below the coastal districts of Los Angeles and the city of Long Beach. The area overlying the field is industrial and residential, and includes Long Beach harbour and a naval shipyard. By the end of 1967, more than one thousand million barrels of oil (184×10^6 m^3) and nearly one Tcf (23.8×10^9 m^3) of gas had been produced over 30 years. However, within a decade of the beginning of major production about 1937, when the importance of the discovery became apparent, surface subsidence began to threaten coastal installations. Figure 8-18 shows the pattern of events up to 1967. Maximum subsidence, over the crest of the field structure, reached 8.8 m (29 ft), and the maximum subsidence rate was about 0.7 m/yr. This vertical movement, with horizontal displacements up to 3 m, caused extensive damage on the surface and to oil wells at depth. Measurements in boreholes (see Mayuga, 1970) showed that almost all the compaction had taken place in the producing zones.

The correspondence between production rate and subsidence rate left no doubt about the cause, so water injection was planned, with a pilot-scheme starting in 1953. Five years later, when the major scheme started, production rates could be increased while the rate of subsidence further decreased. In some areas of greatest water injection, the surface was even elevated by up to 25 cm.

This was not an isolated instance, but its setting made the consequences more serious. M. ap Rhys Price, in his discussion of a paper by Kugler (1933, p. 769), reported that subsidence over the Lagunillas field on the shore of Lake Maracaibo, Venezuela, was found to be "in direct proportion with the production taken out".

The prevention of subsidence, as we have seen, was found to have a beneficial effect on production; so compaction drive is not only undesirable but also inefficient.

Secondary recovery

Production from an oil reservoir leads, in all but those reservoirs with strong water or gas drive, to a gradual loss of pressure with time, and so to a declining production rate. The obvious methods of maintaining production (the word *stimulation* is usually restricted to methods of improving individual well performance by improving the permeability around the well) is to do artificially what Nature does — that is, to replace the oil produced by injecting water close below the oil/water contact, or gas into the gas cap. We will consider water flooding.

The energy of the water below the oil/water contact can be represented by a potentiometric surface. If the water was originally static, the potentiometric surface was originally horizontal; but by the time water injection be-

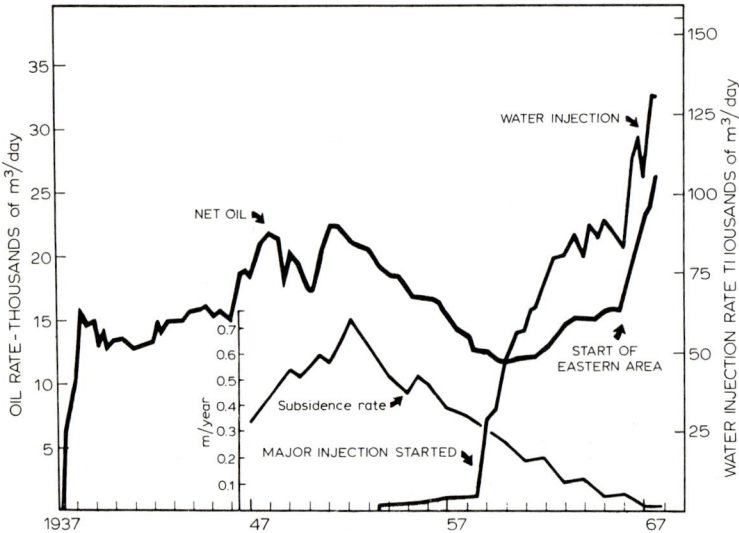

Fig. 8-18. Relationship between oil production rates, subsidence rates, and water injection rates in the Wilmington field, California. (After Mayuga, 1970, p. 177, fig. 16.)

comes necessary, it will be inclined slightly towards the oil reservoir from outside. The effect of an injection well is to impose on this potentiometric surface a *cone of impression* analogous to the cone of depression imposed by a producing well. The shape of this cone of impression is described by an equation similar to 8.19, with the term $\rho g k/\eta = K$, the hydraulic conductivity, and the sign of Q is negative.

The flow of water from an injection well is thus radial from the well, but the symmetry will be disturbed in an injection well close to the oil/water contact due to the low relative permeability to water at low oil saturations about the contact, in the field of flow towards the accumulation.

Heads, being scalar quantities, are additive; and the water-injection well being within the radius of influence of a producing well, the potentiometric surface takes a shape that is the algebraic sum of the total heads of the several components (cf. Fig. 6-5, p. 119).

Pressure maintenance is better than trying to restore pressure lost, so waterflood operations usually begin early in the life of a field, soon after recognition that the natural water drive is inadequate.

The uncomfortable fact that about 2/3 of the oil in place remains in the reservoir, even after secondary recovery operations, has encouraged research into more efficient means of recovery. The problem here is both technical and economic: the cost of the operation must be less than the value of its effect. There are two main factors on the technical side. First, the *immiscibility* of oil and water affects the recovery because of the capillary pressures involved. Secondly, the effects of wells are dominantly radial. Injection of solvents or agents that promote miscibility increases the proportion of oil ultimately recovered, but the solvents or agents must be cheap or readily recoverable. They must also be absolutely clean (as with the water for injection) so that they can be injected at high rates for long periods of time without clogging the injection well.

The radial tendency of the processes leaves large volumes of the reservoir inefficiently drained, like the space between coins placed together on a table, and anisotropy in the reservoir can (and usually does) lead to unstable displacement fronts with "fingering". Once a preferential displacement path reaches a producing well, the greater relative permeability to the solvent or agent (or water in water flooding) ensures that significant proportions of the reservoir will be by-passed and will remain undrainable without drilling new wells.

REFERENCES

Allen, D.R., 1968. Physical changes of reservoir properties caused by subsidence and repressuring operations. *J. Petrol. Technol.*, 20: 23—29.

Bruce, W.A. and Welge, H.J., 1947. Restored-state method for determination of oil in place and connate water. *Oil Gas J.*, 46 (12): 223—238.
Chapman, R.E., 1981. *Geology and water: an introduction to fluid mechanics for geologists.* Nijhoff/Junk, The Hague, 228 pp.
Chapman, R.E., 1982. Effects of oil and gas accumulation on water movement. *Bull. Am. Ass. Petrol. Geol.*, 66: 368—374.
Dake, L.P., 1978. *Fundamentals of reservoir engineering.* Elsevier, Amsterdam, 443 pp.
Darcy, H., 1856. *Les fontaines publiques de la ville de Dijon.* Dalmont, Paris, 674 pp.
Debye, P. and Cleland, R.L., 1959. Flow of liquid hydrocarbons in porous VYCOR. *J. Appl. Phys.*, 30: 843—849.
Geertsma, J., 1973. Land subsidence above compacting oil and gas reservoirs. *J. Petrol. Technol.*, 25: 734—744.
Gibson, H.S., 1948. The production of oil from the fields of southwestern Iran. *J. Inst. Petrol.*, 34: 374—398.
Horner, D.R., 1951. Pressure build-up in wells. *Proc. 3rd World Petrol. Congress*, The Hague, 1951. Sect. 2: 503—521.
Hubbert, M.K., 1940. Theory of ground-water motion. *J. Geol.*, 48: 785—944.
Klinkenberg, L.J., 1942. The permeability of porous media to liquids and gases. *Am. Petrol. Inst. Drilling Product. Practice*, 1941: 200—213.
Kugler, H.G., 1933. Contribution to the knowledge of sedimentary volcanism in Trinidad. *J. Instn. Petrol. Technol.*, 19: 743—760: discussion, 19: 760—772.
Mayuga, M.N., 1970. Geology and development of California's giant — Wilmington oil field. In: M.T. Halbouty (Editor), Geology of giant petroleum fields. *Mem. Am. Ass. Petrol. Geol.*, 14: 158—184.
Morrow, N.R., 1971a. The retention of connate water in hydrocarbon reservoirs. I. A review of basic principles. *J. Can. Petrol. Technol.*, 10: 38—46.
Morrow, N.R., 1971b. The retention of connate water in hydrocarbon reservoirs. II. Environment and properties of connate water. *J. Can. Petrol. Technol.*, 10: 47—55.
Morrow, N.R., 1971c. Small-scale heterogeneities in porous sedimentary rocks. *Bull. Am. Ass. Petrol. Geol.*, 55: 514—522.
Nordberg, M.E., 1944. Properties of some VYCOR-brand glasses. *J. Am. Ceram. Soc.*, 27: 299—305.
Pirson, S.J., 1958. *Oil reservoir engineering* (2nd ed.). McGraw-Hill, New York, N.Y., 735 pp.
Rose, W., 1958. Volumes and surface areas of pendular rings. *J. Appl. Phys.*, 29: 687—691.
Thornton, O.F. and Marshall, D.L., 1947. Estimating interstitial water by the capillary pressure method. *Trans. Am. Inst. Min. Metall. Engrs.* (Petroleum Div.), 170: 69—77.
Versluys, J., 1916. *De capillaire werkingen in den bodem.* Thesis, Versluys, Amsterdam, 136 pp.
Versluys, J., 1917. De beweging van het grondwater. *Water*, 1: 23—25, 44—46, 74—76, 95.
Von Engelhardt, W. and Tunn, W., 1954. Über das Strömen von Flüssigkeiten durch Sandsteine. *Heidelb. Beitr. Mineral. Petrogr.*, 4: 12—25.
Von Engelhardt, W. and Tunn, W.L.M., 1955. The flow of fluids through sandstone. *Ill. State Geol. Surv. Circ.*, 194, 17 pp.
Wyckoff, R.D., Botset, H.G., Muskat, M. and Reed, D.W., 1933. The measurement of the permeability of porous media for homogeneous fluids. *Rev. Sci. Instr.*, New Ser., 4: 394—405.
Wyckoff, R.D., Botset, H.G., Muskat, M. and Reed, D.W., 1934. Measurement of permeability of porous media. *Bull. Am. Ass. Petrol. Geol.*, 18: 161—190.

CHAPTER 9

ORIGIN AND MIGRATION OF PETROLEUM:
GEOLOGICAL AND PHYSICAL ASPECTS

SUMMARY

(1) Petroleum is a product of the diagenesis of fundamental organic compounds in organic matter that accumulated with fine-grained sediment in a low-energy environment deficient in oxygen.

(2) The energy of petroleum in its source rock is greater than that it will have when it reaches the accumulation. Each path of migration from source to accumulation is a path of continuously decreasing energy during migration. The energy is derived largely from the compaction of the petroleum source rock — usually a mudstone, but some may be fine-grained carbonates.

(3) Petroleum exists as a separate phase by the end of primary migration, from source to a permeable carrier bed. Capillary forces then retard migration, and prevent it when the energy of the migrating petroleum is insufficient.

(4) Primary migration may be stratigraphically upwards or downwards, depending on the direction of decreasing energy. Both normally occur in a compacting mudstone that is intercalated between sandstones or other permeable units. The surface dividing upward and downward migration within the mudstone is a perfect physical and chemical barrier to migration.

(5) Secondary migration is lateral within porous and permeable rock units, generally towards the land of the time. Petroleum accumulates when it arrives in a position in which there is insufficient energy to move it further.

INTRODUCTION

The origin and migration of petroleum have been topics of interest for at least a century, since the early days of the industry, but they are still poorly understood. There are several reasons for this, but the main one is that the processes are too slow to model in the laboratory with confidence, and scaling the model introduces doubts regarding the chemical aspects. Migration through a mudstone is a very slow process because of the low permeability. We can speed this up by taking a material of greater permeability — but the material, and so the chemical composition of it, must be changed. Also, we can accelerate the chemical reactions by heating, but it is not certain that in so doing we get the same reaction as that that would have taken place at a lower temperature, with lower energy, over a longer time. Chemical reactions in the laboratory are subject to cosmic radiation: those in the presumed source rock are subject to gamma radiation from some clay minerals, and we cannot say for certain that this difference is unimportant.

It is impossible at present to distinguish in logic between the generation of petroleum and its primary migration. We believe we can recognize petroleum source rocks from the nature of their organic contents: we can identify petroleum accumulations. No migration path has ever been recognized physically with confidence and reported, so the connection between source and accumulation is *inferred* from analyses of the oil and analyses of the organic content of the supposed source rock, and from geological considerations. It is for this reason that we cannot claim to *understand* the origin and migration of petroleum. At best, we can construct plausible hypotheses.

There is another, less creditable, cause of difficulty: the widespread misunderstanding of the physics of fluid movement through porous rocks. Perhaps the commonest error is the assertion that water moves from high pressure to low pressure. Geologists holding this view cannot defend their position because there are artesian basins in which the water is *demonstrably* flowing from low pressures near the intake area to higher pressures in the aquifer at depth. This misconception has bred others, particularly the widely held view that petroleum migration is always upwards (stratigraphically or absolutely). Downward migration over parts of the migration path is not a new concept, nor one that depends on mathematical arguments: King (1899, p. 80 and fig. 9, and p. 99 fig. 14) clearly understood the movement of ground water without the use of mathematical formulation. Others have postulated downward movement in ground-water and petroleum contexts, notably Versluys (1919), Hedberg (1926) and Hubbert (1940).

The growth of petroleum geochemistry over the last decade or so has been spectacular, and the conclusions reached have been widely accepted. It is not easy for geologists to assess geochemical hypotheses because of their general lack of familiarity with chemical arguments. By the same token, it is not easy for chemists to follow geological arguments. There are books, of course, and all can read; but understanding also requires *doing*, because it is in practising our science that we acquire a feel for it.

For these reasons, we shall take the topic of origin and migration of petroleum in three parts. First, the hydrodynamic aspects will be considered with the geological because this is the physical context of petroleum migration, and they put some constraints on the geochemical hypotheses. We shall then consider the geochemical aspects, finishing with a discussion of the whole topic. The only matter that we shall accept uncritically is that petroleum has its origin in organic matter that accumulated with fine-grained sediment in a low-energy environment deficient in oxygen.

HYDRODYNAMIC ASPECTS OF MIGRATION

We have already seen in earlier chapters that fluids, if they move, move from positions of greater energy or potential to positions of lesser energy or potential. Movement involves loss of energy. It is therefore axiomatic that: (1) *the energy of petroleum in its source rock is greater than that it will have when it reaches the accumulation; and* (2) *each path of petroleum migration from source to accumulation is one of continuously decreasing energy during migration.*

These axioms are independent of any hypothesis concerning the nature and state of petroleum during migration. If any part of the total migration path involves transport in solution in water, then the paths of water migration are, of course, those of petroleum migration and the same constraints apply to both water and the petroleum. In any part of the migration that takes place as a separate phase from water, the path of petroleum migration may diverge from that of the associated water, and capillary forces may act in such a way that migration of one phase may be prevented or retarded by the other. But no situation can arise in which either phase can move to a position in which it will have greater energy.

The commonest source rock is considered to be a mudstone. It must be recognized that a mudstone will rarely be entirely source rock: the source will naturally be part of a mudstone. If the physiographic environment of a sand was not favourable for the preservation of organic matter, it is unlikely that the contiguous mudstone facies was favourable for its preservation because that would involve the coincidence of physical and chemical criteria. The concept of diachronous rock units suggests that at any moment of time during the accumulation of a sequence of mudstones and sandstone, for example, some areas accumulate mud and others sand; and part of the area of mud accumulation may also accumulate and preserve a significant organic content (Fig. 9-1). There will then be a source facies within the mudstone facies that will be both laterally and vertically discontinuous.

As this mudstone subsides, and more sediment accumulates on top of it, it

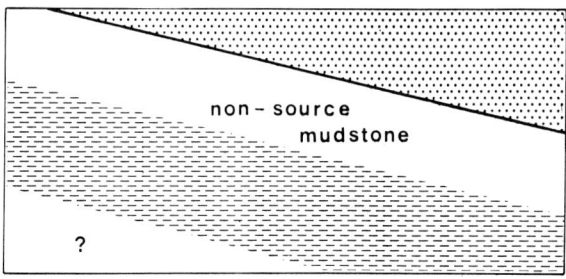

Fig. 9-1. Source rock is a *facies* that is not contiguous with the sandstone facies. (Sandstone, dots; source mudstones, dashes.) View as map or section.

tends to compact. Because compaction can only take place if the pore fluids can be compressed or expelled, a potential gradient is generated in the fluids in the mudstone. When compaction is due to gravity, the direction of this gradient is essentially vertical — upwards and/or downwards to the nearest permeable bed in which the pore fluids are at a smaller potential. If the mudstone is underlain and overlain by such beds, pore fluids in the mudstone will tend to move both upwards and downwards during compaction (Fig. 9-2). There will be a physical surface of maximum potential near the middle of the mudstone that separates the upward tendency from the downward tendency to flow. *This surface is both a hydraulic and a chemical insulator:* no fluid can move across it. If the petroleum source rock is above this insulating surface, any products of organic diagenesis that can move will move upwards towards the overlying bed. If the source rock is below this surface, any products that can move will move downwards towards the underlying bed. If the source rock straddles the surface, movable products of diagenesis will move towards both overlying and underlying beds. And if, as seems likely, petroleum generation involves a net increase in volume, it will also increase pore pressures and the potential gradients. Within a mudstone, this will tend to shift the insulating surface *into* the zone of generation, and so divide it (at least temporarily) into upward and downward zones of migration.

The rate at which the products of organic diagenesis move depends on their state (gas, liquid, or in solution in water), on the potential gradient, and on the effective permeability of the mudstone to that fluid. Those products of organic diagenesis that are taken into solution in the pore water will migrate

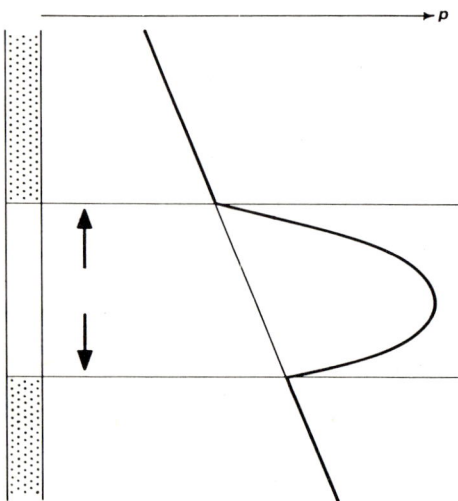

Fig. 9-2. Pressure-depth diagram of upward and downward migration of pore water from compacting mudstone to underlying and overlying sandstones at normal hydrostatic pressures.

at approximately the same rate as the water. Theoretical studies by Bredehoeft and Hanshaw (1968) and Smith (1971) have indicated that during subsidence the abnormal fluid potential in thick mudstones can be maintained for geologically significant periods of time, and that the rates of fluid flow are very slow indeed. Those products of diagenesis that exist as a separate phase will move more slowly because of relative permeability and capillary influences. Products with large molecules may be retained in the mudstone.

We shall postpone consideration of the state of petroleum during primary migration to the next chapter, where chemical aspects will be discussed; but whatever the state may be during most of its primary migration, it is very likely that it exists as a separate phase before primary migration is complete. The main reason for this belief is that the carrier bed is relatively inert chemically, compared with mudstones, and the path to the accumulation may be short.

When petroleum exists as a separate phase in water, two factors affect its migration: water saturations and capillary pressure. It seems certain that the flow of two immiscible fluids in mudstone is similar in principle to such flow in more permeable lithologies. There will be some critical water saturation above which the petroleum can only exist as discrete globules in the pore spaces, and that this petroleum is then virtually immobile. There will also be some irreducible water saturation at which the water is immobile, but at which the effective permeability to petroleum is close to the intrinsic permeability of the mudstone. Such a state does not mean that the petroleum flows: for that, the capillary displacement or injection pressure required for the continuous petroleum phase to move must be less than that existing in the petroleum phase.

Once the continuous petroleum phase reaches the porous and permeable carrier bed, the capillary displacement pressure in the latter is very much less and, as Hubbert (1953, p. 1979) showed, the imbalance of capillary pressure at each end of the volume occupied by petroleum is sufficient to expel it (Fig. 9-3). At the water saturation likely for such a continuous petroleum phase, the effective permeability to petroleum will be relatively large.

The problem of the continuity of petroleum as a separate phase in water is an intriguing one of some interest. We are faced with seemingly incompatible alternatives. When the water saturation is so high that petroleum can only exist as discrete droplets in the pores, the effective permeability to petroleum is zero and it cannot migrate. Within the zone of generation, organic matter is evidently disseminated, so the petroleum generated will also be disseminated. Petroleum must be added to these droplets, or water removed, so increasing the petroleum saturation and decreasing the water saturation (drainage) until a continuous phase is reached. Water will be removed by compaction, so it seems that petroleum disseminated in the pores will only be a transient condition. For petroleum migration in mudstone, the petroleum must either be in solution or in a continuous phase at or close to the irreducible

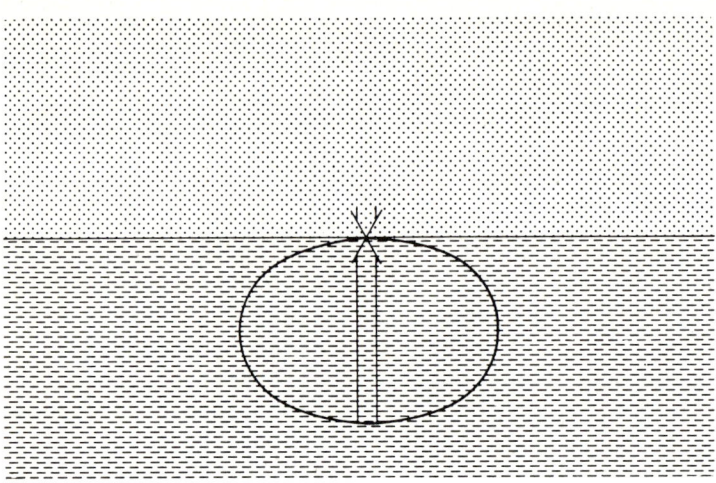

Fig. 9-3. Capillary pressure alone can expel a globule of oil from a mudstone into a sandstone once its interface reaches the sandstone. (After Hubbert, 1953, p. 1978, fig. 14.)

water saturation, at which the effective permeability to water is very small or zero, but the effective permeability to petroleum close to the intrinsic permeability. Migration of petroleum in solution may require less work than that as a separate phase, but the whole migration path must be considered. If we are correct in assuming that petroleum exists as a separate phase before primary migration is complete, then that part becomes virtually impermeable to water, impeding water movement from "upstream" (Chapman, 1972; Hedberg, 1974).

Primary migration takes place in a more rigorous chemical environment than secondary migration, and during it, the fluids are subjected to more severe physical changes. In this connection, upward migration must be distinguished from downward migration. Upward migration above the insulating surface may take all fluids from relatively high pressures to relatively low on expulsion into the overlying carrier bed (the exact amount depending on the rate of upward movement in a subsiding and compacting mudstone). Downward migration into an underlying carrier bed may involve little or no pressure change. Upward migration may involve little temperature change, while downward migration will be to higher temperatures. These influences will be amenable to analysis when we can identify with confidence the source rocks of petroleum accumulations, and so compare the crude oils and gases that have migrated upwards with those that have migrated downwards.

Petroleum migrating as a separate phase almost certainly comes into contact with mineral grain surfaces, or is separated from them by a very thin film of adsorbed water, as it does in the reservoir when the water saturation is sufficiently low. Several common clay minerals are known petroleum catalysts, and it is unlikely that a significant film of water separates the two effectively in the mudstone during primary migration.

Indirect evidence of petroleum as a separate phase during primary migration is given by abnormal pressures and abnormally high resistivities in parts of mudstones in some areas (such as the Bakken Shale of the Williston basin, U.S.A., reported by Meissner, 1978).

We visualize primary migration ending and secondary migration beginning over a large area of the interface between the mudstone that contains the petroleum source rock and the carrier bed.

Secondary migration

In spite of the fact that no secondary migration paths have been recognized with confidence and reported, the conceptual difficulties are not as great as those with primary migration. In the first place, we have petroleum seepages in many parts of the world; and secondly, when we put an oil well onto production, oil demonstrably flows through the reservoir to the well. The main difficulty concerns the role of water movement, so we shall begin with a discussion of secondary migration in an aquifer in which the water is at rest. We shall also simplify the discussion by assuming oil migration in an isotopic, homogeneous, granular, water-wet carrier bed; and regard oil as incompressible, without gas in solution. The principles apply to gas.

Movement in the final stages of primary migration seems to require the petroleum to be in a continuous phase through the pore space, so it is inferred that it remains as a continuous phase initially (at least) in the carrier bed.

Considering upward migration from the mudstone interface first, there is some critical vertical dimension to the oil that will enable it to move upwards under the force of gravity (buoyancy) against the resisting forces, chiefly capillarity. The upward pressure due to buoyancy increases relative to the ambient water pressure by $(\rho - \rho_o)g$ per unit of elevation above the carrier bed interface from which the oil is emerging. Hobson (1954, p. 73) and others have estimated this critical vertical dimension to be of the order of a few metres at most in typical carrier and reservoir rocks.

We can measure the capillary displacement pressure required to move the oil front from one set of pores to the next, so the critical dimension can be estimated from:

$$\Delta h_o = p_i/(\rho - \rho_o)g. \tag{9.1}$$

If we take 10 kPa (1.5 psi) as being a representative maximum carrier bed displacement pressure, and 200 kg m^{-3} as a representative difference of oil and water mass densities, then:

$$\Delta h_o = 10^4/(200 \times 9.8) = 5 \text{ m}.$$

The oil being less dense than the water, the macroscopic water/oil interface is mechanically unstable. As the oil is forced out of the mudstone over a wide area, the water/oil interface will become wavy, with a tendency for "diapiric" oil bodies to form; and the amplitudes of those near the dominant wavelength (see p. 332) will be amplified at the expense of the others (Fig. 9-4). As soon as one oil "diapir" reaches its critical vertical dimension, it will move and tend to drain others. This oil will move vertically upwards, in static water, until it reaches the cap rock.

The criterion with any change of lithology encountered during this vertical migration, and there may be several within real carrier beds, is that the capillary displacement pressure required for further progress must be less than that existing in the oil. Vertical migration ceases as soon as the pressure in the oil is insufficient to overcome the capillary resistance. The cap rock is a fine-grained material of which the capillary displacement pressure exceeds — and usually exceeds by a wide margin — that existing in the oil. The oil is then diverted along this lithological interface, in the direction of decreasing energy, in the up-dip direction literally and strictly.

When the oil is diverted along the base of the cap rock, the situation is comparable with that at the termination of primary migration from an overlying source rock, but not identical to it. The criterion for up-dip migration is the same as before: the difference of vertical elevation within the continuous oil phase must exceed the critical vertical dimension.

For migration downward from an overlying source rock, there is no mechanical instability at the oil/water interface, and the critical vertical dimension depends on the relief of the cap-rock/carrier-bed interface. For source rocks sufficiently rich to generate enough oil to form a commercial accumulation, the critical vertical dimension must be achieved sooner or later, unless the source rock is directly over the trap.

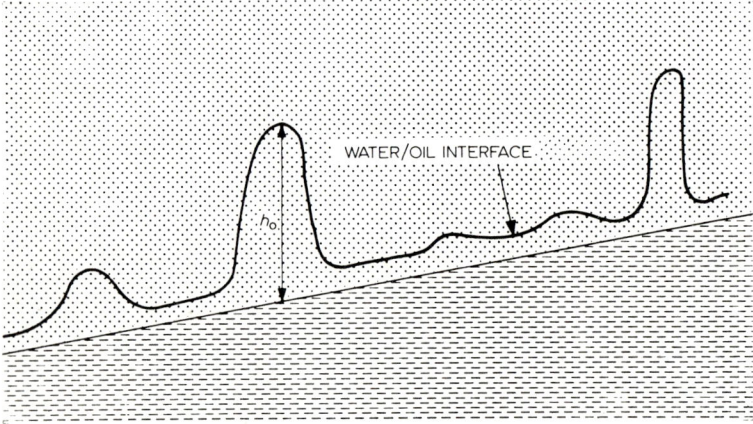

Fig. 9-4. "Diapiric" water/oil interface at the beginning of secondary migration at the bottom of the carrier bed.

Secondary migration up-dip follows a path, or paths, of local minimum potential — much as rivers do — and migration continues until the oil arrives in a position of minimum potential with respect to the physical constraints of the cap rock. Here it accumulates. The quantity that can accumulate depends on the volume of this space of minimum potential, and on the quantities of oil generated. If the space of minimum potential becomes entirely filled with oil, further secondary migration to it will result in overflow from the *spill point*, which is a local position of minimum potential with respect to the accumulation, and so the process will continue until the oil either reaches another space of minimum potential or dissipates at the surface.

The rate of migration of the oil depends not only on the vertical dimension, the physical properties of the oil, and the lithology of the carrier bed, but also on the relative permeability to oil, which is a function of the water saturation (Fig. 9-5).

We noted on p. 168 that experimental curves of relative permeability show a hysteresis effect, depending on whether the initial water saturation was zero or one. The advancing oil front is an injection of oil into 100% water saturation. This part, therefore, corresponds to drainage — from a practical point of view, the drainage curve for water saturations less than the critical oil saturation required for a continuous oil phase. In this part of the relative permeability curves, the relative permeability to oil is very small, and that to water, quite high. We infer, therefore, that much of the pore water is displaced at the oil front so that the water saturation within the migration oil column is such that the relative permeabilities are at least better balanced ($s_w \simeq 0.6-0.7$).

If there is a "tail" or retreating oil front at the lower end, this corresponds to imbibition, and the water saturation increases here at least to the critical saturation at which continuity of the oil phase is lost. The relative perme-

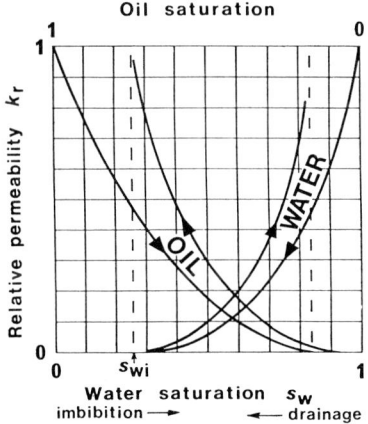

Fig. 9-5. Relative-permeability—water saturation diagram.

ability to oil at the tail is probably less than it is at the head, so there may be a tendency for the oil stream to attenuate. Any oil left behind as a discontinuous phase will be swept up by the next front. All fronts (if there are more than one) will follow the paths of local minimum potential energy on the upper surface of the carrier bed (like an inverted river drainage system), so progress towards the trap will tend to bring migration paths together, and the accumulation will be fed by one or more streams of oil. A detailed contour map of the top of the carrrier bed would indicate the possible paths.

Natural carrier beds are rarely homogeneous or isotropic, so we must consider briefly the main effects of heterogeneity and anisotropy on migration paths.

When two immiscible liquids occupy a single pore, the pressure in the wetting liquid is slightly less than that in the non-wetting liquid, and the non-wetting liquid occupies the position that minimizes its potential energy. The difference of pressure across the liquid/liquid interface is the capillary pressure. We are not concerned here with isolated drops of oil in pores, but with a continuous network of oil through the pore space in a definite volume of the carrier bed. We are concerned, therefore, with the macroscopic upper water/oil interface.

The magnitude of the capillary displacement pressure is a function of the radii of curvature of this interface within each pore along the macroscopic interface, such that the smaller the radius of curvature the greater the capillary displacement pressure. The radii of curvature in the smaller pores are less than those in the larger pores, and the capillary displacement pressures required for displacement through the smaller pores are greater than those for the larger pores. The migrating oil occupies the larger pores preferentially, because these are the paths of least work.

Heterogeneities and anisotropy in carrier beds are generally related to bedding, and so affect the upward migration of oil across the bedding. Migration across a graded carrier bed in which the grain size, and so the pore size, increases upwards is facilitated by the decreasing capillary displacement pressure required and the increasing pressure available within the oil. If the oil is a bubble that is large compared to a single pore, the imbalance of capillary pressure at the leading and trailing surfaces impels the bubble upwards. Grading in the opposite sense retards migration. Beds of alternating fine and coarse grain that are not horizontal lead to refraction of the migration path up-dip (Hubbert, 1953, p. 1972, fig. 10). In the coarser beds, the flow path will tend to deviate up-dip: in the finer, more vertical.

Similarly, the lateral migration will also be along paths of least resistance, favouring the larger pores and perhaps by-passing the smaller.

The water in which petroleum migrates will not always be at rest, but also moving along an energy gradient to positions of lower energy.

Migration in moving water

When secondary migration takes place in a carrier bed in which the pore water is in motion, the petroleum migration paths are affected by this motion, and also the geometry of any accumulation. Rich (1921, 1923) clearly understood this, but present understanding is due to Hubbert (1953), who rationalized and quantified the effects. The reader who wishes a more rigorous analytical argument is referred to Hubbert's paper.

When the pore water is at rest, its potential energy is constant throughout the carrier bed: surfaces of equal pressure are horizontal, and the direction of petroleum migration is determined solely by gravity — that is, when unrestrained by cap rock, it is vertical; when constrained by cap rock, it is up dip. When the pore water is in motion, this has the effect of rotating the hor-

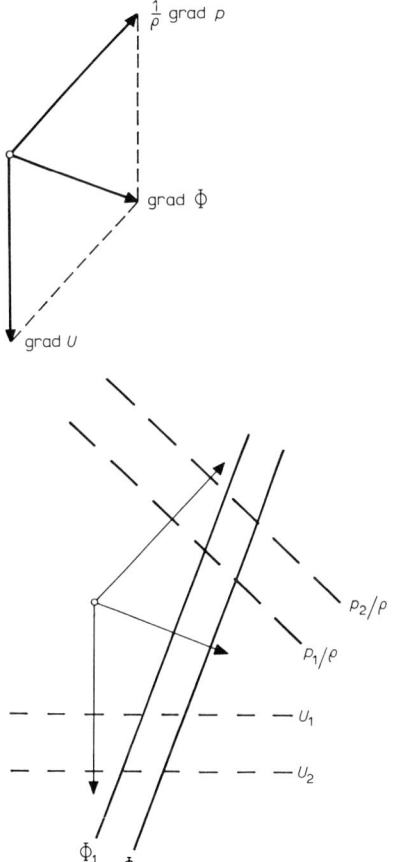

Fig. 9-6. Surfaces of constant pressure, constant gravity potential ($U = gz$), and constant fluid potential (Φ) are normal to their gradient vectors.

izontal reference plane to a plane inclined in the direction of motion, and its normals are similarly rotated, of course, from the vertical. Surfaces of equal potential (the equipotential surfaces) are normal to the direction of motion, surfaces of equal pressure are inclined in the direction of motion, and a component of lateral motion is imparted to migrating petroleum. The magnitude of this component depends on the density difference between the petroleum and the water, in such a way that as the density of the petroleum approaches that of the water, the more nearly do their directions of motion coincide. The direction of gas migration across a carrier bed will be more nearly vertical than that of oil.

The direction of water movement is the resultant of two forces: the force of gravity acting on unit mass of the water, and the force due to pressure acting on unit mass of the water. The resultant is the impelling force acting on unit mass of the water; and, like the others, it has the dimensions of an acceleration (LT^{-2}), and it is the potential gradient (Fig. 9-6).

At any point in the water, *and at any point capable of being occupied by the water*, the water has a potential. When the water is at rest, the potential is constant through the body of water: when the water is in motion, the potential is not constant but decreases in the direction of flow. The water flows in a direction normal to the surfaces of equal potential, which can be mapped through the body of water.

A measure of the water potential at a given point (eq. 8.12, p. 171) is:

$$h = \frac{p}{\rho g} + z \qquad (9.2)$$

where h is the total head, $p/\rho g$ is the pressure head, and z is the elevation head (or simply, elevation) of the point relative to an arbitrary datum level (negative downwards).

Oil in the water also has a potential, and it tends to move in a direction normal to its equipotential surfaces. If we consider a small volume of oil migrating, a measure of this potential is:

$$h_o = \frac{p}{\rho_o g} + z \qquad (9.3)$$

and since the capillary pressure is a very small part of the pressure in the oil in an aquifer at the depths that we are concerned with, we can take the pressure p to be the ambient water pressure that would exist at that point. Solving eq. 9.2 for p, and substituting it into eq. 9.3, we get:

$$h_o = \frac{\rho}{\rho_o} h - \frac{\rho - \rho_o}{\rho_o} z \qquad (9.4)$$

where z is, as before, positive when measured upwards above the arbitrary datum.

Following Hubbert (1953, p. 1991, footnote 3) we divide eq. 9.4 by $(\rho - \rho_o)/\rho_o$:

$$\frac{\rho_o}{\rho - \rho_o} h_o = \frac{\rho}{\rho - \rho_o} h - z. \tag{9.5}$$

Thus, if there is enough data to map h and z, we can map h_o. We shall return to this equation.

Clearly, the projection of the direction of water movement on a horizontal surface is the same as that for oil movement while the oil is migrating without constraint from the cap rock. Both oil and water equipotential surfaces are inclined in the direction of water motion, but by different amounts. If the water flow is directly down-dip, there is some critical dip that equals the "dip" of the oil equipotential surfaces and is normal to the migration path of the oil through the water. A small volume of oil under a cap rock at this critical dip would not move.

Hubbert (1953, pp. 1986—1987) has shown that this critical dip, θ_c, is given by:

$$\tan \theta_c = \frac{\rho}{\rho - \rho_o} \frac{dh}{dx} \tag{9.6}$$

where dh/dx is the slope of the water's potentiometric surface as given, for example, by the contour interval divided by the distance separating two contours on the potentiometric surface (not the hydraulic gradient, in which the length is measured along the aquifer). The coefficient $\rho/(\rho - \rho_o)$ indicates that the heavier the oil, the steeper the slope of the oil's equipotential surfaces and the critical dip. If the oil's density equals that of the water, oil can only accumulate by capillary effects because there is no gravitational effect that will accumulate it.

Gas has a more nearly vertical migration path across a carrier bed by virtue of its smaller density relative to oil and water. The consequences of these effects separately on oil and on gas may well not be the same as the combined effect. If gas alone would take the path G in Fig. 9-7, and oil alone the path O, it is most unlikely that these would be the paths if both were migrating simultaneously. Some intermediate path depending on saturations would be more likely.

Once restrained by the upper surface of the carrier bed, further migration follows paths of local minimum potential. These will not in general be indicated by a detailed contour map of the interface between the cap rock and the carrier bed; but if the contours were drawn relative, not to sea level but to a surface at the critical dip, such a map would indicate the possible paths of migration.

To get an idea of the magnitudes of the critical dip or slope θ_c, consider a potentiometric surface with a slope of 10^{-3}, which is about the steepest regional slope of the Great Artesian basin of Australia. This is a slope of about 3 min of arc. The mass density of most crude oils falls in the range 750—900 kg m^{-3}, so the amplifying factor ranges from four to ten, and the

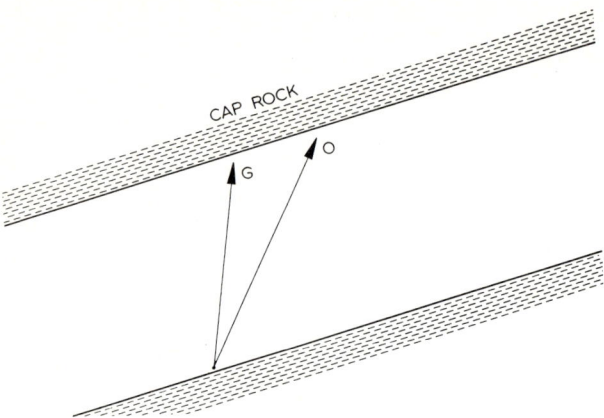

Fig. 9-7. Because of its smaller weight density, gas migration paths are more nearly vertical than oil's under hydrodynamic conditions.

maximum critical slope barely reaches 0.5°. If we take a potentiometric surface with a slope of 10^{-2}, which exists amongst the oil and gas fields of the south German Molasse basin west of Munich (see Chapman, 1981, p. 102, fig. 5-7), the critical slope varies over the range 2°—6°. Locally, greater slopes are possible.

Only in areas of very low structural relief and very large hydraulic gradients or potentiometric slopes will oil and gas migrate other than up-dip qualitatively. If the water is flowing with a component along strike (as it does in the Molasse basin) the oil migration paths will also have a component along strike: the direction of oil migration will be determined by the dip surface relative to the oil's equipotential surfaces, for the oil will migrate along the bedding plane in a direction normal to the line of intersection of the two surfaces. But once a local minimum oil potential has been reached, the migration will be along these axial regions, the plunges of which are usually slight and less than the dips on either side.

Reverting now to eq. 9.5, which we shall simplify by letting $u = h_o \rho_o / (\rho - \rho_o)$ and $v = h\rho/(\rho - \rho_o)$, so that

$$u = v - z \quad \text{(each with dimension of length),} \tag{9.7}$$

we see that a map of v is a map of the water's potentiometric surface amplified so that it becomes a map of the conceptual surface of critical dip and an oil equipotential surface (Fig. 9-8). Its contours are, of course, also equipotential lines as well as lines of constant elevation on this surface. A map of z on the base of the cap rock is a structural map and its contours are lines of constant elevation. Each intersection of a structural contour with a contour of v is a point on a line of constant u — that is, a line of constant h_o — with its value given by eqs. 9.5 and 9.7 (Fig. 9-9).

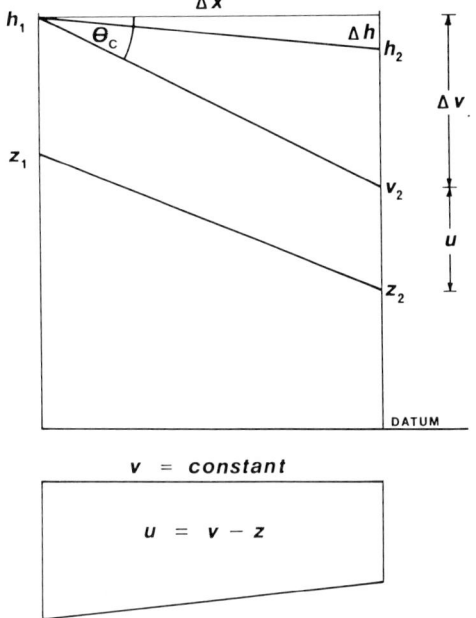

Fig. 9-8. If $\Delta h/\Delta x$ is the slope of the water's potentiometric surface, the slope of the oil's potentiometric surface is obtained by multiplying Δh by the factor $\rho/(\rho - \rho_o)$, the product being Δv. The lower diagram shows the slope of u relative to the oil's equipotential slope.

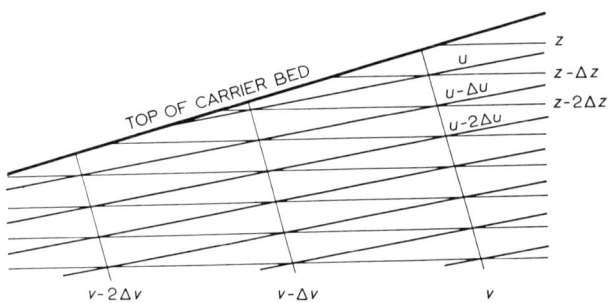

Fig. 9-9. Graphical representation of the interrelationships between surfaces of constant depth (z), u and v, in section.

It is convenient at this point to note that when oil accumulates, its oil/water contact is an equipotential surface of the oil and a boundary to water flow. The slope of the oil/water contact is also given by eq. 9.6, so the map of u is not only a map from which possible oil migration paths can be inferred, but also a map of possible oil/water contacts when enclosing spaces of minimum oil potential. The map of u can also be viewed as an isopach map of the interval between an oil equipotential surface and the top of the reservoir rock or carrier bed. Closed contours of low potential are possible areas of accumulation, and the contours are possible oil/water contacts.

If we take the structural contour interval to be 100 m, then v can be mapped with 100 m contour intervals by choosing Δh so that $\Delta h \rho / (\rho - \rho_o) = 100$ m. Contours constructed from the intersection of the structural contours and the contours of v are oil equipotential lines on the base of the cap rock, the top of the carrier bed.

Fig. 9-10 is a structure map on the top of a carrier bed, and it shows a nose plunging to the south-east. If the water is at rest, all horizontal surfaces are equipotential and each structural contour coincides with an oil equipotential line at the intersection of an equipotential surface with the top of the carrier bed. Oil would migrate up-dip and accumulate in an area of minimum potential defined by closed equipotential contours. This would be off the map to the north-west. The oil/water contact would be horizontal.

Figure 9-11 is a map of the potentiometric surface of the water in the carrier bed, with a slope of 10^{-2}, indicating flow down the nose to the south-east.

Figure 9-12 is a map of $v = h\rho/(\rho - \rho_o)$ assuming a mass density such that $\rho_o/\rho = 0.8$, the contour interval being equal to that of the structure map, superimposed on the structure map.

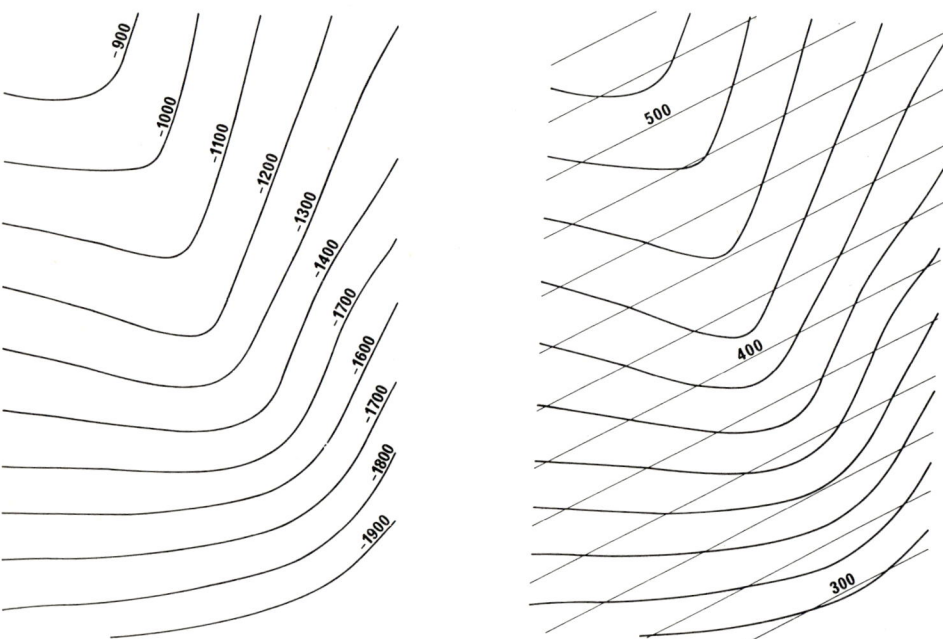

Fig. 9-10. Structure contour map on nose plunging South-East.

Fig. 9-11. Map of potentiometric surface with slope 10^{-2} superimposed on the contour map of nose (Fig. 9-10).

Figure 9-13 shows the construction of a contour map of u from the intersections of the other two sets of contours. Possible migration paths for oil of relative density 0.8 are shown, normal to the equipotential lines of the oil in the direction of smaller potential. Note that one of these is *down* the axial plunge to an area of minimum potential that is enclosed by oil equipotential lines. These closed contours are possible intersections of the oil/water contact with the top of the reservoir rock, and their contour interval is the same as the structure map. The possible volume of accumulated oil is therefore obtained from a map such as that of Fig. 9-13 in the same way it would be obtained from a structure map if the water were at rest.

Heavier oil would accumulate further down the nose, and Figs. 9-14 and 9-15 show part of the configuration for relative densities of 0.85 and 0.9. These show marked shifts of accumulation due to the density changes, and similar shifts would occur if the water flow were stronger.

Since gas is very much less dense than water, it is not displaced much from the structural culmination. Hubbert (1953) demonstrated that oil can be separated from its gas cap under hydrodynamic conditions.

Another form of hydrodynamic trap has interesting possibilities: that caused by changes in the slope of the potentiometric surface due to changes of thick-

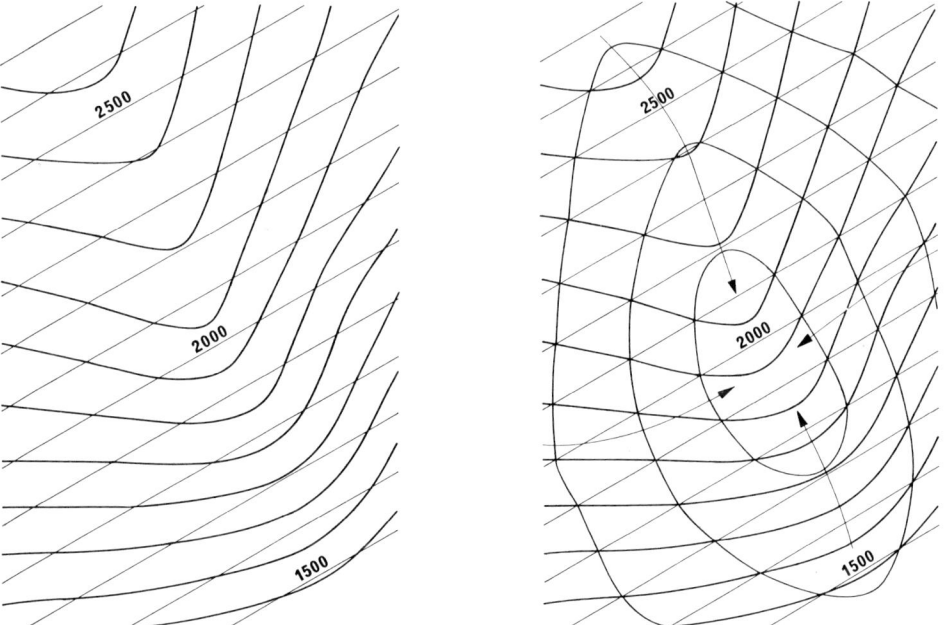

Fig. 9-12. Map of v for $\rho_o/\rho = 0.8$. This is the potentiometric surface of Fig. 9-11 amplified by the factor 5.

Fig. 9-13. Map of u constructed from Figs. 9-10 and 9-12, showing also possible migration paths for crude oil of relative density 0.8.

ness or of permeability in the aquifer. The hydraulic gradient of an aquifer with constant volumetric rate of flow varies with cross-sectional area and with permeability according to Darcy's law:

$$\Delta h/l = Q/K A \tag{9.8}$$

so that if the aquifer thins in the direction of flow, or if the hydraulic conductivity (K) decreases in the direction of flow, the hydraulic gradient, and hence the slope of the potentiometric surface, increases. With down-dip flow, it is possible that this increase in the slope of the potentiometric surface is sufficient to arrest up-dip migration of oil, or to combine with capillary barriers to form an accumulation.

Similarly, when an aquifer is faulted across the flow path with a throw that is less than the thickness of the aquifer, water flow will be impeded and there will be a potential drop across the fault, the amount depending on the permeability of the fault plane and the reduction of area normal to flow at the fault plane (Fig. 9-16). If the dip of the carrier bed/aquifer is greater than the critical hydrodynamic dip, the change in water potential gradient across the fault could act as a hydrodynamic trap irrespective of the permeability of the fault to oil.

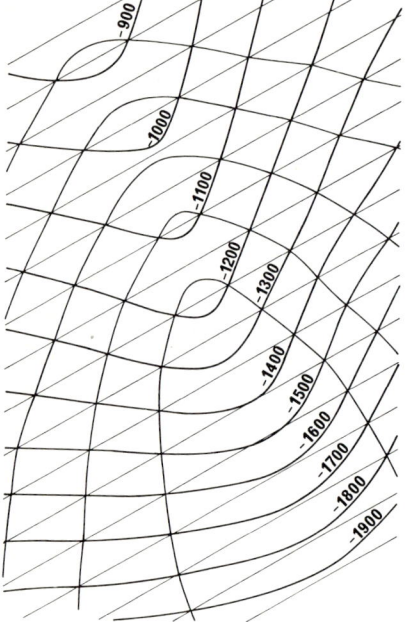

Fig. 9-14. Constructed map of u for $\rho_o/\rho = 0.85$.

Fig. 9-15. Constructed map of u for $\rho_o/\rho = 0.9$.

If the throw of a fault is greater than the thickness of the aquifer, such that a different aquifer is juxtaposed, then the water flow moves to a different stratigraphic level that may have entirely different hydraulic properties (Fig. 9-17). The conservation of matter, assuming negligible compressibility, requires that the volumetric rate of flow, Q, be constant in both aquifers,

$$Q = K_1 A_1 \Delta h_1 / l = K_2 A_2 \Delta h_2 / l \qquad (9.9)$$

so the ratio of their hydraulic gradients is given by $K_2 A_2 / K_1 A_1$ where, for practical purposes, the areas are proportional to the thickness. In such a case there may be an abrupt change of slope of the potentiometric surface (and so of the critical hydrodynamic dip) across a fault, as well as a drop of potential.

We rarely have enough data to map such details, but it is worth noting that the appropriate potentiometric map for hydrodynamic conditions may involve several aquifers in faulted regions. By the same token, if petroleum can pass these faults, its migration path may lie through several stratigraphically distinct horizons. All these complications exist in regions of growth faulting, and the hydrodynamics of the water flow due to compaction changes as the faults move.

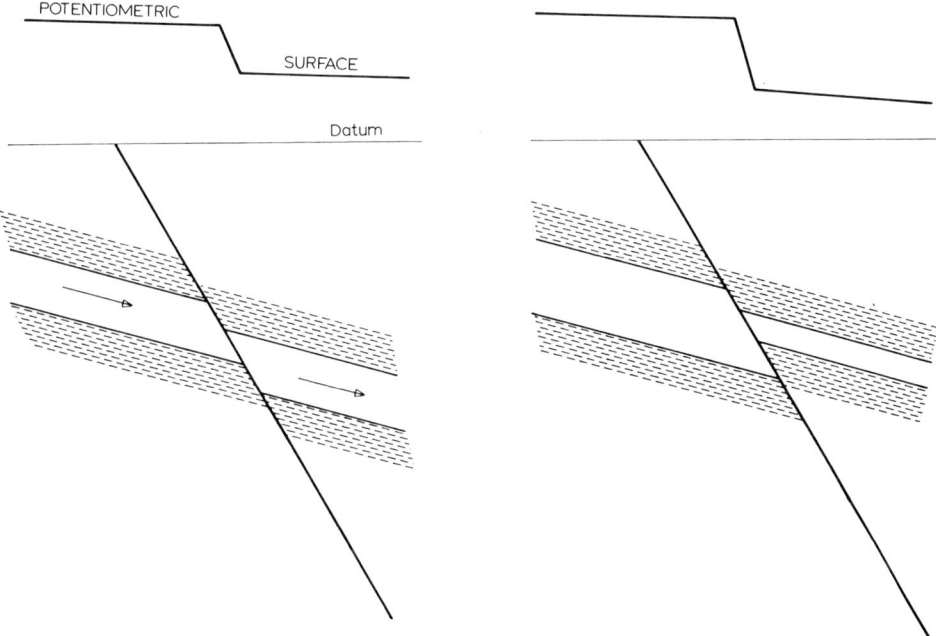

Fig. 9-16. Water flowing through a fault loses energy.

Fig. 9-17. Juxtaposed aquifers will normally have different hydraulic properties.

Discussion of the role of faults in petroleum migration will be postponed to Chapter 11, but the same principles apply to such migration across faults from one block to another: the pressure in the oil at the fault must exceed the injection pressure required to pass the fault. There is evidence that this can happen; but the evidence of fault traps is that faults can seal significant quantities of oil behind them.

It cannot yet be claimed that use of the hydrodynamic approach has led to significant oil discoveries that would not have been found taking the usual hydrostatic approach; but, by the same token, there may be large oil fields waiting to be found in positions that would not be credible under hydrostatic assumptions. Where sufficient data are available, such maps should be drawn because the possible rewards far exceed the labour involved. Furthermore, such maps put some constraints on the directions in which the source rocks of accumulations may lie because they can only lie "upstream" of the possible migration paths.

Of course, such statements are more easily made than justified. They assume that the hydrodynamics of the area has not changed since oil generation, and that the structural relief was much the same. Each area must be judged on its own merits, and it can commonly be assumed that the structural relief was no greater at the time of oil generation and migration.

Rates of secondary migration

We do not have reliable field data from which the rates of migration to individual fields can be determined, but a few sums indicate the order of magnitude required.

A giant oil field with 500 million barrels of recoverable oil has about 1.5×10^9 bbl of oil in place — about 250×10^6 m^3. There are Pliocene/Miocene giants (Halbouty et al., 1970, p. 504, table I), so such quantities cannot take longer than 10—15 m.y. to accumulate. Let us assume that it took one million years to accumulate 250×10^6 m^3, that is, 250 m^3/yr. This is rather less than 0.7 m^3/day, so a single migration path of 1 m^2 cross-sectional area would require a flow of 0.7 m/day, or about 10 μm/s. Migration over a distance of 10 km would involve a transit time of about 40 yrs.

Consider a gas-free crude oil with kinematic viscosity (ν) equal to 6×10^{-6} m^2/s flowing in a carrier bed in which the effective permeability to oil is 100 md (100 μm^2) and the effective porosity is 20%. Assuming that the critical vertical dimension of the oil is exceeded by only one metre, the gradient of total head can be estimated from eq. 9.3 to be about 0.2 in static water for $\rho_o/\rho = 0.8$. So, from Darcy's law:

$$q_o = k_o (g/\nu) \Delta h/l$$
$$= 30 \ \mu m/s.$$

The macroscopic velocity through the carrier bed is approximately q/f, that is, about 160 µm/s, or 14 m/day. There are no obvious difficulties in accumulating enough oil for a giant field in a million years — or even a few hundred thousand years.

The rate of primary migration per unit area of interface between mudstone and carrier bed is undoubtedly very slow; but this is probably compensated by the very large areas of such interfaces.

If gas, and gas only, is migrating, the principles are the same except that as gas moves to positions of different pressures and temperatures (usually lower) its volume changes significantly. If gas and oil are migrating together in separate phases, the principles are the same, but the details become very complex because gas can dissolve in oil and water. If oil has gas in solution . . . there are many variations on the theme. Oil is also slightly soluble in water, so any oil left behind in a discontinuous phase will probably be removed eventually in solution in moving water.

Accumulation of petroleum in a trap

When petroleum, trickling along one or more paths of local minimum potential, arrives in the trap and begins to accumulate, migration ends but a new set of physical changes begins:
— Water is displaced downwards from the top of the reservoir.
— The oil/water or gas/water contact is displaced downwards.
— The pressure in the petroleum increases while petroleum accumulates.
— The petroleum in the accumulation continues to move in response to these changes.

Migrating petroleum has negligible kinetic energy, so the newly-arrived petroleum is added to the accumulation at the interface, without penetration into the accumulation (much as cream poured into a jug, rather than milk). The water contact will be rather lower near the point of entry. Due to the density difference between oil and water, and gas and water, the potential energy of the accumulation will be greater where the water contact is lower, and there will be sympathetic movement within the accumulation in the direction that tends to restore equilibrium.

Within the accumulation, a new physical environment develops. Within the continuous oil phase, the pressure decreases with elevation above the oil/water contact according to the relationship:

$$\Delta p / \Delta z = \rho_o g. \tag{9.10}$$

This rate of decrease is less than that in a continuous water phase, so that, if we take their pressures to be equal at the oil/water contact, the oil at any depth within the reservoir above this is at a higher pressure than the water in continous phase with it. This greater pressure is applied to the water, which acquires the pressure gradient $\Delta p_w / \Delta z = \rho_o g$, and so acquires a downward fluid potential gradient. The oil moves upwards, and displaces the water downwards until hydraulic continuity is lost and the water saturation becomes irreducible. At this stage, we infer, the oil comes into contact with the solid surfaces (or a very thin layer of adsorbed water not more than about 1 nm thick) and isolates the pendular rings.

The reasons for this are not clear. Just prior to the acquisition of irreducible water saturation, the effective permeability to water is very small indeed, while the effective permeability to oil approaches the intrinsic permeability. It is possible, therefore, that the steadily increasing reservoir pressure, concomitantly reducing effective stress and increasing porosity (Terzaghi's relationship) proceed at a rate that cannot be matched by water movement. The inferred size of pendular rings suggests small capillary pressures and so early acquisition of irreducible water saturation. Whatever the cause, a significant proportion of the pore space remains filled with water that is apparently immobile. This water is the original water at the time of petroleum accumulation, and may therefore differ in quality from the water subsequently found below the oil/water contact.

Petroleum accumulation may therefore affect the diagenesis of the reservoir rock within the accumulation. A most interesting study of the Gifhorn Trough in Germany (Phillip et al., 1963) revealed that the reservoir sands contained little cement within the accumulations but were well-cemented outside. The authors drew the conclusion that the accumulation of oil inhibited the deposition of cement, and that therefore the oil accumulated before the processes of cementation had proceeded very far. Similar observations led to similar conclusions for some Nigerian accumulations (Lambert-Aikhionbare, 1982), and in Triassic gas sands in the North West Shelf of Australia (Campbell and Smith, 1982).

It must therefore be noted that if porosity, as determined from the sonic or other log, is found to increase upwards across a water contact, this may well be a real effect and not one induced by the pore fluid's influence on the velocity of sound (or the parameter being logged).

ORIGIN AND MIGRATION OF PETROLEUM IN CARBONATE ROCKS

While most commercial accumulations originate from organic matter in fine-grained argillaceous source rocks, some accumulations seem to require a source in marls or limestones. There are no particular difficulties with this. We assume that the physico-chemical requirements for a carbonate source rock are the same as for argillaceous source rocks — the accumulation of organic matter with sediment in an anoxic environment, and its preservation until buried deep enough for the processes of petroleum generation to begin. There would be few field geologists who have not found dark, dense limestones that give off a bituminous smell on fracture.

Carbonates are prone to chemical alteration and solution (rather than mechanical compaction) and the effects of these diagenetic changes on bulk density trends with depth are not clear. Many carbonates appear to suffer no appreciable mechanical compaction because well-preserved, delicate fossils and original structures are commonly found. On the other hand, some marls apparently compact much as mudstones. McCrossan (1961) found that the bulk density of some Devonian marls in Canada increased with $CaCO_3$ content as well as with depth; and the higher the $CaCO_3$ content, the smaller the relative compaction.

One process is evidently solution at grain contacts and recrystallization in pores and voids. This has the same effect as mechanical compaction — reduction of porosity and increase of bulk density — through chemical rather than mechanical transport. Reduction of porosity can only be effected if the pore fluids are commensurately compressed or expelled; and the expulsion of pore fluids from compacting carbonates involves the removal of soluble components. Not only are there no obvious new difficulties in primary migration from carbonate source rocks, but the recrystallization may actually make the process more efficient provided it does not take place before petroleum is generated. The common occurrence of bituminous limestones may indicate early recrystallization that traps the products of organic diagenesis in the rock.

The matter of porosity and permeability in carbonates is much more varied and complex than in sandstones, for there may be two generations of porosity, primary and secondary, or more, and open fractures and joints are much more common than in sandstones. Secondary migration, therefore, may take place in exactly the same way in calcarenites as in sandstones, and accumulation and production from calcarenite reservoirs will also be the same. But fracture porosity in carbonates is more important than it is in sandstones.

Primary porosity in carbonates may be completely destroyed during diagenesis, the recrystallization leading to porosity of a different sort. Vugs may be formed by the solution of fossils, for example, and considerable permeability may be retained in a bed that consists largely of material of very small porosity and negligible permeability. Recrystallization also increases the strength of the material, in particular, its tensile strength, so that fractured carbonate carrier beds and reservoirs are important in some areas. The prolific Asmari Limestone in Iran and Iraq has a matrix porosity of 9—14% on average in Iran, and an intrinsic permeability averaging about 10 md (Hull and Warman, 1970, p. 431) but it is extensively fractured. These fractures are demonstrably open because some wells have very high production rates on very small penetrations of the reservoir, and production changes in one well are relatively rapidly detected in other wells at a considerable distance away in the same reservoir. Hull and Warman note that where sandstones are present, they appear to be unfractured. They also note that there is a regional hydraulic gradient within the Asmari Limestone from the mountains to the Gulf and oil/water contacts on the two flanks of some oil fields differ by as much as 150 m.

The effective drainage areas of wells in carbonate reservoirs with fracture porosity are large, and productivity of wells is usually large. These are important factors in the economics of oil production because the oil can be produced with relatively few wells. High productivity is one of the characteristics of carbonate reservoirs, including reef reservoirs. Intisar field, Libya, for example, tested one well at 74,867 bbl/day (11,900 m^3/day) clean 37^0 API oil (sp. gr., 0.84) from 223 m (731 ft) of fossil reef (*World Oil*, January 1968).

As regards irreducible water saturations, less is known of carbonate reservoirs than of sandstone reservoirs, and some carbonate reservoirs are thought to be oil-wet. It seems certain, however, that oil and gas will come into close contact with solid carbonate surfaces in the reservoir. In reservoirs with fracture porosity, the irreducible water saturation is probably very low in the fractures, but may be high in the rock itself.

Of particular importance in carbonate provinces is the evidence for differential entrapment of oil and gas. Gussow's (1954) hypothesis of differential entrapment is very simple, and it grew from the observation that in a sequence of reefs in a single trend in the Western Canada basin, the deepest reefs contain gas only; the shallowest, water only; there will be one deep reef with a gas/oil contact, and one shallow one with an oil/water contact, but the others will be full to spill point. Gussow explained this characteristic distribution as follows:

Starting with saturated oil (that is, oil that contains the maximum amount of dissolved gas) migrating up-dip (Fig. 9-18), gas comes out of solution as it moves to lower pressures and temperatures. At the first trap encountered on its migration path, both oil and gas accumulate; but as the trap fills, a gas cap forms by gravity segregation, so when the trap is full, only oil spills out and continues its migration. However, gas can still enter the first trap, displacing oil until the trap is filled with gas only, and filled to the spill point. When this stage is reached, both oil and gas bypass the first trap, and the process is

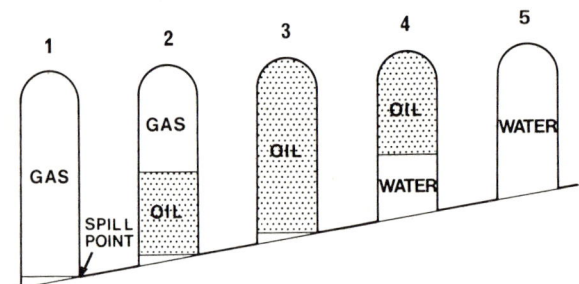

Fig. 9-18. Differential entrapment of oil and gas in traps that are hydraulically connected (Gussow's principle).

repeated at the second trap. When the quantity of gas becomes insufficient to fill a trap, oil with a gas cap is left; and once this trap is full, only oil migrates further and the next trap up-dip contains only oil. When the quantity of oil generated is no longer sufficient to fill a trap, that trap has an oil/water contact, and traps further up-dip will be wet.

Gussow also observed that there is a tendency, not always marked, for the oil to be progressively heavier in traps up-dip. This he attributed to the loss of gas from solution in the oil.

Further subsidence after petroleum generation has largely ceased will result in compression of the accumulated gas, and some may be taken back into solution in the oil, thus raising the water contacts above the spill points. All accumulations in the trend are, of course, hydraulically interconnected, but the water contacts are at different levels depending on the spill points, generally rising up-dip.

A particularly good example of regional differential entrapment is the Silurian pinnacle-reef belt east of Lake Michigan, studied by Gill (1979). This belt is 270 km long and 16—32 km wide (170 by 10—20 miles). Petro-

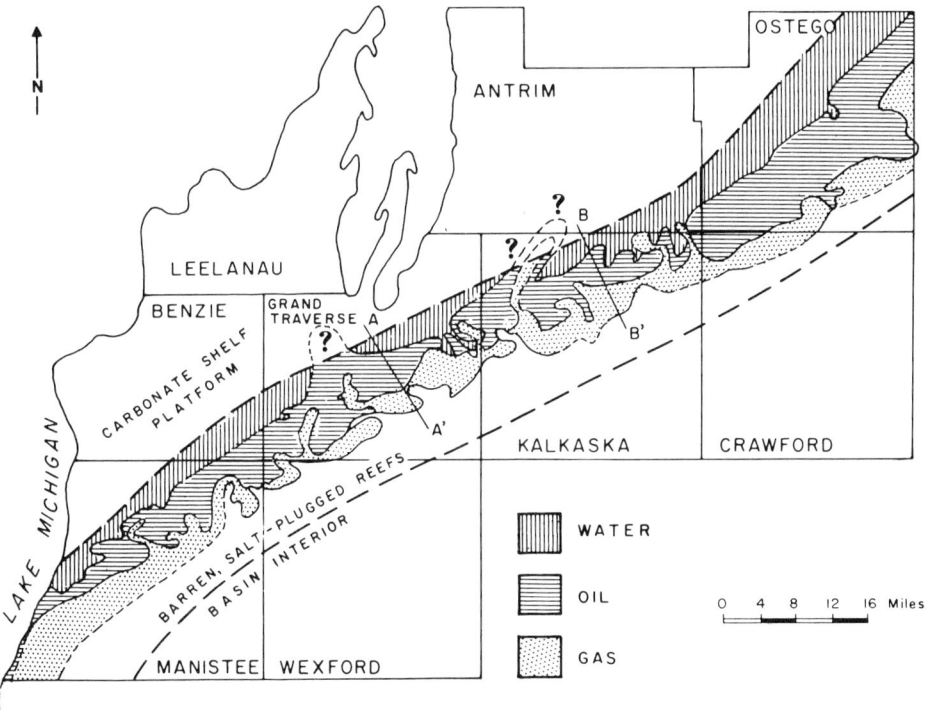

Fig. 9-19. Map of north-western part of Michigan basin showing zones of water, oil and gas accumulations in Silurian reefs. (Reproduced from Gill, 1979, p. 614, fig. 4, with permission.)

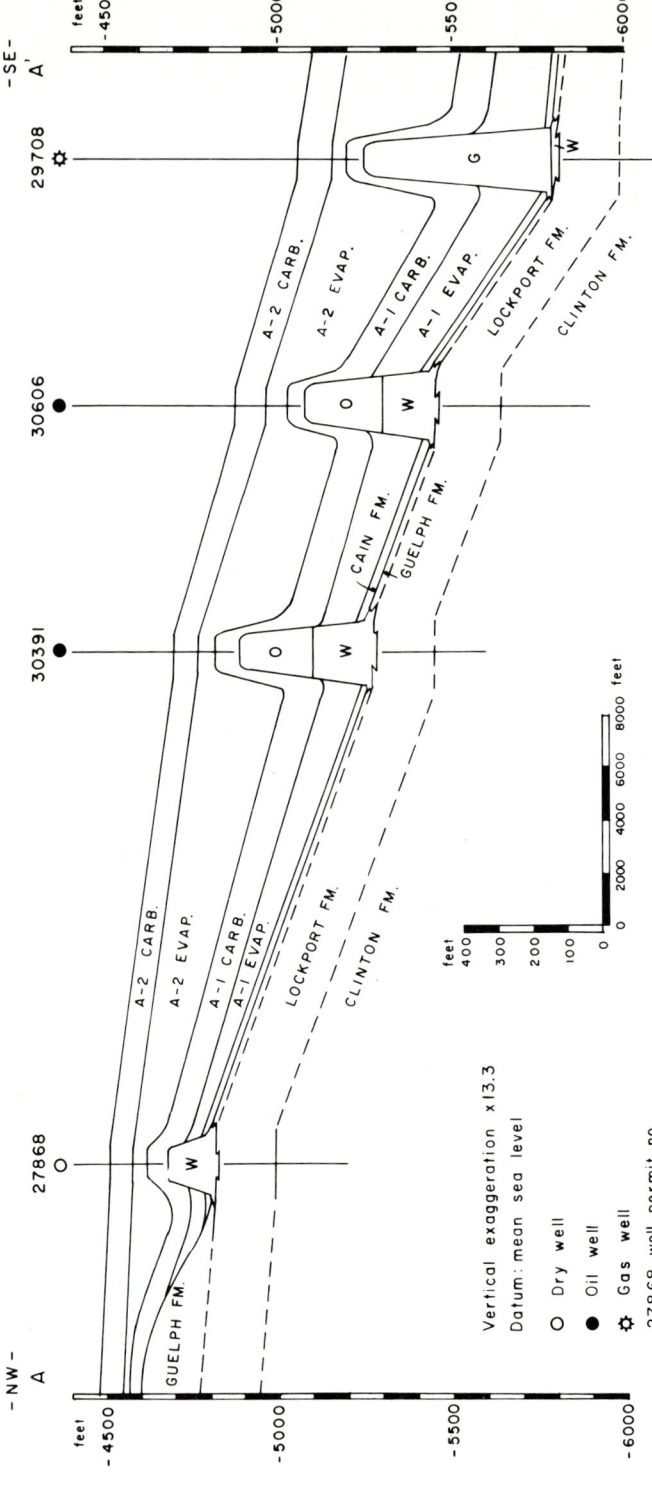

Fig. 9-20. Cross-sections through the reefs of north-western Michigan basin (located in Fig. 9-19). (Reproduced from Gill, 1979, p. 615, fig. 5.)

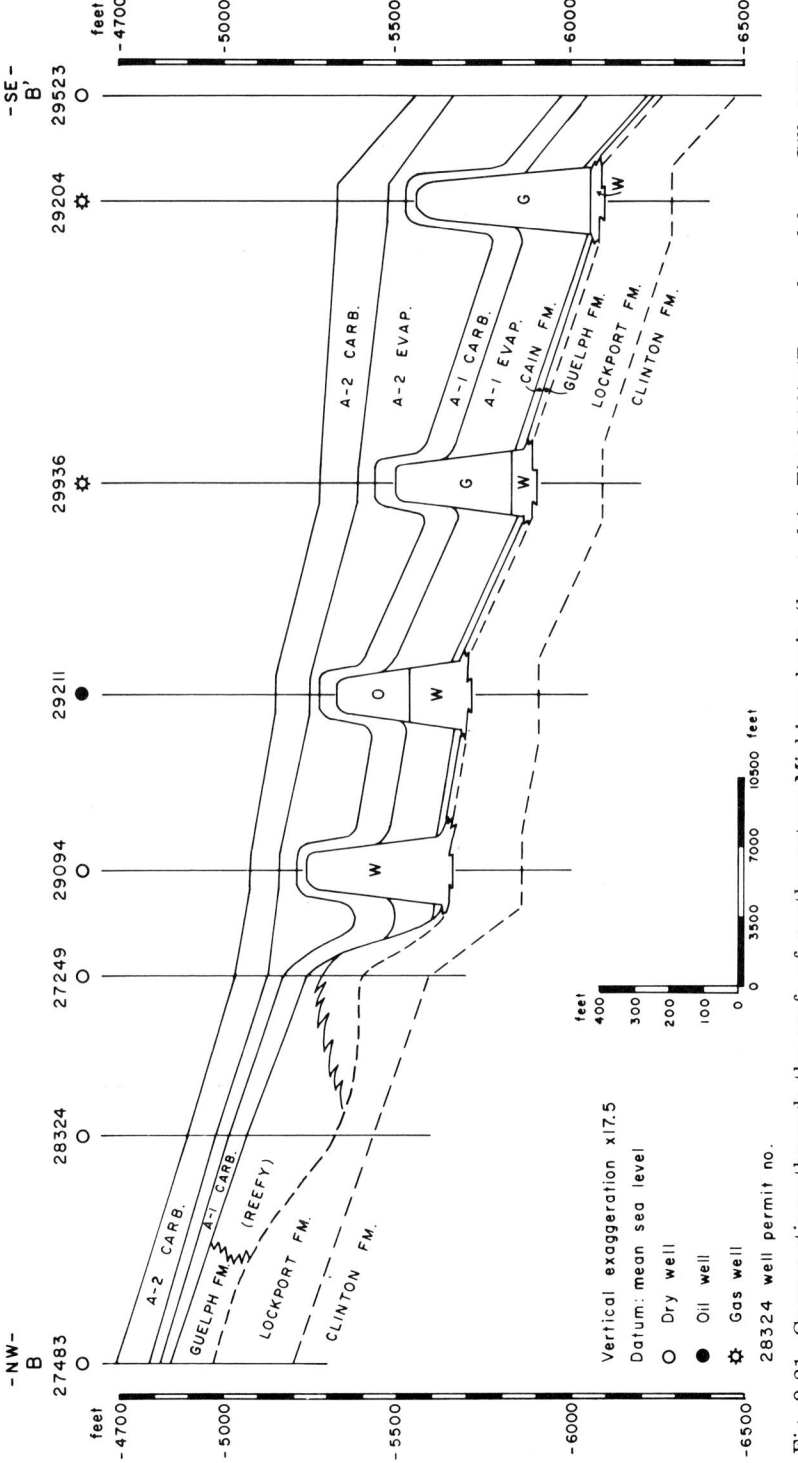

Fig. 9-21. Cross-sections through the reefs of north-western Michigan basin (located in Fig. 9-19). (Reproduced from Gill, 1979, p. 615, fig. 6.)

leum exploration has revealed 432 reefs of which 139 are gas-filled, 221 are oil-filled, and 72 are barren. The cap rock is evaporite. Figure 9-19 shows that the reef belt consists of a down-dip gas zone, an intermediate oil zone, and an up-dip water zone. Figures 9-20 and 9-21 show the cross-sections. The Lockport Formation is the common porous and permeable base to the reefs, and Gill (1979, p. 618) considered that the Cain Formation, *overlying* the Lockport between the reefs and down-dip, is the source rock. The oil gravity increases up-dip from $65°$ to $43°$ API (s.g., 0.72—0.81).

We shall consider these and other reefs in more detail in Chapter 12.

The same processes could, of course, operate in a sequence of sandstone traps; but those in carbonates are clear.

REFERENCES

Bredehoeft, J.D. and Hanshaw, B.B., 1968. On the maintenance of anomalous fluid pressures: I. Thick sedimentary sequences. *Geol. Soc. Am. Bull.*, 79: 1097—1106.

Campbell, I.R. and Smith, D.N., 1982. Gorgon 1 — southernmost Rankin Platform gas discovery. *J. Aust. Petrol. Explor. Ass.*, 22 (1): 102—111.

Chapman, R.E., 1972. Primary migration of petroleum from clay source rocks. *Bull. Am. Ass. Petrol. Geol.*, 56: 2185—2191.

Chapman, R.E., 1981. *Geology and water: an introduction to fluid mechanics for geologists.* Nijhoff/Junk, The Hague, 228 pp.

Gill, D., 1979. Differential entrapment of oil and gas in Niagaran pinnacle-reef belt of northern Michigan. *Bull. Am. Ass. Petrol. Geol.*, 63: 608—620.

Gussow, W.C., 1954. Differential entrapment of oil and gas: a fundamental principle. *Bull. Am. Ass. Petrol. Geol.*, 38: 816—853.

Gussow, W.C., 1968. Migration of reservoir fluids. *J. Petrol. Technol.*, 20: 353—363.

Halbouty, M.T., Meyerhoff, A.A., King, R.E., Dott, R.H., Klemme, H.D. and Shabad, T., 1970. World's giant oil and gas fields, geologic factors affecting their formation, and basin classification. Part I. Giant oil and gas fields. In: M.T. Halbouty (Editor), Geology of giant petroleum fields. *Mem. Am. Ass. Petrol. Geol.*, 14: 502—528.

Hedberg, H.D., 1926. The effect of gravitational compaction on the structure of sedimentary rocks. *Bull. Am. Ass. Petrol. Geol.*, 10: 1035—1072.

Hedberg, H.D., 1974. Relation of methane generation to undercompacted shales, shale diapirs, and mud volcanoes. *Bull. Am. Ass. Petrol. Geol.*, 58: 661—673.

Hedberg, H.D., 1980. Methane generation and petroleum migration. In: W.H. Roberts and R.J. Cordell (Editors), Problems of petroleum migration. *Am. Ass. Petrol. Geol., Stud. Geol.*, 10: 179—206.

Hobson, G.D., 1954. *Some fundamentals of petroleum geology.* Oxford University Press, London, 139 pp.

Hubbert, M.K., 1940. The theory of ground-water motion. *J. Geol.*, 48: 785—944.

Hubbert, M.K., 1953. Entrapment of petroleum under hydrodynamic conditions. *Bull. Am. Ass. Petrol. Geol.*, 37: 1954—2026.

Hull, C.E. and Warman, H.R., 1970. Asmari oil fields of Iran. In: M.T. Halbouty (Editor), Geology of giant petroleum fields. *Mem. Am. Ass. Petrol. Geol.*, 14: 428—437.

Illing, V.C., 1933. The migration of oil and natural gas. *J. Instn. Petrol. Technol.* (now: *J. Inst. Petrol.*), 19 (114): 229—260; discussion, 19: 260—274.

Kawai, K. and Totani, S., 1971. Relationship between crude-oil properties and geology in some oil and gas fields in the Niigata basin, Japan. *Chem. Geol.*, 8: 219—246.

King, F.H., 1899. Principles and conditions of the movements of ground water. *Annu. Rep. U.S. Geol. Surv.* (1897—1898), 19 (2): 59—294.

Lambert-Aikhionbare, D.O., 1982. Relationship between diagenesis and pore fluid chemistry in Niger delta oil-bearing sands. *J. Petrol. Geol.*, 4 (3): 287—298.

McCrossan, R.G., 1961. Resistivity mapping and petrophysical study of Upper Devonian inter-reef calcareous shales of central Alberta, Canada. *Bull. Am. Ass. Petrol. Geol.*, 45: 441—470.

Meissner, F.F., 1978. Petroleum geology of the Bakken Formation, Williston basin, North Dakota and Montana. *In: The economic geology of the Williston basin; Montana, North Dakota, South Dakota, Saskatchewan, Manitoba.* Montana Geol. Soc., Billings, pp. 207—227. (Montana Geol. Soc., 24th Annual Conference; 1978 Williston basin symposium.)

Munn, M.J., 1909a. Studies in the application of the anticlinal theory of oil and gas accumulations. *Econ. Geol.*, 4: 141—157.

Munn, M.J., 1909b. The anticlinal and hydraulic theories of oil and gas accumulation. *Econ. Geol.*, 4: 509—529.

Philipp, W., Drong, H.J., Füchtbauer, H., Haddenhorst, H.-G. and Jankowsky, W., 1963a. Zur Geschichte der Migration im Gifhorner Trog. *Erdöl Kohle Erdgas Petrochem.*, 16: 456—468.

Philipp, W., Drong, H.J., Füchtbauer, H., Haddenhorst, H.-G. and Jankowsky, W., 1963b. The history of migration in the Gifhorn trough (NW-Germany). *Proc. 6th World Petrol. Congress*, Sect. 1: 457—478; discussion: 479—481.

Rich, J.L., 1921. Moving underground water as a primary cause of the migration and accumulation of oil and gas. *Econ. Geol.*, 16: 347—371.

Rich, J.L., 1923. Further notes on the hydraulic theory of oil migration and accumulation. *Bull. Am. Ass. Petrol. Geol.*, 7: 213—225.

Smith, J.E., 1971. The dynamics of shale compaction and evolution of pore-fluid pressures. *J. Int. Ass. Math. Geol.*, 3: 239—263.

Stuart, M., 1910. The sedimentary deposition of oil. *Rec. Geol. Surv. India*, 40: 320—333.

Versluys, J., 1919. De duinwater-theorie. *Water*, 3 (5): 47—51.

CHAPTER 10

ORIGIN AND MIGRATION OF PETROLEUM: GEOLOGICAL AND GEOCHEMICAL ASPECTS

SUMMARY

(1) Petroleum is primarily a product of the diagenesis of fundamental organic compounds contained in organic matter that accumulated with fine-grained sediment in a low-energy environment deficient in oxygen. The diagenesis takes place during burial, under the influences of heat, time and pressure — probably in the presence of clay catalysts. Early diagenesis removes the more volatile and soluble components, leaving kerogen. Some kerogens are considered to be the important source material for petroleum.

(2) Mudstones are important source rocks of petroleum, but not all mudstones are source rocks. Thicker mudstones are probably more important than thinner because greater volumes of water must be expelled from them eventually. Concentrations of residual organic matter may be less in thicker mudstones than in thinner mudstones, but the total quantity must be large for large volumes of petroleum to be generated.

(3) Petroleum is probably generated as a separate phase in the source rock, and migrates as a separate phase in water to the accumulation. During migration, oil may be altered by removal of the more soluble components (water washing); and bacterial degradation may remove normal-alkanes in the range C_{16}—C_{25}, then progressively those with higher carbon numbers.

(4) High-wax crude oils seem to be genetically related to the environment in which the sediments accumulated. They are associated with sandstone/mudstone sequences, both transgressive and regressive, of Devonian age and younger. Not all petroleum provinces have waxy crude oils, and some important carbonate provinces appear to have no waxy crude oils.

INTRODUCTION

The advances in petroleum geochemistry during the last few decades have been spectacular to the point that many regard the basic hypotheses as facts. This may be so; but the topic of origin and primary migration is so complex and contains so many apparent anomalies and contradictions that there is still great danger of error. As before, we shall accept without further scrutiny that *commercial* accumulations of oil and gas have their origin in organic matter that accumulated with fine-grained sediment in a low-energy environment deficient in oxygen, and that this organic matter was derived from the tissues of living plants and animals.

When living tissues die, they are either consumed by living organisms and pass into the food chain, or they decompose with the aid of bacteria. If the environment is sufficiently rich in oxygen, those tissues not consumed by scavengers decompose to CO_2 and water (mainly): if the environment is deficient in oxygen and free of scavengers, the products are more complicated and varied, but are essentially compounds of carbon, hydrogen, oxygen, and nitrogen.

We concluded in the general review that the lack of correspondence between the composition of crude oils and the nature of organic life on earth when the crude oil was presumably formed, revealed by the fossils, indicated a source for hydrocarbons in the fundamental biological molecules: proteins (amino acids), lipids (fats, vegetable oils, waxes), carbohydrates, and, in the higher plants, lignins. Some of these are present in all forms of life, and their initial preservation, which is necessary for their subsequent alteration, depends on their rapid transport to an environment free of scavengers and deficient in oxygen, and their incorporation there into the accumulating sediment. This anoxic environment is not *necessarily* at the sea water/sediment interface, but it must exist at a very shallow depth of burial. What matters is the balance between the amount of material to be oxidized and the available oxygen, because a large supply of organic material can create its own anoxic environment. These environments will be free from scavenging animals, but not free from bacteria.

The physical environment most prone to develop anoxic chemical environments is one of little energy, that is, one in which only fine-grained sediment is present. Thereafter, the preservation of organic matter in quantity depends on its being buried as it is supplied, that is, the area must be one accumulating fine-grained sediment in a sedimentary basin. Note that the accumulation of fine-grained sediment is not in itself sufficient: there must also be a supply of organic matter to an anoxic environment. So true source rock must be distinguished from source rock in a broad sense because not all of a mudstone unit, for example, will necessarily be true petroleum source rock.

The requirement of large supplies of organic matter over considerable periods of time implies that we are concerned with sedimentary basins near continental margins mainly (but not exclusively) because these are areas of large biomass where sediment is commonly accumulating into the stratigraphic record. In such areas, great variations may exist between the relative supply of terrestrial and marine organic matter, and these differ in their chemical compositions. For large supplies of organic matter, organisms such as planktonic algae in the marine environment and higher plants in the terrestrial are more likely progenitors than vertebrates, for example. Marine planktonic algae contain variable amounts of proteins, lipids and carbohydrates, whereas higher land plants contain cellulose and lignin, with some lipids in spores, seeds, pollen and bark (as in cannel coal).

Bacteria themselves can form a significant biomass. ZoBell (1964) found up to 0.5 km m^{-3} in Holocene sediment, but the concentration decreased rapidly with depth.

The two dominant sources of organic matter differ in some general, but important respects. The atomic H/C ratio is higher in marine planktonic organic matter, 1.7—1.9 compared to 1.0—1.5 for terrestrial organic matter. The former is more *aromatic;* the latter is more *aliphatic.* Microbial action may induce further changes, including the liberation of sulphur.

During burial to greater depths, the sediment with its contained organic matter is subjected to increasing temperatures and pressures, and the volatile and more soluble components and products of early diagenesis are lost to the pore water. The portion that is insoluble in organic solvents is called *kerogen,* of which three main types have been recognized (Tissot et al., 1974):

Type-I kerogen: derived from organic matter with lipids, dominantly marine, with H/C ratios greater than 1.5. It is composed dominantly of aliphatic chains, with some aromatic chains. It is considered to be a good source material for oil and gas.

Type-II kerogen: derived from marine organic matter mainly, with H/C ratios smaller than for Type I. It is composed of aromatic and naphthenic rings, and considered to be quite a good source material for oil and gas.

Type-III kerogen: derived from terrestrial higher plants mainly, with H/C ratios usually less than 1.0. Not a good source material, it may generate gas rather than oil.

During burial, these kerogens are altered, with trends to lower H/C and lower O/C ratios (Fig. 10-1) due to the generation and liberation of hydrocarbons, carbon dioxide, and water. The degree of alteration of the kerogens is considered to be the prime measure of *maturity* in a petroleum source rock. A source rock is not mature until the CO_2, H_2O, and some compounds of nitrogen, sulphur and oxygen have been released; and although some hydrocarbons may be released at this stage, most are released later. Geochemists call the processes of the first stage *diagenesis,* and those of the second stage, prior to metamorphism, *catagenesis.*

The rate of the conversion reactions is a function of temperature and, perhaps, pressure. It is also probably affected by minerals with catalytic properties, so the combined process is usually lumped together under the term *thermocatalysis.*

It seems to the author that the matter of catalysis in source rocks and during primary migration has received too little attention in recent years. The early catalytic cracking units of refineries used natural Fuller's Earth (hydrated aluminium silicates) as the catalyst to improve the yield of high-octane-number gasoline and reduce the gas yield. A great deal of laboratory work has shown that most of the clay minerals, and nickel, are catalysts in petroleum reactions and that splitting of hydrocarbon molecules can take place in the

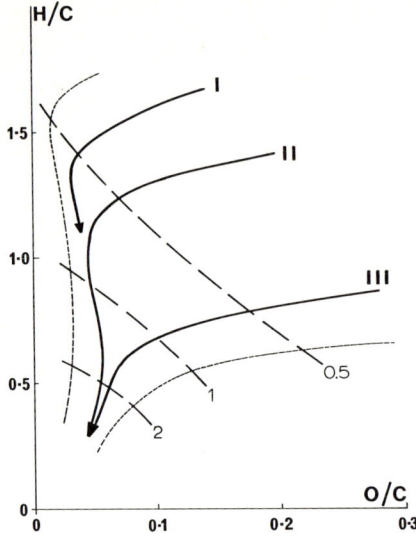

Fig. 10-1. Kerogen evolution during burial in terms of atomic ratios of hydrogen/carbon and oxygen/carbon. The dashed lines crossing the trends are approximate vitrinite reflectance values (in percent). Trends are towards the elimination of hydrogen and oxygen with increasing depths of burial. (After Tissot and Welte, 1978, p. 149, fig. II.5.1.)

presence of catalysts at temperatures well below cracking temperatures (McDermott, 1940; Brooks, 1948; Dobryansky, 1963; Louis, 1966; Hunt, 1967). The objection to catalysis as a geological process has been raised that the minerals are covered by a film of water and so cannot act on the oil. The evidence of petroleum reservoirs (Chapter 8) is that there may be a thin film of adsorbed water, perhaps 2 or 3 molecules thick, separating petroleum from solid surfaces. Adsorbed water, surely present in the catalytic cracking unit, may inhibit catalysis but is unlikely to prevent it. An efficient catalyst from a laboratory point of view is not required in nature: there is plenty of time.

The point here is that different clay minerals may exist along a migration path due to facies changes, and may exist in different petroleum source rocks, and so may contribute to variations in the quality and quantity of petroleum generated that are independent of temperature.

In view of all these variables and unknowns, it has little purpose to seek to make such statements as "the principal zone of oil generation is at 80°C, or 1500 m" or any other temperature or depth. Palaeozoic source rocks could have generated significant quantities of crude oil at 50°C, while Pliocene source rocks might have required 120°C to generate the same amount. Likewise, the depth of intense oil generation can be much deeper than 1500 m in young rocks in areas of low geothermal gradient, and shallower in older rocks in areas of larger geothermal gradient. The role of pressure is not considered to be very sensitive, but there have been hints that abnormally high pore pressures may inhibit maturation, and that high effective stress may facilitate maturation by deforming the kerogen and so accelerating its transformation. There is much work in progress on quantitative modelling of these interactions so that we may be able to estimate the quantities of hydrocarbons generated under any conditions (Lopatin, 1971, 1980; Connan, 1974, Tissot and Welte, 1978, p. 500ff; Waples, 1980).

For the reasons stated on p. 69, we shall not go into details of the chemical transformations of kerogens and other organic matter buried with sediment: for these, the reader is referred to Tissot and Welte (1978) and Hunt (1979). We merely note that empirical data obtained by analysis of extracts from supposed source rocks indicates that low-temperature cracking of large molecules into smaller molecules takes place, and similar results have been obtained by heating laboratory samples in the presence of clay minerals. In addition, the common clay mineral smectite (montmorillonite) can acquire organic compounds by cation exchange, and these, when heated without oxygen, yield hydrocarbons similar to petroleum (Weiss, 1963).

Coal and petroleum

It is natural that one should enquire whether the two great fossil fuels, coal and petroleum, have any significant geological relationship; and this enquiry has been going on for more than a century. Coal results from the diagenesis of vegetable organic matter that accumulated in an environment largely devoid of sediment. Conditions on the actual surface of accumulation may have been reducing or oxidizing; but close below this surface, reducing conditions prevailed. Coal consists largely of carbonized plant tissues, wood and bark, with spores (particularly the more durable spore coatings), leaf cuticles, waxes and resins. Coals form a series, with peat at one end and graphite at the other, and they are *ranked* according to their degree of alteration from lignites to anthracites. The *type* of coal groups coals of similar composition. Cannel coal, for example, is a coal rich in volatiles that burns easily and commonly contains significant proportions of spore coatings. Heating of a "bituminous" coal results in distillation of a gas that consists largely of hydrogen and methane, with numerous other components in small proportions (some being hydrocarbons). The coal-tar residue contains hydrocarbon oils (benzene, toluene, etc.) with other components.

Not surprisingly, early work on the association between coal and petroleum was carried out in Pennsylvania, U.S.A. — a coal mining area in which commercial oil production began in the United States. The early work culminated in White's "Carbon Ratio Theory", in which he related the occurrences of oil and gas to the percentage of "fixed carbon" (see Glossary) in associated coals. He ranked petroleum from low-ranking heavy oils to high-ranking light oils, and noted that oil occurred where the fixed carbon is less than 65%, mostly less than 60%; no commercial accumulation occurred when the percentage of fixed carbon is greater than about 70%; and gas occurred in the intervening zone (White, 1915). There was thus observed a ranking of oil with depth analogous to the ranking of coal with depth in many fields. This theory was considered to be widely applicable.

This line of thought, with the associated idea that petroleum was a distillation product of coal (a demonstrable process to some extent) went out of fashion as a marine origin for oil became fashionable. But it has returned with the increasing evidence that some petroleums can hardly have had a marine source, and that some crude oils, particularly those with high wax content, have an important component of vegetable organic origin (a matter that we shall examine shortly). The diagenesis of vegetable organic matter to petroleum has been well established since the work of Brooks and Smith (1967, 1969). And the Groningen gas field, the largest in the world, with 58 Tcf (10^{12} cubic feet; 1.6×10^{12} m^3) recoverable proven reserves, is believed to be the result of distillation of Upper Carboniferous (Pennsylvanian) coals (Stäuble and Milius, 1970).

Brooks (1970) considered that crude oil generation does not begin until the rank of the coal reaches 80% total carbon, dry mineral matter free; and that oils are formed until the rank reaches about 85%; and light oils with gas, and dry gas, up to about 90% total carbon, dry mineral matter free. (The relationship between fixed carbon and total carbon is not precise: 60% fixed carbon is approximately equivalent to 83% total carbon, dry mineral matter free; 65—86%, and 70—87%.)

White's carbon ratio theory was revived and reclothed in the late 1950s and early 1960s. The coal industry required a measure of rank in sedimentary sequences that contained no coal, and vitrinite (a coal maceral) was found to have a property that closely followed the rank of coals and was used for ranking coals: its capacity to reflect light increased with increasing rank. Largely through the work of M. Teichmüller (1958, 1963), vitrinite reflectance became accepted in the petroleum industry as a measure of the thermal maturity of sedimentary rocks. The mean reflectance of many particles of vitrinite is determined from measurements under standard conditions, and the result recorded as R_o or $\overline{R}_o = x\%$ (the suffix o indicating oil as the medium in which the reflectance was measured). In general terms, it has been found that only biogenic gas occurs where $R_o < 0.5\%$, and the rocks are thermally immature; oil occurs where $0.5 < R_o < 1.3$; wet gas and gas only occur where $R_o > 1.3$.

It is most important for the petroleum geologist to appreciate that vitrinite reflectance is a measure of the thermal maturity of the sedimentary rock in which it was found, whereas petroleum may accumulate at some distance from its source rock. It is therefore important to phrase the statements concluding the previous paragraph with care. It is an *interpretation* of empirical observations that leads to statements such as "oil is to be expected in this area because the vitrinite reflectance is 0.9%", and such statements may be very misleading. The data can be acquired from a single well, so they must be used positively to indicate areas of interest, not negatively to discard areas. Very large volumes of oil, including some of the giant oil fields, are in sedimentary rocks that are regarded as immature by vitrinite reflectance standards.

Palynologists had also noted that the colour of spores and pollen became darker with increasing depth, and interpreted these changes as due to thermal influences (Gutjahr, 1966). The same reservations apply to this index of maturity as for vitrinite reflectance.

PRIMARY MIGRATION

We infer that petroleum source material is disseminated through the source rock sensu stricto and therefore petroleum generated from it is also disseminated. The question then arises: how does this move out of the source rock?

There are three possible processes that each have their adherents: molecular solution, colloidal solution, and as a separate, immiscible phase, in the pore water.

Solution in pore water is attractive because the initial dissemination is no obstacle, and it is the process that requires least work. The difficulties with this hypothesis are twofold. Some process must exist for the exsolution of the petroleum; and the laboratory-measured solubilities appear to be too low by at least an order of magnitude. We will take the latter point first.

Even allowing for some doubt about the actual solubilities of the various hydrocarbons, we can reach an estimate of the solubility required in some areas. Jones (1981, p. 105) took Dow's (1974) data for the Bakken Shale in the Williston basin, Wyoming, U.S.A., and showed that a solubility of at least 15,000 ppm was required to account for the known oil in place assuming that the Bakken Shale compacted from 10 to 5% porosity during generation. (There is no point in making the distinction between weight/weight and volume/volume in such estimates of solubility.) The maximum solubility of oil was taken as 200 ppm (see Price, 1976, p. 220, fig. 7). Even if the compaction range is extended to 25—5% during generation, the solubility required is only reduced to 3000 ppm. No manipulation of these figures seems capable of reducing the requirement to the observed order of magnitude of oil solubility in water.

This argument, it will be noted, does not *eliminate* solution as a process — even as an important process — but it does suggest that if the source rock generates *liquid* petroleum, solution is unlikely to be the main, general process of primary migration to commercial oil fields.

Bonham (1980), on the other hand, modelled fluid migration in a compacting sedimentary basin and, applying it to a "representative" area of the U.S. Gulf Coast, came to the conclusion that a solution process could account for all the known oil and gas accumulations of the area. In a long but interesting argument, he asserted that pore-water movement may generally be upward relative to the stratigraphic levels, but in a subsiding basin it will be downwards in an absolute sense. Accepting the temperature of onset of oil generation as 90°C and that of peak generation as 120°C, he found that a 28°C drop of temperature since the early Pliocene could have exsolved the known quantities of oil and gas in the area. Bonham pointed out that this was not the only possible process. The difficulty with this hypothesis from our point of view is that the depths of onset of oil generation and primary migration in Bonham's area are believed to be below 3000 m, while there is much oil in the U.S. Gulf Coast above 2000 m.

Treating deltaic sequences as a class apart, as Jones (1981) did, does not really help very much. But it must be noted that the regressive sequences with important zones of abnormally pressured mudstones (shales) are common and important in petroleum geology, and the apparent general lack of maturity in the mudstones associated with the reservoirs in such areas will be taken up in the next chapter.

The chemical evidence of accumulations in the context of solubility is confusing to the non-specialist: some authors have claimed that the chemistry of accumulations supports the solution process (Baker, 1959, 1960, 1962), others that it does not.

Price (1976) found that the solubility of hydrocarbons in water increased markedly with increase of temperature above 100°C, and postulated very deep sources with migration in faults upwards in aqueous solution, the petroleum exsolving at shallower, cooler levels.

We conclude that solution *is* a process of primary migration, but that in those areas where a shallow source is indicated by geological or other argument it is not the main process quantitatively (although its indirect importance in other processes may be great).

Colloidal or micellar solution are similar processes that were proposed by Baker (1959) and Meinschein (1959), and revived by Cordell (1973), because of the possibility of solubilizing petroleum in pore water at low temperatures. The difficulties are much the same as for molecular solutions, with two more. Colloids and micelles are larger than the individual molecules because they are disordered and ordered groups of molecules, respectively. There will therefore be an even greater tendency to restrict movement mechanically. The larger micelles solubilize hydrocarbons better than the smaller, but by less than an order of magnitude. The soap concentrations required to form micelles is about two orders of magnitude greater than that found in formation waters.

The size problem is not amenable to numerical assessment as yet because we do not know the sizes of pore *throats* in mudstones, and the pore-*size* estimate of about 5 nm (see Cordell, 1973, pp. 1619, 1623—1625) may be misleading in pores that have a smaller vertical dimension than the lateral.

The conclusion reached by Tissot and Welte (1978, p. 276) and Hunt

(1979, p. 207) is that this process is unlikely to be significant.

Separate phase. The conceptual simplicity of generating liquid or gaseous hydrocarbons directly from source material, and evidence that the liquid or gaseous state exists by the beginning of secondary migration, make primary migration as a separate phase attractive. The difficulties are largely mechanical, and these have been the main reason for regarding migration as a separate phase unlikely if not impossible. If disseminated organic matter generates disseminated droplets of oil, the capillary displacement pressure required to move these from one pore to another is far greater than that available in the pore water. Displacement of oil under these conditions must await mechanical forces during the reduction of pore volume during compaction (Hobson, 1954, pp. 78—80). The problem is reduced to an assessment of possible processes of concentration, because oil in a continuous, separate phase requires, as we saw in Chapter 8 on reservoirs, far less work for its movement.

Dickey (1975) suggested that the pore water in a source mudstone is largely structured water that behaves mechanically more as a solid than as a liquid. Under these conditions, the oil saturation relative to movable water may be high enough to form a continuous phase.

Mudstones are seen under electron microscopes (Dickey, 1975, p. 340, fig. 2) to consist of dominantly platy fragments that are usually slightly deformed but lying generally in the bedding planes. Permeability of mudstones, though difficult to measure with precision, appears to be appreciably anisotropic, with lateral permeability much greater than transverse. We infer that the pore shapes are also "flat", with smaller transverse dimensions ("vertical") than lateral. We assume that during compaction the loss of porosity is achieved largely by reduction of the vertical pore dimension.

As an oil droplet forms, it will tend to occupy the position that minimizes its potential energy, but the spherical shape that also minimizes its potential energy can no longer be maintained once the diameter of the droplet reaches the minimum, vertical, dimension of the pore: it is thereafter distorted, which requires energy. As the droplet grows, any pore throat that comes to be on the upper or lower, flattened surfaces of the droplet will tend to be entered; and if the pore throat diameter is larger than the vertical dimension of the pore, it will enter the throat rather than grow in the pore. It is therefore possible that the anisotropy in the mudstone induces an anisotropy in oil concentration that favours vertical continuity of oil phase at water saturations that are larger than those required in isotropic pores. The water will tend to concentrate in pore space away from throats, analogous to the pendular rings in reservoir rocks.

We concluded on p. 158 that the thickness of an adsorbed water film on VYCOR at room temperature is not thicker than 1 nm because the flow of water, acetone, and *n*-decane obeyed Darcy's law through pores about 4 nm in diameter. The intrinsic permeability of VYCOR was found to be about 10^{-16} cm^2 (Chapman, 1981, p. 65), or 10^{-8} darcies, which is near the lower end of the range of mudstone permeabilities reported by Magara (1971, p. 241, fig. 9), and by Bredehoeft and Hanshaw (1968). It therefore seems sound to infer that the *effective* diameter of the pore/thoat system is usually larger than 5 nm, and, if the tortuosity of mudstone is greater than that of VYCOR, appreciably greater than 5 nm. It is, of course, also evident that n-decane will pass through pores of 4 nm diameter, perhaps rather less.

It now appears that primary migration as a continuous, separate, immiscible phase is entirely possible, irrespective of considerations of structured water, and that the effect would be of multitudinous, sinuous, vertical streams. Furthermore, these streams would be largely in contact with mineral grains, which include clay minerals that are petroleum catalysts, or only separated from them by a very thin film of adsorbed water. Relative permeability to water would be close to zero while oil was migrating, and almost all the compaction energy would be devoted to expulsion of oil.

If this is an important process in primary migration, it suggests that solution would be restricted to early and late stages.

There is no great difficulty in the expulsion of oil or gas as a separate phase from the source rock sensu lato into the carrier bed because, as Hubbert (1953, p. 1979) pointed out, the greater capillary pressure in the finer-grained material is sufficient on its own to effect expulsion once the continuous phase reaches the sedimentary interface (Fig. 10-2).

The main difficulty lies, however, between the source rock sensu stricto and the carrier bed because it is unlikely that true source rock is contiguous with the carrier bed (Chapman, 1974). The reason for this lies in facies considerations. An alternating sequence of sands and mudstones is essentially diachronous, sand accumulating in some areas contemporaneously with muds

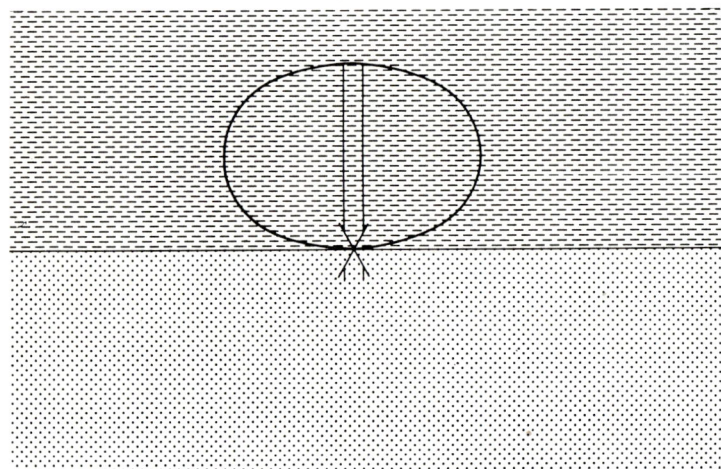

Fig. 10-2. Capillary pressure tends to expel a continuous oil phase from mudstone once part of its interface reaches the sandstone. (After Hubbert, 1953, p. 1978, fig. 14.)

in others. It is most unlikely that the continguous environment in which mud accumulates close to sand is also a favourable environment for the preservation of organic matter because it is probably relatively well oxygenated. The lateral oscillation of these areas leads to the alternating sequence, with true source rock near the middle of the mudstones (Fig. 10-3). To cross this barren mudstone, injection pressures of sufficient magnitude must be generated. During this crossing, the petroleum may come into contact with different clay minerals in the different facies of the mudstone.

If Meissner (1978) is correct in attributing the abnormal pressures in the Bakken Shale of the Antilope field, North Dakota, to oil generation, the conversion of kerogen resulting in larger volumes of oil, it would be evidence of oil generation as a separate phase as well as a process that would contribute to the attainment of a long continuous phase, and provide energy for primary migration.

In the Gulf of Paria, illitic clays are associated with the delta platform of the Orinoco river and waters of lower salinity; montmorillonite (smectite) is associated with more saline environments (Van Andel and Postma, 1954, p. 78). Gibbs (1977) found that the montmorillonite proportion increased away from the Amazon. Likewise, Porrenga (1966) found that the offshore Niger delta has roughly parallel zones of increasing montmorillonite content, with smaller proportions in the sandier facies, to more than 50% offshore. Kaolinite decreased in the seaward direction. Both Porrenga and Gibbs concluded that this was due to physical segregation (i.e., different baselevels), not to chemical changes. If these relationships are typical, a regressive sequence

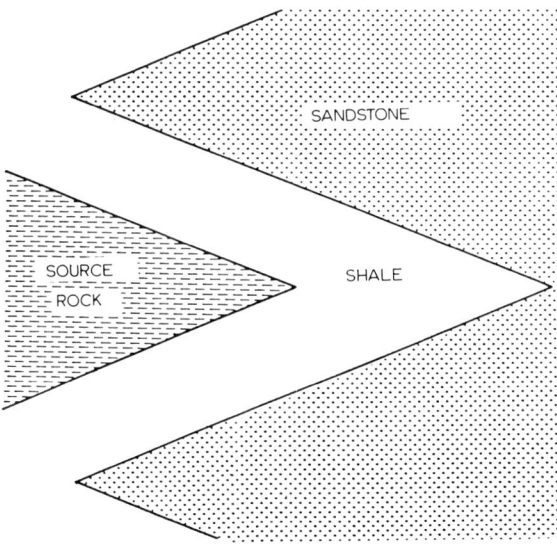

Fig. 10-3. Petroleum source rock is a facies: by Walther's Law, the vertical distribution of facies reflects the horizontal. (Idealized cross-section.)

would originally consist of sands with kaolinitic—illitic mudstones on kaolinitic—illitic mudstones on montmorillonitic mudstones — and a sandy transgressive sequence, perhaps, the reverse. Because these are facies of mudstones and petroleum source rocks are facies of mudstones, it is almost certain that primary migration exposes petroleum to different clay minerals along its path to the nearest sand or sandstone.

CRUDE OILS WITH HIGH WAX CONTENT

Petroleum waxes in crude oil are solid hydrocarbons, mainly normal alkanes in the range C_{22}—C_{30}. Their melting points are above 30—35°C, so they can cause difficulties during production. They have an intrinsic value as a product for the manufacture of candles, impregnated paper, polishes, and water-proofing agents. Most crude oils contain some wax, but some contain up to 15% wax by weight (rarely, more). Crude oils with more than 5% wax are regarded as high-wax oils.

Hedberg (1968) suggested that the high wax content of some crude oils is an original characteristic related to shale-sandstone stratigraphic sequences of non-marine origin or origin in waters of less than normal marine salinity in continental, paralic, or near-shore marine environments, commonly associated with coals, of Devonian to Pliocene age. From these associations he inferred that the wax content reflects a terrigenous organic contribution to the source material. These associations have been widely accepted as causal, because land plants synthesize waxes to C_{37} (see, for example, Tissot and Welte, 1978, pp. 394—396; Hunt, 1979, p. 91).

Reed (1969) tested Hedberg's hypothesis in the western offshore Niger delta, assessing the environment of the sequences in which high-wax oil occurred from the faunal content. He found that 34 out of the 41 samples studied supported the hypothesis. (Note that the hypothesis concerns the environment in which the sequence accumulated and the salinity of its waters, not the salinity of present-day pore waters in the sequences.)

Because wax content is not always determined, reliance is also placed on two properties of crude oil that are normally determined: *pour-point*, which is the temperature at which the crude oil will just flow, is normally (but not always) associated with high-wax crude oils; and *cloud-point*, which is the temperature at which a cloud is formed in reduced pressure distillation fractions by separation of wax on cooling.

The list of high-wax crudes that Hedberg assembled has a number of interesting points. Apart from the associations already mentioned, it seems that there is a positive association between measured wax content and API gravity — the *lighter* the oil, the greater the measured wax content (but such associations must be taken with caution because analyses of the lighter, more valuable crudes may be more readily available than those of heavy crudes that are hard to produce). Hedberg pointed out that high-wax crudes are not ubiquitous; large and important petroleum provinces such as the Middle East, Mexico and West Texas — all mainly carbonate provinces — did not appear to have high-wax crude oils. To this observation we add that there seems to be no preference for regressive or transgressive sequences of mudstone—sandstone lithologies, but in regressive sequences, the heaviest, shallowest crude oils are non-waxy. These heavy oils are typically associated with fresh to brackish *formation* waters, and their properties have been attributed to water-washing and bacterial degradation in the reservoir.

The causal association between terrestrial contribution and high wax content of crude oils seems established beyond reasonable doubt. We therefore conclude that when high-wax crude oils are found in sandstones that accumulated in continental, paralic or near-shore marine environments, *their source may be stratigraphically associated with the accumulation*, and, for the hydrodynamic considerations of mudstone-sandstone sequences discussed on pp. 61—62, we infer that they are intimately associated, with short primary migration paths.

ALTERATION OF CRUDE OIL AFTER GENERATION: WATER-WASHING AND BIODEGRADATION

There are no great difficulties in understanding secondary migration as a separate phase in water, and we shall assume that this is the state of secondary migration and address ourselves to the matter of possible transformations of crude oil during its passage through the carrier bed, and within the accumulation. As always, catalytic transformations are possible when the water saturation is sufficiently small, but the most important processes for the geologist are *water washing* and *biodegradation*.

Crude oil flowing past water, and water flowing past crude oil, can result in the removal of some of the more water-soluble components. In general, hydrocarbons of low molecular weight are more soluble than those of higher molecular weight, so the general effect of water washing is to increase the density, decrease the API gravity, of the crude oil. According to Hunt (1979, p. 390), hydrocarbons up to C_{15} or even higher can be removed. In the extreme, an asphaltic, heavy, unproducible oil is left.

A very similar, perhaps indistinguishable effect is due to microbial action. Both seem to take place under similar geological conditions, so the two processes are commonly taken together.

The role of microbes, of which bacteria are probably the most important, in the transformation of crude oil is almost certainly important, but difficult to assess because of the lack of reliable data. This is due mainly to the enormous difficulties of avoiding contamination during the taking of subsurface samples. Bacteria are normally present in surface waters and drilling fluids.

There is a lack of substantiating literature on subsurface microbial processes, as distinct from assertions that they take place based on comparisons between details of crude oil compositions and those resulting from laboratory degradation by bacteria.

The facts appear to be these (see Davis, 1967): Microbes are 0.2—2 μm in smallest dimension, and they require oxygen, mineral nutrients and water for their existence, generally attaching themselves to a solid surface. They reproduce by cell division, and leave cell walls when they die. Their motility is random, with a bias towards lower temperatures and pressures. They can live at temperatures up to 75°C at least, and some survive pressures of 170 MPa (25,000 psi) corresponding to hydrostatic water pressures at depths to about 3.5 km. Aerobic bacteria at the surface can oxidize crude oil (to alcohols. ketones and acids). They attack n-paraffins in the range C_{16}—C_{25} first, and may later attack the higher range and other components. They convert low molecular weight hydrocarbons to higher molecular weight hydrocarbons, creating long-chain insoluble waxes around C_{40}. Anaerobic bacteria are not *known* to utilize hydrocarbons with this result, but Bailey et al. (1973b) have suggested that they do so more slowly than aerobes. Cell walls have not been found in oil reservoirs.

It is not known how long a bacterium lives, but Davis (1967, p. 51) reports that of the bacteria found at a depth of 2m in the sediment of lake Biserovo, Russia, 70% were dead. This suggests a negligible life-span from a geological point of view. It is not the life span of the individual but the true life span of a viable community of bacteria that is important.

Bacteria were found in artesian water from the Carrizo Formation (Eocene) of south Texas from a depth of 1280 m (4200 ft) at 58°C (136°F) (Davis, 1967, p. 167). This well is 45 km from the outcrop of the aquifer, and the velocity of the water flow is estimated to be between 15 and 30 m/yr. If bacteria were introduced into the sand in its intake area, and were transported at the maximum water-flow rate, the community would have to survive at least 1500 years to be produced from this, the deepest well. It is curious, though, that no bacteria were found in a nearby well.

The motility of bacteria is random, of the order of centimetres a year, with a tendency to follow a decreasing temperature gradient and a decreasing pressure gradient. Because passage into the subsurface is against these tendencies, and because their tendency to cling to solid interfaces and their size would also be retarding influences, it is unlikely that a bacteria colony would reach the deepest Carrizo well in less than 5000 years. During this time, an adequate supply of nutrients would have to be maintained.

With reservations, therefore, concerning the longevity of microbe communities, there is no known parameter that would prevent their existence at depths of one or two kilometres, given the right conditions. Movement of ground water from the surface to the migrating or accumulating crude oil therefore seems to be an essential condition for biodegradation of the crude oil.

Gas chromatograms of crude oil alkanes (Fig. 10-4) indicate the full range of effects observed in laboratory experiments, from the presence of the full range of n-alkanes, to suppression in the range C_{16}–C_{25}, to almost total suppression of the entire range. It is inferred that these changes are due to the activity of aerobic bacteria. The questions arise, how, where, and when?

The widely accepted view is that biodegradation and water washing occur in the accumulation after the crude oil has accumulated, as a result of water flowing through it. Hunt (1979, pp. 382—390) discusses the topic in his chapter on petroleum in the reservoir, and Tissot and Welte (1978, p. 417 and 419) also regard them as alterations that take place in the accumulation.

The evidence of petroleum reservoirs (pp. 156—161) indicates that water

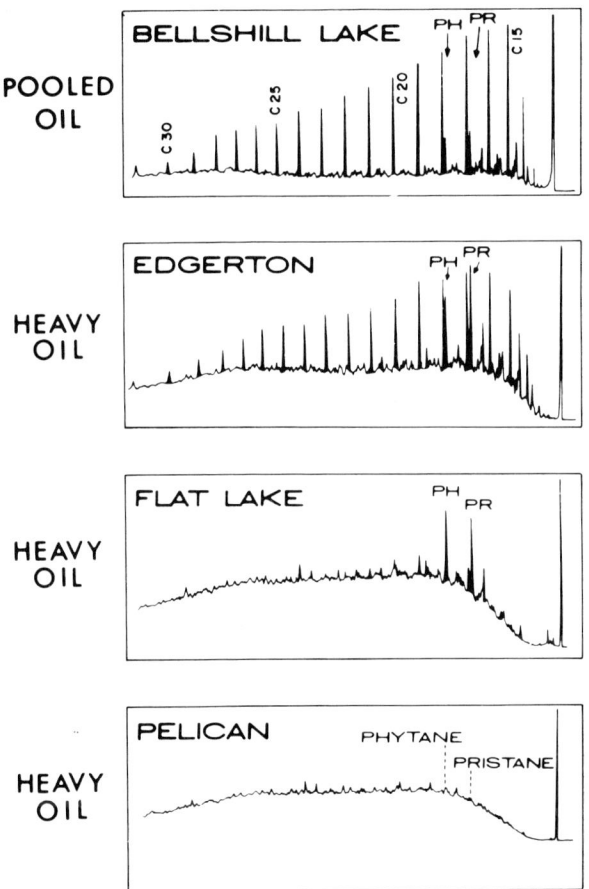

Fig. 10-4. Progressive biodegradation of crude oil in some fields in western Canada, from unaltered crude in Bellshill Lake field to severely altered crude in Pelican field. The gas chromatograms show progressive reduction in n-alkanes and, in Pelican, the elimination of pristane and phytane (after Deroo et al., 1974).

cannot flow upwards through an oil accumulation, and so cannot carry bacteria from the aquifer into the accumulated oil. Water washing is limited to the period of actual accumulation, prior to the acquisition of its irreducible water saturation, while water is being displaced by oil in the trap. Since the latter applies to all oil accumulations, variations of effect would be limited to variations of water composition and temperature, and their capacity to take hydrocarbons into solution. Bacterial degradation of a complete accumulation after accumulation (or a significant thickness of an accumulation after accumulation) could only result from bacteria trapped in the pendular rings at grain contacts. Hydraulic continuity does not exist here: bacteria could not move beyond the pendular ring in which they were trapped, and the requirements of oxygen and nutrients for a community would not be satisfied.

If conditions favourable for biodegradation exist at the time of oil accumulation, they also exist during at least the later stages of secondary migration. Indeed, the passage of an oil stream through water in which there are aerobic bacteria is clearly the most favourable condition for both bacterial degradation and water washing. The movement of oil through the water would expose it to larger bacterial communities, the cell walls of which would remain in the carrier bed (if not transformed). Time is not a constraint because Jobson et al. (1972) and Bailey et al. (1973b) found in the laboratory that dramatic changes took place in days. Long secondary migration paths would enhance water washing.

It is possible that microbes could penetrate an accumulation for a short distance towards the level of irreducible water saturation, and so degrade the oil near the oil/water contact. The presence of such "tar mats" is evidence that the water quality has changed since the oil accumulated, otherwise the whole accumulation would have been degraded.

Likewise, accumulations that result from asphalt seals near the surface (and there are some very large ones, such as Lagunillas, Venezuela, and Coalinga, California) are probably due to the arrival of migrating oil in an environment favourable to degradation. Migration ceases when degradation immobilizes the oil, leaving a gradation to unaltered crude oils down-dip.

In a regressive sequence with multiple reservoirs and sources, the shallowest, heavy oil sometimes shows evidence of biodegradation (see, for example, Hunt, 1979, pp. 388—391). Paraffinic, waxy oils are found in the deeper reservoirs. Hedberg's (1968) tabulation of crude oils from which he inferred a causal association between high-wax crude oils and terrigenous source materials shows a positive association between wax content and API gravity — the lighter the oil, the more wax — with appreciable wax content, in general, in crudes of greater than 30—35°API (s.g., 0.88—0.85). This seems to lend support to the hypothesis of biodegradation of the heavier crudes at shallow depth, with the removal of paraffin waxes. But the deeper reservoirs were once shallow, and were not degraded (Connan et al., 1975, showed that in the laboratory a biodegraded crude oil does not mature to normal crude oil, and is still recognizable). So some change in the water flow must have occurred during the accumulation of the regressive sequence to bring bacteria into the system during secondary migration of the biodegraded oil. This is an *environmental* matter related to the proximity of the regressing land, and may therefore involve a change in source material.

The Bell Creek field in Montana, U.S.A., studied by Winters and Williams (1969), is both interesting and perplexing. There are several sandstone reservoirs in the Cretaceous Muddy Formation, and the petroleum source rocks are considered to be in the enveloping mudstones. The average reservoir depth is about 1370 m, and their temperatures are in the range 35—41°C, so conditions are well within the tolerances of microbes. Within a single reservoir, three types of crude oil are found: unaltered crudes, crudes with all normal alkanes missing, and crudes with normal alkanes missing up to about C_{18}. One curious feature is that the zone in which all normal alkanes are missing separates the zone of unaltered crudes in the south-west from those with up to C_{18} missing in the north-east.

Winters and Williams reported direct evidence of microbial action in the Bell Creek field. They found that water produced with the oil of some wells, when cultured, produced aerobic micro-organisms that consumed both *n*-butane and *n*-hexadecade, while others did not. These wells matched the data of the chromatograms. While this evidence appears to indicate alteration within the accumulations, there may be another explanation because, as Winters and Williams noted, the microbial activity must take place at the oil/water interface.

There is another particularly perplexing problem. There are chromatograms of mudstone extracts that show all the characteristics of biodegradation although the sample came from a position some metres away from the nearest sand*. It seems physically impossible for bacteria to enter a compacting mudstone against the water currents of compaction, highly unlikely that a bacteria community could survive such an environment for millions of years, and chemically implausible that biodegradation of the source material led to results so similar to biodegradation of the products.

Taking all these things into account, the accessibility of bacteria to crude oil during secondary migration seems to require that biodegraded crude oil was formed early, and may have had an inherently different composition from crudes that were not biodegraded in the same field.

---

* I have not found a published example, but have been shown two in the research laboratories of two oil companies.

The Microbe

The microbe is so very small
You cannot make him out at all,
But many sanguine people hope
To see him through a microscope.
His jointed tongue that lies beneath
A hundred curious rows of teeth;
His seven tufted tails with lots
Of lovely pink and purple spots,
On each of which a pattern stands,
Composed of 40 separate bands;
His eyebrows of a tender green;
All these have never yet been seen —
But scientists, who ought to know,
Assure us that they must be so . . .
Oh! let us never, never doubt
What nobody is sure about!

Hilaire Belloc

REFERENCES

Bailey, N.J.L., Jobson, A.M. and Rogers, M.A., 1973a. Bacterial degradation of crude oil: comparison of field and experimental data. *Chem. Geol.*, 11: 203—221.

Bailey, N.J.L., Krouse, H.R., Evans, C.R. and Rogers, M.A., 1973b. Alteration of crude oil by waters and bacteria — evidence from geochemical and isotope studies. *Bull. Am. Ass. Petrol. Geol.*, 57: 1276—1290.

Bailey, N.J.L., Evans, C.R. and Milner, C.W.D., 1974. Applying petroleum geochemistry to search for oil: examples from Western Canada basin. *Bull. Am. Ass. Petrol. Geol.*, 58: 2284—2294.

Baker, E.G., 1959. Origin and migration of oil. *Science*, 129: 871—874.

Baker, E.G., 1960. A hypothesis concerning the accumulation of sediment hydrocarbons to form crude oil. *Geochim. Cosmochim. Acta*, 19: 309—317.

Baker, E.G., 1962. Distribution of hydrocarbons in petroleum. *Bull. Am. Ass. Petrol. Geol.*, 46: 76—84.

Baker, E.G., 1967. A geochemical evaluation of petroleum migration and accumulation. In: B. Nagy and U. Colombo (Editors), *Fundamental aspects of petroleum geochemistry*. Elsevier, Amsterdam, pp. 299—329.

Bonham, L.C., 1980. Migration of hydrocarbons in compacting basins. *Bull. Am. Ass. Petrol. Geol.*, 64: 549—567.

Bredehoeft, J.D. and Hanshaw, B.B., 1968. On the maintenance of anomalous fluid pressures: I. Thick sedimentary sequences. *Geol. Soc. Am. Bull.*, 79: 1097—1106.

Brooks, B.T., 1948. Active-surface catalysts in formation of petroleum. *Bull. Am. Ass. Petrol. Geol.*, 32: 2269—2286.

Brooks, B.T., 1949. Active-surface catalysts in formation of petroleum — II. *Bull. Am. Ass. Petrol. Geol.*, 33: 1600—1612.

Brooks, J.D., 1970. The use of coals as indicators of the occurrence of oil and gas. *J. Aust. Petrol. Explor. Ass.*, 10: 35—40.

Brooks, J.D. and Smith, J.W., 1967. The diagenesis of plant lipids during the formation of coal, petroleum and natural gas. I. Changes in the n-paraffin hydrocarbons. *Geochim. Cosmochim. Acta*, 31: 2389—2397.

Brooks, J.D. and Smith, J.W., 1969. The diagenesis of plant lipids during the formation of coal, petroleum and natural gas. II. Coalification and the formation of oil and gas in the Gippsland basin. *Geochim. Cosmochim. Acta*, 33: 1183—1194.

Chapman, R.E., 1974. Depths of oil origin and primary migration: a geologist's discussion. *Bull. Am. Ass. Petrol. Geol.*, 58: 1853—1857.

Chapman, R.E., 1981. *Geology and water: an introduction to fluid mechanics for geologists*. Nijhoff/Junk, The Hague, 228 pp.

Connan, J., 1974. Time—temperature relation in oil genesis. *Bull. Am. Ass. Petrol. Geol.*, 58: 2516—2521.

Connan, J., 1976. Time-temperature relation in oil genesis: reply. (to Waples, 1976.) *Bull. Am. Ass. Petrol. Geol.*, 60: 885—887.

Connan, J., Le Tran, K. and van der Weide, B., 1975. Alteration of petroleum in reservoirs. *Proc. 9th World Petrol. Congress*, 2: 171—178.

Cordell, R.J., 1972. Depths of oil origin and primary migration: a review and critique. *Bull. Am. Ass. Petrol. Geol.*, 56: 2029—2067.

Cordell, R.J., 1973. Colloidal soap as proposed primary migration medium for hydrocarbons. *Bull. Am. Ass. Petrol. Geol.*, 57: 1618—1643.

Cordell, R.J., 1974. Depths of oil origin and primary migration: a geologist's discussion — reply. (to Chapman.) *Bull. Am. Ass. Petrol. Geol.*, 58: 1857—1861.

Davis, J.B., 1967. *Petroleum microbiology*. Elsevier, Amsterdam, 604 pp.

Deroo, G., Tissot, B., McCrossan, R.G. and Der, F., 1974. Geochemistry of the heavy oils of Alberta. In: L.V. Hills (Editor), Oil sands, fuel of the future. *Mem. Can. Soc. Petrol. Geol.*, 3: 148—167.

Dickey, P.A., 1975. Possible primary migration of oil from source rock in oil phase. *Bull. Am. Ass. Petrol. Geol.*, 59: 337—345.

Dobryansky, A.F., 1963. La transformation du pétrole brut dans la nature. *Rev. Inst. Fr. Pétrole*, 18: 41—49.

Dow, W.G., 1974. Application of oil-correlation and source-rock data to exploration in Williston basin. *Bull. Am. Ass. Petrol. Geol.*, 58: 1253—1262.

Dow, W.G., 1978. Petroleum source beds on continental slopes and rises. *Bull. Am. Ass. Petrol. Geol.*, 62: 1584—1606.

Dufour, J., 1957. On regional migration and alteration of petroleum in south Sumatra. *Geol. Mijnbouw*, (Nw. Ser.), 19: 172—181.

Gibbs, R.J., 1977. Clay mineral segregation in the marine environment. *J. Sediment. Petrol.*, 47: 237—243.

Gutjahr, C.C.M., 1966. Carbonization measurements of pollen-grains and spores and their application. *Leidse Geol. Meded.*, 38: 1—29.

Hedberg, H.D., 1968. Significance of high-wax oils with respect to genesis of petroleum. *Bull. Am. Ass. Petrol. Geol.*, 52: 736—750.

Hilt, C., 1873. Die Beziehungen zwischen der Zusammensetzung und den technischen Eigenschaften der Steinkohlen. *Z. Ver. Dtsch. Ing.*, Band XVII, Heft 4: 193—202. (Reprinted in 1873 in *Dinglers Polytech. J.*, 208: 424—434.)

Hobson, G.D., 1954. *Some fundamentals of petroleum geology*. Oxford University Press, London, 139 pp.

Hood, A., Gutjahr, C.C.M. and Heacock, R.L., 1975. Organic metamorphism and the generation of petroleum. *Bull. Am. Ass. Petrol. Geol.*, 59: 986—996.

Hubbert, M.K., 1953. Entrapment of petroleum under hydrodynamic conditions. *Bull. Am. Ass. Petrol. Geol.*, 37: 1954—2026.

Hunt, J.M., 1967. The origin of petroleum in carbonate rocks. In: G.V. Chilingar, H.J. Bissell

and R.W. Fairbridge (Editors), *Carbonate rocks*. Elsevier, Amsterdam, pp. 225—251.
Hunt, J.M., 1979. *Petroleum geochemistry and geology*. Freeman, San Francisco, Calif., 617 pp.
Jobson, A., Cook, F.D. and Westlake, D.W.S., 1972. Microbial utilization of crude oil. *Appl. Microbiol.*, 23: 1082—1089.
Jones, R.W., 1981. Some mass balance and geological constraints on migration mechanisms. *Bull. Am. Ass. Petrol. Geol.*, 65: 103—122.
Laplante, R.E., 1974. Hydrocarbon generation in Gulf Coast Tertiary sediments. *Bull. Am. Ass. Petrol. Geol.*, 58: 1281—1289.
Lopatin, N.V., 1971. Temperatura i geologicheskoye vremya kak faktory uglefikatsii. *Izvest. Akad. Nauk SSSR*, Ser. Geol., 3: 95—106.
Lopatin, N.V., 1976. Istoriko-geneticheskiy analiz nefteobrazovaniya s isopol'zovaniyem modeli ravnomernogo nepreryvnogo opuskaniya neftematerinskogo plasta. *Izvest. Akad. Nauk SSSR*, Ser Geol., 8: 93—101.
Lopatin, N.V., 1980. Historic-genetic analysis of oil generation using a model of uniform continuous subsidence of the oil-source layer. *Int. Geol. Rev.*, 22: 193—200.
Louis, M., 1966. Essais sur l'évolution de pétrole à faible température en présence de mineraux. In: G.D. Hobson and M.C. Louis (Editors), *Advances in organic geochemistry 1964*. Pergamon Press, Oxford, pp. 261—278.
Magara, K., 1971. Permeability considerations in generation of abnormal pressures. *J. Soc. Petrol. Engrs.*, 11: 236—242.
McDermott, E., 1940. Geochemical exploration (soil analysis) with some speculation about the genesis of oil, gas, and other mineral accumulations. *Bull. Am. Ass. Petrol. Geol.*, 24: 859—881.
Meinschein, W.G., 1959. Origin of petroleum. *Bull. Am. Ass. Petrol. Geol.*, 43: 925—943.
Meinschein, W.G. 1961. Significance of hydrocarbons in sediments and petroleum. *Geochim. Cosmochim. Acta*, 22: 58—64.
Meissner, F.F., 1978. Petroleum geology of the Bakken Formation, Williston basin, North Dakota and Montana. In: *The economic geology of the Williston basin; Montana, North Dakota, South Dakota, Saskatchewan, Manitoba*. Montana Geol. Soc., Billings, pp. 207—227. (Montana Geol. Soc. 24th Annual Conference; 1978 Williston basin symposium.)
Milner, C.W.D., Rogers, M.A. and Evans, C.R., 1977. Petroleum transformations in reservoirs. *J. Geochem. Explor.*, 7: 101—153.
Philippi, G.T., 1965. On the depth, time and mechanism of petroleum generation. *Geochim. Cosmochim. Acta*, 29: 1021—1049.
Porrenga, D.H., 1966. Clay minerals in Recent sediments of the Niger delta. *Clays Clay Mineral.; Proc. 14th Natl. Conference*, pp. 221—233.
Price, L.C., 1976. Aqueous solubility of petroleum as applied to its origin and primary migration. *Bull. Am. Ass. Petrol. Geol.*, 60: 213—244.
Reed, K.J., 1969. Environment of deposition of source beds of high-wax oil. *Bull. Am. Ass. Petrol. Geol.*, 53: 1502—1506.
Rodionova, K.F. and Maximov, S.P., 1981. *Geokhimiya organicheskogo veshchestva i neftematerinskiye porody fanerozoya*. Nedra, Moscow, 367 pp.
Saikia, M.M. and Dutta, T.K., 1980. Depositional environment of source beds of high-wax oils in Assam basin, India. *Bull. Am. Ass. Petrol. Geol.*, 64: 427—430.
Smoot, T.W. and Narain, K., 1960. Clay mineralogy of pre-Pennsylvanian sandstones and shales of the Illinois basin, Part II — Clay mineral variations between oil-bearing and non-oil bearing rocks. *Ill. State Geol. Surv. Circ.*, 287: 1—14.
Stäuble, A.J. and Milius, G., 1970. Geology of Groningen gas field, Netherlands. In: M.T. Halbouty (Editor), Geology of giant petroleum fields. *Mem. Am. Ass. Petrol. Geol.*, 14: 359—369.

Teichmüller, M., 1958. Métamorphisme du charbon et propection du pétrole. *Rev. l'Indust. Minér.*, Num. Spéc. (Pétrologie Charbons), pp. 99—113.

Teichmüller, M., 1963. Die Kohlenflöze der Bohrung Münsterland 1 (Inkohlung, Petrographie, Verkokungsverhalten). *Fortschr. Geol. Rheinland Westfalen*, 11: 129—178.

Teichmüller, M., 1971. Anwendung kohlenpetrographischer Methoden bei der Erdöl- und Erdgasprospektion. *Erdöl Kohle Erdgas Petroch.*, 24: 69—76.

Teichmüller, M. and Teichmüller, R., 1968. Geological aspects of coal metamorphism. *In:* D. Murchison and T.S. Westoll (Editors), *Coal and coal-bearing strata.* Oliver and Boyd, Edinburgh, pp. 233—268.

Tissot, B., Durand, B., Espitalié, J. and Combaz, A., 1974. Influence of nature and diagenesis of organic matter in formation of petroleum. *Bull. Am. Ass. Petrol. Geol.*, 58: 499—506.

Tissot, B.P. and Welte, D.H., 1978. *Petroleum formation and occurrence.* Springer, Berlin, 538 pp.

Van Andel, Tj. and Postma, H., 1954. Recent sediments of the Gulf of Paria. Reports of the Orinoco Shelf Expedition, vol. 1. *Verh. K. Ned. Akad. Wetensch.*, Afd. Natuurk., Eerste Reeks, XX (5), 245 pp.

Waples, D., 1976. The time-temperature relation in oil genesis: discussion. *Bull. Am. Ass. Petrol. Geol.*, 60: 884—885.

Waples, D.W., 1980. Time and temperature in petroleum formation: application of Lopatin's method to petroleum exploration. *Bull. Am. Ass. Petrol. Geol.*, 64: 916—926.

Weiss, A., 1963. Organic derivatives of mica-type layer-silicates. *Angew. Chem. Int. Edit.*, 2: 134—144.

White, D., 1915. Some relations in origin between coal and petroleum. *Wash. Acad. Sci.*, 5 (6): 189—212.

Winters, J.C. and Williams, J.A., 1969. Microbiological alteration of crude oil in the reservoir. *Preprints of symposia*, Division of Petroleum Chemistry, Am. Chem. Soc., 14 (4): E22—E31.

Zhang Yi-gang, 1981. Cool shallow origin of petroleum — microbial genesis and subsequent degradation. *J. Petrol. Geol.*, 3 (4): 427—444.

ZoBell, C.E., 1964. Geochemical aspects of the microbial modification of carbon compounds. *In:* U. Colombo and G.D. Hobson (Editors), *Advances in organic geochemistry 1962.* Pergamon, Oxford, pp. 339—356.

CHAPTER 11

ORIGIN, MIGRATION AND ACCUMULATION OF PETROLEUM: DISCUSSION

SUMMARY

(1) High-wax crude oil that is found in sediments of inner neritic, paralic or continental facies was probably generated in source rocks stratigraphically close to the reservoir.

(2) Fields in which the composition of the crude oils and water varies from reservoir to reservoir probably received their oil from source rocks stratigraphically adjacent to the reservoir.

(3) Fields and provinces in which petroleum occurrences and petroleum quality are associated with the sand/shale ratio probably derived their petroleum from source rocks interbedded with the reservoirs.

(4) Biodegradation of crude oil is evidence of early generation, dominantly in rocks of continental and paralic environments, rather than marine. The association of heavier crude oils with fresher formation waters suggests that such crudes either have an inherently different composition or tend to be immature.

(5) Density of crude oils is determined largely by facies of the source rock, rather than by temperature and time.

(6) If crude oil generated early is heavier than the later, secondary migration directions may be dominated by water motion because the critical slope may be 10 or 15 times the slope of the potentiometric surface.

(7) Fields with similar crude oils in reservoirs of different facies probably had a source stratigraphically removed from the reservoirs.

(8) Faults generally act as barriers to lateral migration, but can leak if the vertical dimension of the pooled crude oil or gas is sufficient to generate a sufficiently large pressure differential across the fault to another permeable rock unit.

(9) Fields with fault traps and petroleum in the same rock unit in both blocks were either sourced from local source rocks or migration took place during fault movement.

(10) Faults do not generally act as conduits for petroleum migration.

EVIDENCE OF STRATIGRAPHIC POSITION OF SOURCE FOR ACCUMULATIONS

It is a matter of considerable importance to petroleum geology and the industry at large to note that geochemical hypotheses concerning the source rocks of some oil fields and provinces are at variance with geological and logical arguments. Our purpose in this chapter is to concentrate on the discrepancies because it is from these that true relationships and processes will most convincingly be demonstrated eventually. We cannot be satisfied until a coherent geological, geochemical and hydrogeological synthesis for each field emerges. We start on a positive note.

Probably the most convincing case in which geological argument concerning the source, migration and accumulation of significant quantities of oil and gas is not contradicted by geochemical or hydrogeological arguments lies in the North American Silurian reef province of the Michigan basin, and the Devonian reef province of the Western Canada basin.

There are two reasons for stating that the source rock of reef accumulations does not lie in the reef itself:

(1) The contemporary reef environment is one of high energy, rich in oxygen, with much of the organic matter in the food chain. All these are considered inimical to the preservation of large quantities of organic matter.

(2) In groups of reefs, some contain gas, some contain oil and gas, some contain oil and some only water. Groups of reefs probably shared closely comparable environments while they grew, and have since had almost identical geological histories, so it seems inconceivable that the variety of petroleum content could arise from an internal source.

Gussow's (1954) hypothesis of differential entrapment, which arose from his observations in the Western Canada basin, has since been found to apply convincingly to the Michigan basin (Gill, 1979) and to other areas (e.g. West Irian: Trend Exploration Technical Staff, 1973). This hypothesis *requires* for those areas where differential entrapment has occurred that the source is down-dip from the deepest, gas-filled reef, and that the secondary migration path to the shallowest, oil-bearing reef extends at least to the deepest. The progression from lighter to heavier crudes from the deepest to the shallowest oil-bearing reefs may also be a function of the length of the secondary migration path (not necessarily exclusively), with the removal of the more water-soluble components during migration, as well as any effect of de-gassing.

These geological observations require for those areas with differential entrapment that the source of the petroleum is stratigraphically closely related to the accumulations: the stratigraphy suggests that the source rock is the fine-grained mudstones or marls that *over*lie the permeable carrier bed — the permeable carbonate "platform" on which the reefs grew (Fig. 11-1). These source rocks accumulated while the reefs grew, but they accumulated in the still, deep water at some distance from the reefs. Generation, migration and accumulation *could* have begun as soon as the reefs were drowned and closed. In any case, generation and migration probably took place over a considerable span of time as subsidence gradually took the source rocks to temperatures at which generation could begin.

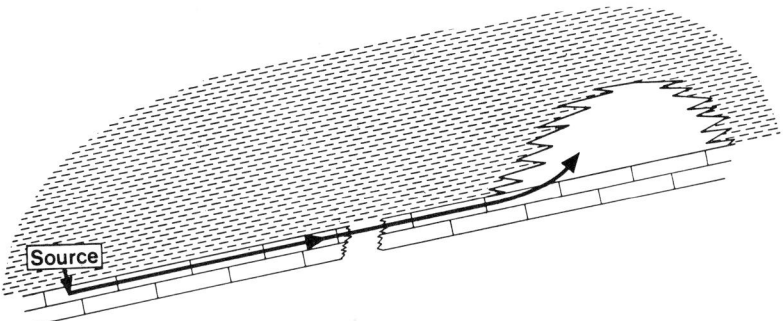

Fig. 11-1. Probable source—accumulation relationship in fossil reefs.

It has long been observed that in fields with multiple sandstone reservoirs heavier oil tends to be at shallower depth and associated with fresher formation waters; lighter oils tend to be at greater depth, associated with more saline formation waters. Bunju field, on the island of that name in northern Kalimantan, Indonesia, is a particularly interesting example of that association.

The shallowest reservoirs, in common with many fields of this type, contain asphaltic crude oil, and the transition to paraffinic crudes occurs at about 800 m. The paraffinic crudes become lighter with depth, and at about 950 m they are 29°API (s.g., 0.84). Below this, they become heavier again until at 1900 m they are 20°API (s.g., 0.88) with an increase in waxy residue from 20 to 45%. Concomitant with this decrease in API gravity, the salinity of the formation waters decreases from 10 to 1.8 g/l, with an increase in carbonate content (Weeda, 1958, pp. 1345—1346). Weeda does not give the temperatures of these reservoirs. The data of Kenyon and Beddoes (1977) suggests that the geothermal gradient is unlikely to be much less than 30°C/km, so we may suppose that the temperature at 1900 m is at least 80°C.

While we cannot eliminate bacteria as a possible cause of the changes of gravity, the facts as presented permit no deep source for the oil. The association of crude oil gravities and formation water salinities strongly suggests an interbedded source, closely related stratigraphically to the reservoirs, with short primary migration paths. The variation in crude oil gravity appears to be due to variation in source material due to variation of facies and the environment in which the source rocks accumulated.

In eastern Venezuela, oil has accumulated in numerous Tertiary sandstone reservoirs, in stratigraphic and fault traps, in several fields. Hedberg et al. (1947), in a classic paper that should still be read, described the area and examined the geological evidence for the position of the source of the oil. Most of the crude oil comes from the Oficina Formation, but not all the sands contain oil or gas. The oil varies greatly in composition, and some of it is waxy. The oil gravities vary also, both within and between reservoirs. The range within reservoirs is probably due to gas de-asphalting (Renz et al., 1958, p. 596; Evans et al., 1971, pp. 154—159). The total ranges of API gravity are listed in Table 11-1: these are for the Oficina Formation in West Guara field, from top to bottom over a depth range from about 1500 to 2500 m. The temperature of the M and N reservoirs, some 240 m above the deepest horizon listed in Table 11-1, is about 88°C (Hedberg et al., 1947, p. 2132, table IX).

TABLE 11-1

Total ranges of API gravity in successive reservoirs in the Oficina Formation, West Guara field, E. Venezuela. (Data from Renz et al., 1958, p. 587, fig. 21)

OF.1	20°—21°	OF.9	33°—51°	D	14°—44°	J	14°—35°	P	10°—15°
OF.5	10°—37°	OF.11	24°—56°	E	13°—43°	L	10°—31°	R	10°—18°
OF.6	32°—48°	OF.13	44°—47°	F	13°—46°	M	10°—26°	S	10°—16°
OF.7	36°—47°	AB	21°—46°	H	14°—44°	N	11°—22°	T	11°—14°
OF.8	22°—50°	C	12°—47°	I	12°—47°	O	11°—22°	U	11°—14°

The salinity of the formation waters also varies, and there is a tendency for the heavier crude oils to be associated with the less saline waters. The increase in crude oil gravity, decrease in API gravity, towards the base of the Oficina Formation coincides with a general decrease in salinity and a trend towards less-marine environments of deposition — that is, locally transgressive.

Hedberg et al. (1947, pp. 2136—2137) are well worth quoting in full:

The writers believe that the multiple-sand character and intricate system of segments and reservoirs in the fields of this area present an exceptional opportunity for the study of matters concerning place of origin, migration, and accumulation of petroleum, not always afforded by simpler single-reservoir fields. Evidence for the origin of the Oficina oil at stratigraphic horizons very close to those in which it is at present found may be summarized as follows.

1. In spite of essentially conformable deposition, essentially identical age, and a common geological history, there is a marked variation in the character of the oils found in different sands within the same trap segment. This variation includes a range in gravities from less than 10° to 57°API, differences in wax content from a negligible amount to more than 15 per cent, differences in dissolved gas values, variations from undersaturated oil to all gas, and differences in color, sulphur content, and other qualities. Moreover, these variations show no more systematic relation to depth than is called for by the general changes in environment of deposition from bottom to top of the Oficina Formation, and marked differences in these qualities are commonly found to occur from one sand to the next in the stratigraphic succession. A striking example is the presence of the highly waxy D and E sand production of the Oficina field in the middle part of the productive section, overlain and underlain by sands producing non-waxy oil.

2. Differences in the salinity of water associated with different sands in any one structural trap are not compatible with extensive migration from one bed to another or mixing of fluids across stratigraphic boundaries.

3. The persistent shale bodies which lie above and below the various sands of the Oficina Formation should theoretically have served as effective barriers to vertical migration between sands.

4. The faults of the Greater Oficina area, while numerous, definitely have acted as barriers rather than avenues of migration (West Guara cross section, Figure 15).

5. Sands of the basal Freites Formation immediately above the Oficina Formation in the present fields invariably are barren of oil. These sands rest conformably on the Oficina Formation and are separated by no greater shale barriers from the Oficina producing sands than separate these latter from each other. These basal Freites sands occupy the same relation to structural traps, are of essentially the same degree of continuity, and are covered by the same seal (Freites shale). It is hard to find any explanation of the absence of petroleum in the Freites Formation in the present fields other than that source material was lacking.

6. The general tendency towards heavier oil with depth (although broken by numerous exceptions) appears to accord with the general tendency toward more brackish and less marine environment of deposition toward the base of the Oficina Formation.

7. Certain sands within the oil-producing section of the Oficina Formation are locally barren of oil, although overlain and underlain by productive sands. The only reason for these differences, provided there is no leak in the fault barrier forming the trap, seems to be in the environment of deposition of the sands or of the adjacent shales.

Evidence for the geographically local origin of oil found at any particular horizon (limited lateral migration) is as follows.

1. There are marked changes in gravity and character of oil in individual sands from one area to another. For example, as mentioned in the previous section, the S sand produces $16°API$ oil in the segment just north of the Oficina fault, $41°API$ oil in the OG-116 area 7 kilometers distant, $30°API$ oil in the OG-187 area, 7 kilometers farther west, $25°API$ oil in YS-17, 13 kilometers still farther west, and $44°API$ wax-oil in the Agua Clara field. Yet electrical-log correlation shows that essentially the same sand is represented in all these cases.

2. Sands in the upper part of the Oficina Formation are barren of oil in the southern and western parts of the Greater Oficina area in spite of favorable structural traps. These same sands contain oil farther north and east where more marine conditions apparently prevailed during the deposition of these beds.

3. Southward and westward from the Greater Oficina area where the whole Oficina Formation becomes definitely less marine, heavy tar-oil replaces the lighter oils known in the Greater Oficina fields.

4. Oil is commonly found in the Greater Oficina area in lenticular sands of small lateral extent. It is difficult to see how it could have reached these sands by any extensive lateral migration.

Considering the evidence outlined, the writers are inclined to believe that the oil produced from the Greater Oficina area had its origin largely in the shales of the Oficina Formation immediately above and below each of the productive sands and that it migrated laterally only moderate distances in these sands within the Greater Oficina area ...

They later note (Hedberg et al., 1947, p. 2138) that some lateral migration is required, and that the fact that almost all accumulations are on the basinward side of faults and other barriers to migration suggested to them that this had been up the regional dip, radially away from the centre of the Eastern Venezuelan basin. Although some of the faults are growth faults (Hedberg et al., 1974, pp. 2115, 2156, 2168 and fig. 18), the main faulting that caused the accumulations occurred later, when there was a kilometre or

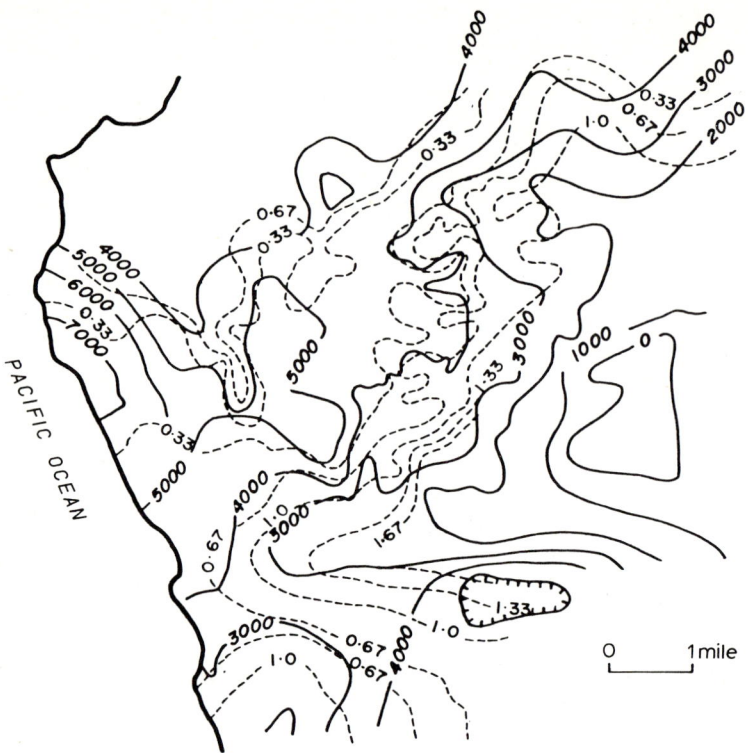

Fig. 11-2. Map of La Brea—Pariñas, Peru, showing relationship between structural elevation in feet (solid lines) and sand/shale ratio (dashed lines). (After Youngquist, 1958, pp. 700—701, figs. 4 and 5.)

so of overburden on the productive Oficina Formation.

These are strong arguments for the source-accumulation relationships of eastern Venezuelan crude oil, and are worth studying because similar arguments are applicable to other areas.

Youngquist (1958) concluded that migration had been short in the La Brea—Pariñas field, Peru, and that the faults, which moved contemporaneously (or nearly so) with sediment accumulation, acted as seals. The oil and gas are found in Eocene sandstone reservoirs at depths between 1000 and 2000 m, but there is no discernable reason for the occurrence of oil, gas, or water, in a particular sand because there is no discernable pattern to their distribution. Nor does structural relief have its usual influence because, in general, the sands in the highest structural positions are wet.

Most remarkable of his findings (and one well worth investigating in other areas) is that the sand/shale ratio has a controlling influence (Fig. 11-2 and 11-3). Sand/shale ratios greater than 1.2 in La Brea—Pariñas are not conducive to the generation and pooling of petroleum, and he concluded that sedimentary facies was the most important control on petroleum occurrence, and that structure was important because it controlled facies. He concluded that the optimum sand/shale ratio in La Brea—Pariñas is about 0.6: at this ratio the reservoirs appear to be just large enough to pool all the oil generated. At smaller ratios, he suggested, the reservoirs are not large enough to pool the oil generated.

The interpretation of the role of the sand/shale ratio is more easily understood in the context of the discussion of Walther's law in Chapter 1 (p. 8) and the assertion made on p. 181 that the facies of a mudstone contiguous with a sandstone is unlikely to be a true source rock. Thus the true source rock, laterally more distant from the reservoir facies, can only appear in the sequence above or below a potential reservoir when the proportion of "shale" is large enough to include it (Fig. 11-4).

Hunt (1953, p. 1862) found a correlation between sand/shale ratios and the gravity of crudes in Wyoming, U.S.A., the lighter crude oils being associated with smaller sand/shale ratios, and concluded from this and other considerations that the major differences between Wyoming crudes are due to source material and the environment of deposition.

Recently, Saikia and Dutta (1980) reported that high-wax crude oils in Tertiary sandstone reservoirs in fields of the Assam basin at depths between two and three kilometres have consistently high specific gravities. There are wide variations in wax content both vertically and laterally, and the variations from field to field could be related to depositional environment. The wax content varies from 8 to 16% by weight, and the API gravities from 22° to 38° (s.g., 0.92—0.83). The sandstone bodies are not persistent, and they concluded that there had been little or no migration, and that the effect of temperature on the quality of the crudes has been "insignificant".

All these examples serve to illustrate the geological arguments that have been applied to areas around the world concerning the source and migration paths of crude oil found in traps. However, there are some important petro-

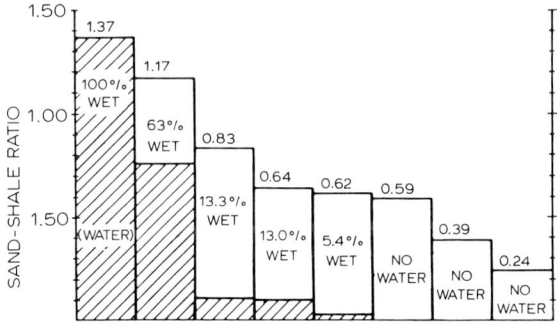

Fig. 11-3. Sand/shale ratios appear to influence the content of reservoirs in La Brea—Pariñas, Peru. (After Youngquist, 1958, p. 711, fig. 10.)

leum provinces in which the geological evidence and the geochemical are in conflict in the matter of source and migration of crude oil.

A striking example of conflicting geochemical and physical evidence is to be found in the Ekofisk oil field in the Norwegian North Sea. The Paleocene/Upper Cretaceous (Danian/Maastrichtian) chalk reservoir is overlain by abnormally pressured Tertiary mudstones, and the reservoir itself is overpressured. Byrd (1975, p. 443, fig. 2) illustrated the pressure regime in the mudstones with the sonic transit time plotted against depth, and concluded that the pressure drop in the mudstones to the reservoir not only reinforces the seal but suggests that the source of the accumulated oil is in the overlying mudstones. Van den Bark and Thomas (1981, p. 2361, fig. 24) also feature this pressure regime, differing from Byrd only in putting the top of abnormal pressures in the Tertiary mudstones rather shallower, at 1036 m (3400 ft).

However, Van den Bark and Thomas (1981, p. 2353) state that the Paleocene mudstones had been postulated as the source because of the similarity of oil extracted from them and the accumulated oil; but that geochemical study had later disqualified them from being petroleum source rocks on the grounds of immaturity (R_o = 0.59—0.62%, at the shallow threshold of thermal maturity). They state in support of this conclusion that the gas chromatography and mass spectral data of the saturate fraction of these extracted oils, from

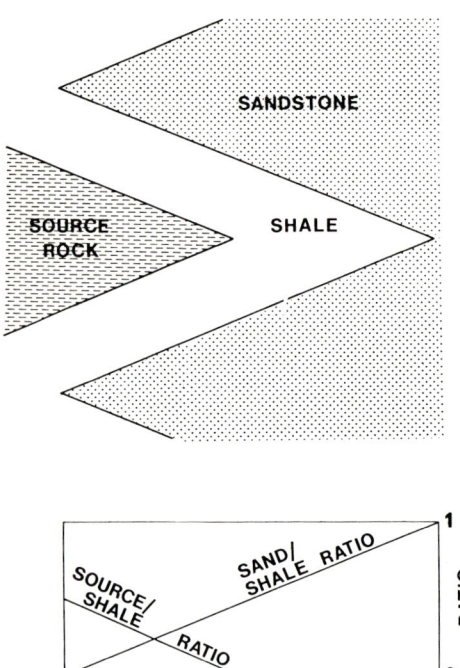

Fig. 11-4. Relationship between sand/shale ratio and source/non-source mudstone ratio.

Paleocene mudstones off-structure, showed both rearranged and unrearranged steranes. This relationship, according to Seifert and Moldowan (1978) is characteristic of relatively immature organic matter. Van den Bark and Thomas drew attention to the Jurassic (Kimmeridge) mudstones encountered in Ekofisk wells, in which the vitrinite reflectance values range from $R_o = 0.93$ to 1.16%, indicating the peak range of thermal maturity for hydrocarbon generation. They also produced the results of other analyses that support the conclusion that the accumulated crude oil has a Jurassic source, showing that Jurassic extracted oil has several features in common with the accumulated oil, while these features are not shared with Paleocene extracted oils. The manner of migration of the Jurassic hydrocarbons "is not yet clear", but they suggested that faults in other parts of Ekofisk have acted as conduits. They attributed the oil in the Paleocene mudstone cap rock to "oil migrating through the section" and concluded that it is not indigenous.

This is a good example of a conclusion that was based on published data being later retracted for reasons that are not published in detail. We are not told of the difficulties in accounting for oil migration from the Jurassic source to the accumulation, but a consequential hypothesis — faults acting as conduits — is introduced to overcome the difficulties. Nor are we told why the "definite evidence" of dissimilarities between accumulated oil and oil in the Tertiary mudstones supports the other consequential hypothesis that the latter oil is migrating through the section.

The physical reasons for postulating the overlying Paleocene source originally have not been refuted, and are worth further examination. The facts appear to be these:

— There is a downward energy gradient from the overlying mudstones to the top of the reservoir,

— There is oil in these mudstones that has "the same general characteristics" as the oil in the accumulation,

— The pressures in the reservoir are well above normal hydrostatic.

If these are indeed facts, they seem to require the following conclusions:

— Crude oil from the reservoir cannot enter the cap rock because it has insufficient energy. This was the conclusion of Byrd (1975, p. 443) and, curiously, also the conclusion of Van den Bark and Thomas ("such a downward pressure drop serves to strengthen the shale caprock . . .": Van den Bark and Thomas, 1981, p. 2357).

Therefore:

— The oil in the Paleocene mudstone cannot be oil migrating *upwards* through the section;

— The downward energy gradient suggests that the Paleocene mudstone is the source of at least some of the accumulated oil, and that this oil could be in primary migration to the accumulation;

— The abnormally high pore pressures in the chalk reservoir indicate that lateral permeability through the chalk unit as a whole is poor, and that therefore a source at higher energy than the accumulation is required in the area of the permeable reservoir.

The importance of resolving this conflict of evidence is obvious, because *either* the geochemical approach is right *or* the physical approach is right: they cannot both be right. By far the more serious of the two possibilities of error is the first, because exploration takes geochemistry more seriously into account than fluid mechanics. If the geochemical concept of maturity of source rock is in error, as seems possible from the Ekofisk data, areas may be dismissed as unprospective when they should not be. Some questions that need answers for Ekofisk are:

(1) If the analytical techniques that discriminated against a Paleocene source for the accumulated oil are valid, why do they not also discriminate against migration "through the section"? Is this migration through the section seen as leakage from the reservoir, or by some other path?

(2) Where is the oil in the Paleocene mudstone going?

(3) If faults are *not* conduits for oil migration, do physical arguments similar to those above preclude a Jurassic source? Does the postulated Jurassic source satisfy the requirement of possessing greater energy than the oil at the oil/water contact, with a path of continuously decreasing energy between them?

(4) Are there two sources for this oil, above and below?

The most important contemporary problem of petroleum geology is, in my view, the reconciliation of geochemistry and fluid mechanics because, once petroleum has been generated, its movement is governed by physical laws within geological constraints. The most fundamental of these is that fluids lose energy while in motion. If the source postulated for an accumulation has less energy than the accumulation, then it must be established that it could have had higher energy at the time of migration, or that there was a source of energy. No real progress can be made in understanding petroleum generation and migration until this central problem is understood.

Ekofisk is not an isolated example. About 400 km north of Ekofisk lies the Frigg gas field, on the Norwegian and United Kingdom boundary. This field, one of the largest offshore gas fields in the world, was found in what has been interpreted as a submarine fan (Héritier et al., 1979, 1980) of Paleocene age, sealed by middle Eocene mudstones. Recoverable reserves are estimated at 7 Tcf (200×10^9 m^3), and the gas is 95.5% methane. There is a 10 m naphthenic oil zone that cannot be economically produced on account of its density (23—24°API, s.g. 0.91—0.92). The pressures are normal hydrostatic.

The crude oil of the main Frigg accumulation is described as "anomalous, suggesting biodegradation caused by bacteria" (Héritier et al., 1979, p. 2018; 1980, p. 78), compounds with carbon numbers less than C_{17} being a minor fraction of the crude oil, and *n*-alkanes are almost absent. This quality is not shared with other accumulations in the area: the East Frigg pool, for example, also has an oil leg, but it contains a significant proportion of *n*-alkanes. If we are correct in our conclusion that such alterations took place during secondary migration, they must have taken place during the last stages before or during accumulation. The anomaly is also a geological one.

Geochemical analyses suggest a Jurassic source in the Dogger and Lias, not the Kimmeridge Clay; and the shallower possibilites are eliminated because "lower Tertiary and Cretaceous source rocks of the Frigg area show only a low degree of diagenesis" (Héritier et al., 1979, p. 2019; 1980, p. 78).

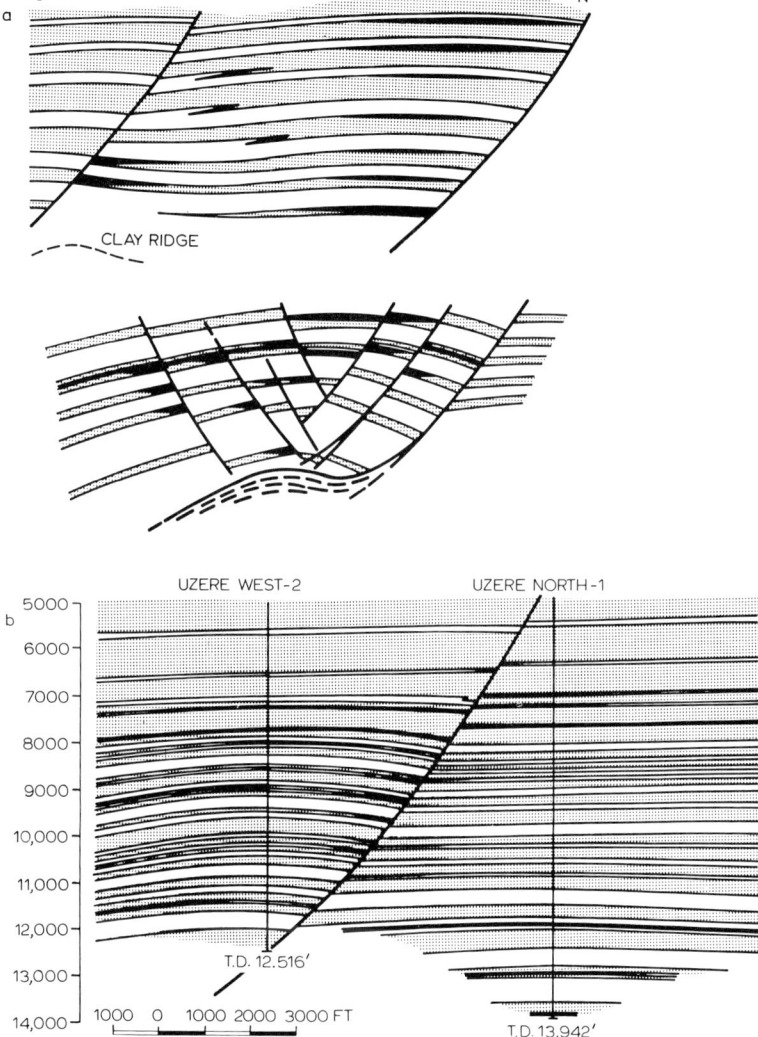

Fig. 11-5. (a) Cross-section through typical Nigerian fields. (After Weber, 1971, p. 561, fig. 3.); (b) Cross-section through Uzere field, Nigeria. (After Short and Stäuble, 1967, p. 777, fig. 8.)

It is very difficult to understand on the data available how the oil and gas migrated to the reservoirs through the Kimmeridge Clay as well as more than one kilometre of Cretaceous mudstones and marls. The pressure regime has not been published, but Héritier et al. (1979, p. 2003; 1980, p. 64) report "soft gumbo clays of early Oligocene to middle Eocene age" overlying the reservoir, and "208 m of dark grey undercompacted shales of Early Cretaceous age" underlying it, so there would appear to be abnormally pressured mudstones both above and below the accumulation. Again, the source must not only have been at higher energy than the accumulation, but also there must have been a path of continuously decreasing energy between the two. There is the further complication in the Frigg accumulations that if Frigg and East Frigg have the same source, the differences in their crude oils must surely have their cause between the two.

The Niger delta is a thick dominantly regressive sequence that, in the simplest terms, has been a delta since Africa and South America parted, and it grew through the Tertiary. Intense exploration activity has discovered about 150 fields, many with multiple reservoirs. These fields are dominantly faulted anticlines, the anticlines being generally roll-over structures on the downthrown side of growth faults (Fig. 11-5). The trends of the faults (as usual with growth faults) are parallel to the depositional strike, reflecting the shape of the growing delta (Fig. 11-6), and they shifted seaward with the regression so that the younger faults are on the downthrown sides of the older.

The stratigraphic sequence has been divided into three formations: the sandy, terrestrial Benin Formation on top, underlain by alternating sands and mudstones of the Agbada Formation from which almost all the production comes, and then the dominantly marine mudstone of the Akata Forma-

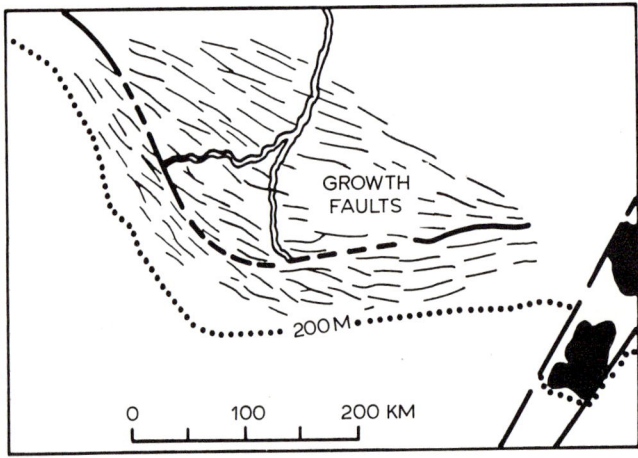

Fig. 11-6. Growth faults in the Niger delta reflect the growth of the delta, younger growth faults being seaward of the older. (After Weber, 1971, p. 560, fig. 2a.)

tion. These three formations are recognized over most of the delta, but of course their ages vary with position, being younger in the directions of progradation. The Akata Formation, a thick mudstone, is generally undercompacted and overpressured.

Fränkl and Cordry (1967) and Short and Stäuble (1967) observed that the crude oils vary from field to field, and also within fields, and concluded that the source rocks are stratigraphically associated with their reservoirs, and that migration has been short. Weber and Daukoru (1975) noted that the faults appeared to be the spill point of many reservoirs, and concluded that the main growth faults had acted as conduits for the migration of oil, and that the main source beds were the deeply buried Akata mudstones lying below the productive Agbada Formation. Evamy et al. (1978), in a detailed study of the delta, came to the conclusion that the interbedded mudstones of the Agbada Formation were the sources in the western delta, but that the main source beds in the east were within the continuous mudstone section of the Akata Formation. This conclusion was based on evidence for a threshold temperature of 115°C (240°F) for main oil generation. Recently, Ekweozor and Okoye (1980), in a preliminary report on a geochemical study, concluded that the main source beds are in the "deeply buried shales" at about 3375 m (11,000 ft) in the onshore delta, and 2900 m (9500 ft) in the offshore delta, below the productive sands, at a threshold temperature of 95°C (203°F) for intense oil generation.

We see that in the Niger delta, geochemical argument has advocated a source to the petroleum that is much deeper than the accumulations, and consequential hypotheses have had to be proposed for the migration of the oil.

Weber and Daukoru's paper, read at the 9th World Petroleum Congress, did not get unqualified support during the panel discussion that followed (*Proc. 9th World Petrol. Congress*, 2: 264—265). Welte queried the assumption of the source's lying in the overpressured Akata Formation mudstones because of the immature nature of the shallower crude oils. Ovanesov said that the consensus of opinion in the USSR favoured interbedded source rocks, and Magnier thought that the Mahakam delta source rocks (East Kalimantan, Indonesia) were probably adjacent to the reservoirs because of the haphazard distribution of oil and gas, and the absence of faults.

We must, of course, accept that it is possible that the Niger delta may have characteristics that differ from those of other deltas, but the fundamental questions are these:

(1) Is it possible to generate from a single source bed crude oils that migrate up faults and accumulate as distinctive crudes in some of the sands but not all of them?

(2) Is our *chemical* concept of maturity sufficiently well established that identification of source rocks on this basis is more reliable than the *geological* reasoning for interbedded source rocks when these are chemically "immature"?

(3) Can petroleum migrate up faults?

The Bomu field, for example, in the Niger delta produces waxy crude oil from a sand at about 2200 m, the oil being of 34.4°API (s.g., 0.853) with 5.1% wax (Fränkl and Cordry, 1967, pp. 204—206). There is a small shallower accumulation, but all the other numerous sands are wet. Taking Ekweozor and Okoye's (1980) figure of 3375 m as the depth of main oil generation (determined in Cawthorne Channel, some 30 km to the WSW from Bomu), the question is, how did oil generated at this depth in Bomu, which is in the Agbada Formation, bypass all the intervening sands, including those adjacent to the source, to reach that at 2200 m?

Evamy et al. (1978, p. 23) reported that crude oils that have a character attributed to biodegradation are separated from lighter, paraffinic crudes consistently by a *temperature* of 65°—80°C. If these crude oils migrated up faults into reservoirs, and if they cannot be biodegraded once they accumulate, then they must have been biodegraded during migration up the fault (unless, as we noted earlier, sufficient bacteria are trapped in the connate water to effect the changes and leave no cell walls).

The two opposing hypotheses, which appear to be mutually exclusive, may be summarized thus:

— In favour of deeper origin and migration up faults in the Niger delta is the chemical evidence of maturity and the coincidence of oil/water contacts with the intersection of faults and the upper surface of many reservoirs,

— In favour of the interbedded sources is the apparent haphazard distribution of crude oils and oil-bearing sands amongst the wet sands, the simplicity of migration and the possibility of biodegradation. Of lesser weight, perhaps, are: the more likely source beds for waxy crudes are in the paralic environment (as Reed, 1969, found), and the undercompacted, overpressured mudstones have yet to expel much of their water (but they may have expelled the oil). In connexion with the latter point, Weber, in the panel discussion mentioned above, said that few commercial accumulations had been found with pore-pressure gradients of more than 0.5 psi/ft (11 kPa/m) from the surface.

There are similar conflicts of views in the U.S. Gulf Coast, the North Sea, and elsewhere, and it is probably true to say that the chemical view of maturity prevails, and the consensus of opinion now favours deep sources in overpressured mudstones, with migration up faults. There are serious hydrodynamic difficulties with sources removed stratigraphically from the accumulations (as discussed on p. 245ff), and the matter of faults as conduits for migration will be taken up below.

The importance of resolving this problem is obvious: if the source is in fact deep, variability in the crudes must be created either during migration or within the reservoir, and the light thrown on this aspect would be of great interest in other contexts. If in fact the source is interbedded, then the concept of maturity — *and so many of the geochemical concepts of petroleum generation* — may be wrong, or seriously misleading. These concepts are widely used to assess areas of petroleum potential, so we may be discarding areas on account of a false impression of immaturity.

Meanwhile, it is essential to understand clearly that petroleum can and does occur in sedimentary sequences that have chemical indications of immaturity. As Hunt (1979, p. 528) wrote, "Considerable research still needs to be done to define the immature—mature boundary with different types of both marine- and land-derived organic matter".

PETROLEUM MIGRATION AND FAULTS

Geochemical evidence for deep source rocks, stratigraphically removed from the accumulations, required an explanation of the migration from the one to the other. During the last decade, it has become almost universally accepted that faults are the conduit for this migration. This hypothesis is a consequence of the geochemical, to a great extent, because the consensus of opinion up to the mid-1960s was that faults generally acted as barriers to petroleum migration. In the mid-1970s, great interest was shown in the hypothesis proposed by Price (1976; and in verbal presentations before that) that petroleum was generated at depths of 10 km or more, and migrated in solution in water up faults. Such ideas are stimulating in themselves, and stimulate further research into the possible processes. But the physical difficulties of fluid migration in fault planes has received little attention.

We must be careful to distinguish between lateral migration *through* faults, from one block to the other, and *in* faults, up and down fault planes (Fig. 11-7). As regards lateral migration through faults, there is good field evidence that many faults do not permit this migration, and that some do. As regards migration in faults, up or down (but mostly up), there is also good field evidence that many do not permit this migration, and that some do. To simplify the discussion, we shall take these two dichotomies separately, starting with lateral migration through a fault.

Any fault trap is evidence that lateral migration is prevented (note the tense) by the fault, but we must distinguish between fault traps that are contained by juxtaposed porous and permeable rock units that could have been carrier beds, from those that are contained by fine-grained, relatively impermeable juxtaposed rock units that could not be carrier beds. We are not concerned for the moment with the latter, but with a fault plane that is itself a barrier to migration due to the fine-grained material or gouge in it.

In general, movement on such faults must have caused a reduction in mean pore size so that the capillary displacement pressure required for further migration exceeds that available in the migrating oil. We can therefore visualize a range of fault-plane textures varying from a clayey fault gouge or smear with very high capillary displacement pressures to one with no significant change of texture from one side to the other. We infer that the trapping capacity of a fault is variable, and over a range of textures the capacity to accumulate will be limited so that only some critical vertical dimension of oil or gas can accumulate.

We do not know what happens when this critical vertical dimension is exceeded, and the pressure within the petroleum exceeds the capillary displacement pressure. Once a continuous oil phase, for example, exists through a fault plane, it is possible that the whole accumulation will drain through, leaving some residual oil in the original reservoir. But if oil migration away from the fault is intermittent, breaking the oil-phase continuity in the fault plane, it is possible (as Smith, 1966, suggested) that the reservoir will be maintained at about its critical vertical dimension, with the surplus passing through. The existence of several fault traps in one field (e.g., Seria field, Brunei, shown in Fig. 15-13) suggests that the latter occurs. In either case, no generalizations can be made except that in areas rich in fault traps, it is probably safe to say that all faults are sealing.

Smith reports evidence that depletion of reservoirs on one side of a fault by production led to the breakthrough of undepleted juxtaposed reservoirs when the differential pressure across the fault reached 1 or 2 MPa (150—300 psi)

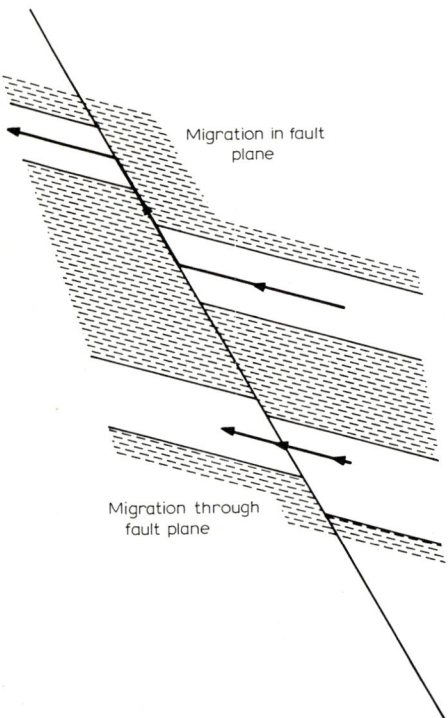

Fig. 11-7. Migration through faults must be distinguished from migration in faults.

(Smith, 1980; and I know of a field with similar evidence). This suggests that there is indeed some critical vertical dimension to the oil accumulation, but that it is relatively large — 500—1000 m — for these faults.

Now, zero effective permeability to oil or gas in the fault plane does not mean zero permeability to water in those parts of the fault plane where aquifers are juxtaposed (but it does mean zero effective permeability to water in the accumulation because the water saturation in the accumulation will be at irreducible minimum). We conclude from this that no large water-pressure discontinuity can be generated in juxtaposed aquifers, so that the effective stress in the aquifer on one side of the fault is not significantly different from that on the other side.

Migration of petroleum (or indeed water) *in* a fault plane, up or down, can only take place in significant quantities under certain conditions:

(a) The fault plane must have permeability.

(b) There must be a potential gradient up or down the fault plane *and in the fault plane.*

(c) The beds on either side of the fault plane must have very low permeability, so that the product $KA\Delta h/l$ within the fault plane is large compared to that in the enclosing beds in a direction parallel to the fault. Note carefully the factor of area, because the area of a fault normal to the direction of flow within the fault is very small compared to the area of sedimentary rocks on either side of the fault.

Evidently these criteria require a tendency for the fault blocks to separate in beds of very low permeability; and that migration of water or petroleum in the fault plane, up or down, will terminate at the first permeable bed in which the pore fluids have less energy than the migrating fluids.

A normal fault is the commonest fault in which there could be a tendency for the blocks to separate, and the stress field around a normal fault under flat topography is such that the greatest principal stress is vertical, and the intermediate and least principal stresses are horizontal. Hubbert (1951, p. 367) showed that the strength of normal sediments is such that the least principal stress is compressive below a few hundred metres at most, and the least compressive stress exceeds the cohesive strength of most sedimentary rocks. Thus it is quite conceivable that faults act as conduits for fluids at shallow depth. Indeed, seepages along faults are common. The question arises, are there any conditions that can lead to important fluid migration in fault planes at the depths of most oil and gas fields?

Secor (1965) concluded that there are. Taking a composite failure envelope (Fig. 11-8) that satisfies both theory and experiment reasonably well, he derived the following expression for the maximum depth at which open fractures could exist in regions where the greatest principal stress is vertical:

$$z_{max} = 8\sigma_0/\rho_b g(1-\lambda) \tag{11.1}$$

where σ_0 is the tensile strength of the material, ρ_b is the mean bulk mass den-

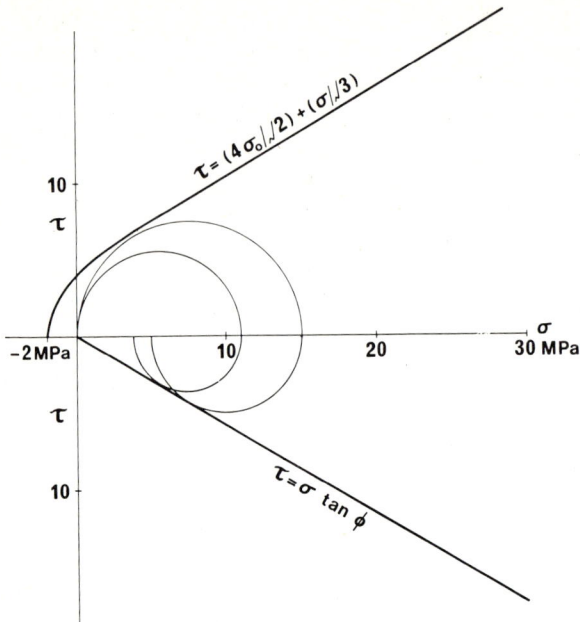

Fig. 11-8. Secor's composite failure envelope for rocks with 2 MPa tensile strength (above the axis); conventional, with zero tensile strength, below.

sity of the overburden, λ is the ratio of pore-fluid pressure to overburden pressure, and the factor 8 is the largest difference between σ_1 and σ_3 in units of σ_0 at which σ_3 can be zero. The requirement for an open fracture is that the greatest and least principal stresses must be such that the circle on which they lie must reach the failure envelope while the least principal stress is negative.

It must be remembered, as Secor was careful to point out, that Mohr diagrams are simplifications of a very complex process, and nowhere is the process more complex than in the region of very small effective stresses. Secor concluded that in areas in which the difference between the greatest and the least principal stresses is small (small circles on the Mohr diagram) and the ratio of pore-fluid pressure to overburden pressure is large, open fractures could be created and exist at great depths, and that previous shear fractures could be opened up.

The existence of open fractures at depth, such as in the Asmari Limestone reservoirs of Iran (Hull and Warman, 1970) leaves no doubt that the argument is qualitatively correct. The difficulty with quantitative assessment lies in determining the value of the tensile strength of rocks in situ.

In the Midland gas field, Louisiana, studied by Fowler et al. (1971), each fault block in Oligocene sedimentary rocks has its own pressure regime, and the values of λ are known quite accurately from reservoir data. We assume that the existence of these separate pressure regimes indicates that the faults in this field are not open conduits, but there could be leakage through the faults.

In the central block, the Hayes gas sand has a value $\lambda = 0.78$ close to a fault at a depth of about 11,300 ft (3445 m). Substituting $z_{max} = 3445$ m, $\rho_b = 2,300$ kg m^{-3}, and $\lambda = 0.8$ into eq. 11.1, we find that σ_0 must be less than 2 MPa (20 bars). [Similarly, Magara's (1968) data lead to values of σ_0 less than 2 MPa for Miocene mudstone in Japan.]

In areas of normal faulting, such as the U.S. Gulf Coast, the stress field is such that the greatest principal stress is vertical, the least and intermediate, horizontal. In the Midland field, the greatest principal effective stress (taking pore-fluid pressure into account) at a depth of 3445 m is:

$$\sigma_1 = \rho_b gz (1 - \lambda) = 15 \text{ MPa}$$

and at about 1000 m, where the pore-fluid pressures on both sides of the fault are normal hydrostatic, $\lambda = 0.5$ and $\sigma_1 = 11$ MPa, approximately.

Figure 11-8 shows the Mohr diagram for the two possible conditions of zero and 2 MPa tensile strength, plotted on either side of the axis. It is clear that smaller differential stresses are required for shear failure without absolute

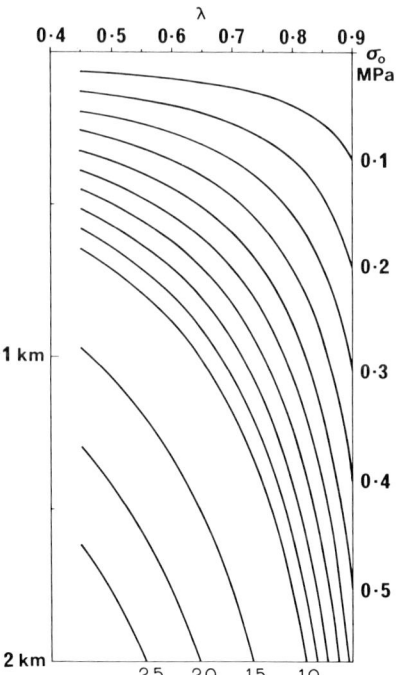

Fig. 11-9. Depths to which open fractures could occur in materials of tensile strengths to 2.5 MPa (25 bars) for various values of λ.

tension (below the line) than for an open fracture to occur. Because fractures exist in this field that are closed, we infer that the tensile strength of the rocks is close to zero.

This inference is not inconsistent with Secor's conclusions. Our conclusion is that in *young* rocks with normal faulting, or incipient normal faulting, the differential stress ($\sigma_1 - \sigma_3$) is too large and the tensile strength of the rocks too small for faults to open and form conduits, or for open faults to be created. If this seems a sweeping conclusion from a single case study, it must be remembered that the many known fault traps also support this conclusion.

Figure 11-9 shows the depths to which open fractures could occur in materials of tensile strengths to 1 MPa (10 bars). It is a matter of scale. Figure 11-10 shows the critical region of the Mohr diagram plotted in units of σ_0. When σ_0 is very small, the value of σ_1 required for open fracture is also very small.

Open fractures, then, appear to be restricted to areas in which the sediments are consolidated, the stress field is close to hydrostatic (in the structural sense of $\sigma_1 \simeq \sigma_3$), and the pore-fluid pressures are close to the overburden pressures. We cannot invoke this process in primary or secondary migration without evidence of these criteria. The commonest geological context of the first two criteria is probably stratigraphic traps in transgressive sequences; but the geological context of the last criterion is generally structural traps in regressive sequences. That the tensile strength of the rocks is a critical parameter is indicated by Hull and Warman's observation that whereas the Asmari limestone has open fractures, sandstones that occur with it do not appear to be fractured (Hull and Warman, 1970, p. 431).

Returning to the problem of source rocks in the Niger delta, we see that there are serious geological and physical objections to a deep source, stratigraphically removed from the accumulations. It seems geologically implausible that the source rocks are far removed stratigraphically from the accumulations, and physically impossible for petroleum to migrate in fault planes

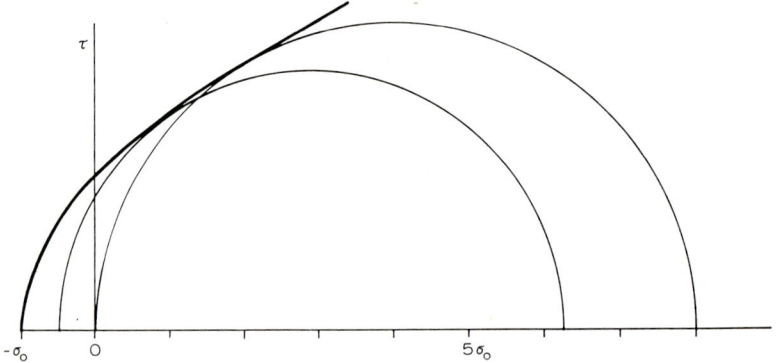

Fig. 11-10. Critical region of the Mohr diagram, plotted in units of σ_0.

past several potential reservoir sands to accumulate in a select few. The coincidence of oil/water contacts with fault intersections at the top of reservoirs, to make a spill point, could perhaps be due to migration through a fault when in juxtaposition with another sand, the oil column exceeding its critical vertical dimension.

Lambert-Aikhionbare (1982) studied the relationship between pore-fluid chemistry and diagenesis in some of the oil-bearing sands of the Agbada Formation, and found that the diagenesis of the oil-bearing part of the sands differed from that of the water-bearing part of the same sands. The sands are dominantly quartz-rich, fine to medium grained, with less than 10% feldspars, clays and other minerals. Both authigenic and detrital clay minerals are present, of which kaolinite is dominant among the authigenic clays. Within the oil-bearing part of a sand, kaolinite only is found, whereas in the water-bearing part the authigenic minerals include kaolinite, siderite, pyrite and calcite, as well as minor amounts of smectite and illite. He concluded, as others have done in other areas, that early entry of the crude oil into the sands inhibited diagenesis. Most of the pore water was displaced, and the connate water that remained could not sustain production of kaolinite, let alone other authigenic minerals.

On present evidence, therefore, the geochemical concepts that led to the postulate of a deep source in the Niger delta may well be incomplete or incorrect — and, by analogy, the same may apply to other areas in which the same concepts have been used with the same result. If this is found to be the case, there will be no *need* to postulate faults as conduits for petroleum migration.

The issue is of immeasurable importance to petroleum geology, and must not be prejudged. Areas in which this conflict of evidence appears to exist must be scrutinized for evidence that will discriminate between two alternatives. It seems impossible that both can be right, and much will be learnt by finding out which is right and which is wrong.

REFERENCES

Byrd, W.D., 1975. Geology of the Ekofisk field, offshore Norway. In: A.W. Woodland (Editor), *Petroleum and the continental shelf of north-west Europe*, Vol. 1, Geology. Institute of Petroleum, London, pp. 439—444.

Campbell, I.R. and Smith, D.N., 1982. Gorgon 1 — southernmost Rankin Platform gas discovery. *J. Aust. Petrol. Explor. Ass.*, 22: 102—111.

Deroo, G., Powell, T.G., Tissot, B. and McCrossan, R.G., 1977. The origin and migration of petroleum in the Western Canada sedimentary basin, Alberta. A geochemical and thermal maturation study. *Bull. Geol. Surv. Can.*, 262, 136 pp.

Ekweozor, C.M. and Okoye, N.V., 1980. Petroleum source-bed evaluation of Tertiary Niger delta. *Bull. Am. Ass. Petrol. Geol.*, 64: 1251—1259.

Evamy, B.D., Haremboure, J., Kamerling, P., Knaap, W.A., Molloy, F.A. and Rowlands, P.H., 1978. Hydrocarbon habitat of Tertiary Niger delta. *Bull. Am. Ass. Petrol. Geol.*, 62: 1—39.

Evans, C.R., Rogers, M.A. and Bailey, N.J.L., 1971. Evolution and alteration of petroleum in western Canada. *Chem. Geol.*, 8: 147—170.

Fowler, W.A., Boyd, W.A., Marshall, S.W. and Myers, R.L., 1971. Abnormal pressures in Midland field, Louisiana. *In:* Houston Geological Society. *Abnormal subsurface pressure: a study group report 1969—1971*. Houston Geological Society, Houston, Texas, pp. 48—77.

Fränkl, E.J. and Cordry, E.A., 1967. The Niger delta oil province: recent developments onshore and offshore. *Proc. 7th World Petrol. Congress*, 2: 195—209.

Gill, D., 1979. Differential entrapment of oil and gas in Niagaran pinnacle-reef belt of northern Michigan. *Bull. Am. Ass. Petrol. Geol.*, 63: 608—620.

Gussow, W.C., 1954. Differential entrapment of oil and gas: a fundamental principle. *Bull. Am. Ass. Petrol. Geol.*, 38: 816—853.

Gussow, W.C., 1968. Migration of reservoir fluids. *J. Petrol. Technol.*, 20: 353—363.

Hedberg, H.D., 1964. Geologic aspects of origin of petroleum. *Bull. Am. Ass. Petrol. Geol.*, 48: 1755—1803.

Hedberg, H.D., 1967. Geologic controls on petroleum genesis. *Proc. 7th World Petrol. Congress*, 2: 3—11.

Hedberg, H.D., Sass, L.C. and Funkhouser, H.J., 1947. Oil fields of Greater Oficina area central Anzoategui, Venezuela. *Bull. Am. Ass. Petrol. Geol.*, 31: 2089—2169.

Héritier, F.E., Lossel, P. and Wathne, E., 1979. Frigg field — large submarine-fan trap in Lower Eocene rocks of North Sea Viking graben. *Bull. Am. Ass. Petrol. Geol.*, 63: 1999—2020.

Héritier, F.E., Lossel, P. and Wathne, E., 1980. Frigg field — large submarine-fan trap in Lower Eocene rocks of the Viking graben, North Sea. *In:* M.T. Halbouty (Editor), Giant oil and gas fields of the decade 1968—1978. *Mem. Am. Ass. Petrol. Geol.*, 30: 59—79.

Hubbert, M.K., 1951. Mechanical basis for certain familiar geologic structures. *Bull. Geol. Soc. Am.*, 62: 355—372.

Hull, C.E. and Warman, H.R., 1970. Asmari oil fields of Iran. *In:* M.T. Halbouty (Editor), Geology of giant petroleum fields. *Mem. Am. Ass. Petrol. Geol.*, 14: 428—437.

Hunt, J.M., 1953. Composition of crude oil and its relation to stratigraphy in Wyoming. *Bull. Am. Ass. Petrol. Geol.*, 37: 1837—1872.

Hunt, J.M., 1979. *Petroleum geochemistry and geology*. Freeman, San Francisco, Calif., 617 pp.

Kenyon, C.S. and Beddoes, L.R., 1977. *Geothermal gradient map of southeast Asia*. Southeast Asia Petroleum Exploration Society, Singapore, and Indonesian Petroleum Association, Jakarta.

Lambert-Aikhionbare, D.O., 1982. Relationship between diagenesis and pore fluid chemistry in Niger delta oil-bearing sands. *J. Petrol. Geol.*, 4 (3): 287—298.

Magara, K., 1968. Compaction and migration of fluids in Miocene mudstone, Nagaoka Plain, Japan. *Bull. Am. Ass. Petrol. Geol.*, 52: 2466—2501.

Philipp, W., Drong, H.J., Füchtbauer, H., Haddenhorst, H.-G., and Jankowsky, W., 1963a. Zur Geschichte der Migration im Gifhorner Trog. *Erdöl Kohle Erdgas Petroch.*, 16: 456—468.

Philipp, W., Drong, H.J., Füchtbauer, H., Haddenhorst, H.-G., and Jankowsky, W., 1963b. The history of migration in the Gifhorn trough (NW-Germany). *Proc. 6th World Petrol. Congress*, Sect. 1: 457—478; discussion, 1: 479—481.

Price, L.C., 1976. Aqueous solubility of petroleum as applied to its origin and primary migration. *Bull. Am. Ass. Petrol. Geol.*, 60: 213—244.

Price, L.C., 1980a. Utilization and documentation of vertical oil migration in deep basins. *J. Petrol. Geol.*, 2 (4): 353—387.

Price, L.[C.], 1980b. Shelf and shallow basin oil related to hot-deep origin of petroleum. *J. Petrol. Geol.*, 3 (1): 91—116.

Reed, K.J., 1969. Environment of deposition of source beds of high-wax oil. *Bull. Am. Ass. Petrol. Geol.*, 53: 1502—1506.

Renz, H.H., Alberding, H., Dallmus, K.F., Patterson, J.M., Robie, R.H., Weisbord, N.E. and MasVall, J., 1958. The Eastern Venezuela basin. In: L.G. Weeks (Editor), *Habitat of oil*. Am. Ass. Petrol. Geol., Tulsa, Okla., pp. 551—600.

Saikia, M.M. and Dutta, T.K., 1980. Depositional environment of source beds of high-wax oils in Assam basin, India. *Bull. Am. Ass. Petrol. Geol.*, 64: 427—430.

Sarkisyan, S.G., 1972. Origin of authigenic clay minerals and their significance in petroleum geology. *Sediment. Geol.*, 7: 1—22.

Secor, D.T., 1965. Role of fluid pressure in jointing. *Am. J. Sci.*, 263: 633—646.

Seifert, W.K. and Moldowan, J.M., 1978. Applications of steranes, terpanes and monoaromatics to the maturation, migration and source of crude oils. *Geochim. Cosmochim. Acta*, 42: 77—95.

Short, K.C. and Stäuble, A.J., 1967. Outline of geology of Niger delta. *Bull. Am. Ass. Petrol. Geol.*, 51: 761—779.

Smith, D.A., 1966. Theoretical considerations of sealing and non-sealing faults. *Bull. Am. Ass. Petrol. Geol.*, 50: 363—374 (Errata in 51: 1427).

Smith, D.A., 1980. Sealing and nonsealing faults in Louisiana Gulf Coast salt basin. *Bull. Am. Ass. Petrol. Geol.*, 64: 145—172.

Smoot, T.W. and Narain, K., 1960. Clay mineralogy of pre-Pennsylvanian sandstones and shales of the Illinois basin, Part II — Clay mineral variations between oil-bearing and non-oil-bearing rocks. *Ill. State Geol. Surv. Circ.*, 287: 1—14.

Trend Exploration Technical Staff, 1973. Reef exploration in Irian Jaya, Indonesia. *Proc. Indon. Petrol. Ass.*, 2: 243—278.

Van den Bark, E. and Thomas, O.D., 1980. Ekofisk: first of the giant oil fields in western Europe. In: M.T. Halbouty (Editor), Giant oil and gas fields of the decade 1968—1978. *Mem. Am. Ass. Petrol. Geol.*, 30: 195—224.

Van den Bark, E. and Thomas, O.D., 1980. Ekofisk: first of the giant oil fields in western Europe. *Bull. Am. Ass. Petrol. Geol.*, 65: 2341—2363.

Webb, J.E., 1974. Relation of oil migration to secondary cementations, Cretaceous sandstones, Wyoming. *Bull. Am. Ass. Petrol. Geol.*, 58: 2245—2249.

Weber, K.J. and Daukoru, E., 1975. Petroleum geology of the Niger delta. *Proc. 9th World Petrol. Congress*, 2: 209—221.

Weeda, J., 1958. Oil basin of East Borneo. In: L.G. Weeks (Editor), *Habitat of oil*. Am. Ass. Petrol. Geol., Tulsa, Okla., pp. 1337—1346.

Weeks, L.G., 1958. Habitat of oil and some factors that control it. In: L.G. Weeks (Editor), *Habitat of oil*. Am. Ass. Petrol. Geol., Tulsa, Okla., pp. 1—61.

Wilson, H.H., 1975. Time of hydrocarbon expulsion, paradox for geologists and geochemists. *Bull. Am. Ass. Petrol. Geol.*, 59: 69—84.

Youngquist, W., 1958. Controls of oil occurrence in La Brea-Pariñas field, northern coastal Peru. In: L.G. Weeks (Editor), *Habitat of oil*. Am. Ass. Petrol. Geol., Tulsa, Okla., pp. 696—720.

Yurkova, R.M., 1970. Comparison of post-sedimentary alterations of oil-, gas- and water-bearing rocks. *Sedimentology*, 15: 53—68.

PART 2. PETROLEUM GEOLOGY OF TRANSGRESSIVE SEQUENCES

CHAPTER 12

FOSSIL CORAL REEFS

SUMMARY

(1) Dominant transgressive sequences tend not to be folded, so this is the geological context of stratigraphic traps, of which fossil reefs are probably the purest form.

(2) Reefs become large petroleum traps when their vertical dimension has been extended by upward growth with a rising sea level, relatively. It seems necessary for them to have grown on a permeable platform that acted as a carrier bed, because the reef organisms themselves were not the source of the petroleum. Ultimately they must be exterminated by drowning and/or smothering by muddy sediment.

(3) Diagenesis has a marked influence on reservoir properties. Dolomitized reefs tend to be more porous and permeable than limestone reef reservoirs, and they are also more common.

(4) Fossil reefs commonly contain very large quantities of recoverable petroleum and are capable of producing it at a great rate. Crude oil quality tends to be rather heavier than that from sandstone reservoirs, with relatively high sulphur content, but no wax.

INTRODUCTION

The most remarkable feature of dominantly transgressive sequences is that they tend not to be folded. *The corollary for petroleum geology is that this is the habitat of stratigraphic traps.* Fossil coral reefs are arguably the purest form of stratigraphic trap; they are strongly associated with transgressive sequences that have almost invariably not been folded, whatever their age, with regional dips commonly less than 5°, many less than 1°. When fossil reefs become petroleum reservoirs, they can contain very large quantities of oil and gas, and individual wells are usually very productive.

We must be careful with terminology because the detailed terminology of present-day reefs is rarely applicable to fossil reefs. There are two separate influences on comparisons between modern reefs and subsurface reefs. First, the dimension of time may exaggerate or blur the morphology that existed in the living reef; and secondly, drilling is unlikely to yield all the information required for a rigorous classification or recognition of features.

The morphological distortions imposed by the dimension of time are of two broad types. Subsidence, which is one essential ingredient of a transgressive sequence, requires the living reef to grow upwards or sideways to maintain its living community at or near sea level (Fig. 12-1). Lateral growth will leave in the stratigraphic record a rock unit that would probably be called a biostrome, although at any time during its development it would have been called a reef. Lateral growth cannot take place once the sea floor around the reef is deeper than the depth tolerance of the colonial organisms. Vertical growth, which is particularly favourable for subsequent petroleum accumulation, distorts the vertical dimension. As noted in Chapter 1, the vertical dimension of many fossil coral reefs, commonly 150—200 m, far exceeds the presumed depth tolerance of the reef-building organisms, and so is evidence of a deepening sea and transgression.

Because petroleum-bearing reefs are commonly prolific, well spacing tends to be larger than for sandstone reservoirs, with the result that the density of information may be rather sparse. Moreover, if the water contact is high in the reef, few wells will penetrate the water-bearing part.

Lowenstam (1950) stipulated that a reef must properly contain organisms that were frame builders, and that these organisms grew to produce a wave-resistant structure, with sediment retention and sediment binding playing an important part. This criterion of biological potential to build a wave-resistant structure, clearly satisfactory for reefs studied in outcrop, may not be determinable in the subsurface (although it may be implied by the morphology). If we insist on the information required to satisfy a definition of reefs in biological, ecological, and morphological terms, many of the so-called reefs are not true reefs (for example, the Capitan Formation in the Guadelupe Mountains of New Mexico; Achauer, 1969).

The strict definition of *coral* reef must, of course, require that corals were important frame-builders. Again, the information obtained from the subsurface may not be sufficient to establish this. Dolomitization and recrystallization may destroy the evidence, and corals are rarely the dominant frame builder. Playford (1969) assessed corals as third in importance as frame builders after algae and stromatoporoids in the Devonian reefs of Alberta and those of Western Australia. Girty's early work on the Guadelupian fauna in New Mexico did not reveal corals (Girty, 1908).

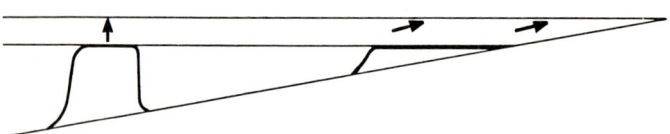

Fig. 12-1. A subsiding reef maintains its living community near sea level by vertical or lateral growth.

These difficulties account for the rather loose usage of the term "reef" in subsurface geological work, with the common term "reef complex" allowing ignorance of detail. It must never be forgotten that the terminology of petroleum geology must take the practical side into account: a "lump" revealed by seismic reflection survey may be termed a reef prospect long before it is drilled. We shall therefore use the term reef for any reef or reef-like carbonate body, acknowledging that careful studies can lead to better definitions in some areas — notably in the Devonian reefs of the Western Canada basin, studied by so many workers (e.g., Barss et al., 1970; Hemphill et al., 1970; Hriskevich et al., 1980).

Reefs are divided into three main zones: fore reef, reef, and back reef (Fig. 12-2). These three zones are broadly both depositional and ecological zones and are recognizable by both lithology and fossil content. The *reef* is the wave-resistant structure. The *fore reef* is the apron of detritus on the outside, or weather side, of the reef and is characterized by an original dip of 30° or more. The *back reef*, on the sheltered side, is contiguous with the reef and consists of organic, bioclastic, and sometimes terrigenous sediments.

Reefs on the continental shelf fall into two main classes: *barrier reefs* and *random* or *patch reefs*. *Fringing reefs* around continental islands or peninsulas also occur in the stratigraphic record. *Pinnacle reefs* are subsurface patch reefs of exaggerated height without inter-reef facies (but the criterion usually

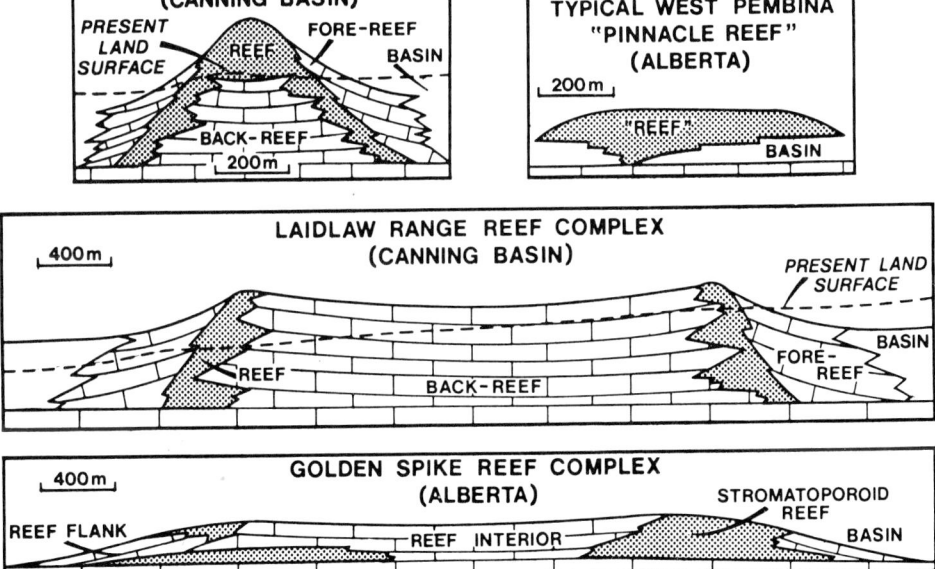

Fig. 12-2. Typical reef forms and their terminology. (Reproduced from Playford, 1982, p. 266, fig. 12, with permission.)

seems to be morphology). The stratigraphic record of reefs is virtually confined to reefs of these types because oceanic reefs on seamounts and around volcanic islands, by their very nature, are unlikely to be preserved in the stratigraphic record — or be recognized if they are.

The foundations of fossil reefs show the same general characters as those of the present day. Devonian reefs in Alberta, Canada, grew on carbonate banks or shelves; some as fringing reefs around the Peace River arch. Some Silurian reefs in the Great Lakes area of North America grew on bioclastic calcarenites and some on limy muds into which the reef mass settled (Shrock, 1939; Lowenstam, 1950). The petroleum potential of a reef seems to be very largely dependent on the existence of a permeable base on which it grew. Reefs without a permeable base rarely contain commercial quantities of petroleum.

The termination of a reef is also important because, to become a petroleum reservoir, it must be covered by a fine-grained, relatively impermeable cap rock. In many cases this occurred by simple drowning of the reef, when reef growth (for one reason or another) failed to keep up with sea level. Mudstones or marls then accumulated in due course, and differential compaction may have draped them over the reef like an anticline (for example, the Cretaceous reefs of Mexico). Some reefs grew to modify their own environment to the point of self-extermination by the creation of hypersaline conditions in which evaporites were deposited and accumulated.

The final ingredient required is a petroleum source rock in hydraulic continuity with the reefs. The environment of a living reef, as discussed earlier, is not favourable for the preservation of organic matter because it is richly oxygenated and has relatively high water energy. Small quantities of petroleum may be generated in situ, as found in Silurian reefs quarried for commercial dolomite near Chicago, but the evidence for a more distant source is compelling. The main source material seems to be situated in the areas of deepening water between reefs or reef trends, or outside them, where euxinic (anoxic) conditions may develop during the transgression. Such source rocks may well be contemporaneous with the reefs, but they will normally accumulate *on top of* the lateral equivalent of the permeable base on which the reefs grew, because this is the nature of the accumulation of sediment in a transgressive sequence. Primary migration will be downwards to the permeable carrier bed, and secondary migration will then be up-dip towards the reef (see Fig. 11-1).

The area of interface between source rock in the loose sense and the carrier bed may be very large and, with the slight relief of the carrier bed, there may be many migration paths. Not all of these will necessarily reach a reef, and not all reefs will necessarily lie on a migration path. In view of the slight relief inferred from the present slight relief, the migration paths may be greatly affected by water flow in the carrier bed; but this will generally be from the deeper parts receiving the waters of compaction to the shallower parts towards the land of the time, and so be towards the reefs in general.

SILURIAN REEFS OF THE GREAT LAKES AREA

Lowenstam (1950) demonstrated that the Silurian (Niagaran) reefs of the Great Lakes area could properly be called reefs because the stromatoporoids not only had the potential to build a wave-resistant structure, but clearly did so. Silurian reefs are exposed at the surface around Lake Michigan and around the north shore of Lake Huron to the Bruce Peninsula, from which the outcrop extends south-east (see Fig. 12-3). Chamberlin (1877) provided the first description of fossil coral reefs in North America from the Silurian reefs of Wisconsin. The subsurface reef trends are to the east of Lake Michigan, around the eastern and southern shores of Lake Huron, in Michigan and Ontario.

In this broad area during Niagaran and earliest Salinian times, there was a general carbonate accumulation with associated evaporites. Along the margin of the Michigan basin, the Silurian reefs seem to have developed into effective barrier reefs, leading to the accumulation of evaporites in the back-reef areas.

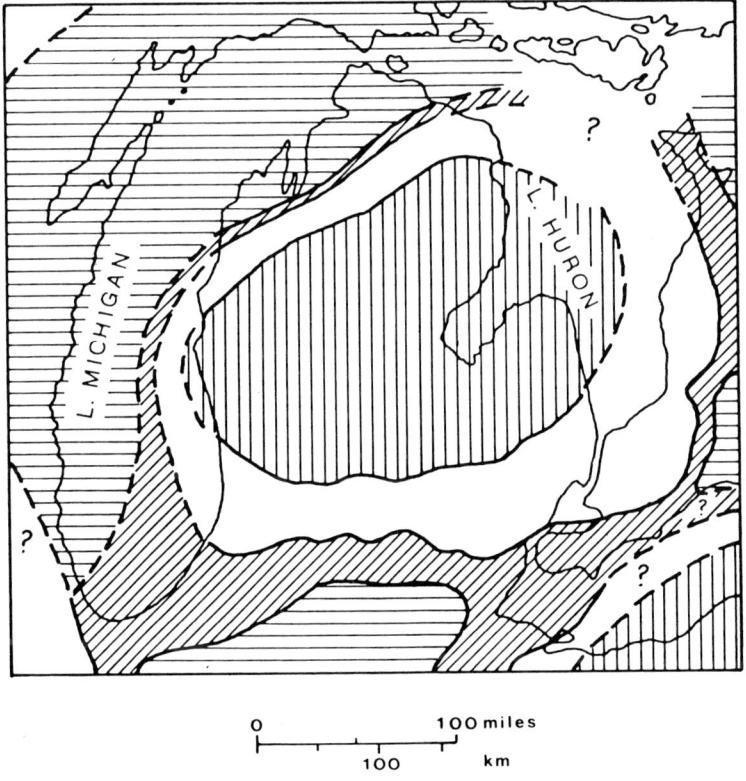

Fig. 12-3. Palaeogeographic map of north-western Michigan basin. Horizontal ornament: platform-shelf carbonates. Diagonal: platform-margin barrier reef. Blank: platform slope pinnacle reefs. Vertical: basinal carbonates. (After Gill, 1979, p. 609, fig. 1.)

Jodry (1969) has pointed out that these lateral facies changes may not have been strictly contemporaneous if the relationship has been obscured by differential compaction. The reefs and associated carbonates have been extensively dolomitized.

A well-documented Niagaran reef is the Thornton reef complex in Illinois (Ingels, 1963) about 15 km south-west of the southern tip of Lake Michigan, which is exposed in a commercial dolomite quarry. It is roughly circular in plan, with an area at the surface of about 5 km^2; and it is about 175 m thick at its maximum development. The foundation of the reef on its southern flank is a bioclastic calcarenite, and Ingels suggested that coral colonies coalesced to form the initial reef growth. Subsequently, rising sea-level led to upward and outward growth. The water depth was inferred to have been about 60 m outside the reef, which implies subsidence of about 120 m relative to sea level during the growth of the reef. The prevailing winds were west-south-westerly (by present geography) and this led to ecological zoning with biotopes and lithotopes trending south-east to north-east.

There are hundreds of subsurface reefs, but these cannot of course be known in the same detail as those that are exposed at the surface, and quarried. First oil production from a reef in Illinois came from the Marine Pool, and this reef was studied by Lowenstam (1948, 1950). Drilling revealed it to have a horseshoe shape, open to the north; and Lowenstam interpreted it to be a reef with detrital bars caused by prevailing southerly winds. Only one well penetrated the reef itself. Around the reef are terrigenous clastics of the Mocassin Springs Formation. The clays in these contiguous sediments are usually red, and Lowenstam suggested that they accumulated below the euphotic ceiling. Near the reef, however, they are green: here they probably accumulated above the euphotic ceiling. The Mocassin Springs sediments are thicker to the north of the reef, suggesting redistribution by wave action to the lee of the reef.

Perhaps the most interesting reef trend is that studied by Gill (1979) to the east of Lake Michigan, in the north-west of the Michigan basin.

The Michigan basin is remarkable in that it is a relatively undeformed area of Middle Silurian carbonates and evaporites, with many hundreds of reefs, that occupies the general area of the Great Lakes of North America. In the central part of the basin (Fig. 12-3) the environments of the carbonates formed concentric rings, one of which consisted of almost continuous reefs — like a barrier reef. Within this ring was another of patch reefs, which was to develop into an area of pinnacle reefs in which petroleum would accumulate. In the central area was deeper water in which carbonates, mainly, accumulated.

The basin was subsiding during the Middle Silurian, the central part more rapidly than the margins, with the result that the height of the pinnacle reefs tends to increase towards the centre of the basin (Fig. 12-4). Despite the subsidence, the ring of continuous reefs appears to have isolated the interior of the basin from the sea because the period of reef growth was followed by the

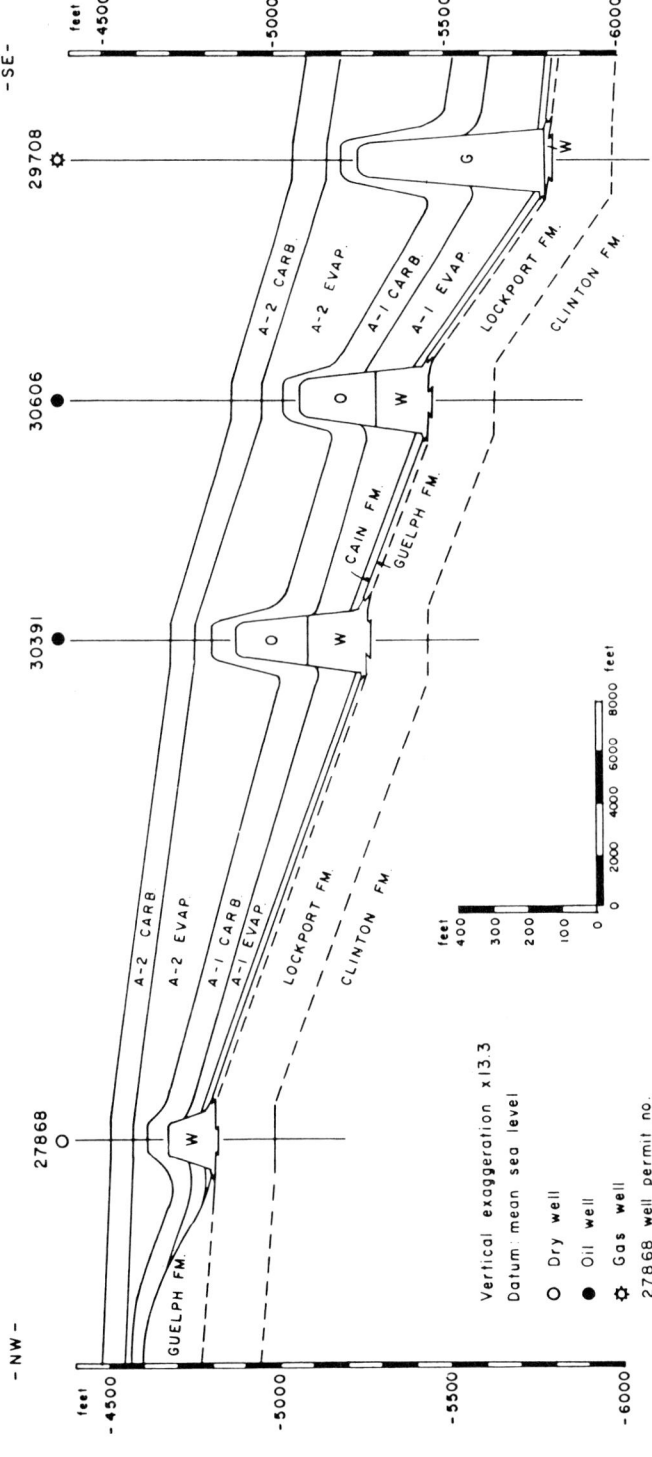

Fig. 12-4. Section through north-western Michigan basin showing tendency for reefs to increase in height with depth. (Reproduced from Gill, 1979, p. 615, fig. 5, with permission.)

accumulation of evaporites. This environmental change seems to be due largerly, if not entirely, to the growth of the encircling reefs.

The north-western belt of pinnacle reefs, studied by Gill (1979) and referred to in Chapter 9 in the context of differential entrapment, grew on a carbonate platform — the Lockport Formation — and euxinic conditions developed in the deeper water in which the Cain Formation accumulated. The Cain Formation is only 7—8 m thick, and consists of a basal dark grey to black argillaceous mudstone with a mean total organic carbon content of 0.3%, followed by cyclic calcite and anhydrite with halite. The carbonate components are rich in bituminous matter.

The reefs were dolomitized and secondary porosity development made them good potential reservoir rocks. They were then sealed by evaporites. The regional dip of the Lockport Formation, on which the pinnacle reefs grew, is less than 2°.

Thus, before the end of the Silurian period all the ingredients of a petroleum province existed: source rock, carrier bed, cap rocks and stratigraphic traps. The size of the traps, had it not been for subsidence, would have been very small because the average basal area of a pinnacle reef here is only 0.3 km^2 (Gill, 1979, p. 612); but subsidence gave them vertical dimensions of 90 to 180 m at depths from 900 to 2100 m. The fluid capacity of the reefs, even then, might have been quite small had it not been for the diagenetic development of porosities averaging about 6%.

The source of the petroleum is clearly not in the reefs themselves because the very strong evidence of differential entrapment and the existence of gas-bearing, oil-bearing and water-bearing reefs of the same type indicate the main source to be down-dip of the petroleum-bearing reefs, and that it generated both oil and gas. The evidence of migration and differential entrapment requires a common carrier bed and a common source that had more energy than the carrier bed and the deepest reservoir. There may have been other contributary sources in some areas, with these energy relationships to particular parts of a carrier bed and reservoir, but the main source must have been in hydraulic continuity with the Lockport Formation. Sharma (1966, p. 347) suggested the underlying Clinton Formation as source rock for the Peters reef in St. Clair county, Michigan; but Gill (1979, p. 618) pointed out that it does not appear to be in a favourable facies in positions that satisfy the requirements for the main source, while the Cain Formation overlying the Guelph and Lockport Formations does.

DEVONIAN REEFS OF WESTERN CANADA

The prolific and widespread reefs of the Silurian Period were followed in the Devonian Period by perhaps even more prolific and widespread reefs around the world (although they were, perhaps, more concentrated during

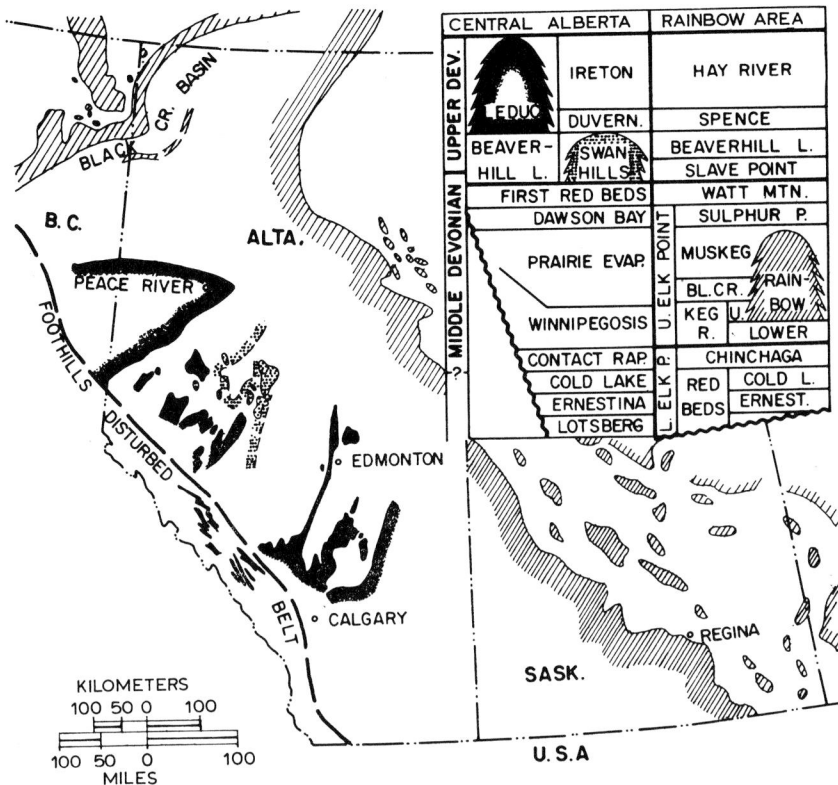

Fig. 12.5. Devonian reefs of the Western Canada basin. (Reproduced, with permission, from Barss et al., 1970, p. 20, fig. 1.)

these periods before the opening of the Atlantic ocean). Of particular importance to petroleum geology are the Devonian reefs of the Western Canada basin. The discovery of oil at Leduc in 1947 was an event of great significance not only to Alberta and Canada, but also to the petroleum industry at large, because it reminded the industry of the great potential of fossil coral reefs and encouraged a wave of research into ancient and modern reefs around the world.

The Western Canada basin (Fig. 12-5) lies east of the Rocky Mountains and extends from the Arctic Islands in the north, through the Northwest Territories and Alberta, to the Williston basin in north-central United States. A prominent feature of the basin is the Peace River arch, in British Columbia and Alberta, which influenced sediment and its accumulation throughout most of the Palaeozoic, forming in late Devonian seas an island of Precambrian granodiorites, schists and gneisses. Around this grew fringing reefs, and cycles of reef rock, detritus and clastic limestones (commonly dolomitic) accumulated (De Mille, 1958).

Throughout the late Middle Devonian and Late Devonian, the basin accumulated a dominantly transgressive sequence of sediments, the transgression progressing generally from north to south. North of the Peace River arch, in middle Devonian times, grew the reefs of the Rainbow—Zama area. These reefs grew to the south of the great Presqu'ile barrier reef along which, to the north-east, occur the lead-zinc deposits of Pine Point on the southern shore of the Great Slave Lake. South of the Peace River arch, in latest Middle and early Late Devonian times, grew the reefs of the Swan Hills area. Further to the south, and a little later, grew the reefs of the Leduc area. In all three areas, important petroleum reserves exist in the reefs themselves, with some in structures formed by differential compaction over them.

This whole area has suffered very little deformation. Dips are generally less than $1°$; and faulting is relatively rare to the east of the foothills, with normal faults usually throwing less than 30 m. Decades of meticulous research have failed to reveal any structural control on the reef trends.

The Presqu'ile barrier reef was a strong influence on the environment, separating an area of terrigenous sediment with some reefs to the north and west (the Mackenzie basin) from the Rainbow—Zama area of carbonates and evaporites with many reefs to the south and east.

To the north and west of the Presqu'ile barrier reef there are isolated reefs with diameters up to about 12 km, associated with marine shales. These have been disappointing from the petroleum point of view (Hriskevich, 1967).

To the south and east of the barrier reef, detailed facies analysis enabled Langton and Chin (1968) to recognize two distinct types of these Middle Devonian (Givetian) reefs: they are either pinnacles (without lagoonal or back-reef facies) or atolls. They grew on a platform of carbonate and argillaceous carbonate of the Keg River Formation, crinoids apparently forming the nucleus to reef growth. Stromatoporoids appear to have been the dominant frame builder, with corals in a subsidiary role. The reef structures are up to about 250 m high, and the rim development suggests that the prevailing wind was from the present north-east. Some original porosity is present, but most is attributable to diagenesis, local differences of which have led to variations in reservoir characteristics. As with the Silurian reefs of the Great Lakes area, the reefs of the Rainbow—Zama area were terminated by environmental changes, probably brought about by the growth of the reefs themselves, that resulted in the deposition and accumulation of evaporites and carbonates that acted as caprocks (Fig. 12-6).

Normal faults with throws up to 25 m are common in the Rainbow area, and the faulting took place during or soon after the reef growth because the faults appear to be confined to the reefs — and could have been caused by their own weight (Barss et al., 1970, pp. 42—44).

In the Rainbow area, the pattern of accumulations has no obvious indication of differential entrapment: there are reefs with a little oil or a little gas, reefs that have oil with associated gas caps, reefs with oil and no gas, reefs

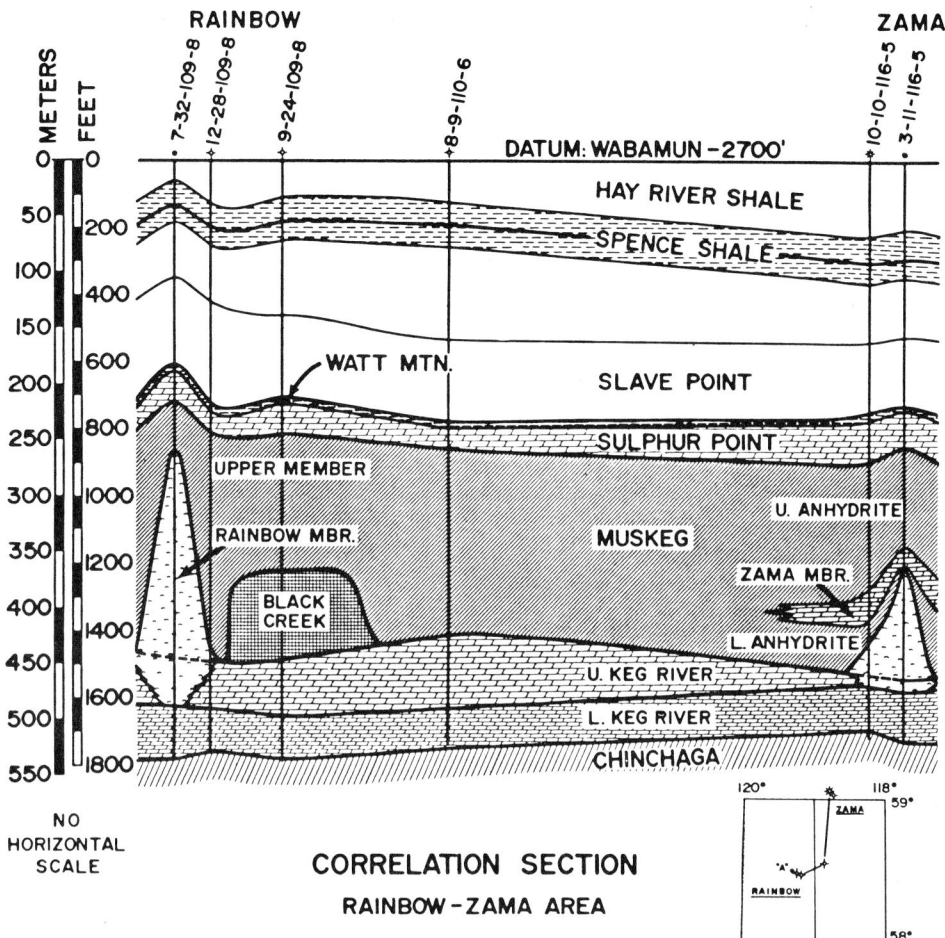

Fig. 12-6. Stratigraphic section through Rainbow-Zama area, Alberta. (Reproduced from Barss et al., 1970, p. 42, fig. 20, with permission.)

nearly full to spill point, and reefs with oil/water contacts above spill point (Fig. 12-7). The reservoirs have a wide range of properties. The source of the oil and gas is probably in contemporaneous strata in euxinic, deeper water facies outside the reef area, with various migration paths.

Some 400 km to the south, and south of the Peace River arch, are the reefs of the Swan Hills area (Fig. 12-8), which have been described by Murray (1966) and Hemphill et al. (1970). These reefs, of latest Middle Devonian to early Late Devonian age (rather younger than those to the north) also grew on a carbonate shelf in normal marine waters, and stromatoporoids were the main frame builders. Some fringed carbonate banks. They grew in a transgressive sea to reach vertical thicknesses up to rather more than 100 m, and their growth was terminated by drowning, probably rather earlier in the north than in the south (Murray, 1966). Naturally, the transgression was locally to the north against the Peace River arch. Deeper-water sediments of the Waterways Formation lie to the east and the north of the Swan Hills reefs, and are thought by some to be the source of the petroleum. The reefs and their associated banks have not been dolomitized. The area is virtually devoid of structural deformation, and the regional dip is about ½° to the south-west.

The reefs of the Swan Hills area form two parallel trends, each with gas in the down-dip reef. This suggests differential entrapment, and therefore a source down-dip to the south-west of the area. In the Kaybob—Snipe trend, the oil gravity changes progressively from 43°API in Kaybob, through 40° in Goose River, to 37° in Snipe Lake. No marked change of gravity is found in the other trend, but a point of interest is that both pools in the Carson Creek North have a small gas cap on undersaturated oil (Hemphill et al., 1970, p. 83). The crudes are all paraffin base, undersaturated, with a low sulphur content (< 5%).

Porosity and permeability are both rather low in these fields, the average porosity being between 6 and 10%, and permeability between 5 and 170 md.

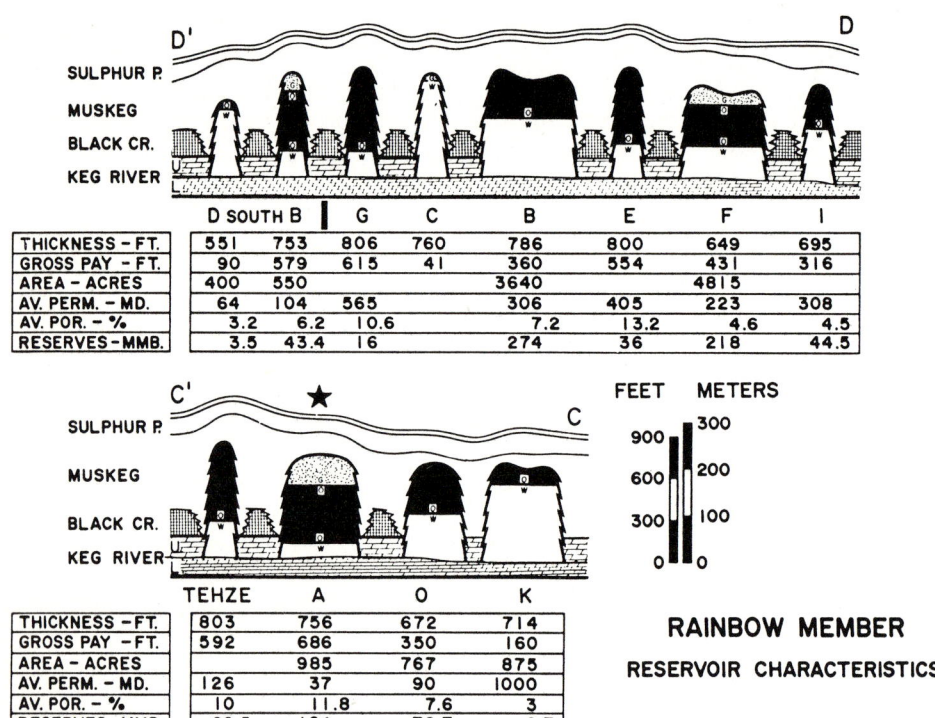

Fig. 12-7. Reservoir variability in the Rainbow—Zama area. (Reproduced from Barss et al., 1970, p. 47, fig. 25, with permission.)

The reservoirs showed a rapid pressure decline, indicating poor communication and a low primary recovery of about 16% of the oil in place (which secondary recovery is expected to double).

The youngest reefs, and the first to be discovered, occur in the area between Calgary and Edmonton. These are of Late Devonian (Frasnian) age, in three major trends running more or less parallel to each other (and to those in the Swan Hills area) in a NNE—SSW direction. The stratigraphic relationships are shown in Fig. 12-9, the production coming almost exclusively from the reef limestones (mostly dolomitized) of the Leduc Formation. Minor, but still important production comes from dolomites of the D-2 reservoir of the Nisku Formation, overlying the Woodbend Group, in anticlines formed by differential compaction and draping over the reefs. Dolomitization has destroyed the original diagnostic features of the reefs, both texture and organic content.

These reefs grew in shallow seas that transgressed towards the south-east over a contemporaneous biostromal carbonate platform (Cooking Lake Formation) that has not been dolomitized. Prevailing winds were from the present north-east. The reefs themselves are generally flat-topped ("table" reefs of Andrichuk, 1958) ranging in size up to about 30 km in greatest horizontal dimension. Due to the transgression and to subsidence, the reef complexes grew to thicknesses between 180 and 300 m. Their growth was terminated

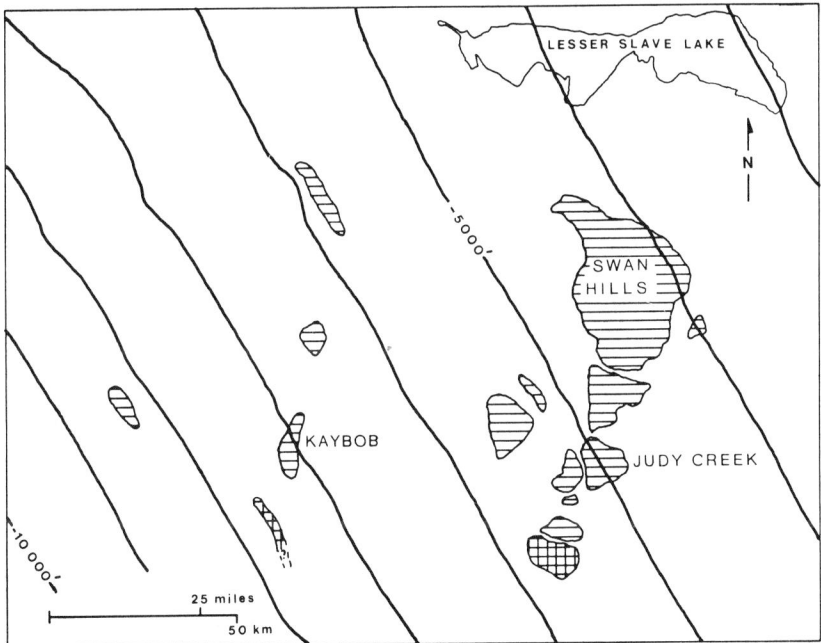

Fig. 12-8. Structure contour map on top of Beaverhill Lake Group, Swan Hills area, Alberta. Contours in feet below sea level. Horizontal ornament, oil; crossed, gas. (Simplified from Hemphill et al., 1970, p. 82, fig. 21.)

by drowning, and they were subsequently covered by grey marine shales and argillaceous limestones of the Ireton Formation.

The reefs seem to have influenced the environment to the south-east of the Stettler area because evaporites become more important in the Duvernay Formation and the equivalents in that direction.

No obvious structural feature caused the trends, and the Leduc is lithologically indistinguishable from the Cooking Lake under the reefs. But the trends themselves are of interest because, like the other reef areas of the Western Canada basin, there is very little deformation.

The Redwater field, discovered in 1948 by seismic reflection survey, produces oil of 34—36°API gravity from the most imporant undolomitized reef. It lies at a depth of about 1000 m, but, because only the top 50 m or so contains oil, little is known of the complex as a whole. The main frame builders in the oil-bearing part of the reef were stromatoporoids, and the biosomes and lithosomes show a general parallelism with the reef trend from NW to SE (Andrichuk, 1958; Klovan, 1964).

Klovan distinguished seven facies: a *Megalodon* (pelecypod) facies along the outer edge, and then successively the facies are, tabular stromatoporoid reef, massive stromatoporoid detritus, skeletal calcarenite, and a back reef facies that encloses an *Amphipora* limestone facies. Corals (*Phillipsastrea* and *Alveolites*) occur in most areas of the reef and appear to have been important frame builders locally. *Alveolites* is found commonly in all but the back-reef facies and so seems to have been able to adjust to the various environments. *Syringopora* and *Syringoporella* are found in the fore-reef, in which original dips of 20° have been recorded (Klovan, 1964).

Golden Spike (west of Leduc) is also undolomitzed, and Duhamel is partly dolomitized. We cannot be sure that the dolomitized reefs were similar in

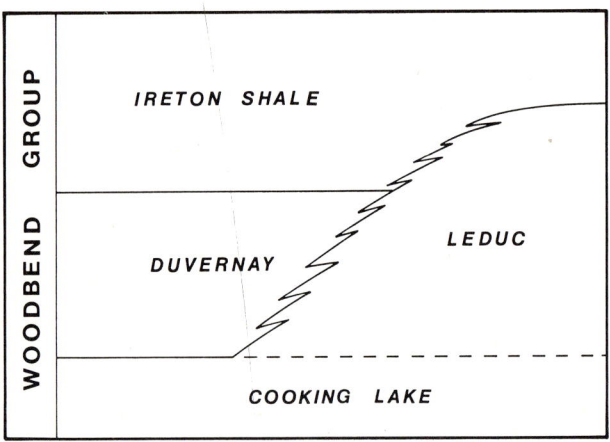

Fig. 12-9. Stratigraphic relationships of formations of the Woodbend Group in the Leduc trend.

organic composition to the undolomitized, but it seems likely that they were, and that while reefs seem to have been prone to dolomitization, it is not related to particular reef environments.

Not all the southern reefs contain petroleum; and the distribution of gas, oil and gas, and oil fields has a pattern that suggests differential accumulation. This was one of the areas that Gussow (1954, 1968) used to illustrate his hypothesis, and was probably the one that suggested it to him. Both the observation of variety and the argument for differential entrapment lead to the conclusion that the source of the petroleum lies outside the reef trends themselves, down-dip, probably in the contemporaneous fine-grained sediments that accumulated in quieter, deeper water (a conclusion supported by Deroo et al., 1977, from geochemical evidence).

The overlying Nisku Formation of the Winterburn Group also contains important reserves of oil (with some associated gas) in anticlines formed by differential compaction over the deeper reefs. Recently these have been added to by the discovery of oil in a reef facies of the Nisku, called the Zeta Lake Member, in the Pembina area to the south-west of Edmonton (Chevron Standard Limited, 1979). These reefs are small in area, flat-topped, and up to 110 m thick. Stromatoporoids appear to be subsidiary to corals and algal encrustations. The pools are small, with recoverable reserves varying from about 160,000 m^3 (10^6 bbl) to about 6.4×10^6 m^3 (40×10^6 bbl). Diagenesis has had a marked influence on reservoir properties. Limestone reservoirs average 3.5% porosity and 300 md permeability, but dolomitized reservoirs average 14.5% porosity and 2.3 darcies permeability. Most pools will be produced from a single well, with another well for pressure maintenance. The source of the oil is considered to be local, either in the Ireton or in the off-reef Nisku strata.

The presence of Nisku reef oil suggests that the other Nisku accumulations have their own source, rather than resulting from spillage from the Leduc Formation reservoirs underlying them.

The Devonian reefs of the Western Canada basin have been important for petroleum geology both in providing a natural laboratory for study and in stimulating studies that have been, and will be applied elsewhere.

CRETACEOUS REEFS OF MEXICO

Mexico was one of the early countries to produce important quantities of crude oil, and her petroleum history is particularly interesting. Their first well, drilled near oil seeps in 1869, ten years after Drake's well in Pennsylvania, found some oil at 28 m. Serious production began in 1904, also as a result of drilling near seeps, and by 1908 it could be said that La Faja de Oro — the Golden Lane — had been discovered (Guzmán, 1967; Viniegra-O. and Castillo-Tejero, 1970). In the following year, a number of prolific wells was drilled, some of which ran wild, culminating in 1916 with the completion of Cerro Azul 4 for nearly 41.500 m^3/day (260,858 bbl/day), setting a world record that probably still stands. In 1919 the first warning signs appeared, with water encroachment in some fields. Mexico had been the world's second largest oil producer after the United States of America, and production peaked in 1921 at 840,240 m^3/day, with a severe decline after that. By 1930, Mexico had fallen to 6th largest oil producing country, and the expropriation of foreign and domestic oil companies in 1938 was followed by 20 years of difficulties (see Hewins, 1957; Young, 1966; Owen, 1975; for interesting accounts of these events).

Although Mexico soon ceased to be a major oil producing country, she was not inactive. The application of improved geological and geophysical techniques led to a number of significant discoveries in the years following the War of 1939—1945, and the Golden Lane was found to extend to the southeast, the extension being called the New Golden Lane. Late in the 1950s, exploration moved offshore, and the discovery of the Marine Golden Lane began in 1963. During the 1970s, Mexico's successful exploration onshore and offshore raised her once again to be one of the world's largest producers of crude oil, with the added advantage of geographical position close to North America and far from the Middle East. By the end of 1980, proven recoverable reserves were estimated at 44×10^9 bbl (7×10^9 m^3), ranking 5th by reserves after Saudi Arabia, Kuwait, USSR and Iran — and well ahead of the U.S.A. Production in 1980 averaged 1.96×10^6 bbl/day (312,000 m^3/day), giving her a reserves/production ratio of 62 : 1 (data from *Oil and Gas Journal*, 78 (52), p. 79).

Important production comes from middle Cretaceous limestones, probably reefal, that form a huge atoll 70 km wide in east—west direction, 130 km long in north—south direction (Fig. 12-10), with a perimeter of about 350 km.

The Old Golden Lane is remarkable not only for its production, but also for the shallow depths from which it was obtained: from 500 m in Cerro Azul to 800 m in San Isidro. Production comes from the El Abra limestone (Albian—Cenomanian) in a series of culminations that are reefal in origin, or erosional. The top surface tends to be karstic, and the limestone is fractured and cavernous (several caves having been found when drilling). The productive trend is very narrow, less than 3 km. The oil columns are thin relative to the 1 km relief on the top of the El Abra, so little is known about the deeper parts of the reef. The cap rock is shale that ranges in age from Early Cretaceous in the central and southern areas to Late Oligocene in the northern Golden Lane. These overlying formations are draped over the reef trend to form an anticline, and this originally led to the belief that the Golden Lane was a tight fold.

The New Golden Lane fields are rather deeper than the old, below 2 km in the south-east, but their character is the same. However, the largest onshore field in Mexico, Poza Rica, discovered in 1930, lies to the west of the New Golden Lane on what is known as the Tamabra trend. "Tamabra" is a corruption of Tamaulipas and Abra, and refers to the facies that is intermediate between the El Abra reefal facies to the east and the dense, micritic Tamaulipas equivalent to the west. This facies is either a fore-reef detrital apron or a local reef growth that did not persist. The general loss of permeability towards the west and the variable permeability within the producing sequence both influence oil accumulation and account for reservoir variation. The Poza Rica accumulation lies off the culmination of its trend, on the north-east flank and south-east down a nose. It is not clear whether the apparently inclined oil/water contact in this field is due to these facies changes or to a strong hydrodynamic flow to the south-east.

The Marine Golden Lane fields lie rather deeper still, between two and three kilometres, and discoveries are still being made. These fields also occur in local culminations of the top surface of the El Abra, but few details have been published.

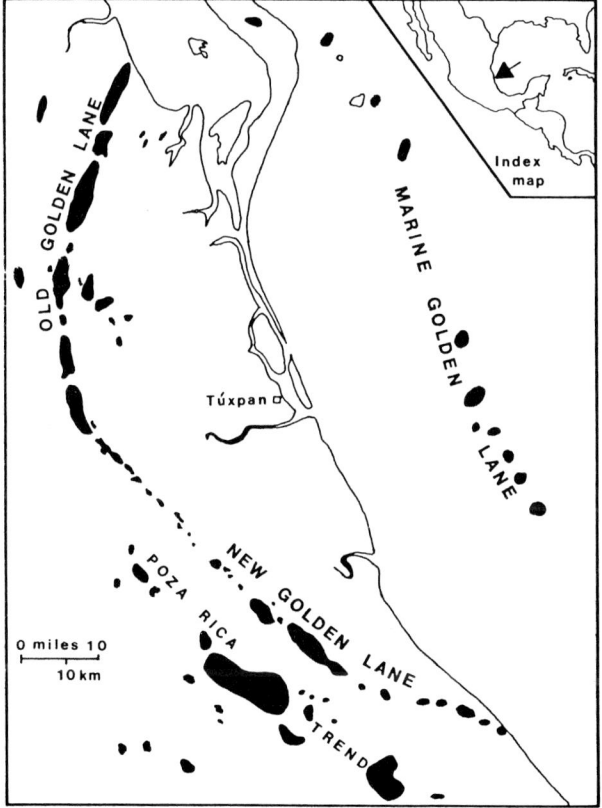

Fig. 12-10. Map of the Old, New, and Marine Golden Lanes, Mexico, showing the large atoll form. (Adapted from Coogan et al., 1972, p. 1422, fig. 2.)

Assuming general contemporaneity around this huge atoll, the present configuration (Fig. 12-10) indicates tilting to the east with a maximum regional dip of about 5° — less on average. There are local patterns to the petroleum distributions, but no general pattern*, so there are probably several source rocks.

PALEOCENE REEFS OF LIBYA

The discovery of oil in a Paleocene reef province of Sirte basin, Libya (Fig. 12-11) in 1967 once again illustrated that fossil reefs could trap significant quantities of crude oil that is producible with relatively few wells. The Idris "A" discovery well, subsequently renamed Intisar, tested 6916 m^3/day (43.500 bbl/day) 43.5° API oil; the Intisar "C" discovery well tested 2798 m^3/day (17,600 bbl/day) [37.5°] API oil; and the Intisar "D" discovery well tested 11,903 m^3/day (74,867 bbl/day) 37.2° API oil from 223 m of reef reservoir (Heatzing and Michel, 1968, p. 1500, table XIV). Of the six reefs found in the area, two (A and D) were found to be giants with 1500 and 1200 million barrels of recoverable oil (240 × 10^6 and 190 × 10^6 m^3), respectively, ranked 76 and 91 among the giants of the world by Halbouty et al. (1970, p. 504, table 1). Three had small accumulations: Intisar "C" was commercial, and "B" and "E" subsequently became so, producing 50° and 34.7° oil, respectively. One was dry. Their location is shown on Fig. 10-14. All were discovered from seismic reflection surveys.

The Intisar "A" discovery well (Terry and Williams, 1969) penetrated the top of the reef at 2870 m and found an oil column of 292 m in an Upper Paleocene reef complex 364 m thick. Appraisal and development drilling showed it to be an oval structure, 5 × 4 km, with a NW—SE orientation. Its recoverable reserves of 240 × 10^6 m^3 were expected to be producible from 42 wells (with about 20 water injection wells for pressure maintenance) from depths between 2700 m and the oil/water contact at about 3065 m below sea level.

The stratigraphy is shown in Fig. 12-12. Terry and Williams (1969) recognized two stages in the growth of the bioherm: first, the development of a foraminiferal bank, loosely bound by algae; then, a period of coral-reef growth that was terminated by drowning or by smothering with argillaceous sediment. The sequence is dominantly transgressive, with two transgressive pulses. The first resulted in the accumulation of a thin calcareous shale on top of the Heira (lower Sabil) dolomite. This shale grades upwards into the Lower Zelten (upper Sabil) carbonates (dense biomicrite) on which formed foraminiferal

---

* See "Mexico Report" in *Oil and Gas Journal*, 77 (34), p. 73.

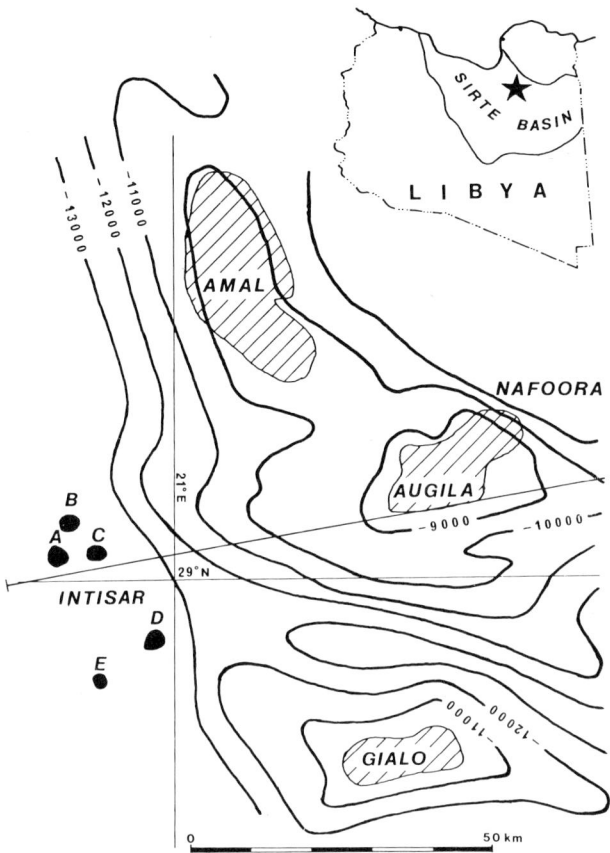

Fig. 12-11. Regional contour map of base Upper Cretaceous sedimentary rocks in the east Sirte basin (contours in feet). The line of the cross-section of Fig. 13-1 passes through Intisar and the southern margin of Augila. (Adapted from Williams, 1972, p. 625, fig. 2, and Brady et al., 1980, p. 544, fig. 1.)

banks up to 200 m thick. The second transgressive pulse motivated the growth of coral reefs, and their ultimate extermination.

Terry and Williams also report that the coral reef member of the bioherm has two mappable units — a coralline biomicrite overlain in general by reef limestone. The coralline biomicrite forms a lens-like body that thins to the edges of the reef from 90 to 105 m in the centre. It has about 15% porosity, but low permeability because much of the porosity is due to solution of coral skeletons and is not effective porosity. The reef limestone is 100—150 m thick. Detrital biomicrite both in coral chambers and coral framework indicates that the corals had a capacity to retain sediment. The porosity of this unit is intergranular and averages about 22%. The reef is overlain by a dense biomicrite, overlain in its turn by the Upper Kheir marl.

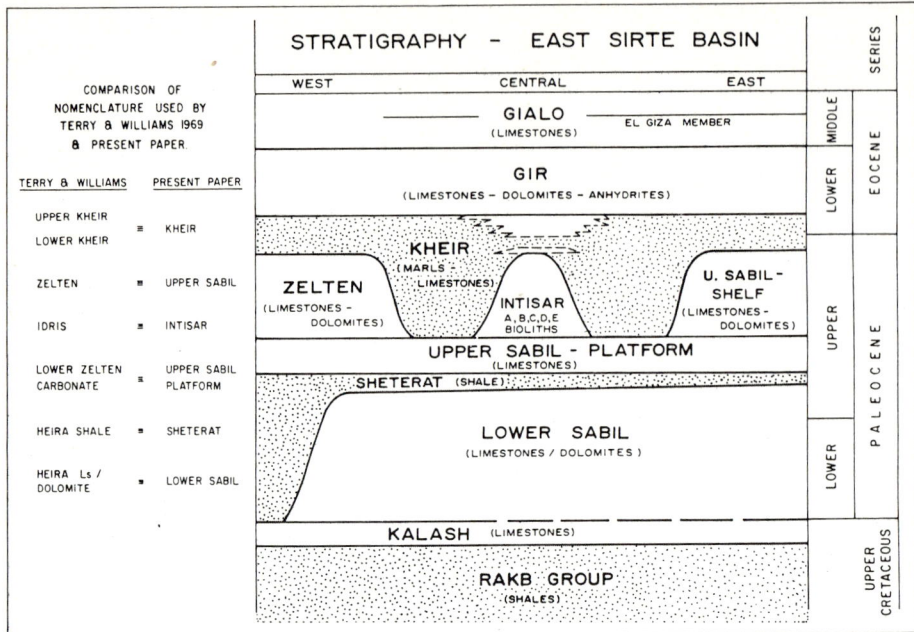

Fig. 12-12. Stratigraphy of the east Sirte basin. Dotted lines are facies variations: heavy lines are generalized time lines. (Reproduced from Brady et al., 1980, p. 547, fig. 4, with permission.)

The Intisar "D" reef (Brady et al., 1980) is of similar size at similar depth, roughly circular in plan with a diameter of about 5 km, and a thickness of 385 m. It differs from the other reefs in the area in that it has a lithologically homogeneous reservoir. However, its permeability varies greatly, from 4 to over 500 md, with an average of at least 200 md indicated by well performances (compared to 87 md arithmetic average of measured horizontal permeablities). Porosities range from 4 to 35%, with a field average of 22%.

The "D" reef is full to spill point with 40°API* undersaturated paraffinic crude, which is rather heavier than the oil in reefs to the north-west, and lighter than that in "E" to the south-west. The oil/water contact is at 2880 m (9450 ft) below sea level, giving a maximum oil column of 291 m (995 ft). Original oil in place is estimated to be 1.8×10^9 bbl (286×10^6 m^3), and the recovery factor may be as high as 75—80% with secondary recovery programmes.

---

* There is usually some variation in reported gravities: the gravities of individual wells and production zones may differ from each other and from the field average of produced crude oil.

An interesting, but unexplained observation of this reef area is that markers are found on the gamma-ray log that are independent of gross lithology yet correlate from reef to reef.

Terry and Williams (1969) considered that the Intisar "A" reef never formed a relief on the sea floor greater than 60 m, which would suggest that transgression and subsidence accounted for about 300 m of the reef's vertical thickness. However, Brady et al. (1980, p. 563) concluded from detailed electrical log correlation between the "A" and "D" reefs that the relief was approximately 240 m (800 ft) when the reefs were terminated. This figure, though large, is not out of proportion to the size of the reefs, and certainly does not require a similar depth tolerance of the reef-building organisms. The combined thinning and differential compaction of the Kheir Formation has given it a relief of about 140 m, and the very varied lithologies of the Kheir suggest local influences. The accumulation of mudstone and marl into the stratigraphic record does not necessarily mean that the water was muddy at shallow depth, because most of this material would have been in the traction load (as we saw in the discussion of growth faults on p. 28). Perhaps the gamma-ray logs provide evidence that has not been reported.

Although some reefs in this province remain to be described, we see that the character of these Paleocene reefs in Libya is much the same as those we have already discussed from other continents and other ages. There is apparently no significant deformation: they are pure stratigraphic traps on a carbonate base.

As regards the source rocks for this oil, Terry (in the discussion that followed the reading of his paper) said that the Kheir marls probably sourced the Intisar "A" oil from an off-reef position. Brady et al. (1980, p. 564) report that source rocks are plentiful, and regard the Sheterat shales, underlying the reefs and platform, as the main source rock, with some contribution from the Kheir marls and Kakb shales. The distribution of undersaturated crudes, lightest in the north and heaviest in the south, is reminiscent of the Western Canada basin, and suggests a northerly source with southerly migration.

The weak water drive of the "A" and "D" accumulations only indicates poor regional permeability on a short time scale. The development of both required water injection for pressure maintenance, and a point of practical interest was the drilling of "dump flood" wells in addition to normal water injection wells. A dump-flood well connects an aquifer to the reservoir directly so that the water drive can be augmented by enlarging the volume of water in hydraulic continuity with the reservoir. This technique requires, of course, that the aquifer potential is at least equal to the undepleted potential of the reservoir, and the vertical distance between the two shall not be very large or temperature reduction may fill the pipe with scale (if a pipe is used) and clog the formation.

REFERENCES

Achauer, C.W., 1969. Origin of Capitan Formation, Guadalupe Mountains, New Mexico and Texas. Bull. Am. Ass. Petrol. Geol., 53: 2314—2323.
Andrichuk, J.M., 1958. Stratigraphy and facies analysis of Upper Devonian reefs in Leduc, Stettler, and Redwater areas, Alberta, Canada. Bull. Am. Ass. Petrol. Geol., 42: 1—93.
Barss, D.L., Copland, A.B. and Ritchie, W.D., 1970. Geology of Middle Devonian reefs, Rainbow area, Alberta, Canada. In: M.T. Halbouty (Editor), Geology of giant petroleum fields. Mem. Am. Ass. Petrol. Geol., 14: 19—49.
Brady, T.J., Campbell, N.D.J. and Maher, C.E., 1980. Intisar "D" oil field, Libya. In: M.T. Halbouty (Editor), Giant oil and gas fields of the decade 1968—1978. Mem. Am. Ass. Petrol. Geol., 30: 543—564.
Chamberlin, T.C., 1877. The geology of eastern Wisconsin. Geol. Wisc., 2 (2): 91—405.
Chevron Standard Limited, 1979. The geology, geophysics and significance of the Nisku reef discoveries, west Pembina area, Alberta, Canada. Bull. Can. Petrol. Geol., 27: 326—359.
Coogan, A.H., Bebout, D.G. and Maggio, C., 1972. Depositional environments and geologic history of Golden Lane and Poza Rica trend, Mexico, an alternative view. Bull. Am. Ass. Petrol. Geol., 56: 1419—1447.
De Mille, G., 1958. Pre-Mississippian history of the Peace River arch. J. Alberta Soc. Petrol. Geol., 6: 61—68.
Deroo, G., Powell, T.G., Tissot, B. and McCrossan, R.G., 1977. The origin and migration of petroleum in the Western Canada sedimentary basin, Alberta. A geochemical and thermal maturation study. Bull. Geol. Surv. Can., 262, 136 pp.
Gill, D., 1979. Differential entrapment of oil and gas in Niagaran pinnacle-reef belt of northern Michigan. Bull. Am. Ass. Petrol. Geol., 63: 608—620.
Girty, G.H., 1908. The Guadalupian fauna. Prof. Pap. U.S. Geol. Surv., 58, 651 pp.
Gussow, W.C., 1954. Differential entrapment of oil and gas: a fundamental principle. Bull. Am. Ass. Petrol. Geol., 38: 816—853.
Gussow, W.C., 1968. Migration of reservoir fluids. J. Petrol. Technol., 20: 353—363.
Guzmán, E.J., 1967. Reef type stratigraphic traps in Mexico. Proc. 7th World Petrol Congress, 2: 461—470.
Halbouty, M.T., Meyerhoff, A.A., King, R.E., Dott, R.H., Klemme, H.D. and Shabad, T., 1970. World's giant oil and gas fields, geologic factors affecting their formation, and basin classification. Part I. Giant oil and gas fields. In: M.T. Halbouty (Editor), Geology of giant petroleum fields. Mem. Am. Ass. Petrol. Geol., 14: 502—528.
Heatzig, G. and Michel, R., 1968. Petroleum developments in North Africa in 1967. Bull. Am. Ass. Petrol. Geol., 52: 1489—1511.
Hemphill, C.R., Smith, R.I. and Szabo, F., 1970. Geology of Beaverhill Lake reefs, Swan Hills area, Alberta. In: M.T. Halbouty (Editor), Geology of giant petroleum fields. Mem. Am. Ass. Petrol. Geol., 14: 50—90.
Hewins, R., 1957. Mr. Five Per Cent; the biography of Calouste Gulbenkian. Hutchinson, London, 254 pp.
Hriskevich, M.E., 1967. Middle Devonian reefs of the Rainbow region of northwestern Canada exploration and exploitation. Proc. 7th World Petrol. Congress, 3: 733—763.
Hriskevich, M.E., Faber, J.M. and Langton, J.R., 1980. Stachan and Ricinus West gas fields, Alberta, Canada. In: M.T. Halbouty (Editor), Giant oil and gas fields of the decade 1968—1978. Mem. Am. Ass. Petrol. Geol., 30: 315—327.
Ingels, J.J.C., 1963. Geometry, paleontology, and petrography of Thornton reef complex. Silurian of northeastern Illinois. Bull. Am. Ass. Petrol. Geol., 47: 405—440.

Jodry, R.L., 1969. Growth and dolomitization of Silurian reefs, St. Clair County, Michigan. *Bull. Am. Ass. Petrol. Geol.*, 53: 957—981.

Klovan, J.E., 1964. Facies analysis of the Redwater reef complex, Alberta, Canada. *Bull. Can. Petrol. Geol.*, 12: 1—100.

Langton, J.R. and Chin, G.E., 1968. Rainbow Member facies and related reservoir properties, Rainbow Lake, Alberta. *Bull. Can. Petrol. Geol.*, 16: 104—143 and *Bull. Am. Ass. Petrol. Geol.*, 52: 1925—1955.

Lowenstam, H.A., 1948. Marine pool, Madison County, Illinois, Silurian reef producer. In: J.V. Howell (Editor), *Structure of typical American oil fields*, Vol. 3. Am. Ass. Petrol. Geol., Tulsa, Okla., pp. 153—188.

Lowenstam, H.A., 1950. Niagaran reefs of the Great Lakes area. *J. Geol.*, 58: 430—487.

Murray, J.W., 1966. An oil producing reef-fringed carbonate bank in the upper Devonian Swan Hills Member, Judy Creek, Alberta. *Bull. Can. Petrol. Geol.*, 14: 1—103.

Owen, E.W., 1975. Trek of the oil finders: a history of exploration for petroleum. *Mem. Am. Ass. Petrol. Geol.*, 6, 1647 pp.

Playford, P.E., 1969. Devonian carbonate complexes of Alberta and Western Australia: a comparative study. *Geol. Surv. W. Aust. Rep.*, 1, 43 pp.

Playford, P.E., 1982. Devonian reef prospects in the Canning basin: implications of the Blina oil discovery. *J. Aust. Petrol. Explor. Ass.*, 22: 258—271.

Sharma, G.D., 1966. Geology of Peters reef, St. Clair County, Michigan. *Bull. Am. Ass. Petrol. Geol.*, 50: 327—350.

Shrock, R.R., 1939. Wisconsin Silurian bioherms (organic reefs). *Bull. Geol. Soc. Am.*, 50: 529—562.

Terry, C.E. and Williams, J.J., 1969. The Idris "A" bioherm and oilfield, Sirte basin, Libya — its commercial development, regional Palaeocene geologic setting and stratigraphy. In: P. Hepple (Editor), *The exploration for petroleum in Europe and North Africa.* Institute of Petroleum, London, pp. 31—48.

Viniegra-O, F., 1981. Great carbonate bank of Yucatán, southern Mexico. *J. Petrol. Geol.*, 3 (3): 247—278.

Viniegra-O, F., and Castillo-Tejero, C., 1970. Golden Lane fields, Veracruz, Mexico. In: M.T. Halbouty (Editor), Geology of giant petroleum fields. *Mem. Am. Ass. Petrol. Geol.*, 14: 309—325.

Williams, J.J., 1972. Augila field, Libya: depositional environment and diagenesis of sedimentary reservoir and description of igneous reservoir. In: R.E. King (Editor), Stratigraphic oil and gas fields — classification, exploration methods, and case histories. *Mem. Am. Ass. Petrol. Geol.*, 16: 623—632.

Young, D., 1966. *Member for Mexico. A biography of Weetman Pearson, First Viscount Cowdray.* Cassell, London, 280 pp.

CHAPTER 13

PALAEOGEOMORPHIC AND UNCONFORMITY TRAPS

SUMMARY

(1) Palaeogeomorphic traps are traps that resulted from the accumulation of sediment over a pre-existing topography, the physiographic expression of which led to facies that could generate reservoir rock and, usually contemporaneously, facies that would eventually act as cap rock. These are almost invariably diachronous, transgressive sequences that include petroleum source rock.

(2) Unconformity traps are those that resulted from the truncation of reservoir rocks and the subsequent sealing of the subcrop by an unconformable, relatively impermeable, fine-grained, rock unit. The source rocks may be within the pre-unconformity sequence, or in the immediate post-unconformity cap rocks. The timing of secondary migration is not, of course, earlier than the time of sealing of the subcrop. There may be a lapse of 50 m.y. or more between the accumulation of the petroleum source rock and the accumulation of its petroleum.

(3) Of particular importance to geology is the occurrence of both types of trap in a world-wide geological context related to the Mesozoic development of rift margins to continents (e.g. North-West Europe, Alaska, Australia) and the development of rift basins that are not parallel to a continental margin (e.g. North Africa). These areas show very similar geological histories, beginning with rifting and growth faulting, followed by transgression and subsidence with little or no fault movement. Commonly the development began in the Permian or Triassic and continued well into the Tertiary, at least. Not all such areas have strictly contemporaneous events, but most show a prolonged "tensional" regime that began long before the opening of the adjacent ocean, and continued while the ocean opened wider.

PALAEOGEOMORPHIC, PALAEOTOPOGRAPHIC TRAPS

About 60 km to the east-north-east of the Intisar fields of Libya, with which we finished the previous chapter, lies the Augila field (see Fig. 12-11) and a good example of a different form of stratigraphic trap that depends on marine transgression over an irregular, usually subsiding, topographic surface. The deepening sea generates sediment from the migrating shore line, the sediment accumulating eventually to form a porous, diachronous reservoir rock that overlies the old topography. As the water deepens, fine-grained material accumulates diachronously over its coarser equivalent; and if the finer-grained material also accumulates petroleum source material, it comes to act both as source rock and cap rock to the carrier bed — the underlying coarser unit. As the transgression proceeds, islands are formed from the hills, and the amount of terrigenous material available decreases; but once an island becomes submerged below some baselevel of energy, the coarse sediment can no longer be generated, and fine-grained sediment from elsewhere (or carbonates, if the environment is right) accumulate over the immersed island. From this point on, the trap is closed. The Augila field is such a trap.

Much of the world experienced a late Cretaceous transgression, an event of great geological significance that is not yet properly understood, and one of great importance to the petroleum reserves of the world. Suess (1888; 1906, p. 290ff) observed that the Cretaceous System has a worldwide transgressive tendency and discussed "the Cenomanian transgression" at some length. This is still a topic of great interest because it suggests a world-wide eustatic event. Detailed study (e.g., Matsumoto, 1977, 1980) has shown that "the Cretaceous transgression" was not a synchronous event around the world. In Libya (and some other parts of the world) it continued well into the Tertiary. In the Sirte basin (see Fig. 12-11) it began in Albian—Cenomanian times and transgressed southwards over land comprising granitic and other igneous rocks, and Palaeozoic sediments, with a topographic relief of some hundreds of metres at least. The Sirte basin was also subsiding, and subsiding irregularly due to contemporaneous movement of faults, so that considerable thicknesses of Upper Cretaceous and Tertiary sediments (mainly carbonates) accumulated (Fig. 13-1). This basin was to become one of the major oil provinces of the world, with estimated proven recoverable reserves of the order of 25×10^9 bbl (4×10^9 m^3) of oil in 1975, only 15 years after the first discovery. Production in the same year averaged 1.47×10^6 bbl/day (235,000 m^3/day), mainly from carbonate reservoirs.

The Augila field (Fig. 13-2) was discovered in 1966 from seismic reflection surveys in an area in which previous work, including drilling and the discovery of the Amal field (see Roberts, 1970), had indicated a large basement high that had probably existed from the early Palaeozoic (Williams, 1972).

The first test found some oil. The second (D1-102) was drilled on another prospect nearly 20 km to the west and tested 2350 m^3/day (14,800 bbl/day) of 36° API oil from a porous limestone 2598—2612 m below the surface. D2 found no limestone on top of the basement, but tested 1213 m^3/day (7627 bbl/day) from devitrified rhyolite and fractured, weathered granophyre. Of the first 11 wells drilled to the D prospect, only two failed to find commercial production (one being wet). The basement was productive in three wells in addition to D2: D5 tested 2248 m^3/day (14,140 bbl/day) from two intervals, 12 m of granite and 18 m of carbonate; D8 tested 2860 m^3/day (18,000 bbl/day) from an open-hole completion in basement and 11 m of

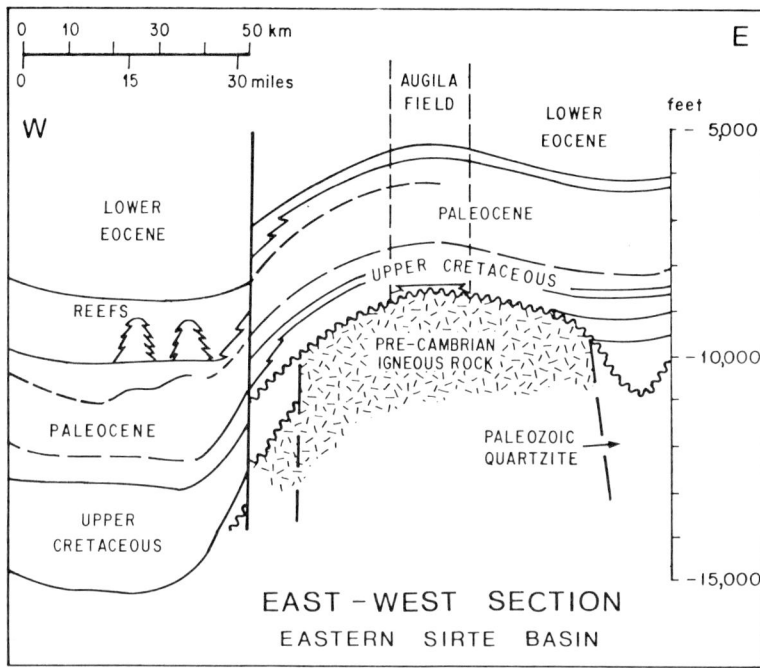

Fig. 13-1. Schematic regional east—west cross section through the Intisar and Augila-Nafoora fields (location shown on Fig. 12-11; after Williams, 1972, p. 626, fig. 3.)

perforations opposite carbonates (most of the oil coming from the basement, according to a flow meter log); and D9 found no sedimentary reservoir but tested 140 m³/day (1500 bbl/day) from weathered granite.

The appraisal drilling of Augila field revealed it to be an Upper Cretaceous stratigraphic, palaeogeomorphic or palaeotopographic trap against a basement high that had formed an area of more than 2600 km² with topographic relief greater than 600 m in late Cretaceous times (see Fig. 12-11). The subaerial relief was steadily reduced by subsidence and the late Cretaceous transgression, so that the ridge became islands that were eventually submerged and covered with fine-grained sediments that became the cap rock.

Nafoora field to the north is similar to Augila, and these two were unitized in 1971 and produced as one field. Halbouty et al. (1970, p. 504, table 1) grouped these with the Amal field and ranked them 27th in the world, with estimated recoverable reserves of 5.2×10^9 bbl (827×10^6 m³).

The diachronous sedimentary rock units of the Augila area consist of the Rachmet shale overlain by the lower Rakb carbonates, with a basal clastic unit that can be regarded as the littoral facies of both the other units. The basal unit is a sandstone composed of material derived from the basement; it is of variable thickness up to 185 m, but 3—10 m in the field (Williams, 1972).

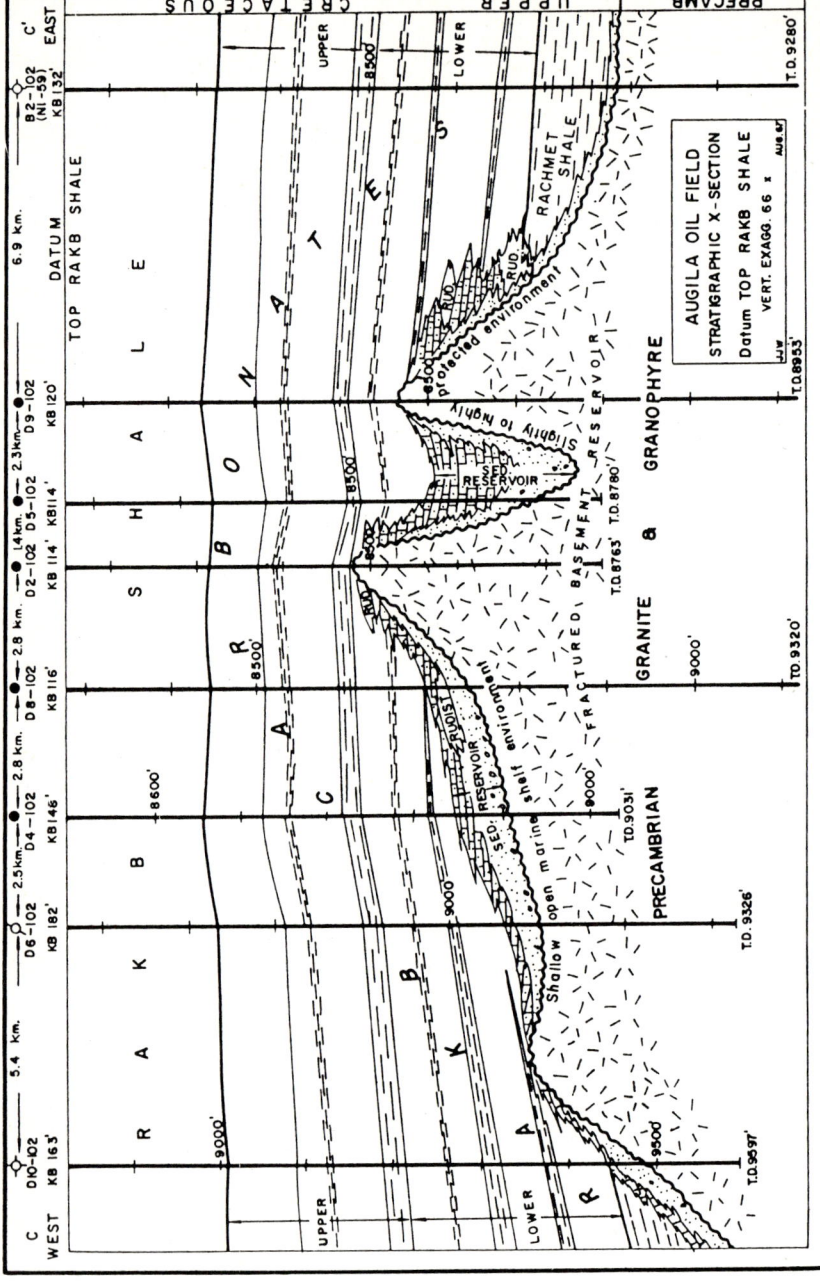

Fig. 13-2. Stratigraphic cross-section through the Augila oil field, Libya. (Reproduced from Williams, 1972, p. 627, fig. 6, with permission.)

Its porosity is small due to poor sorting and clay and carbonate matrix. Its lateral gradation into sandy limestones of the lower Rakb carbonates is, of course, also a vertical one (Walther's law), and the two lithologies form a single reservoir of which the terrestrial component decreases upwards. Three distinct environments are recognized in the carbonates of the reservoir, each relating to its position amongst the islands prior to their submergence and the depth of water prevailing. Porosity in the low-energy zones is due to faunal pelleting of micrites and to solution during diagenesis. In the higher-energy zones, porosity is due to winnowing, with some increase due to solution.

The Rachmet shale did not accumulate in the field area. It is mainly a glauconitic shale with sandstone stringers, with fine-grained sandstone above and below. It is probably the source rock for the oil in areas that were deeper and more tranquil, with primary migration downwards into basal sands and the basement, and secondary migration through them to the traps closed by the accumulation of the upper Rakb carbonates over the whole area. These are argillaceous micrites and calcarous shales of low porosity and permeability. Oil saturation suggests that these may be the source rocks for some of the oil, or for oil that accumulated elsewhere, e.g. Amal (Roberts, 1970, p. 445).

We see from this description of the Augila field, which is based on the informative and perceptive paper by Williams (1972), that the stratigraphic evolution of this trap is entirely consistent with subsidence of an existing subaerial surface, and marine transgression over it. This transgression, while not strictly global at this time, was sufficiently widespread to suggest that subsidence of the Sirte basin was not the only cause. Accepting an organic origin of oil, we can assert that the source rock is in sediments that accumulated while the surface was being inundated or, at the latest, in the immediately overlying sediments. The trap was closed when the Rachmet shale was buried about 100 m. The similarity between Amal, Nafoora and Augila oil suggests that the source for all lay in sediments that accumulated in an euxinic environment — not necessarily in very deep water because Williams (1972, pp. 628 and 632) considered that the Rakb could have accumulated in water as shallow as 35—45 m — and that this favourable environment continued through most of late Cretaceous time. Parts of the upper Rakb are oil-stained, and pyrite is a common filling of foraminifera.

These three fields are not identical, though we may assume that Nafoora and Augila are very similar. Amal differs mainly in that its reservoirs are Palaeozoic fractured quartzose sandstones of the Amal Formation lying unconformably beneath the transgressive, Cretaceous, Maragh Formation and the Rakb Formation that oversteps the Maragh. Sandstones of the Maragh Formation itself form reservoirs in those areas where it exists above the oil/water contact. The basal Maragh is demonstrably derived from the Amal Formation in some areas, so the principal is the same as for Augila. The source rock for the petroleum almost certainly lies stratigraphically above the reservoirs.

On a rather broader scale, favourable conditions for petroleum source rock accumulation continued into the Paleocene for the nearby Intisar fields — probably with some deepening of the water.

UNCONFORMITY TRAPS

The beauty of an unconformity trap is that it exists because of a rather long and sometimes complicated succession of events, each of which is essential for the entrapment of the petroleum. First, an alternating sequence of potential reservoir rocks and cap rocks must accumulate. Then they must be deformed and truncated by an erosion surface. Then the erosion surface must be covered by relatively fine grained sediment that will become the cap rock, and may also be the petroleum source rock. This unconformity must then subside to depths at which petroleum in its source rock (wherever it is) is generated and can be expelled to migrate to the trap. There is no real time limit on these events, but they are unlikely to occur in less than 10 m.y., and may well take 100 m.y. In addition to all these events, the erosion surface must have a configuration that closes the reservoirs into spaces of minimum potential beneath the unconformity, or faults must exist, otherwise any petroleum generated would migrate along the subcrop of the permeable carrier bed and dissipate, or pass through the unconformity to a different type of trap.

Little thought seems to have been given to unconformities since the days of Hutton — largely, perhaps, because in outcrop there is relatively little to see. But subsurface geology provides a clearer picture of both stratigraphy and structure around them, so the study of unconformity traps is of interest to both "pure" and applied geologists.

The word "unconformity" suggests to many geologists subaerial erosion of sedimentary rocks that had suffered orogeny (i.e. uplift) followed by the accumulation of another sequence of sediments, usually marine. This sort of unconformity, though real, is rarely a petroleum trap because the beds immediately overlying the unconformity surface are usually sufficiently porous and permeable for any oil or gas to escape, or accumulate in a different sort of trap. What is required for a petroleum trap is that the post-unconformity sediments shall be fine-grained and very extensive. This requires transgression over a surface of very low relief by a sea with little energy — less than that required, in general, for erosion and removal of the products of erosion — yet a supply of mud. These conditions are also conducive to the preservation of organic matter in the muds. Once these muds have accumulated, further subsidence results in the completion of the post-unconformity sequence.

These conditions were fulfilled over wide areas of the world during the Cretaceous, and more locally at other times (particularly during the late Jurassic and during the early Tertiary), and they are sufficiently peculiar to have deserved more attention that they have received.

The question of petroleum source rocks for unconformity traps is intriguing. There are, of course, only three possibilities: either they are in the pre-unconformity sequence, or in the post-unconformity sequence, or both.

If pre-unconformity source rocks exist, there is one requirement that must be met: generation and primary migration of oil or gas from them must be delayed, at least in part, until the sealing post-unconformity bed has accumulated. This is not a great difficulty provided the erosion leading to the unconformity is not so deep that deep burial before uplift and erosion would be implied. A relatively thin post-unconformity seal will be sufficient to trap the petroleum.

A post-unconformity source cannot, of course, generate any petroleum until the trap already exists, and such generation will occur after substantial subsidence and burial of the unconformity. Primary migration will then take place downwards through the subcrop of porous and permeable beds, and secondary migration will be against the subcrop to the trap. When post-unconformity beds contain a petroleum source rock, they must be rather thicker — at least so that the zone of downward migration embraces the petroleum source rock sensu stricto. The source rock sensu lato may therefore also be the seal. This is not a strict requirement, because the source rock may exist anywhere over the subcrop, but it is very likely because of the requirement of low relief on the unconformity surface. This requirement of low relief also means that very extensive source rocks may exist that will generate very large quantities of petroleum.

There are geological arguments that suggest the location of the source rock relative to the accumulations. First, if there is oil of similar quality in several reservoirs of differing rock type in one trap, there is good reason to postulate a post-unconformity source. Similarity of formation water will reinforce this conclusion. If, on the other hand, there is a variety of crude oils and formation waters, whether in similar or dissimilar reservoir rocks, different sources are indicated, at least one being a pre-unconformity source rock. Of course there will be ambiguities. From a post-unconformity source, petroleum will only enter those permeable rock units that subcrop against it. This need not be in the trap itself, but the corollary is also important: a reservoir that does not subcrop in the trap is not proof of a pre-unconformity source because it may subcrop anywhere where the secondary migration paths will lead to the trap.

The surface of the unconformity is also important in relation to the subcrop because this, in the absence of faulting, determines the trap. It also determines the probable directions of secondary migration, but this is less important. The very size of some unconformity traps suggests a great area of petroleum genesis, but the migration paths are restricted stratigraphically. If the surface of unconformity does not form a perfect seal, this will usually be indicated by concordant oil/water contacts, or gas/water contacts, in the subcropping reservoirs. In such a case, similarity of crude oils is to be expected, and must be taken into account when reasoning back to the source rock.

Any post-unconformity folding will, of course, fold the unconformity and the pre-unconformity sequence. Such deformation varies from the mild "baldheaded" anticline (some of which are demonstrably due to continued diapiric movement, as will be seen in Part 3) to more complicated deformation. But the old saying "You don't find much oil in steep structures" is as true of unconformity traps as of structural traps.

Prudhoe Bay oil field

The Prudhoe Bay oil field on the North Slope of Alaska is of interest not only because it is the largest oil field in North America and situated 400 km (250 miles) north of the Arctic Circle, with more than 600 m of permafrost, but also because it illustrates several of the points discussed above. In its simplest terms, it is an unconformity trap with several Palaeozoic and lower Mesozoic reservoirs of different rock types sealed by a Lower Cretaceous mudstone, the whole gently folded.

Prudhoe Bay oil field was discovered in 1968, but production and development were delayed for nearly ten years by social and political controversy over the construction of the pipeline (see Jamison et al., 1980). It was found to contain 23.5×10^9 bbl of oil in place, with 9.6×10^9 bbl recoverable (3.7×10^9 m³ and 1.5×10^9 m³, respectively) for a recovery factor of about

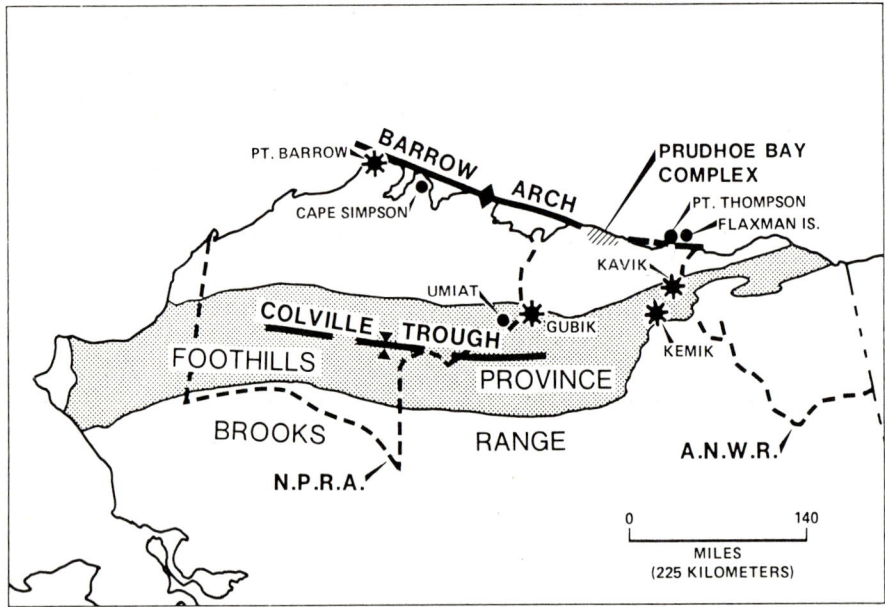

Fig. 13-3. Major structural elements of the North Slope of Alaska. (Reproduced from Jamison et al., 1980, p. 290, fig. 1, with permission.)

40%. Recoverable gas reserves were estimated to be 26 Tcf (736×10^9 m^3). These reserves are over the nose of a major "high" known as the Barrow Arch (Figs. 13-3 and 13-4), which brings Lower Palaeozoic basement rocks to within 800 m of the surface at Point Barrow (Morgridge and Smith, 1972; Jones and Speers, 1976). The accumulations are due not so much to the arch as to the unconformable Cretaceous mudstones that seal the truncated pre-Cretaceous reservoir rocks far down the plunge of the Barrow Arch.

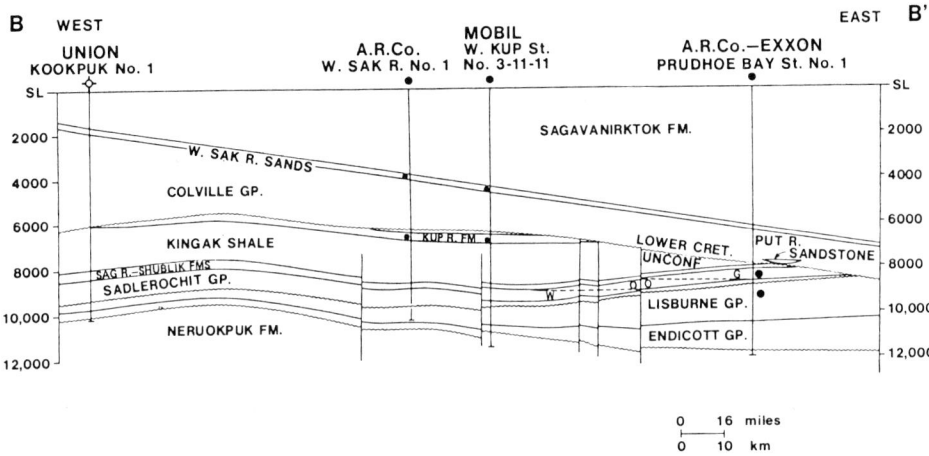

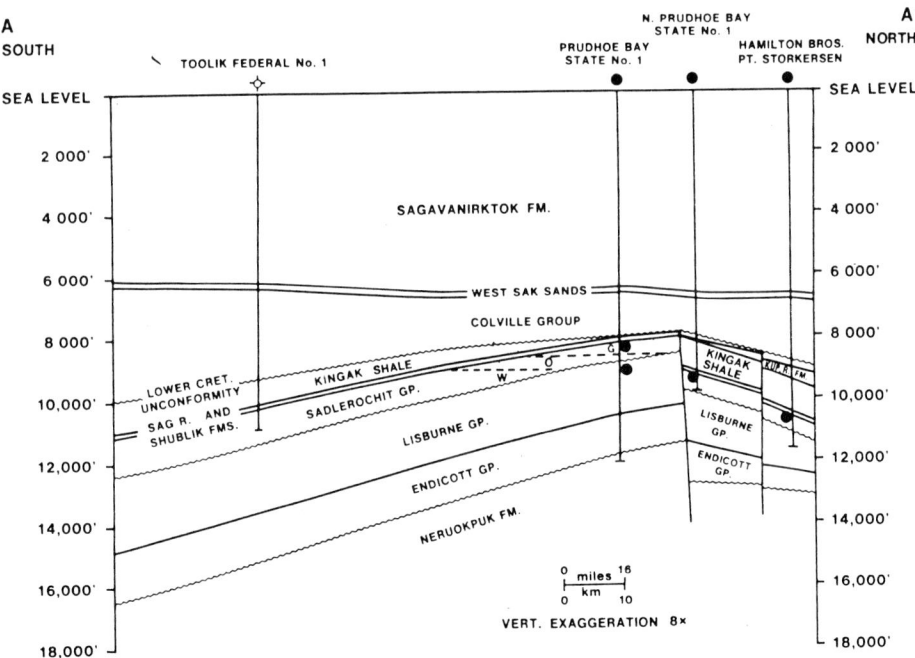

Fig. 13-4. Cross-sections through Prudhoe Bay field. (Reproduced, with permission, from Jamison et al., 1980, p. 296, figs. 7 and 8.)

288

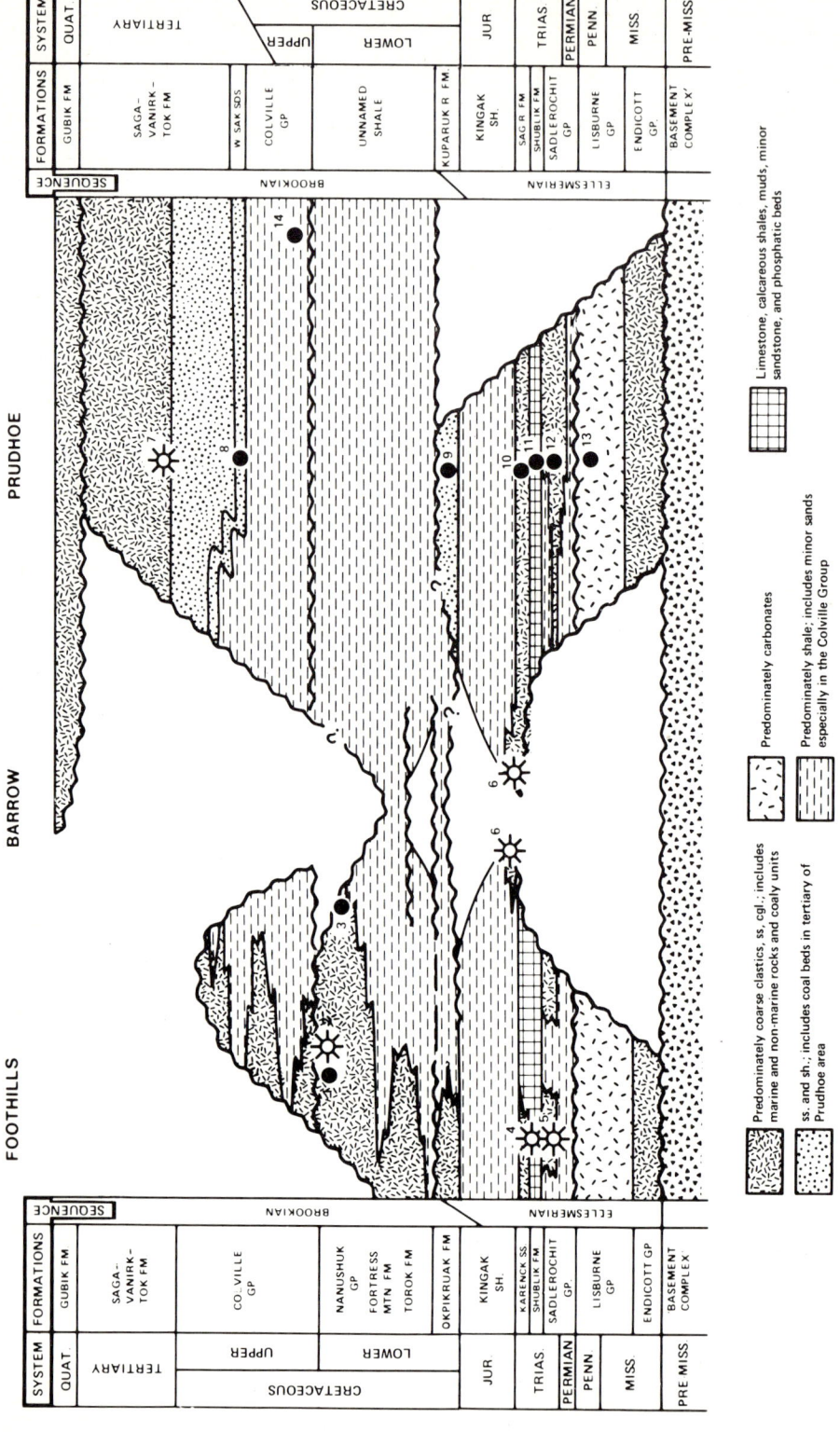

Fig. 13-5. Generalized stratigraphic chart showing significant oil and gas occurrences, North Slope, Alaska. *1* Umiat; *2* Gubik; *3* Simpson; *4* Kemik; *5* Kavik; *6* Point Barrow; *7–13* Prudhoe Bay field; *14* Flaxman Island. (Reproduced from Jamison et al., 1980, p. 294, fig. 5, with permission.)

Much of the North Slope of Alaska is a Naval Petroleum Reserve, and the United States Navy (through the U.S. Geological Survey) spent nearly 10 years exploring it (1944—1953). They drilled on and round the Barrow Arch looking for pre-Cretaceous subcrops (finding one small gas field); and further south, looking for Lower Cretaceous paralic sands in closed anticlines (finding a small oil field and a small gas field). This was intelligent exploration, and the information gained was to lead to more successful drilling by others later. In 1964, after a quiet decade, Atlantic Richfield (ARCO) — as it was soon to become — and Humble joined the search.

They drilled first in the south, about 100 km south of Prudhoe Bay, where steep, narrow anticlines formed long and rather sinuous trends, separated by broad, gentle synclines (a structural style that will be discussed in Chapter 15). These anticlines were found to have sheared cores, and the Upper Cretaceous sands had poor porosity and permeability. The Cretaceous here is dominantly regressive. They then turned their attention to the north, partly because the geology seemed more attractive, and partly because the State leasing in the north was more favourable for exploration in allowing larger areas than the Federal leasing in the south (Morgridge and Smith, 1972; Jamison et al., 1980).

In the Prudhoe Bay area, the Barrow Arch brought the pre-Cretaceous sequence to depths much shallower than in the Colville trough to the south, where at least 9 km of sediment accumulated. The discovery well was drilled on the local culmination of the structure, and penetrated the gas cap of the main reservoir. The next well, drilled 11 km to the south-east, on the flank of the structure, had an oil column of 120 m.

Dips on the south flank were found to be very gentle, about $2°$. There are normal faults in the pre-unconformity sequence in both flanks, and the density of them obscures the dip on the north flank. The stratigraphy is shown in Fig. 13-5. The most important reservoir, the Sadlerochit (Permian—Lower Triassic), is the consequence of a regressive episode in a generally transgressive pre-unconformity sequence. This regression was southwards from a land area to the north (where the Beaufort Sea now is). Thinning of the Sadlerochit onto the structure from the south indicates structural growth beginning in the Permian (probably). Faulting was no later than early Cretaceous. The post-unconformity sequence is regionally regressive, but Lower Cretaceous (Barremian) mudstones seal the subcrops in the Prudhoe Bay area (Fig. 13-6). Whereas the sediment source for the pre-unconformity sequence lay to the north, that of the post-unconformity sequence lay to the south or southwest, with the evolving Brooks Range.

There is general consensus that the source of the petroleum in the Prudhoe Bay field is in the Lower Cretaceous marine mudstones on the unconformity (Morgridge and Smith, 1972; Jones and Speers, 1976). Geochemical evidence does not preclude a pre-Cretaceous contribution. The *total organic carbon* content, which is considered to be a guide to source rock potential, was found

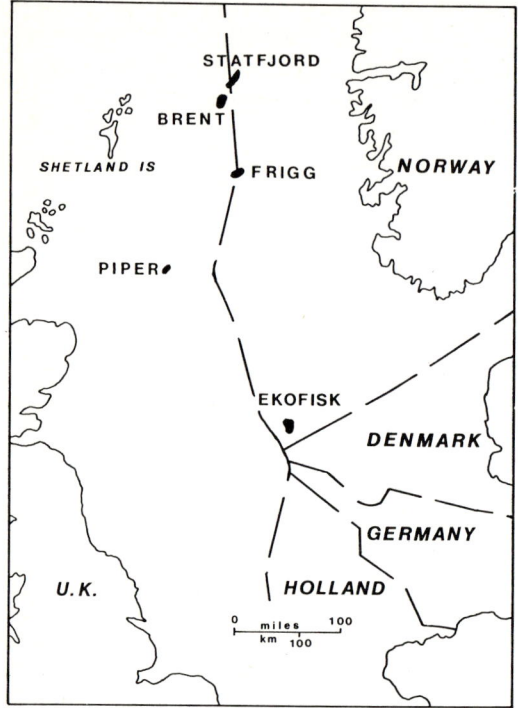

Fig. 13-6. Location map, North Sea. (Adapted from Kirk, 1980, p. 96, fig. 1.)

to be high in the Jurassic marine mudstones, and very high in the Cretaceous marine mudstones on the unconformity. The C_{15+} hydrocarbon content (i.e., hydrocarbons with 15 and more carbon atoms in the molecules), considered by some to be an indication of oil (rather than gas) generative capacity, is also high in both these mudstones. The post-unconformity Barremian mudstones are favoured as the petroleum source rock because they are on the subcrop of all the known reservoir rock units. This conclusion was supported by Jones and Speers (1976), who found that crude oils from the Lower Cretaceous Kuparuk River Formation (which underlies the unconformity in some areas) shares various chemical characteristics with those from the Permo-Triassic Sadlerochit reservoir.

Details of the oil/water contacts of the various reservoirs have yet to be published, but Jones and Speers (1976) report that in the main reservoir, the Sadlerochit, the oil/water contact is deeper in the east and the north-east, and that faults do not appear to affect the oil/water contacts significantly.

The geological history of Prudhoe Bay since the sealing of the subcrops has been one of continued, irregular subsidence without faulting. It is possible that the thickness of the permafrost, more than 600 m in places, indicates continued subsidence. The presence of ice layers buried in sediments supports this view.

There is no difficulty in accepting a post-unconformity source for the Prudhoe Bay oil. Compaction of the Barremian mudstones would have expelled fluids downwards into any subcropping permeable formation in which the water was at lower potential, but the position of the source rocks is restricted to the areas of subcrop. If evidence of intercommunication between reservoirs is obtained, the requirement is reduced to source rocks' being in the area of one subcropping formation, or overlying one of the reservoir rock units.

The North Sea

The first exploration well was drilled in the North Sea in 1964, the first discovery was made in 1965 (British Petroleum's West Sole gas field), and within 15 years the North Sea has become a major petroleum province. Many of the very large accumulations are in stratigraphic traps, and unconformity traps are typical of the northern accumulations.

The *Statfjord oil field* lies mainly on the Norwegian side of their boundary with the United Kingdom, in the Viking graben (Fig. 13-6). Its essential features are shown in Fig. 13-7. It is the largest single oil field in the North Sea, with recoverable reserves of 3×10^9 bbl (477×10^6 m^3) of 38°/41°API oil. The volume of oil in place is estimated at 4.8×10^9 bbl (763×10^6 m^3). Most of the oil is found in the Brent (sandstone) Formation, some in the Statfjord (sandstone) Formation, and a small pool is in a sandstone in a member of the intervening Dunlin Formation. There are several points of interest.

The Brent Formation (which is the main producing horizon of the Viking graben) is unconformably overlain by transgressive Upper and Middle Jurassic shales that are less than 50 m thick (Kirk, 1980). These thin to zero on the eastern flank where the later Jurassic and early Cretaceous unconformity cuts down the reservoir sequence. Above this second unconformity lie transgressive Cretaceous mudstones and marls, with some limestones (Fig. 13-8).

The crude oil in the Statfjord Formation is of poorer quality than that in the Brent Formation, and it is a smaller accumulation with an oil/water contact 220 m (725 ft) deeper than that in the Brent. It is therefore reasonable to conclude that there are two sources. The small pool in the Dunlin Formation has the same oil/water contact as the Brent Formation accumulation: there is therefore evidence of communication between the two, probably at the unconformity surface. The reworked Jurassic sand found in places at the subcrop is the obvious connexion.

Geochemical source rock evaluation indicates the Kimmeridge Clay Formation (Upper Jurassic) as the main source rock and its position overlying the Brent Formation and its subcrop, with the intervening mudstone of the Heather Formation, lends geological support to this (but the reservoirs are over-pressured, and detailed study of the pressure regime in the area would also be required). It appears from published information (Kirk, 1980, p. 100,

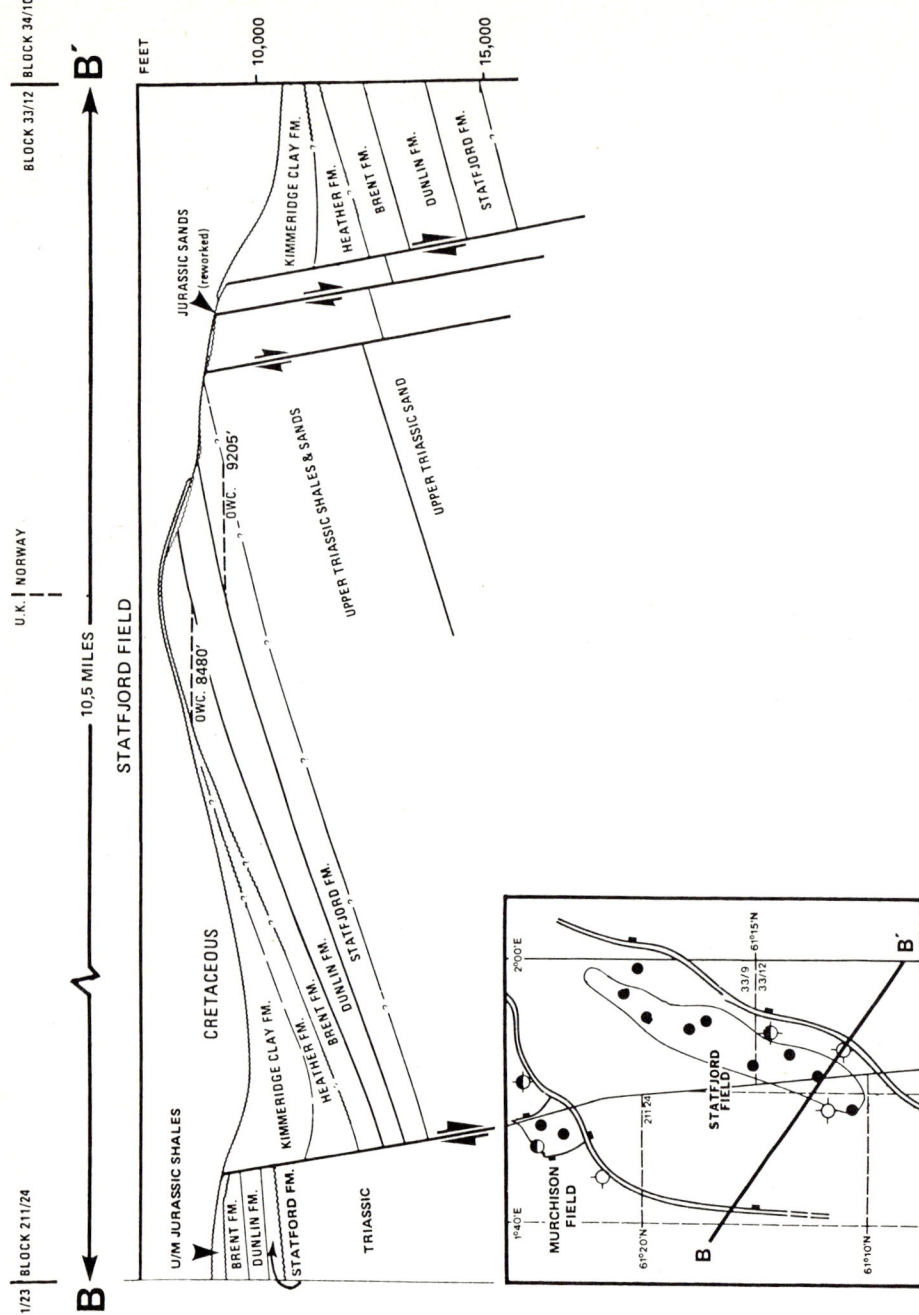

Fig. 13-7. Cross-section through Statfjord field. (Reproduced from Kirk, 1980, p. 100, fig. 5, with permission.)

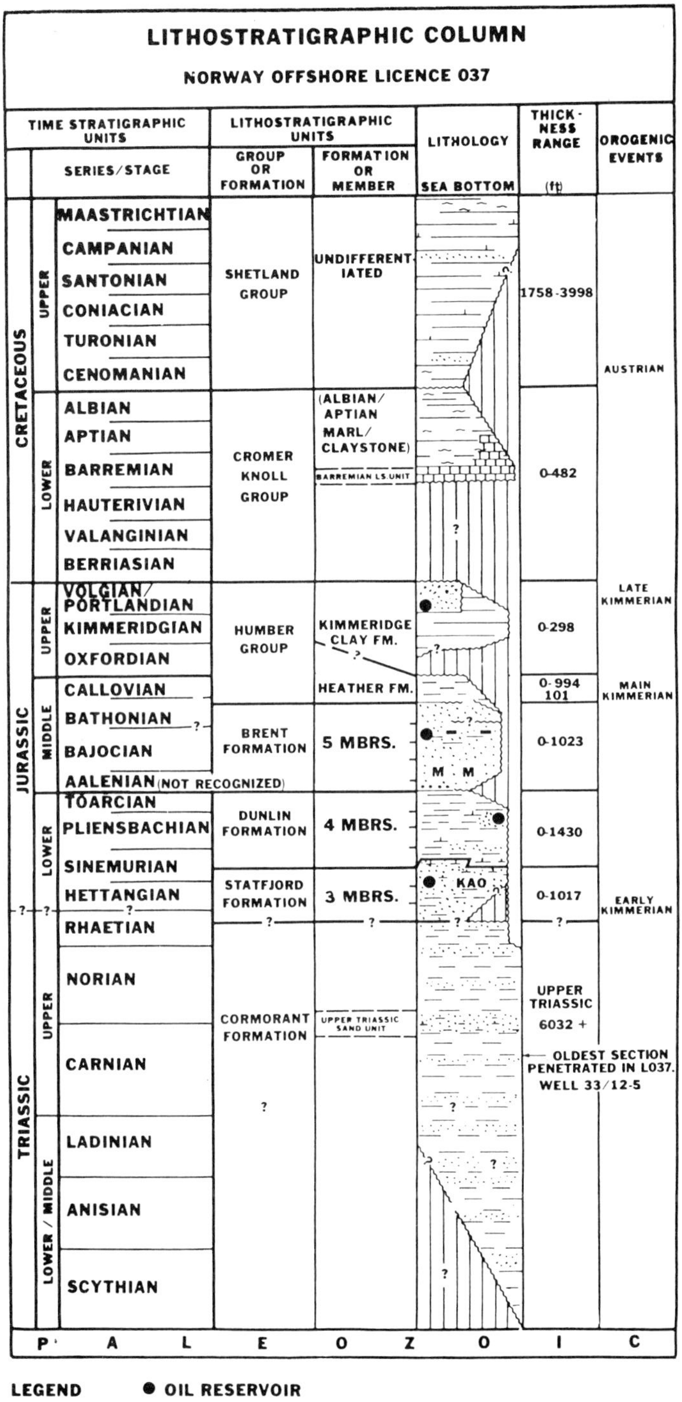

Fig. 13-8. Stratigraphic chart, Statfjord area, North Sea. (Reproduced from Kirk, 1980, p. 102, fig. 7, with permission.)

fig. 5) that these Jurassic mudstones or shales do not reach the Statfjord Formation subcrop, which is overlain by Cretaceous mudstones. If this is true over the pertinent area around the accumulation, it could account for the different quality of the Statfjord Formation crude oil.

The fact that the reservoirs are abnormally pressured requires that the source rock not only was in a position with greater energy than the accumulating oil, but almost certainly still is, and migration is still taking place. The ultimate condition of mechanical stability is with normal hydrostatic pressures in the water around the accumulation, so there must be potential gradients in the fluids, and so a tendency for them to migrate towards the beds with fluids with lower energy. These are either above or below, or both.

Growth faulting also influenced the development of the traps before they were closed in Cretaceous times, so they probably did not affect the accumulation of oil while they moved. The growth faulting, however, did affect the relief of the unconformity surface (not necessarily the topography to the same degree) and so affected the stratigraphic relationships between the Heather and Kimmeridge Clay Formations on the one hand, and the Brent, Dunlin and Statfjord Formations on the other. The very size of the accumulation and the thinness of the apparent source rock over the accumulation suggests that the main source lay outside the present accumulation, and that migration could pass through, or bypass, the faults.

The *Brent oil field* (Bowen, 1975) is very similar to Statfjord. It also has two reservoirs: the Brent sandstone with 36°API oil and the Statfjord with 38.5°API oil, and the oil/water contact of the latter is about 200 m deeper. The similarities go even further because Bowen reports Kimmeridge Clay on the Brent, partly covering the subcrop, but Albian—Aptian marls on the Statfjord. The relief on the unconformity surface is reported to be 900 m, but much of this could be due to differential subsidence: 900 m of topographic relief at the time is not implied. The source of the oil is thought to be the deep Upper Jurassic shales in flanking troughs. The qualities of the crude oils are not reported, but the unconformity surface is effectively sealed.

These two fields illustrate an important point in petroleum geology. The similarities between the two are *similarities of two highs*, and the stratigraphic relationships here are probably not representative of the area as a whole.

The *Piper oil field* in the Moray Firth basin is a folded and faulted unconformity trap with crude oil in a marine sandstone of Oxfordian to early Kimmeridgian age (Williams et al., 1975; Maher, 1980). The main field has an oil/water contact at 2594 m (8512 ft), unaffected by faults. To the south-east, however, there is a pool on the downthrown side of a growth fault that moved during the Cretaceous; and this pool has a deeper oil/water contact at 2804 m (9199 ft), 210 m deeper than that in the main field. This fault accumulated thicker Coniacian and Santonian marls (Fig. 13-9) on the downthrowing side, and movement continued through the Campanian, marls of which cap the subcrop in the main field area. So, by the time the main trap was closed, this fault already had a throw of about 300 m on the base of the productive sand.

Maher (1980, p. 162) points out that at the time of closure, the Kimmeridge Clay had already been buried to more than 1500 m in the adjacent Witch Ground graben, so oil generation and migration could have begun before the trap was closed. Maher (1980, p. 159) also reports better porosity in the oil-bearing sandstones than in the wet sandstones, so early accumulation is not only possible but likely. The point will not be missed that indices of thermal maturation in the reservoir may be quite irrelevant, and so misleading.

These three accumulations of oil have several points in common, and all are traps because of disconformities and unconformities. It is important for the understanding of these and similar accumulations that the nature of such disconformities and unconformities shall be properly understood.

It is commonly assumed that there was general erosion on such a surface, but this is not necessarily true. When there is erosion, the products are transported elsewhere, and accumulate where the energy is insufficient to move

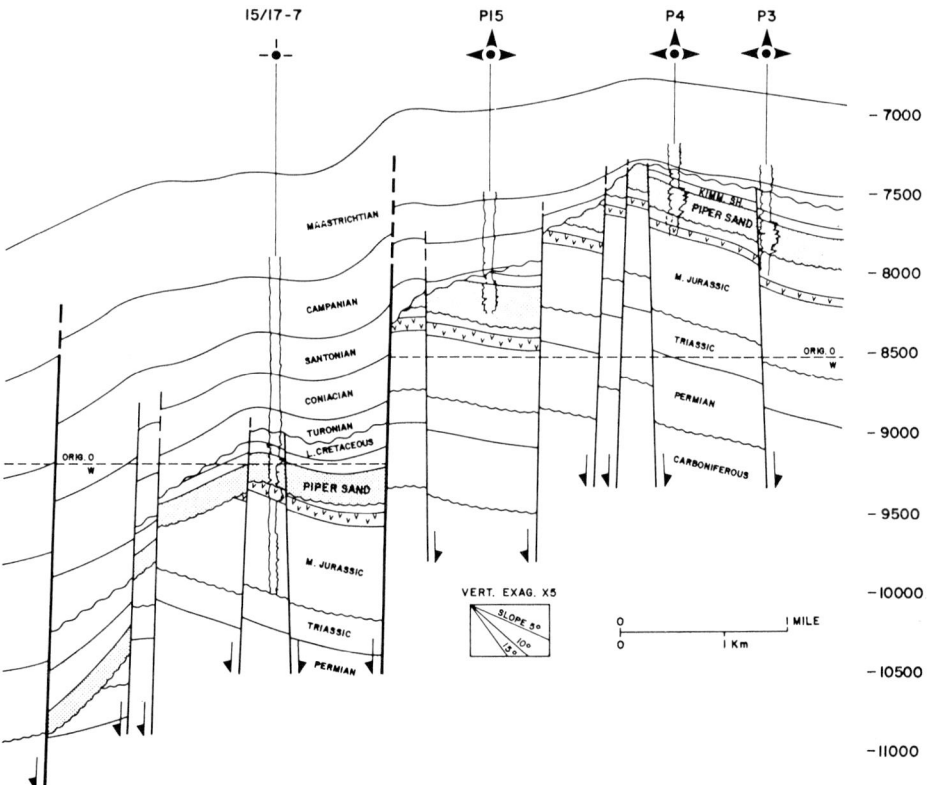

Fig. 13-9. Cross-section through Piper oil field, Moray Firth basin, North Sea. (Reproduced from Maher, 1980, p. 142, fig. 9, with permission.)

them further. In the Piper field, the accumulation of Coniacian and Santonian marls only on the downthrowing side of a southern fault in places suggests low energy. It is inconceivable that subsidence at the rate of 50,000 yrs/m could maintain a balance between accumulation and non-accumulation of marls for 5 m.y.: this must be the net result. Hence we infer that sediment did accumulate from time to time on both sides of the fault (which was moving very slowly indeed), but that erosion removed more of this sediment from the upthrowing block from time to time, rather than cutting into the older sediments. The areas of erosion and truncation of older strata are probably limited to the highs.

The present relief of an unconformity surface is not necessarily a good guide to the contemporary relief of the surface. The transgression that completes the unconformity is necessarily diachronous, and accumulation early in the lows tends to maintain a more subdued physiographic relief than that of the erosion or non-accumulation surface; and much of the present relief may have been created by subsequent differential subsidence.

Bass Strait oil and gas fields, Australia

In the south-east corner of mainland Australia, offshore from the state of Victoria, lies the Gippsland basin (Fig. 13-10). In 1965, the first offshore

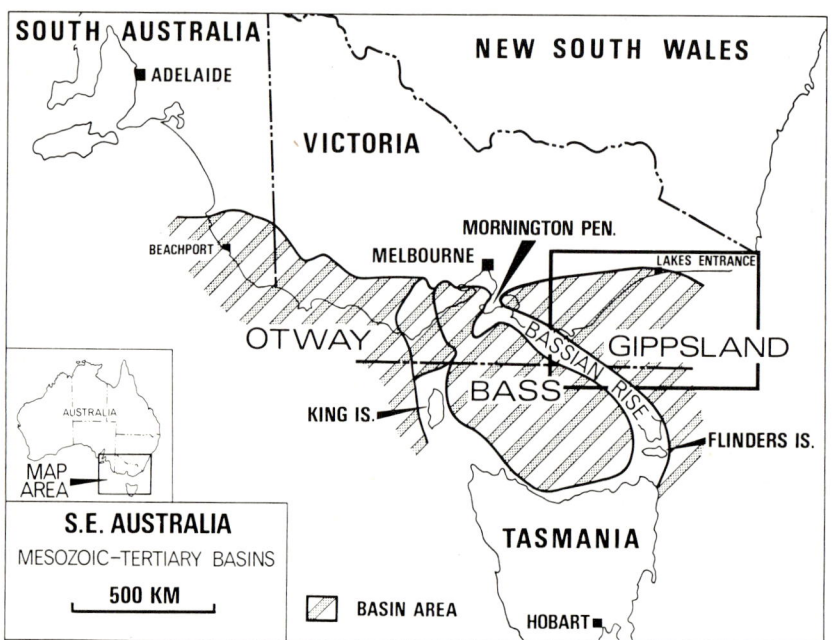

Fig. 13-10. The Mesozoic–Tertiary basins of south-east Australia. (Courtesy of Esso Australia Ltd.)

well to be drilled in this basin discovered the Barracouta gas field in a large anticline, 20 × 4 km, that had been revealed by reflection seismic surveys. The gas was found at a depth of 1021 m in Eocene sandstones of the Latrobe Group, below a disconformity that separates the older alluvial—deltaic sequence of sandstones, mudstones and coals from the younger marine mudstones and marls of the Lakes Entrance Formation of Oligocene age. The cap rock of this accumulation, however, is not the mudstone above the disconformity but a mudstone just below it (Griffith and Hodgson, 1971; James and Evans, 1971; Threlfall et al., 1976). Within three years, another large gas accumulation was found (Marlin), and two large oil fields (Kingfish and Halibut). Subsequent discoveries fell into a pattern of gas fields in shallower water north-west of oil fields in rather deeper water (Fig. 13-11). The stratigraphy was found to be as shown in Fig. 13-12.

To the east of Barracouta, the Marlin accumulation was found to be in several reservoir rocks of the Latrobe Group, but the trap was formed by the unconformity, marine mudstones of the Lakes Entrance Formation sealing the westerly dipping reservoirs at their subcrops (Fig. 13-13). The gas/water contact was found to be constant at 1564 m in the main reservoirs, about 200 m below the culmination of the trap, which is determined by the shape of the unconformity surface. The pre-unconformity sequence is faulted by normal faults that do not, in general, affect the accumulations, but one such fault has trapped a small accumulation of oil (with associated gas). There is also a deep Paleocene gas reservoir.

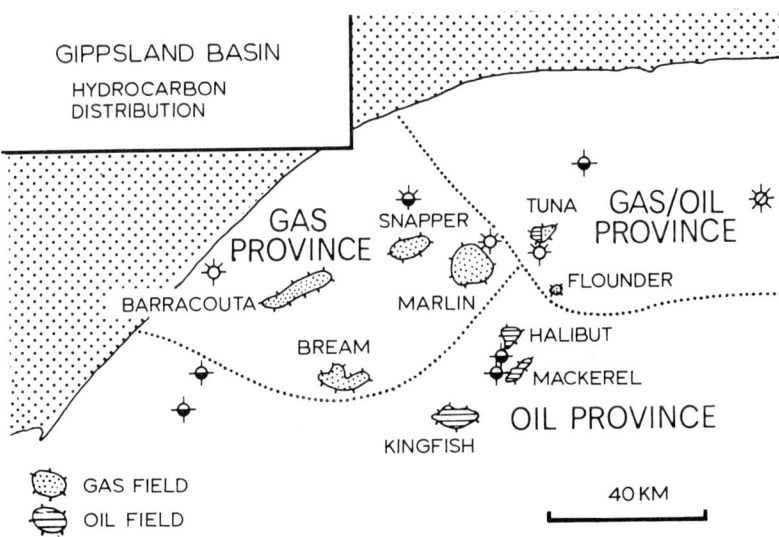

Fig. 13-11. Hydrocarbon distribution in the Gippsland basin. (Courtesy of Esso Australia Ltd.)

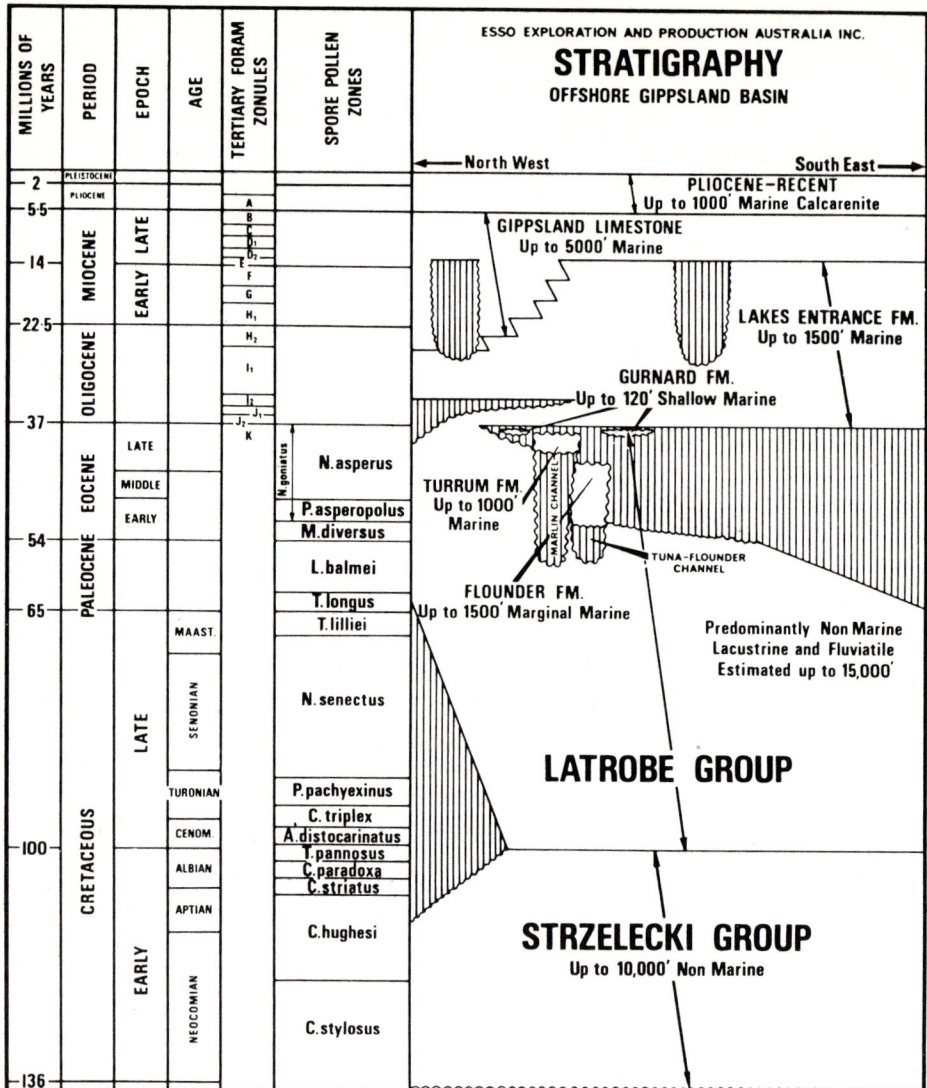

Fig. 13-12. Stratigraphic chart, Gippsland basin (offshore). (Courtesy of Esso Australia Ltd.)

To the south of Marlin lies the Halibut oil field. It too is an unconformity trap in westerly-dipping Latrobe Group reservoirs, sealed at the subcrops by the Lakes Entrance Formation. A constant oil/water contact was found at 2396 m, unaffected by a fault that terminates at or below the unconformity surface. Close to the west of Halibut, a new accumulation of oil was found (Thornton et al., 1980) in younger Latrobe Group reservoirs that are also

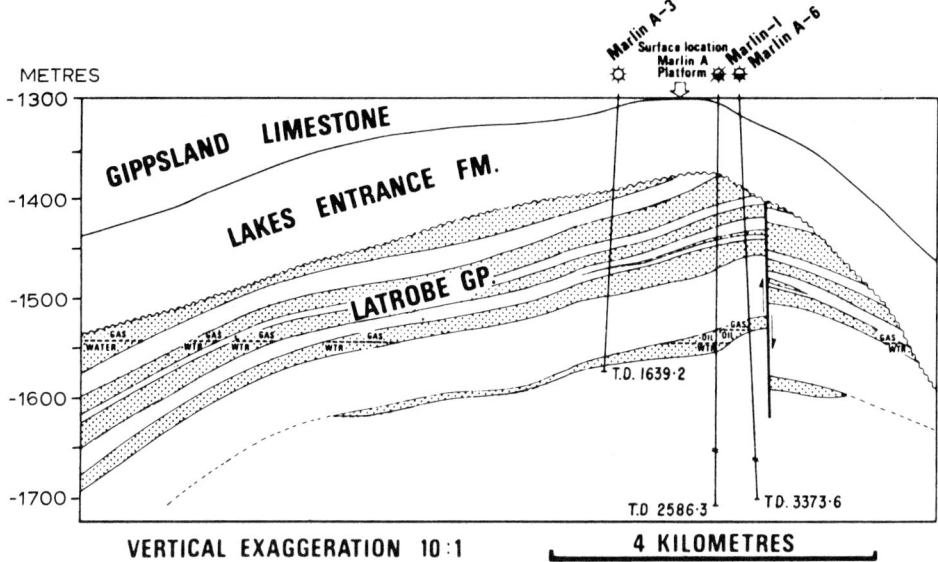

Fig. 13-13. Cross-section through Marlin field, Gippsland basin. (Courtesy of Esso Australia Ltd.)

sealed at their subcrops by the Lakes Entrance Formation. This new field, Fortescue, has its oil/water contact 24 m deeper than that in Halibut (Fig. 13-14).

The reservoirs of all these fields are fluvial—deltaic sands with good porosity and very high permeability in a predominantly non-marine sequence that includes coals. The crude oils vary from field to field, but are typically paraffin-base of about 45°API. Halibut crude has a very high wax content of 26.8%; Kingfish has 13% wax, and so is also classified as a high-wax crude. The small oil accumulation in Marlin is a 50°API crude with 2.7% wax. The gas compositions are typically 85% methane, 6% ethane and 3% propane. Pressures are normal hydrostatic.

The concordant water contacts within fields indicates that either the Latrobe unconformity surface (as it is called) is an imperfect seal in itself, but seals by virtue of its configuration, or that the reservoir sands themselves are interconnected. The evidence of the Halibut and Fortescue fields, with their separate oil/water contacts within a single culmination of the unconformity surface, suggests that the reservoir sands are interconnected within a field because here the unconformity surface has effectively sealed the two accumulations. This conclusion is supported by the chemistry of the Halibut and Fortescue crudes. These are similar, and were probably generated from similar source rocks; but they are considered to have some differences that would not exist if they were a single accumulation (Thornton et al., 1980).

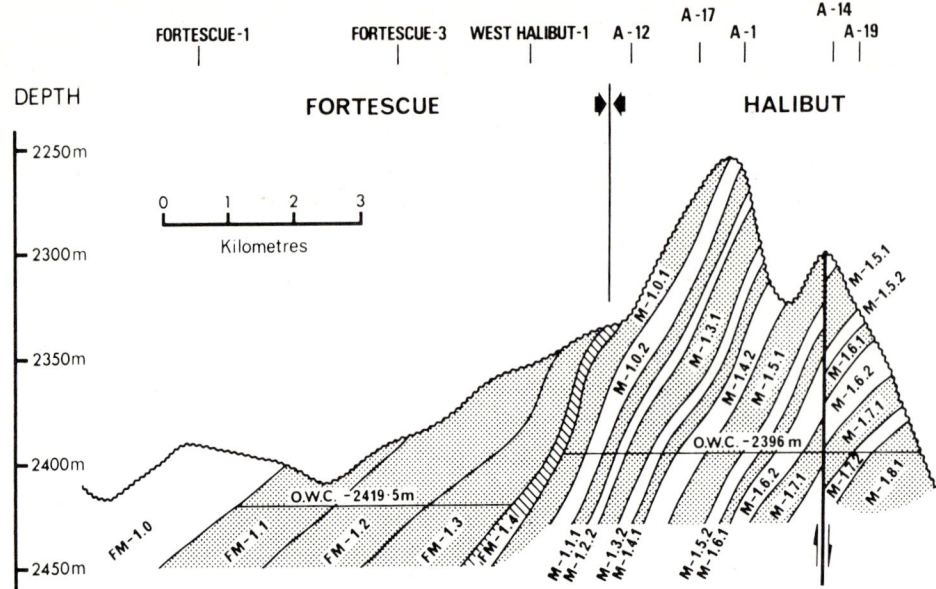

Fig. 13-14. Cross-section through Fortescue and Halibut fields, Gippsland basin. (Courtesy of Esso Australia Ltd.)

The general similarities of the accumulations — their stratigraphy and trap configurations — suggests that these fields have similar source-reservoir relationships. The choice, therefore, is between non-marine pre-unconformity source rocks and marine post-unconformity source rocks. Although the post-unconformity Lakes Entrance Formation contains a small quantity of heavy, 15.7°API, asphaltic crude oil near the coast at Lakes Entrance itself (from which the formation name came), which was exploited from 1925 to 1956 for a total of only 1300 m^3, there is little evidence that it was the source of the offshore oil and gas. This small heavy oil accumulation occurs where the Lakes Entrance Formation lies nonconformably on Devonian granite, and has been severely degraded during secondary migration. It could have been sourced from contiguous formations down-dip, offshore. Thomas (1982) reported that geochemical studies of the offshore Latrobe Group had revealed abundant potential source rocks, rich in oil- and gas-generating macerals (resinite, cutinite, sporinite). Given the right conditions, these could have generated oil in some places, gas in others. These studies found the post-unconformity mudstones organically barren.

There seems little doubt, therefore, that the Gippsland basin petroleum was generated from pre-unconformity, non-marine source rocks. These are likely to be close stratigraphically to the reservoirs. A certain amount of migration across stratigraphic boundaries is likely from the evidence of the fields themselves, but the occurrences of small oil accumulations in the gas fields seems to preclude a deep source with significant migration across stratigraphic boundaries. Threlfall et al. (1976, p. 54) noted that the gas province is where the reservoir rocks of the Latrobe Formation are of Eocene age, while the oil province is where they are of Paleocene and late Cretaceous age.

REFERENCES

Adiwidjaja, P. and DeCoster, G.L., 1974. Pre-Tertiary paleotopography and related sedimentation in south Sumatra. *Proc. Indon. Petrol. Ass.*, 2 (for 1973): 89—103.
Bowen, J.M., 1975. The Brent oil-field. *In:* A.W. Woodland (Editor), *Petroleum and the continental shelf of north-west Europe*, Vol. 1. *Geology.* Applied Science Publishers/Institute of Petroleum, Barking, pp. 353—361.
Clifford, H.J., Grund, R. and Musrati, H., 1980. Geology of a stratigraphic giant: Messla oil field, Libya. *In:* M.T. Halbouty (Editor), Giant oil and gas fields of the decade 1968—1978. *Mem. Am. Ass. Petrol. Geol.*, 30: 507—524.
De Coster, G.L., 1975. The geology of the Central and South Sumatra basins. *Proc. Indon. Petrol. Ass.*, 3 (for 1974): 77—110.
Fraser, W.W., 1967. Geology of the Zelten field, Libya, North Africa. *Proc. 7th World Petrol. Congress*, 2: 259—264.
Gillespie, J. and Sanford, R.M., 1967. The geology of the Sarir oilfield, Sirte basin, Libya. *Proc. 7th World Petrol. Congress*, 2: 181—193.
Griffith, B.R. and Hodgson, E.A., 1971. Offshore Gippsland basin fields. *J. Aust. Petrol. Explor. Ass.*, 11 (1): 85—89.
Halbouty, M.T., Meyeroff, A.A., King, R.E., Dott, R.H., Klemme, H.D. and Shabad, T., 1970. World's giant oil and gas fields, geologic factors affecting their formation, and basin classification. Part. 1. Giant oil and gas fields. *In:* M.T. Halbouty (Editor), Geology of giant petroleum fields. *Mem. Am. Ass. Petrol. Geol.*, 14: 502—528.
Hancock, J.M. and Kauffman, E.G., 1979. The great transgressions of the Late Cretaceous. *J. Geol. Soc. London*, 136: 175—186.
James, E.A. and Evans, P.R., 1971. The stratigraphy of the offshore Gippsland basin. *J. Aust. Petrol. Explor. Ass.*, 11 (1): 71—79.
Jamison, H.C., Brockett, L.D. and McIntosh, R.A., 1980. Prudhoe Bay — a 10-year perspective. *In:* M.T. Halbouty (Editor), Giant oil and gas fields of the decade 1968—1978. *Mem. Am. Ass. Petrol. Geol.*, 30: 289—314.
Jenkins, D.A.L. and Twombley, B.N., 1980. Review of the petroleum geology of offshore northwest Europe. *Trans. Instn. Min. Metall.*, Spec. Issue on Petroleum, pp. 6—23.
Jones, H.P. and Speers, R.G., 1976. Permo-Triassic reservoirs of Prudhoe Bay field, North Slope, Alaska. *In:* J. Braunstein (Editor), North American oil and gas fields. *Mem. Am. Ass. Petrol. Geol.*, 24: 23—50.
Kent, P.E., 1975. Review of North Sea basin development. *J. Geol. Soc. London*, 131: 435—468. (See also Discussion, printed in "General index to volume 131" distributed with volume 132 (5) for September 1976, concerning fig. 18B.)
Kent, P.E., 1976. Major synchronous events in continental shelves. *In:* M.H.P. Bott (Editor), Sedimentary basins of continental margins and cratons. *Tectonophysics*, 36: 87—91.
Kirk, R.H., 1980. Statfjord field — a North Sea giant. *In:* M.T. Halbouty (Editor), Giant oil and gas fields of the decade 1968—1978. *Mem. Am. Ass. Petrol. Geol.*, 30: 95—116.

Maher, C.E., 1980. Piper oil field. *In:* M.T. Halbouty (Editor), Giant oil and gas fields of the decade 1968—1978. *Mem. Am. Ass. Petrol. Geol.*, 30: 131—172.

Matsumoto, T., 1977. On the so-called Cretaceous transgressions. *Spec. Pap. Palaeontol. Soc. Japan*, 21: 75—84.

Matsumoto, T., 1980. Inter-regional correlation of transgressions and regressions in the Cretaceous Period. *Cret. Res.*, 1: 359—373.

Miller, E.G., 1972. Parkman field, Williston basin, Saskatchewan. *In:* R.E. King (Editor), Stratigraphic oil and gas fields — classification, exploration methods and case histories. *Mem. Am. Ass. Petrol. Geol.*, 16: 502—510.

Morgridge, D.L. and Smith, W.B., 1972. Geology and discovery of Prudhoe Bay field, eastern Arctic Slope, Alaska. *In:* R.E. King (Editor), Stratigraphic oil and gas fields — classification, exploration methods, and case histories. *Mem. Am. Ass. Petrol. Geol.*, 16: 489—501.

Roberts, J.M., 1970. Amal field, Libya. *In:* M.T. Halbouty (Editor), Geology of giant petroleum fields. *Mem. Am. Ass. Petrol. Geol.*, 14: 438—448.

Sanford, R.M., 1970. Sarir oil field, Libya — desert surprise. *In:* M.T. Halbouty (Editor), Geology of giant petroleum fields. *Mem. Am. Ass. Petrol. Geol.*, 14: 449—476.

Suess, E., 1885—1909. *Das Antlitz der Erde.* Temsky, Prague, 3 vols. in 4.

Suess, E., 1904—1924. *The face of the Earth (Das Antlitz der Erde).* (Transl. by H.B.C. Sollas under direction of W.J. Sollas) Clarendon Press, Oxford, 5 vols.

Thomas, B.M., 1982. Land-plant source rocks for oil and their significance in Australian basins. *J. Aust. Petrol. Explor. Ass.*, 22 (1): 164—178.

Thornton, R.C.N., Burns, B.J., Khurana, A.K. and Rigg, A.J., 1980. The Fortescue field — new oil in the Gippsland basin. *J. Aust. Petrol. Explor. Ass.*, 20 (1): 130—142.

Threlfall, W.F., 1981. Structural framework of the central and northern North Sea. *In:* L.V. Illing and G.D. Hobson (Editors), *Petroleum geology of the continental shelf of north-west Europe.* Heyden/Institute of Petroleum, London, pp. 98—103.

Threlfall, W.F., Brown, B.R. and Griffith, B.R., 1976. Gippsland basin, off-shore. *In:* R.B. Leslie, H.J. Evans and C.L. Knight (Editors), Economic geology of Australia and Papua New Guinea 3. Petroleum. *Australas. Inst. Min. Metall., Monogr. Ser.*, 7: 41—67.

Vail, P.R. and Todd, R.G., 1981. Northern North Sea Jurassic unconformities, chronostratigraphy and sea level changes from seismic stratigraphy. *In:* L.V. Illing and G.D. Hobson (Editors), *Petroleum geology of the continental shelf of north-west Europe.* Heyden/Institute of Petroleum, London, pp. 216—235.

Van Hinte, J.E., 1976a. A Jurassic time scale. *Bull. Am. Ass. Petrol. Geol.*, 60: 489—497.

Van Hinte, J.E., 1976b. A Cretaceous time scale. *Bull. Am. Ass. Petrol. Geol.*, 60: 498—516.

Williams, J.J., 1972. Augila field, Libya: depositional environment and diagenesis of sedimentary reservoir and description of igneous reservoir. *In:* R.E. King (Editor), Stratigraphic oil and gas fields — classification, exploration methods, and case histories. *Mem. Am. Ass. Petrol. Geol.*, 16: 623—632.

Williams, J.J., Conner, D.C. and Peterson, K.E., 1975. The Piper oil-field, UK North Sea: a fault-block structure with Upper Jurassic beach-bar reservoir sands. *In:* A.W. Woodland (Editor), *Petroleum and the continental shelf of north-west Europe,* Vol. 1. *Geology.* Applied Science Publishers/Institute of Petroleum, Barking, pp. 363—377.

PART 3. PETROLEUM GEOLOGY OF REGRESSIVE SEQUENCES

CHAPTER 14

ABNORMAL PRESSURES

SUMMARY

(1) Abnormal pressures are, for practical purposes, pore pressures that are sufficiently greater than normal hydrostatic to have a noticeable effect when drilling, and to require special precautions. They are stratigraphically related to thick mudstones, usually but not invariably in regressive sequences, with sand/shale ratios less than 10%.

(2) Abnormally high pore pressures are also related to growth structures that are also stratigraphically related — that is, to those that occur in regressive sequences. The top of abnormal pressures usually lies below the level of maximum growth-rate on growth faults, close to (above or below) the level at which growth faulting began. These relationships appear to be causal.

(3) The cause of abnormal pressures is largely mechanical loading of a relatively impermeable mudstone, and it seems likely that the pore pressures in such mudstones have never been normal hydrostatic. In their turn, they cause mechanical instability in such regressive sequences through the retention of relatively large porosity, low bulk density and low equivalent viscosity while the sandier part of the regressive sequence accumulates.

(4) Generation of petroleum in the source rock may also generate abnormal pressures (and resistivities higher than normal).

OBSERVATIONS

There is a great deal of evidence from boreholes around the world that pore-fluid pressures are, in general, normal hydrostatic — that is, the pressures measured in water reservoirs and the water legs of oil reservoirs are sufficient to support a column of that water to close to the land surface (or sea level on the continental shelves). This implies that the pore water is in physical continuity to the surface, however tortuous the paths may be, and that during burial this continuity has been maintained. Were it not for the ever-increasing depths of boreholes drilled for petroleum, it might reasonably have been assumed that this was the universal rule.

The drilling of "gushers" in the cable-tool days does not constitute an early exception to these observations because, as we saw in Chapter 8, the smaller weight density of oil and of gas compared to that of water means that the pressure at the top of the reservoir is higher than the pressure that would have been found if the reservoir had been full of water only. This is the initial driving force of flowing wells, enlarged when the well itself is full of oil or gas, or oil from which solution gas is liberated as it rises to lower pressures.

That reservoir geometry can lead to higher pressures than the normal hydrostatic can be seen in Fig. 14-1, which, although schematic, is representative of several Iranian fields. A thousand metres of oil column of mass density 850 kg m^{-3}, as compared with water of mass density 1010 kg m^{-3}, gives rise to an excess pressure of $(1010-850) g \times 1000 = 1.57$ MPa (= 227 psi). Such a column of gas of mass density 200 kg m^{-3} would lead to an excess pressure of 7.94 MPa (1150 psi) at the top of the reservoir. If these pressures were encountered at a depth of 1500 m, for example, they would lead to excess pressure gradients from the surface of about 1 kPa/m (0.04 psi/ft) and 5.3 kPa/m (0.23 psi/ft). If mud is lost to the formation, and oil fills the hole to the surface, the excess pressure there will be about 3.9 MPa (570 psi). For gas, the excess pressure would be about 19.8 MPa (2900 psi). These pressures are normal in the sense that they arise from normal physical causes in a water environment that has normal hydrostatic pressures.

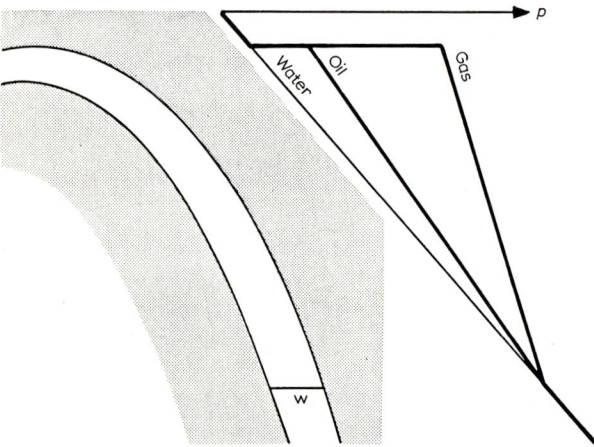

Fig. 14-1. Reservoir geometry can lead to "abnormal" pressures within an accumulation.

Drilling in Trinidad and the U.S. Gulf Coast around 1940 led to some blowouts in which the pore pressures apparently greatly exceeded anything that could reasonably be explained in terms of a normal hydrostatic water regime. These difficulties were encountered more widely by the 1950s — in Brunei, Indonesia, Burma, for example, and then the Niger delta. In the U.S.S.R. similar experiences occurred (see Fertl, 1976, p. 334). Abnormal these pressures might be, but they were becoming very common.

Typically, the experience was this: having drilled 1.5 km or so without difficulty, penetration rate increased and soon afterwards the mud was observed to be flowing from the well at a greater rate over the shale shaker, and the mud tanks started filling up. When drilling was stopped and the mud pumps shut down, mud continued to flow from the well at an increasing rate, so the blowout preventers were closed. It was commonly found that mud of specific gravity 1.5—1.8 was required to control the influx into the borehole. If drilling was continued, the specific gravity of the mud had to be continually increased until the practical limit of about 2.2 was reached. The borehole then had to be abandoned. Such heavy muds were needed to control pressures that approached those of the total overburden, corresponding to specific gravities of 2.2—2.4.

In some boreholes, the mudstone tended to squeeze into the hole, making it tight and tending to stick the tools. These beds were called "heaving shales". They showed up on caliper logs, and it was quite common to find that one could not get back to bottom after a round trip without reaming or redrilling the last part.

If there was much open hole when the mudweight was increased, there was an increased tendency for the pipe to stick — not on bottom, but well up the hole in the normally pressured part of the sequence. This was called "wall-sticking". The excess pressure of the mud over the formation fluids held the drill pipe to the wall of the hole with considerable force. (See Thomeer and Bottema, 1961, for some histories of drilling abnormal pressures.)

The practical solution of these problems is not without interest. In the mid-1950s it was found that the best manner of drilling in areas in which abnormal pressures were expected was to use the *lightest* possible mud from the beginning. This was a remarkable conclusion in the face of the normal practice of weighting up the mud in anticipation of abnormal pressures. With lighter mud, the normally pressured part of the hole drilled faster; and careful control of the drilling parameters resulted in early detection of the increasing drilling rate (called the "drilling break").

As soon as the drilling break was detected, the bit was pulled back above where the drilling break started, pumps were stopped and the mud-level in the borehole watched. The mud was then circulated "bottoms up". It could thus be determined if there was influx into the borehole, and, by analysing the mud from bottom, the nature of the influx. If the borehole flowed, the well was said to have kicked; the blowout preventers were closed and the pressure on them measured. From this the proper mudweight for further drilling could be determined.

At the first kick, or when satisfied that the top of abnormal pressures had been reached, casing was set and drilling proceeded with a properly weighted mud, but as light as possible, in greater safety and with greatly reduced risk of wall-sticking. The same techniques were used to decide when to increase the mudweight, and by how much. By drilling for kicks, the rare combination of increased safety at reduced cost was achieved.

These experiences indicated that at some depth the pore-fluid pressures departed from the normal hydrostatic and approached the overburden pressures in what came to be called the *transition zone*. A few very high pressures equal to, or slightly in excess of, the overburden pressures were reported, but the common limit seemed to be about 90% of the overburden pressures. Almost all the sequences with abnormal pressures were Tertiary in age in the early years, but later such pressures were also found in sequences of Mesozoic and even late Palaeozoic age, although these were rarely of great severity.

For many years the cause was thought to lie in the depth, because there was a remarkable constancy of depth to top of abnormal pressures in the U.S. Gulf Coast (about 3 km). It was thought that the loss of permeability with compaction during burial sooner or later was such that the pore fluids could no longer be expelled. The pore fluids then bore some of the load. It was tacitly assumed that once the top of abnormal pressures had been reached, they would continue indefinitely to greater depths. It was George Dickinson who, with a geological approach, demonstrated in a classic paper that should still be read, that the cause of abnormal pressures is stratigraphic (Dickinson, 1951, 1953).

INTERPRETATION OF OBSERVATIONS

The essence of Dickinson's geological observations in the Louisiana Gulf Coast is that the age of the sediments in which the top of abnormal pressures occurs is younger the father south towards the Gulf, and that the lithology is mudstone below the massive regressive sands (Fig. 14-2). That the top of abnormal pressures tended to occur at depths of about 3 km (10,000 ft) was a coincidence of the geology of the area. Dickinson noted that high pressures existed in sand lenses that were isolated by mudstones, and sandstones that were isolated by faults. The permeability of these could lead to massive water or petroleum influx into the borehole. He attributed these abnormal pressures to the compaction of the mudstones under the gravitational load of the overburden. Numerous measurements indicated that the degree of abnormality increased with depth on a gradient ($\Delta p/\Delta z$) greater than that of the overburden (Dickinson, 1953, p. 420, fig. 6).

Subsequently, the main development in the topic was in recognizing ab-

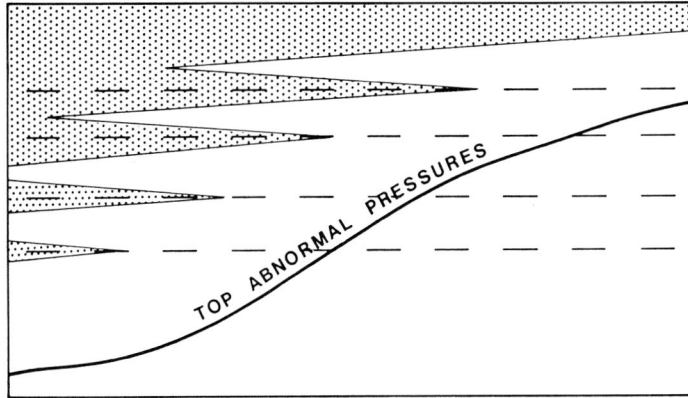

Fig. 14-2. Stratigraphic diagram of relationship of abnormally pressured mudstones to stratigraphy. (Time lines horizontal.)

normal pressures, and estimating the actual pressures, from their effects on electrical and acoustic borehole logs (Hottman and Johnson, 1965). It was found (Fig. 14-3) that both resistivity and sonic transit time showed departures from the normal trends of increasing resistivity and decreasing sonic transit time below the top of abnormal pressures. But these did not provide a *predictive* tool, merely a method of locating and quantifying the abnormality. The drilling break remains the only reliable indicator of impending abnormal pressures, although the great improvements in seismic technology enable qualitative prediction of abnormal pressures, and the depths with an accuracy of 200 or 300 m, from interval velocity analyses. The drilling break was also quantified and, through automation, provided a continuous *computed* pore pressure. This was called the D-exponent, or d-exponent (see Fertl, 1976, pp. 122—130). Computers can, however, be dangerous because the printed figures tend to be accepted uncritically*.

The development of indirect, geophysical methods (including borehole logging) of quantifying abnormal pressures, coupled with the ever-widening areas of the world in which these phenomena were found, led to increasing awareness of the habitat of abnormal pressures — but this interest, naturally, was confined to those in or connected with the petroleum industry. Whereas Dickinson had worked with measured pressures and with pressures estimated from the mudweight needed to contain a pressure, the borehole log responses indicated that the mudstones themselves were abnormally pressured (as had indeed been inferred earlier) although no pressure measurements could be

---

* I have been on an offshore Louisiana rig when the toolpusher ordered a reduction of the mudweight on the grounds that the computer showed a reduced pore pressure. The mud weight had been raised earlier for a part of the section that was still in open hole, and he was surprised when the well kicked.

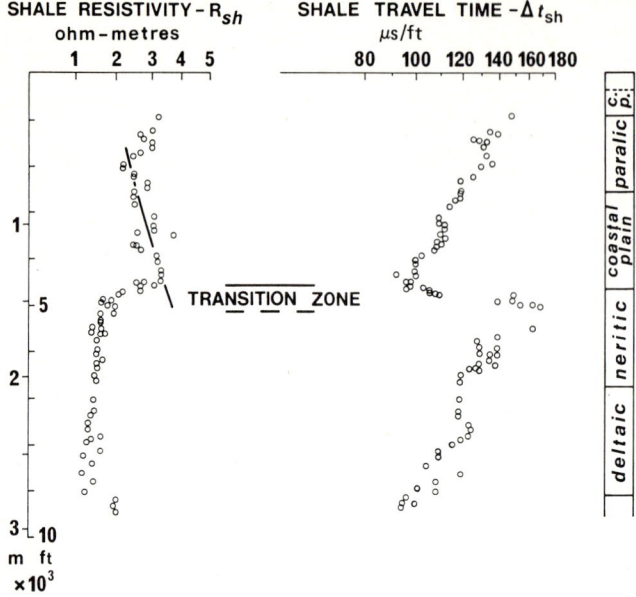

Fig. 14-3. Plot of logarithm of shale resistivity and logarithm of shale transit time in mudstones against depth in a well in Borneo. (Data courtesy of the Royal Dutch/Shell Group.)

made on their pore fluids because their permeability is too small to allow equilibration in the borehole. Such mudstones came to be called "undercompacted" because they showed the properties that were characteristic of mudstones at much shallower depths.

Harkins and Baugher (1969) found that abnormal pressures were associated with those parts of the sequence in which the sand/shale ratio was less than 5—10%.

In some areas, pressures were found to return to normal hydrostatic at greater depths, or a zone of smaller abnormality was found. Examples of the former are to be found in eastern Venezuela (Funkhouser et al., 1948, p. 1891) and north Sumatra (Mulhadiono and Marinoadi, 1977) and some other examples are given by Frederick (1967). A particularly interesting example of reduction in abnormality followed at greater depth by increasing abnormality is described by Fowler (1970) in a study of pressures measured in reservoirs, their petroleum accumulations, and the water salinities in a field in Texas, U.S.A.

During the late 1960s, the mechanical hypothesis (which had been generally accepted) began to be questioned. The dehydration of smectite to illite, which Powers (1967) had suggested could be important for primary migration, was thought to be a possible cause of abnormal pressures. This diagenesis was observed at depths below about 1800 m (5900 ft) (Burst, 1969; Perry and

Hower, 1970, p. 171; Weaver and Beck, 1971, p. 18). The difficulty with this hypothesis is that abnormal pressures occur in several areas at depths much shallower than that of clay-mineral diagenesis. Some of the shallow occurrences of abnormal pressures were in areas of apparent or evident tectonic activity, such as California, Trinidad and New Zealand, and a tectonic cause had some adherents.

Smith and Thomas (1971) and Barker (1972) revived the thermal hypothesis. It had long been known that thermal expansion of water with depth in the subsurface on a geothermal gradient is greater than its compression under the increasing load of overlying water (Versluys, 1932, pp. 924—925) (Fig. 14-4). Assuming negligible permeability to water, heating of the water by burial down a geothermal gradient would lead to expansion that would be resisted by the weight of the overburden. Pressures greatly in excess of overburden pressures could theoretically be developed, but they would lift the overburden and the mudstones would release the excess pressure over the overburden pressure through fractures. The aquathermal hypothesis, as Barker called it, dominated the topic for the next decade (and may continue to do so).

Chapman (1980) found that subsidence at the greatest known rate in the

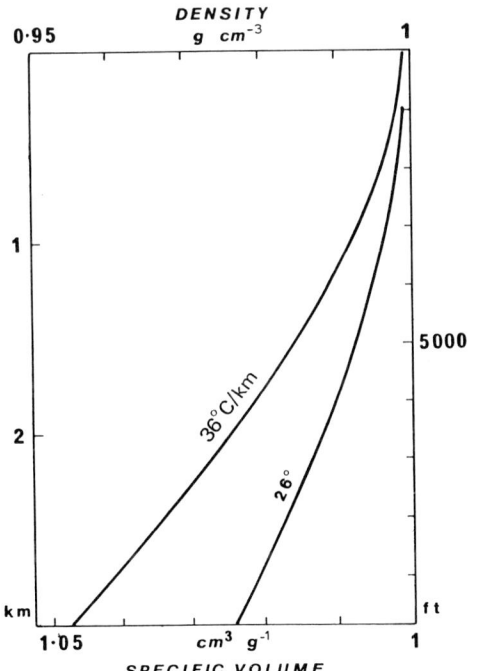

Fig. 14-4. Density-depth diagram for pure water in the subsurface at normal hydrostatic pressures and geothermal gradients of 26 and 36°C/km.

U.S. Gulf Coast down a larger-than-normal geothermal gradient would lead to a rate of water expansion that could be dissipated by a mudstone permeability close to the smallest measured, and concluded that thermal expansion could not be more than a minor component of abnormal pressures. Barker and Horsfield (1982) disputed this conclusion, but did not convince Chapman (1982).

At the time of writing, therefore, it cannot be said that a true cause of abnormal pressures has been identified. A vote would strongly favour the thermal school — but science is not a matter of voting. The importance of correctly understanding this topic is great because error in this leads to errors in the related topics of primary migration and its timing in regressive sequences, and so to error in exploration philosophy. We shall therefore examine the components of this topic in some detail.

THE NATURE OF THE TRANSITION ZONE

The characteristics of the transition zone from normal to seriously abnormal pore pressures are typically these: (a) the penetration rate of drilling increases; (b) the pore pressures measured in sandstones or inferred for mudstones increases with depth, usually more rapidly than the overburden pressures; (c) the geothermal gradient increases; and (d) the sonic log indicates transit times in mudstones increasing with depth (Fig. 14-3):
— the resistivity logs show a decrease in resistivity in mudstones with depth (Fig. 14-3); and
— logs that measure bulk density indicate a decrease in mudstone bulk density with depth, and this is supported by cutting density measurements (see Fertl, 1976, p. 150).

These are all reversals of the normal trend.

These features are not seen in every well, but most that penetrate the transition zone above depths of about 3 km (some deeper) in the U.S. Gulf Coast, Niger delta, Mackenzie delta (Canadian Arctic), and south-east Asia show them all. We are concerned here with the common observations, and must accept that there will be wells and areas that do not conform to these observations.

In seeking a geological understanding of the transition zone, we must clearly understand that the stratigraphic position of the top of abnormal pressures, the top of the transition zone, means that there is probably a mudstone-facies trend that may contribute to these features (see p. 219).

(a) The increase in drilling rate is the *effect* of drilling undercompacted mudstones, and when heaving shales are encountered with the drilling break, it is clearly reasonable to interpret them as such. However, there is an hydraulic effect that is not related directly to undercompaction. Drilling rate is sensitive to the relationship between mudweight and pore pressures: it has long been observed that holes drilled with too heavy a mud drill more slowly than those drilled with a lighter mud. Part of the drilling break is evidently due to the reduction and/or reversal of the potential gradient in the fluids across the bottom of the hole while drilling.

(b) The significance of the rate of increase of pore pressure with depth is that if it exceeds the overburden gradient, an explanation is required. Two explanations have been offered: upward flow of pore water through the relatively impermeable mudstone, and thermal expansion of the pore water in almost totally impermeable mudstone during burial down a geothermal gradient. We shall return to this question later.

(c) The increase in geothermal gradient, while not well documented in the literature (but see Lewis and Rose, 1970; Fertl, 1976, p. 144), is a real effect because it has been general practice in several areas (including the U.S. Gulf Coast) to monitor the temperature of the mud returns. Anomalous increases warn of abnormal pressures. The physical explanation has been offered that the thermal conductivity of the abnormally-pressured mudstone is decreased by the abnormality, so that the geothermal gradient is increased to maintain constant heat flow.

(d) The log responses are all consistent with an increase in mudstone porosity with depth, but there are other effects as well. Any increase in temperature, other things being equal, reduces the resistivity of the mudstone. Temperature alone cannot account for the great relative decreases observed. There is also some evidence (see Magara, 1978, p. 222) that abnormally pressured pore water in mudstone is less saline than that in adjacent sands.

Perhaps of greater significance is the observation that formation density, resistivity, and sonic transit time plots do not always indicate the same depth of top of abnormal pressures (in the author's experience, commonly in that order of increasing depth). Drilling experience tends to support the sonic log as being the most accurate from a practical point of view: it indicates accurately the depth at which an appreciable abnormality has developed, but the others may well indicate the real divergence from normal trends.

It seems quite certain that these anomalies indicate that the porosity of mudstone in the transition zone increases with depth while the pore pressures increase at a greater rate (usually) than the overburden pressures. We cannot understand abnormal pressures without understanding this major anomaly.

The alternative hypotheses concerning the cause of the features of the transition zone involve the permeability of the mudstone and the reversibility of its compaction. The mechanical hypothesis rests on small but real permeability to water or petroleum, and the irreversibility of compaction: the thermal hypothesis rests on near-zero permeability and reversibility of compaction due to expansion of the pore water during burial down the geothermal gradient. These differing requirements lead to totally different conclusions concerning the timing of generation of abnormal pressures and the quantities of pore fluids expelled, so it is essential for our understanding of the petro-

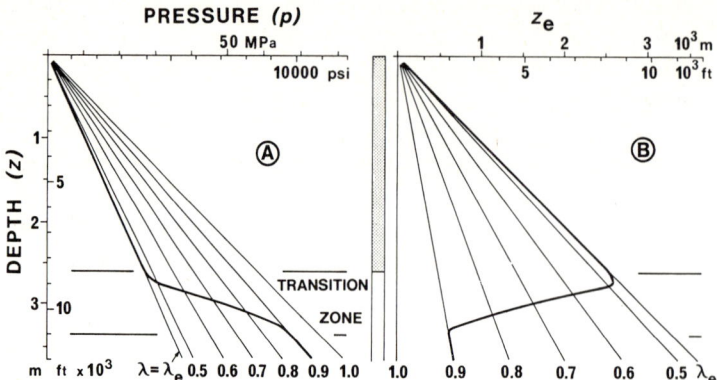

Fig. 14-5. The equilibrium compaction depth, z_e ($= \delta z$), decreases in the transition zone below normal hydrostatic pore-fluid pressures.

leum geology of regressive sequences that we shall discover the true causes of abnormal pressures and their relative importance.

The *mechanical hypothesis* attributes abnormal pressures to compaction of mudstones that do not have sufficient permeability to permit pore water to escape as fast as the other mechanical influences would require for normal compaction and normal hydrostatic pore-fluid pressures. The increase in pressure with depth in the transition zone means that the equilibrium compaction depth decreases throughout the transition zone (Fig. 14-5). Compaction of mudstone is regarded as irreversible, so the history of the equilibrium compaction curve cannot have followed the form in Fig. 14-5, but must have developed along different paths from a much shallower depth, each path tending to increase the effective compaction depth with increasing depth (as shown schematically in Fig. 14-6) because of leakage. Undercompaction therefore requires an early origin, the greater the undercompaction, the earlier the origin.

The transition zone pressure gradient is caused by the overlying permeable sand or sandstone with normal hydrostatic pressures. The potential gradient that this creates leads to upward water flow through the mudstone in the transition zone. The small permeability of the mudstone to water causes large energy losses during flow, with a consequent steep gradient of pressure loss. This pressure gradient, following Terzaghi's principle (p. 57), leads to a compaction gradient, and so a permeability gradient. The slope of the transition zone pressure gradient is therefore also due to the permeability of the transition zone as a whole: the smaller the permeability, the larger the pressure gradient (larger $\Delta p/\Delta z$). Eventually, all the expellable water will be expelled and the mudstones will become normally compacted, with pore fluids at normal hydrostatic pressures.

The *thermal hypothesis* postulates an "isolation depth" at which mud-

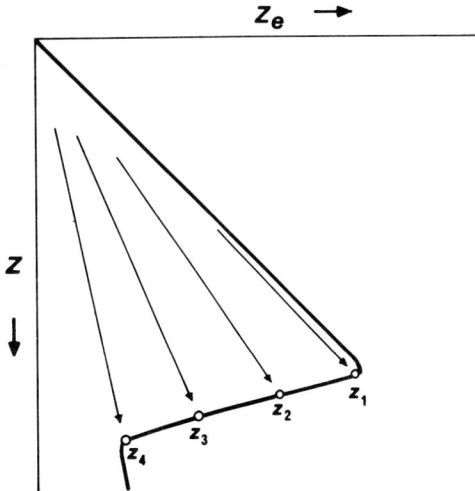

Fig. 14-6. Postulated development of the z–z_e relationship during burial, assuming irreversibility of mudstone compaction.

stones become virtually impermeable to water. Further burial increases the temperature of the pore water and the water's tendency to expand, resisted by the overburden, results in elevated pressures. These have a theoretical capacity to exceed the overburden pressure, but when that pressure is reached, fractures develop along which excess water flows, dissipating some of the excess pressure but still leaving it close to the overburden pressure.

Figure 14-7 shows the relationship between pressure, temperature, and specific volume of pure water. On this has been plotted the normal hydrostatic line for a geothermal gradient of 36°C/km; and the isopycnic lines have been terminated at the approximate overburden line. If, once the isolation depth and temperature are reached, the pore volume remains constant and no more pore water escapes, then further burial will lead to pressure/temperature and pressure/depth gradients parallel to the isopycnic lines. These are similar to some observed gradients.

The thermal hypothesis requires relatively late generation of abnormal pressures, after the isolation depth has been reached, and limits the volumes of pore water expelled to those generated by thermal expansion.

Discussion

The question is not whether the thermal or the mechanical hypothesis is correct, but which is the dominant process, and what are the conditions that favour one or the other in particular circumstances. The critical parameter is the permeability of the mudstones because this affects the thermal hypothesis much more than the mechanical. So it could well be that the importance of

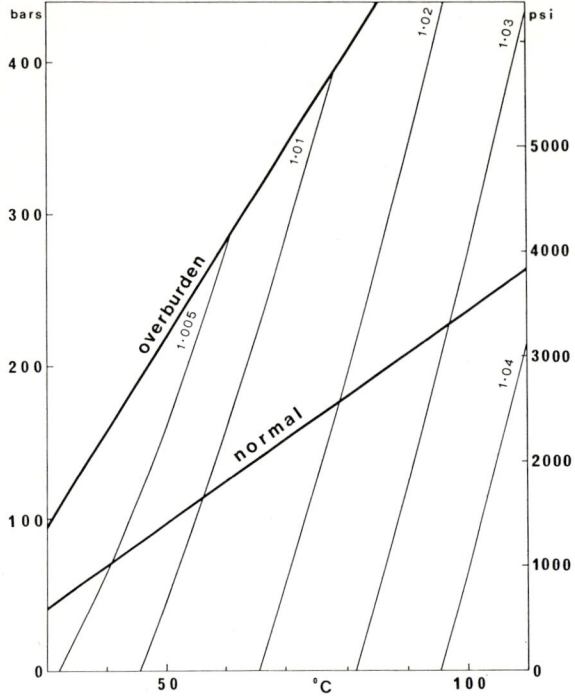

Fig. 14-7. Pressure—volume—temperature (PVT) diagram for pure water. Data of Kennedy and Holser, 1966, table 16-1. Specific volume is in cubic centimetres per gram, or 10^{-3} m^3 kg^{-1}.

the thermal process increases with depth to the top of abnormal pressures — the deeper the top of abnormal pressures, the more important the component due to thermal expansion — and increases as the mechanical process reduces the porosity and permeability of the mudstone. In brief, the thermal process may become more important with time.

Chapman (1980) calculated the permeability required in a mudstone to dissipate the volume of water created by thermal expansion during burial at the maximum rate known in the U.S. Gulf Coast (500 yrs/m) down the maximum geothermal gradient known there (36°C/km). A mudstone 500 m thick, with 20% porosity, would require a hydraulic conductivity of about 10^{-13} m/s, or an intrinsic permeability of about 5×10^{-17} cm^2 (5×10^{-6} md) to dissipate all the water of expansion. This is near the lower limit of Tertiary mudstone permeabilities measured in Japan (reported in Magara, 1971, fig. 9). Thus the volume of expansion under these extreme conditions can be dissipated by very small permeabilities, and we cannot assign a major role to the thermal process. If the thermal process becomes important at depth, it is only after the mechanical process has largely run its course.

The depth of generation of abnormal pressures is also important in understanding abnormal pressures because, as we shall see in the next chapter, the relative incompetence of abnormally pressured mudstones can lead to deformation of the regressive sequence due to mechanical instability, and this deformation can lead to the creation of petroleum traps. The mechanical hypothesis requires a shallow depth of initiation of abnormal pressures, and this depth is estimated by the effective compaction depth, z_e ($= \delta z$). The effective compaction depth is therefore given a real meaning: it indicates the maximum depth at which compaction equilibrium was lost, and the overburden at that time is represented by the sequence of sediments of thickness z_e above the mudstone being considered. This is found to be typically about 600 m or 2000 ft near the bottom of the transition zone, and to extend upwards into the sandy part of the overburden.

This interpretation is supported by occurrences of abnormal pressures at such depths. In their classic study of Pedernales in the Orinoco delta, Kidwell and Hunt (1958, p. 805, fig. 8) found that mudstones above 45 m (140 ft), less than 10,000 yrs old from radio-carbon dating, had pore pressures above normal hydrostatic. In the East Coast basin of North Island, New Zealand, the well Rotokautuku 1 encountered abnormal pressures at 356 m (1168 ft) and the well had to be abandoned at 627 m (2057 ft) because of the difficulties they caused (Katz, 1974, p. 469). Spinks (1970) reported abnormal pressures at 640 m (2057 ft) in the Gulf of Papua. Figure 14-8 shows the measured

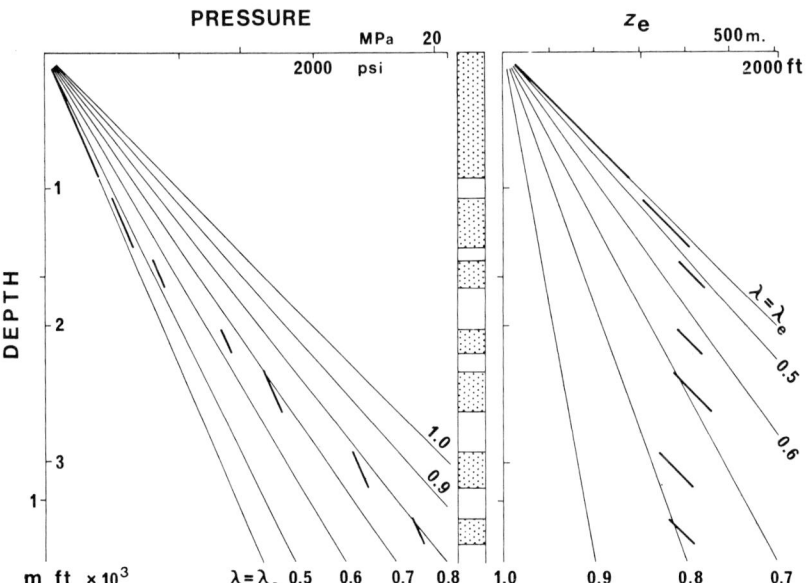

Fig. 14-8. Pressure-depth plot of shallow oil reservoirs in Trinidad, with their corresponding equilibrium depths. (Data courtesy of Royal Dutch/Shell Group.)

pressures in a sequence of reservoirs in Trinidad. The effective compaction depth of all of them is less than 500 m, and the top of abnormal pressures is a little shallower than 500 m (1640 ft).

The evidence of growth faults also supports such shallow depths. Dickinson (1953) found that the age of the mudstones at the top of abnormal pressures in the Louisiana Gulf Coast became younger towards the south, towards the Gulf, in the direction of the regression. Thorsen (1963) made an interesting study of growth faults in western Louisiana and found that the age of the beds showing maximum thickness contrast — that is, the age of maximum rate of growth fault movement — became younger towards the south, in the direction of the regression. Comparison of Thorsen's map (1963, p. 107, fig. 4) with Dickinson's (1953, pp. 416—417, fig. 3) shows that regionally the periods of maximum growth-fault movement occurred within a couple of biostratigraphic subzones above the top of abnormal pressures. Dickey et al. (1968) found that some growth faults in south-western Louisiana appeared to control the tops of abnormal pressures, and Fowler et al. (1971) found in the Midland field, Louisiana, that the top of abnormal pressures was stratigraphically higher and at shallower depth in the downthrown block of a growth fault than in the upthrown block, and the maximum rate of growth fault movement took place within the abnormally pressured sequence.

Thorsen (1963) observed that sand percentage is most closely related to contemporaneous structural growth near "the basinward limit of sand deposition, that is, in those areas of ten per cent or less sand". Harkins and Baugher (1969) observed that the top of abnormally pressured mudstone normally occurs regionally where there is less than five to ten per cent sand. It is difficult to avoid the conclusion that the two are causally related.

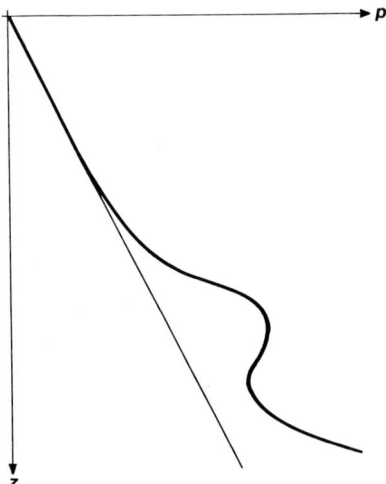

Fig. 14-9. Some transition zones contain zones of lesser abnormality.

Perhaps the most interesting phenomenon in abnormally pressured mudstones is the occurrence of zones of lesser abnormality (Fig. 14-9; see Fowler, 1970, p. 413). Such zones are necessarily below the "isolation depth" postulated by the thermal hypothesis, but the zone above the anomaly cannot be a zone of constant water density. If the thermal process is important, it is perhaps a zone of fractures. But it is logically unsatisfying to have different causes for the two sides of such an anomaly. The mechanical hypothesis would regard the anomaly as a drain with downward flow above it and upward flow below it, and lateral migration of the expelled fluids within the relatively permeable bed (see p. 61).

Finally, if the thermal process is dominant, there should be a tendency for the thickness of the transition zone to increase with increasing depth to the top of abnormal pressures. No world-wide study of this has been published, but experience of several important areas with abnormal pressures lends no support to such a relationship — rather the reverse. Transition zones may tend to be thinner with increasing depth to the top of abnormal pressures. Dickinson (1951, 1953, p. 420, fig. 6) plotted measured reservoir pressures against depth for six Louisiana Gulf Coast wells, and these suggest thinner transition zones with increasing depth to top of abnormal pressures. And in the Midland field, mentioned above, the thickness of the transition zones varies from block to block in one field. This is consistent with the mechanical hypothesis, which attributes the thickness of the transition zone to its permeability.

We therefore conclude, on present evidence, that at depths down to about 3 km at least, the mechanical process is dominant in most areas, and therefore that abnormal pressures are usually generated at shallow depth and, in thick mudstones, retained for significant periods of time.

We have concentrated on these two hypotheses because of their importance to petroleum geology. They are not just "academic", because several aspects of petroleum generation and migration depend on a proper understanding of the processes of development of abnormal pressures. If the thermal process is dominant in an area, the volume of fluids migrating from the mudstone is limited approximately to the volume created by thermal expansion, and the mudstone is, by the requirements of the hypothesis, an impermeable barrier separating the sequence below from the sequence above. If, therefore, a petroleum source rock is postulated in or below an abnormally pressured mudstone for an accumulation above it, consequential hypotheses have to be developed to account for primary migration. This seems to be the origin of the renewed interest in faults as conduits for petroleum migration, and their wide acceptance in that role. This topic has been discussed at length on pp. 245—251, but it is worth noting here that Dickinson's observation of abnormal pressures in sandstones isolated by faults (Dickinson, 1953, fig. 10, pp. 415 and 418 and p. 422 and fig. 11) indicates that these faults, at least, do not act as conduits. Many areas of the world are affected, and the problem must be resolved.

There are, of course, other possible causes of abnormally high pore pressures in mudstones and in permeable strata associated with mudstones.

OTHER POSSIBLE CAUSES OF ABNORMAL PRESSURES

Reservoir geometry

We have already seen (p. 161) that a column of oil or gas of great vertical extent can create high pressures near the top of the reservoir due to the weight density's of oil and gas being generally less than that of water. These are normal; but if the reservoir itself is abnormally pressured, reservoir geometry will add to the pressures. We shall not discuss this further*.

Clay-mineral diagenesis

The diagenesis of smectite (montmorillonite) to illite involves the release of inter-layer water. Powers (1967) suggested that this release of water could aid primary migration, and the hypothesis that this process could lead to abnormal pressures grew from this suggestion. *If* the density of inter-layer water is greater than that of free water, there is expansion on liberation that would result in a net decrease of bulk density of the mudstone, and abnormal pressures, unless the water can escape.

There is ample evidence that this diagenesis is real, but little that it contributes significantly to abnormal pressures. Powers regarded 1800 m (6000 ft) as the ceiling of this diagenesis. Burst (1969) concluded that it did not operate between depths of 800 and 2500 m in the U.S. Gulf Coast. Comparable depths were found by Perry and Hower (1970, p. 171), and rather greater depths were found by Weaver and Beck (1971, p. 18).

Magara (1978, pp. 100—109) has shown that the observed degree of undercompaction cannot be accounted for by smectite diagenesis because it cannot account for the bulk density decrease observed within the transition zone.

This diagenesis cannot contribute, it seems, to those abnormal pressures that lie above its ceiling, so it cannot be a general cause. If it can be established that inter-layer water has significantly greater density than free water, it is possible that release of this water to the pore spaces will contribute to abnormal pressures, and primary migration, at depths below two or three kilometers. As with thermal processes, the rate at which the diagenesis takes place is im-

---

* It is interesting, however, to read the discussion of Dickinson's 1951 paper to the World Petroleum Congress because some of those who spoke had Middle East experience of high pressures that were due to reservoir geometry.

portant: it must generate a volume of water greater than that that can be dissipated, so that the rate of generation exceeds the rate of dissipation.

Tectonic

Various authors have suggested *uplift* as a cause of abnormal pressures. A pressure regime that is normal for one depth would, if elevated and preserved, be abnormal for shallower depths. There may be areas in which this has occurred, but the unambiguous evidence of the major regressive sequences with abnormal pressures, such as the U.S. Gulf Coast, the Niger delta and several sedimentary basins of South-east Asia, is that the abnormal pressures were generated during subsidence and the accumulation of the regressive sequence, and that these areas are still subsiding (p. 352).

Likewise, tectonic *compression* cannot be a general cause because the Niger delta and the U.S. Gulf Coast are tectonically passive, and there is unambiguous evidence from growth faults that the stress field is and was one with a component of horizontal tension.

Nevertheless, tectonic compression could cause abnormal pressures in mudstones, and Berry (1973) has described a zone in California 650—800 km long and 40—130 km wide (400—500 × 25—80 miles) associated with the San Andreas fault.

There is little doubt that on a local scale, abnormal pressures may be associated with faulting, the pressures being due to mechanical deformation of the mudstone and the consequent tendency to reduce porosity.

Osmosis

When a semi-permeable membrane separates two liquids of different salinities, the less saline liquid moves through the membrane into the more saline liquid. This tendency to equalize chemical potentials across the membrane can lead to pressure differences. As a process that could generate abnormal pressures in thick mudstones of regressive sequences, it is unconvincing. Magara (1978, pp. 283—284) has argued that the osmotic gradient in a mudstone opposes the generation of abnormally high pore pressures in mudstones by assisting the expulsion of pore water. There is laboratory evidence (McKelvey and Milne, 1962; Von Engelhardt and Gaida, 1963) that water expelled early from a mudstone is more saline than that expelled later. And there is field evidence (Schmidt, 1973) that the salinity of mudstone pore water is less than that of the adjacent sandstones when the latter are normally pressured. When abnormally pressured, the salinity of sandstone pore water tends to be less than that of normally pressured sandstones, and more comparable to that of the abnormally pressured mudstone.

Hill et al. (1961) suggested that osmosis might be the cause of abnormally low pore pressures in the Mesaverde sandstone in central U.S.A. and the Viking

Sandstone in Canada, both of which not only have pressures below normal hydrostatic but have waters that are fresh. The same conditions are found in the Molasse basin of southern Germany.

Some pore pressure anomalies appear to be related to osmosis (see Fertl, 1976, pp. 33—36), but there is no evidence that the generality of abnormal pressures is so caused, and some that it is not so caused.

Petroleum generation

Petroleum generation could affect mudstone pore pressures in two ways: the generation of petroleum itself could involve an increase in net volume, and the products could reduce the relative permeabilities to water and to petroleum. We shall take the latter first.

Chapman (1972) noted that upward flow of water through the transition zone is towards lower pressures, and concluded that any petroleum exsolution during this migration would reduce the effective permeability of the mudstone to water and so reduce the rate of expulsion of pore fluids. Such a process cannot eliminate permeability, but if relative permeability diagrams determined for sandstone reservoirs (p. 168, Fig. 8-12) are generally applicable to mudstones, the permeability to both fluids together (i.e., the sum of their relative permeabilities) may be reduced to about 1/10th of its value for a single fluid.

Visser and Hermes (1962, p. 228) described the drilling of the Gesa anticline in the Mamberamo delta, Irian Jaya, and the problems encountered with abnormal pressures. Methane and some carbon dioxide were found in the mudstones only, and so could have contributed to these pressures by reducing the permeability to the fluids. Gas is sometimes (but not always) detected in the mud while drilling the transition zone (see Fertl, 1976, pp. 137—140, for a fuller discussion).

Illing (1938, p. 227) stated that the generation of oil from organic matter involved an increase in volume, and that this could give rise to considerable pressures in compacted rocks. Hedberg (1974, 1980) suggested that the generation of methane and other hydrocarbons of low molecular weight is an important source of energy for primary migration, abnormal pressures, mudstone diapirs, and mudvolcanoes. Such pressures are enhanced and retained by the effects on effective permeabilities.

In the Lower Mississippian Bakken Shale of the Williston basin in North America, Meissner (1978) found abnormal pressures in certain areas over certain intervals. The prevailing pressure regime in the sequence is even a little below normal hydrostatic, and Meissner argued that the abnormality is directly due to oil generation in a relatively well-compacted source rock.

Similarly, the Kimmeridge Clay (Jurassic) of the northern North Sea is overpressured and considered by many to be the source rock of much of the North Sea oil. This example is confused, but not irrelevantly to our purpose, by the fact that it is unusually radioactive, and also hot. Héritier et al. (1981, p. 381) reported that the Frigg discovery well penetrated 77 m of oil shale of Kimmeridgian age that registered 250 API units of gamma radiation at about 4200 m (but they do not regard it as the source rock for the Frigg petroleum because the vitrinite reflectance is only 1%).

From these examples, it is clear that attention must be paid to the geophysical properties of the fine-grained rocks that might be petroleum source rocks. Source rocks generating petroleum may also generate local abnormality of pressure, resistivity, sonic velocity, and their temperatures may well provide more direct evidence of the requirements for oil and gas generation.

REFERENCES

Barker, C., 1972. Aquathermal pressuring — role of temperature in development of abnormal-pressure zones. *Bull. Am. Ass. Petrol. Geol.*, 56: 2068—2071.
Barker, C. and Horsfield, B., 1982. Mechanical versus thermal cause of abnormally high pore pressures in shales: discussion. *Bull. Am. Ass. Petrol. Geol.*, 66: 99—100.
Berry, F.A.F., 1973. High fluid potentials in California Coast Ranges and their tectonic significance. *Bull. Am. Ass. Petrol. Geol.*, 57: 1219—1249.
Burst, J.F., 1969. Diagenesis of Gulf Coast clayey sediments and its possible relation to petroleum migration. *Bull. Am. Ass. Petrol. Geol.*, 53: 73—93.
Burst, J.F., 1976. Argillaceous sediment dewatering. *Annu. Rev. Earth Planet. Sci.*, 4: 293—318.
Chapman, R.E., 1972. Primary migration of petroleum from clay source rocks. *Bull. Am. Ass. Petrol. Geol.*, 56: 2185—2191.
Chapman, R.E., 1980. Mechanical versus thermal cause of abnormally high pore pressures in shales. *Bull. Am. Ass. Petrol. Geol.*, 64: 2179—2183.
Chapman, R.E., 1982. Mechanical versus thermal cause of abnormally high pore pressures in shales: reply. *Bull. Am. Ass. Petrol. Geol.*, 66: 101—102.
Dickey, P.A. and Cox, W.C., 1977. Oil and gas reservoirs with subnormal pressures. *Bull. Am. Ass. Petrol. Geol.*, 61: 2134—2142.
Dickey, P.A., Shriram, C.R. and Paine, W.R., 1968. Abnormal pressures in deep wells of southwestern Louisiana. *Science*, 160: 609—615.
Dickinson, G., 1951. Geological aspects of abnormal reservoir pressures in the Gulf Coast region of Louisiana, U.S.A. *Proc. 3rd World Petrol. Congress*, Sect. 1: 1—16; discussion, 1: 16—17.
Dickinson, G., 1953. Geological aspects of abnormal reservoir pressures in Gulf Coast Louisiana. *Bull. Am. Ass. Petrol. Geol.*, 37: 410—432.
Fertl, W.H., 1976. *Abnormal formation pressures*. Elsevier, Amsterdam, 382 pp.
Fowler, W.A., 1970. Pressures, hydrocarbon accumulation, and salinities —Chocolate Bayou field, Brazoria County, Texas. *J. Petrol. Technol.*, 22: 411—423.
Fowler, W.A., Boyd, W.A., Marshall, S.W. and Myers, R.L., 1971. Abnormal pressures in Midland field, Louisiana. In: *Abnormal subsurface pressure: a study group report 1969—1971*. Houston Geological Society, Houston, Texas, pp. 48—77.
Frederick, W.S., 1967. Planning a must in abnormally pressured areas. *World Oil*, 164 (4): 73—77.
Funkhouser, H.J., Sass, L.C. and Hedberg, H.D., 1948. Santa Ana, San Joaquín, Guario,

and Santa Rosa oil fields (Anaco fields) central Anzoátegui, Venezuela. *Bull. Am. Ass. Petrol. Geol.*, 32: 1851—1908.

Harkins, K.L. and Baugher, J.W., 1969. Geological significance of abnormal formation pressures. *J. Petrol. Technol.*, 21: 961—966.

Hedberg, H.D., 1974. Relation of methane generation to undercompacted shales, shale diapirs, and mud volcanoes. *Bull. Am. Ass. Petrol. Geol.*, 58: 661—673.

Hedberg, H.D., 1980. Methane generation and petroleum migration. In: W.H. Roberts and R.J. Cordell (Editors), Problems of petroleum migration. *Am. Ass. Petrol. Geol., Stud. Geol.*, 10: 179—206.

Héritier, F.E., Lossel, P. and Wathne, E., 1979. Frigg field — large submarine-fan trap in Lower Eocene rocks of North Sea Viking graben. *Bull. Am. Ass. Petrol. Geol.*, 63: 1999—2020. Also in: M.T. Halbouty (Editor), Giant oil and gas fields of the decade 1968—1978. *Mem. Am. Ass. Petrol. Geol.*, 30 (1980): 59—79.

Héritier, F.E., Lossel, P. and Wathne, E., 1981. The Frigg gas field. In: L.V. Illing and G.D. Hobson (Editors), *Petroleum geology of the continental shelf of north-west Europe.* Heyden/Institute of Petroleum, London, pp. 380—391.

Hill, G.A., Colburn, W.A. and Knight, J.W., 1961. Reducing oil-finding costs by use of hydrodynamic evaluations. In: *Economics of petroleum exploration, development, and property evaluation.* Prentice-Hall, Englewood Cliffs, N.J., pp. 38—69.

Hottmann, C.E. and Johnson, R.K., 1965. Estimation of formation pressures from log-derived shale properties. *J. Petrol. Technol.*, 17: 717—722.

Illing, V.C., 1938. The origin of pressure in oil pools. In: A.E. Dunstan (Editor), *The science of petroleum*, Vol. 1. Oxford University Press, London, pp. 224—229.

Katz, H.R., 1974. Recent exploration for oil and gas. In: G.J. Williams (Editor), Economic geology of New Zealand. *Australas. Inst. Min. Metall., Monogr. Ser.*, 4: 463—480.

Kennedy, G.C. and Holser, W.T., 1966. Pressure—volume—temperature and phase relations of water and carbon dioxide. In: S.P. Clark (Editor), Handbook of physical constants (rev. ed.). *Geol. Soc. Am., Mem.*, 97: 371—384.

Kidwell, A.L. and Hunt, J.M., 1958. Migration of oil in Recent sediments of Pedernales, Venezuela. In: L.G. Weeks (Editor), *Habitat of oil.* Am. Ass. Petrol. Geol., Tulsa, Okla., pp. 790—817.

Lewis, C.R. and Rose, S.C., 1970. A theory relating high temperatures and overpressures. *J. Petrol. Technol.*, 22: 11—16.

Magara, K., 1971. Permeability considerations in generation of abnormal pressures. *J. Soc. Petrol. Engrs.*, 11: 236—242.

Magara, K., 1978. *Compaction and fluid migration: practical petroleum geology.* Elsevier, Amsterdam, 319 pp.

McKelvey, J.G. and Milne, I.H., 1962. Flow of salt solutions through compacted clay. *Clays Clay Minerals, Proc. 9th Natl. Conf.*, pp. 248—259.

Meissner, F.F., 1978. Petroleum geology of the Bakken Formation, Williston basin, North Dakota and Montana. In: *The economic geology of the Williston basin; Montana, North Dakota, South Dakota, Saskatchewan, Manitoba.* Montana Geol. Soc., Billings, pp. 207—227.

Mulhadiono and Marinoadi, 1977. Notes on hydrocarbon trapping mechanism in the Aru area, north Sumatra. *Proc. Indon. Petrol. Ass.*, 6: 95—115.

Perry, E. and Hower, J., 1970. Burial diagenesis in Gulf Coast pelitic sediments. *Clays Clay Minerals.*, 18: 165—177.

Perry, E.A. and Hower, J., 1972. Late-stage dehydration in deeply buried pelitic sediments. *Bull. Am. Ass. Petrol. Geol.*, 56: 2013—2021.

Powers, M.C., 1967. Fluid-release mechanisms in compacting marine mudrocks and their importance in oil exploration. *Bull. Am. Ass. Petrol. Geol.*, 51: 1240—1254.

Schmidt, G.W., 1973. Interstitial water composition and geochemistry of deep Gulf Coast shales and sandstones. *Bull. Am. Ass. Petrol. Geol.*, 57: 321—337. (See also discussion by Osmaston, 1975, in *Bull. Am. Ass. Petrol. Geol.*, 59: 715—720.)

Smith, N.E. and Thomas, H.G., 1971. Origins of abnormal pressures. *In: Abnormal subsurface pressure: a study group report 1969—1971*. Houston Geological Society, Houston, Texas, pp. 4—19.

Spinks, R.B., 1970. Offshore drilling operations in the Gulf of Papua. *J. Aust. Petrol. Explor. Ass.*, 10 (1): 108—114.

Thomeer, J.H.M.A. and Bottema, J.A., 1961. Increasing occurrence of abnormally high reservoir pressures in boreholes, and drilling problems resulting therefrom. *Bull. Am. Ass. Petrol. Geol.*, 45: 1721—1730.

Thorsen, C.E., 1963. Age of growth faulting in southeast Louisiana. *Trans. Gulf-Coast Ass. Geol. Socs.*, 13: 103—110.

Versluys, J., 1932. Factors involved in segregation of oil and gas from subterranean water. *Bull. Am. Ass. Petrol. Geol.*, 16: 924—942.

Visser, W.A. and Hermes, J.J., 1962. Geological results of the exploration for oil in Netherlands New Guinea. *Verh. K. Ned. Geol. Mijnbouwk. Genootsch. Nederland Kolonien*, Geol. Ser., 20: 1—265.

Von Engelhardt, W. and Gaida, K.H., 1962. Concentration changes of pore solutions during the compaction of clay sediments. *J. Sediment. Petrol.*, 33: 919—930.

Weaver, C.E. and Beck, K.C., 1971. Clay water diagenesis during burial: how mud becomes gneiss. *Spec. Pap. Geol. Soc. Am.*, 134, 96 pp.

CHAPTER 15

DIAPIRS, DIAPIRISM AND GROWTH STRUCTURES

SUMMARY

(1) Diapirs, in the context of petroleum geology, are intrusions of sedimentary rocks, primarily salt or mudstone, into the overlying sedimentary sequence. Incipient diapirs are salt pillows and the analogous mudstone pillows or "shale masses". Deformation of the sedimentary rocks around and above diapirs and incipient diapirs creates potential petroleum traps.

(2) Diapirs are initiated by unequal loading of a layer of material of small equivalent viscosity. The common diapiric materials — salt and abnormally pressured mudstone — may be less dense than the normally compacted sedimentary rocks overlying them. Hence, once a diapir has been initiated (particularly a salt diapir), the forces of buoyancy tend to elongate the deformation vertically.

(3) The upward movement of a diapir is relative to the surrounding sedimentary rocks. The accumulation of a sedimentary sequence over a diapir indicates that it was subsiding with the development of the sedimentary basin. The upward movement is only absolute if the relative movement is faster than the subsidence of the surrounding sedimentary sequence.

(4) This differential subsidence may influence the accumulation of sediments, contributing to the variations of loading on the diapiric mother bed.

(5) The mechanical properties of the diapiric material change with time and position. Salt becomes less viscous with increasing temperature. Mudstone viscosity is a function of pore pressure and depth as well as temperature.

(6) A diapir commonly, but not invariably, shows a gravity minimum. This indicates a deficiency of mass.

(7) Failure of the overburden by faulting may accompany diapiric development; but diapiric development may also inhibit subsidence locally at the surface of accumulating sediment and so lead to a local stratigraphic hiatus.

(8) Diapirism is necessarily contemporaneous with the expulsion of pore fluids from compacting mudstones. Diapiric mudstone may also be a petroleum source rock.

DIAPIRS

Diapirs are essentially intrusions of deeper material into the overlying material of the Earth's crust. The processes of diapirism are dynamic, and lead to structures that range from minor displacements of plastic material to major intrusions of large volumes of material through considerable thicknesses of overlying rocks. Not all the structures that are called diapiric are intrusive, but the development of a true diapir probably passes through stages of *incipient* diapirism in which the mobile material forms ridges and low domes (Fig. 15-1). In plan, diapirs tend to acquire a more or less circular outline: in section, the amplitude may achieve dimensions of several kilometres. The scale of diapirs ranges from kilometres down to centimetres (e.g., load casts). They commonly occur in groups, in lines, or in lines of groups. They may be intimately associated with folding and faulting, and they demonstrate that certain rocks under stress will flow as a quasi-fluid or viscous solid. The materials of diapirs include ice, peat, evaporites (especially salt), mudstones and marls, occasionally sands, and some igneous rocks.

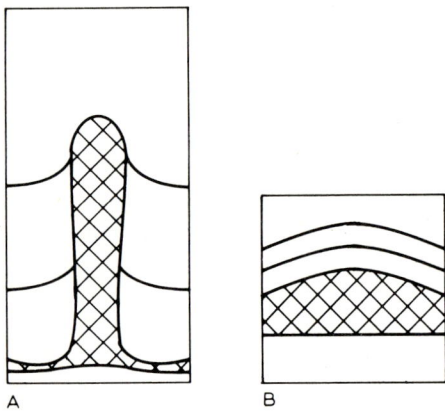

Fig. 15-1. Diapir (*A*) and incipient diapir (*B*).

Confining our attention to diapirs of sedimentary rocks, we find them only in sedimentary basins, of course, in rocks of most geological ages from Proterozoic to Holocene, and in all continents except Antarctica (so far). They are common in the petroleum provinces of the Gulf Coast of the United States of America and offshore in the Gulf of Mexico, in the Middle East, the Caucasus and adjoining regions to the north of the Caspian Sea, and in north-west Europe (Braunstein and O'Brien, 1968). They may be equally common in non-petroleum provinces that have not received the same intensity of geological and geophysical investigation. Geophysical investigation has revealed salt diapirs in sedimentary sequences off the continental shelf of West Africa in water depths to 4 km (13,000 ft) (Beck, 1972; Beck and Lehner, 1974), and in the submarine parts of most of the major deltas.

Diapirs commonly, but not invariably, occupy areas of gravity minima. A local gravity minimum over a diapir indicates a deficiency of mass despite the intrusion of deeper material to shallower depths.

Salt diapirs

Salt diapirs occur in large areas of the Gulf Coast province of North America, north-west Europe, Russia, and around the Arabian or Persian Gulf. Some have reached the surface, or are at very shallow depths, from a mother layer more than 5 km deep. Others take the self-explanatory shape of *salt pillows*, forming the core of diapiric anticlines. The truly intrusive forms are known as salt *domes, plugs,* or *stocks.*

The mining of salt near the surface has shown that the salt is intensively deformed, with complicated flow patterns (rather than folds) but very rare faults. The external form, however, tends to be more regular. Many salt domes, particularly those at shallow depth, have developed a cap rock that consists mainly of anhydrite or gypsum. The cap rock is the less soluble residue from leaching of the salt by ground water. Salt domes may be sheathed in a thin skin of anhydrite or mudstone "gouge". Some cap rock contains sulphur that was commercially extractable until the cleaning of sour gas provided a cheaper and more abundant source. The sulphur of salt domes is associated, probably biogenically, with petroleum-bearing diapiric structures.

Although a salt-dome is usually roughly circular in plan, the cap rock commonly extends laterally over a wider area than the main part of the dome, which may be more or less cylindrical, or narrowing downwards to the mother layer of salt from which the dome was supplied. Characteristically, the upper surface of the mother layer is deformed near the salt dome into a *rim syncline,*

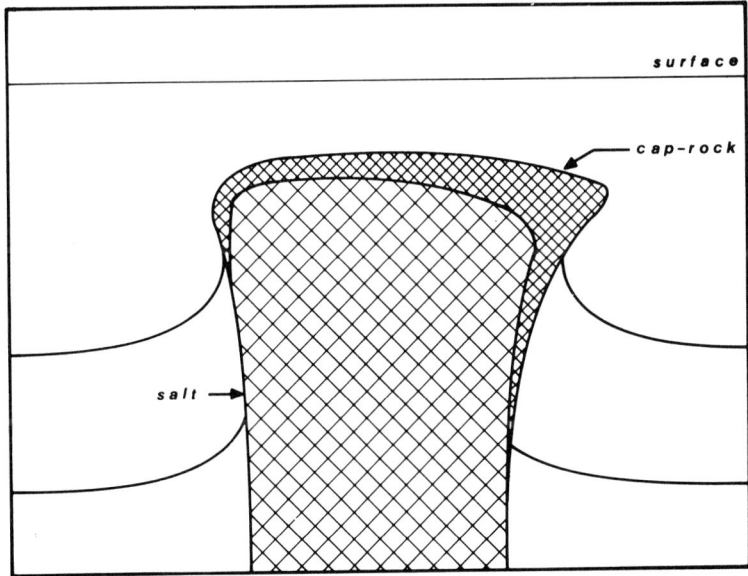

Fig. 15-2. Diagrammatic cross-section through top of well-developed salt diapir.

rim sink, or peripheral sink. A well-developed salt dome may therefore have a profile as depicted in Fig. 15-1A and 15-2.

Structurally, salt domes are important for two reasons. First, the deformation of the overlying sedimentary rocks and the rocks through which the dome has penetrated may lead to traps for petroleum either in the overlying anticlinal form or by the truncation of potential reservoirs by the relatively impermeable salt. Accumulations in such traps may also be affected by faulting that resulted from the deformation. Secondly, three or more salt domes in a group may form an anticlinal trap by virtue of their rim synclines. The structure of the sedimentary rocks over and around a salt dome is usually extremely complex. Strata tend to be variably inclined, and faulted with predominantly radial faults that die out away from the dome (Fig. 15-3). There is no means of knowing whether lateral continuity of particular rock units existed prior to their penetration by salt; but local hiatus are common in the strata overlying domes, and they are commonly associated with features that suggest that the growth of the dome affected the accumulation of sediment.

Some salt diapirs in the Gulf Coast province of North America are also intimately associated with mudstone diapirs, both as a single structure and as separate structures.

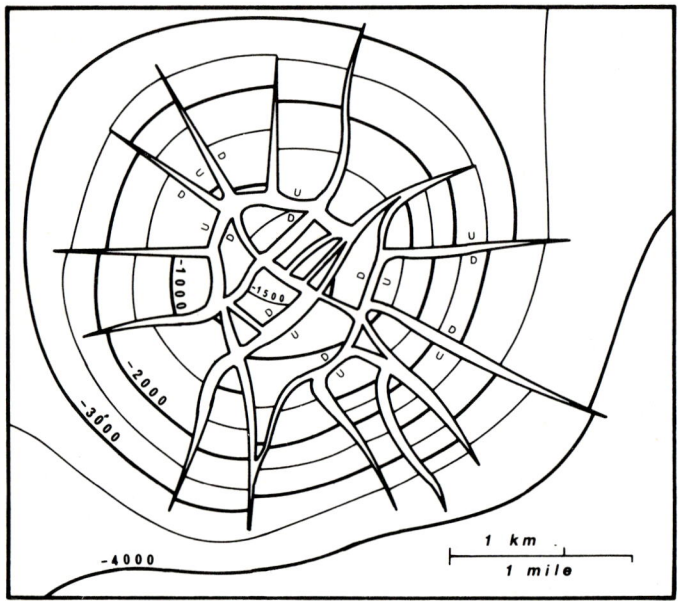

Fig. 15-3. Stratum contour map on Wilcox sand, just above salt of Clay Creek dome, Texas. Contours in feet; contour interval, 500 ft (150 m approximately). (After Parker and McDowell, 1953, p. 2085, fig. 8.)

Mudstone (shale) diapirs

Mudstone does not seem to develop into such clearly defined diapirs as salt. The expressions of mudstone diapirism are typically two:
— True mudstone diapirs that are penetrative stocks. These may have a surface expression in the form of mud volcanoes.
— Incipient diapirs, forming the cores of anticlines, similar to salt pillows.

True mudstone diapirs, and incipient diapirs, typically contain fluids at abnormally high pressures. The true diapirs are sheathed in compacted mudstone or shale. Dips measured in boreholes that penetrate the diapir are usually found to be steep but regular, the result of flow rather than folding. Geophysically, they are low-velocity, low-density features; and drilling has found them to be undercompacted, with small mechanical strength and small equivalent viscosity. These indicate, of course, similar properties for the mother layer. The preservation of these properties in the diapir suggest that the mudstone has not been intruded upwards from great depth, and the accumulation of sediment over the diapir is strong evidence that the whole diapir was subsiding with the sedimentary basin.

Mud volcanoes occur commonly in younger sedimentary basins around the world, and seem to be associated strongly with regressive sequences. They are reported from Trinidad and northern South America, the Gulf Coast province of North America, Asia Minor, Pakistan, Burma, Indonesia, Borneo, New Guinea and New Zealand. The principle of uniformitarianism requires us to postulate that they also occurred in older sedimentary basins, but subsequent geological events have obscured them.

Salt or brackish water is the main fluid of mud volcanoes, but gas (mainly methane and carbon dioxide) and oil also occur. The fluid is usually warm, and the activity intermittent. The fluid is clearly the fluid of expulsion from the mudstone, with some perhaps coming from other material incorporated into the flow.

In the Mississippi delta, mounds and small islands appear from time to time, due to mud volcanism with less than 100 m of overburden. These mud-lumps, as they are called, have long been known, and their intermittent activity (which can be a hazard to navigation) has been attributed to the variable depositional patterns of the delta (see Lyell, 1867, pp. 447—454 for an early description and interpretation of them). Investigation with core holes revealed folds and overthrusts that clearly have a purely gravitational origin (Morgan et al., 1968).

Most mud volcanoes can be considered as mudstone diapirs that have reached the surface, and their activity will continue until sufficient fluid has been expelled from the mudstone to halt the diapiric tendency. The bedding of the mudstone is commonly destroyed, and rock material may be included that has demonstrably been brought from another formation at depth. Some mud volcanoes have apparently resulted from intrusion up a fault plane (Fig.

15-4) — but it is not easy to determine the causal relationship. It seems likely that the diapir caused the fault in many cases, but the two are closely related.

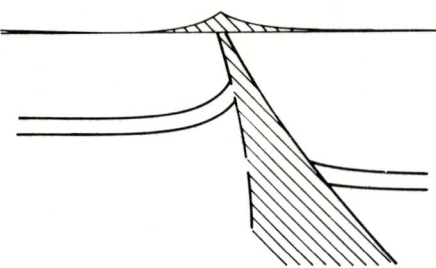

Fig. 15-4. Mud volcanism and mudstone diapirism associated with a fault (diagrammatic).

Generalizations

Diapirs, of whatever sedimentary material, are characteristically overlain by a sequence of sedimentary rocks that is, of course, younger than the material of the diapir. The accumulation of these sedimentary rocks must be taken as clear evidence that they were *subsiding* during periods of sediment accumulation. If the growth of a diapir is reasonably accurately depicted by Fig. 15-5, which is consistent with the observations reported in an extensive literature, most of the strata now penetrated by the diapir were once continuous across it, like those that have not yet been penetrated. If the growth of a diapir is accelerating, there may come a time when there is absolute upward movement at a sufficient rate to inhibit sediment accumulation, and stratigraphic continuity will be broken. This is essentially the concept of "downbuilding" that Barton (1933) proposed.

DIAPIRISM

Diapirism is a dynamic process that takes place under the force of gravity during the accumulation of sediment in a developing sedimentary basin. Its significance for the petroleum geologist is that it is a process that deforms the sedimentary strata while they are compacting, during fluid expulsion from the more compactible lithologies. Mudstone diapirism is probably more significant than salt diapirism because the mudstone itself may be a petroleum source rock, and mudstone diapirism is essentially contemporaneous with fluid expulsion from the mudstone.

Diapirism involves the flow of diapiric material. When referring to the flow of rocks, we use the term "equivalent viscosity" because the term "viscosity" may suggest that Newtonian viscosity is involved, with the velocity of flow at a point proportional to the distance of the point from a static boundary. Carey (1954) argued that the conventional classification of matter into

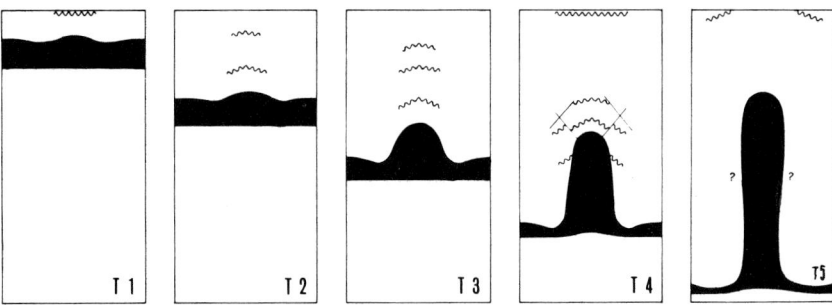

Fig. 15-5. Growth of a salt diapir from a subsiding mother layer with time. The question marks at T5 denote the writer's uncertainty about the deformation from T4 to T5.

fluids and solids has a dividing line that shifts towards the solid side as the dimension of time increase, so that forces acting on a material for a short time may produce results that indicate that the material is a solid, while lesser forces acting on the same material for a longer time produce results that indicate that the material is a fluid. The concept of rock deformation by quasi-fluid or quasi-plastic processes is familiar to all geologists. It is evident in folded strata, and in many common materials. Pitch, for example, behaves as a brittle solid when struck with a hammer, but deforms as a viscous solid when subjected to small stresses over a longer time.

These properties, which are inferred for all natural materials in some degree, were embodied by Carey (1954) into the concept of *rheidity*, which he defined as the ratio of absolute viscosity (a measure of resistance to flow, with dimensions $M\ L^{-1}\ T^{-1}$) to rigidity (a measure of resistance to elastic deformation, with dimensions $M\ L^{-1}\ T^{-2}$) multiplied by an arbitrary factor of 1000 to eliminate the trivial. This ratio has the dimension of *time*. The rheidity of a given material decreases with increasing temperature. Too little is known about the physical properties of salt and mudstone in the subsurface to use this concept quantitatively, but the concept is nevertheless useful in emphasizing the dimension of time in the deformation of sedimentary rocks.

The dynamic forces in the sedimentary column may act for periods of time greatly in excess of the rheidity of the materials. These materials may then be regarded, both generally and mathematically, as fluids. This concept is implicit in the scaling of physical models (Hubbert, 1937) because viscosity has a dimension of time. Properly scaled model materials resemble fluids much more than the materials they represent*.

---

* The resemblance is not complete. For example, turbulence in a water pipe is found to be characterized by a Reynolds number greater than about 2000. The Reynolds number is the ratio Dq/ν, where D is the internal diameter of the pipe, q is the mean velocity of the water (volumetric rate divided by cross-sectional area), and ν is the kinematic viscosity of the water. No reasonable figures for salt and salt domes gives a Reynolds number anywhere near 1, let alone 2000, yet the internal evidence seems to indicate turbulent flow.

The formation of a diapir involves the displacement of diapiric material from the mother layer to the diapir under dynamic forces acting over long spans of time. The flow lines in the mother layer around a well-developed diapir are centripetal, and the material moves down an energy gradient analogous to the fluid potential gradient around a producing oil or water well. If the mother layer is horizontal, the pressure in it adjacent to the diapir is less than that further away, because this is a necessary condition of horizontal flow. The rim syncline, or peripheral sink, around a diapir is therefore to be regarded as an expression of the potential energy of the mother layer analogous to the drawdown of the water table around a producing water well (Ramberg, 1981). This sink is terminated on the inside by the upward drag of the diapir. This is clear for a well-developed diapir.

During stages of incipient diapirism, unequal loading of a potential mother layer (with low equivalent viscosity) creates a disequilibrium that may be restored by flowage from the more heavily loaded areas to the less heavily loaded areas. With mudstone diapirism in mind, this is generally away from the marine margin of the physiographic basin. Locally, however, differences will exist, some of which will be minor, others more important. Because subsidence may locally increase the capacity to accumulate sediment, a further inequality of loading may follow consequentially on initial inequality.

Diapirism, like many geological processes, is not easily reduced to simple statements of cause and effect. There is, however, general agreement on the main factors that contribute to diapirism, even if there is disagreement on the relative importance of each. The main factors are: (1) low equivalent viscosity in the material that contributes to, and forms the diapir; (2) the load on the mother layer, and the variations of the load in space and time; (3) the bulk density of the diapiric material relative to that of the overburden; and (4) the thickness of the mother layer (but perhaps this is only a matter of scale).

None of these factors by itself necessarily leads to diapirism. Not all salt layers feed diapirs, for example, and there are many density inversions with depth in the geological column that do not lead to diapirism. But physical and mathematical models have been constructed that reproduce the essential features of diapirs, both individually and collectively. In physical models there are problems of scaling, but a wide variety of materials leads to structures that resemble real diapirs and incipient diapirs closely.

At the onset of instability, the interface between the overburden and mother layer becomes wavy, and both types of models indicate that some wave lengths become more strongly amplified than others (see, for example, Biot and Odé, 1965). The more strongly amplified wavelength, called the dominant wave length, is affected by the viscosity ratio and the thickness ratio of the overburden and mother layer. Overburdens of larger equivalent viscosity tend to be deformed with a longer dominant wavelength, and the rate of diapiric growth is slower. As the thickness ratio is increased, Biot and

Odé found that the rate of growth was significant for ratios with values of 1 to 5 (overburden/mother layer) and that the dominant wavelength was about 10 to 20 times the thickness of the mother layer. The relative density contrast affects the rate of development, but not the dominant wavelength.

Most physical models have to be started artificially; and mathematical models must assume an initial deformation of the interface between the mother layer and the overburden. But this need not detain us, because rock units in a sedimentary sequence must not be regarded as a Dickensian geological cake, and the interfaces between rock units are rarely, if ever, horizontal planes. Furthermore, diapirism is a dynamic process taking place during the evolution of a sedimentary basin, and the variation of all the parameters with space and time cannot easily be incorporated into models. Nevertheless, both physical and mathematical models lead to results that are consistent with each other and with Nature.

Consider now a vessel in which a liquid is overlain and contained by two identical, frictionless pistons (Fig. 15-6). The interface between the pistons and the liquid is a plane because the vertical forces acting through each piston are equal. If one piston is now loaded with a weight W and the other with a weight $2W$ (merely a device to remind one that in nature the whole area may be loaded, but unequally) the more heavily loaded piston will sink relative to the less loaded piston, displacing a weight W of liquid to the volume under the less loaded piston. If this is a valid, though simple, model of diapiric process with a relatively rigid overburden, low equivalent viscosity and unequal loading are the two essential factors in incipient diapirism. The density of the liquid relative to that of the pistons is irrelevant.

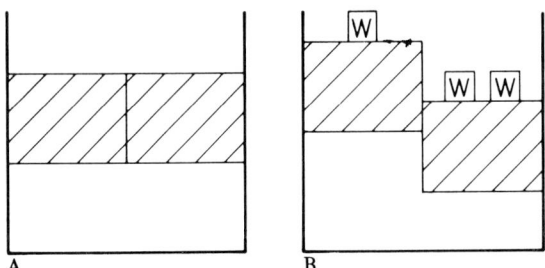

Fig. 15-6. Simple model of diapirism as a result of loading.

However, the common materials with low equivalent viscosity — salt and, under certain circumstances, mudstone — are also less dense than the average overburden.

The mean mass density of the overburden has been assumed to be 2300 kg m^{-3}, for a weight density of 22.6 kPa/m (1 psi/ft), and this is a useful figure for general purposes. However, bulk density is a linear function of porosity and mean bulk grain density, and the relationship:

$$\gamma_{bw} = \gamma_s - f(\gamma_s - \gamma_w) \tag{15.1}$$

indicates that the assumed mean weight density implies a mean porosity close to 22%. Interbedded mudstones and sandstones may have such a mean density but, as we saw in the discussion of abnormal pressures, the porosity and bulk density of abnormally-pressured mudstones correspond to normally compacted mudstone at a shallower depth. Dickinson (1953, p. 429) appreciated the difficulties and showed that the error in taking the mudstone compaction curve for estimating the overburden *pressure* is negligible (at least in the Louisiana Gulf Coast).

However, we are concerned more with the relative densities of the sediments involved in mudstone diapirism, and the model chosen for this is a regressive sequence of sedimentary rocks in which the permeable facies loads the compactible.

Mudstone density depends on its state of compaction (with minor variation on account of the mineralogy). The formula 3.5a relating porosity, depth, and pore-fluid pressure (through the parameter δ) is:

$$f = f_0\, e^{-\delta z/b}. \tag{15.2}$$

Substituting this into eq. 15.1:

$$\begin{aligned}\gamma_{bw} &= \gamma_s - f_0\, e^{-\delta z/b}\,(\gamma_s - \gamma_w) \\ &\simeq 26.0 - 7.9\, e^{-\delta z/b}\ \text{kPa/m}\end{aligned} \tag{15.3}$$

where the scale length b can be estimated from the sonic log. However, the sonic transit time in mudstone is a linear function of porosity (p. 49), so we may use the sonic log with a porosity scale:

$$\begin{aligned}\gamma_{bw} &= \gamma_s - f_0 \left(\frac{\Delta t - \Delta t_{ma}}{\Delta t_0 - \Delta t_{ma}}\right)(\gamma_s - \gamma_w) \\ &\simeq 26.0 - 0.072\,(\Delta t - 55)\ \text{kPa/m}.\end{aligned} \tag{15.4}$$

Figure 15-7 shows a density inversion from about 23.3 kPa/m (ρ_{bw} = 2380 kg m^{-3}) at 1350 m to about 20.6 kPa/m (ρ_{bw} = 2100 kg m^{-3}) by 1800—2000 m, with the density increasing below this. (The very long transit times on the diagram are probably due to hole caving and are spurious.) All the ingredients for diapiric deformation exist — density inversion, reduced equivalent viscosity, unequal loading — and this is a mechanically unstable sequence. The mechanical boundary between the overburden and the potential mother layer is not stratigraphic-lithologic, but rather the top of abnormal pressures.

The vertical, buoyant forces acting on an incipient diapir by virtue of the differing weight densities of the diapiric and overburden materials are of the form:

$$S' = (\gamma_{diapir} - \gamma_{overburden})\,h \tag{15.5}$$

where S' is the vertical component of deviatoric stress in the diapir, and h is the mean height of the diapir (in units consistent with the weight density). The total stresses are all compressive. As a diapir grows, so the deviatoric stress increases, tending to accelerate the growth of the diapir. But in mudstone diapirs, all the quantities in eq. 15.5 change with time. In particular, the mudstone tends to compact, and eventually mechanical equilibrium will be reached. The *effect* of these forces depends on the size of the forces and the time rate of their application, and on the mechanical properties of the rocks on which they act.

While on the topic of densities, it is worth noting that a large gas, or even oil, accumulation constitutes a local density anomaly within the sedimentary sequence that may encourage structural growth. A sandstone with 30% porosity, for example, has a mass density:

$$(2650 \cdot 0.7) + (1040 \cdot 0.3) = 2167 \text{ kg m}^{-3}$$

where it is water-saturated. Within the gas accumulation, assuming 20% water saturation, the bulk density may be:

$$(2650 \cdot 0.7) + (1040 \cdot 0.3 \times 0.2) + (200 \cdot 0.3 \times 0.8) = 1965 \text{ kg m}^{-3}$$

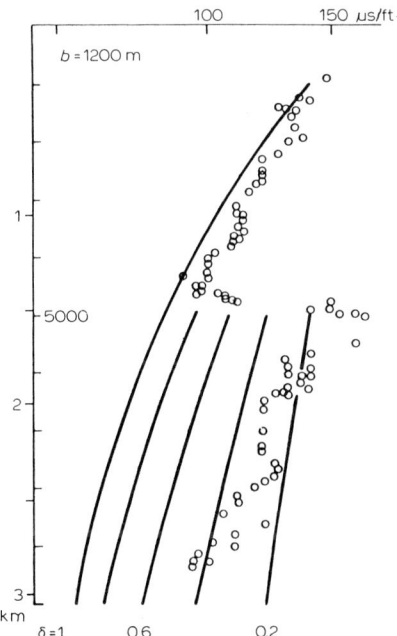

Fig. 15-7. The shale-transit-time plot of mudstone in a well in Borneo indicates a density inversion below about 1350 m, down to nearly 2000 m. (Data courtesy of the Royal Dutch/Shell Group.)

a reduction of 9%. The effect of this on the whole sedimentary column will be very small, but sustained for a great length of time — and perhaps reinforcing a general diapiric tendency — its effect may be real.

DISCUSSION

Salt diapirism is a process that is relatively well understood: it can be modelled satisfactorily, and the features of the models correspond qualitatively, at least, with geological observations. Conversely, geological analysis of areas of diapirism can reveal in some detail the evolution of the diapirs.

A particularly good example of this is the beautiful work of Trusheim (1957, 1960; the name is Trus-heim, not Trush-eim) and Sannemann (1968) in northern Germany (Fig. 15-8). Here, up to 1 km of Zechstein salt (Permian) accumulated, and was followed by Triassic clastic sediments that also accumulated to a thickness of about 1 km. The development of the salt diapirs is

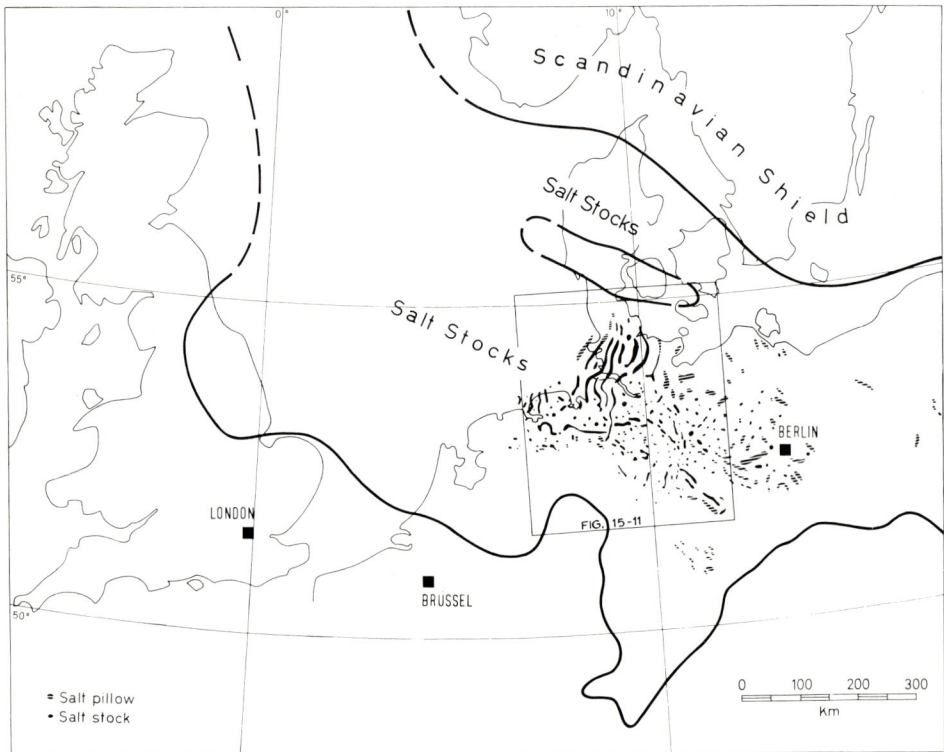

Fig. 15-8. Zechstein (U. Permian) salt-stock area, north-western Germany. (Reproduced from Sannemann, 1968, p. 262, fig. 1, with permission.)

recorded in the stratigraphy, which is well known from over 10,000 boreholes and extensive seismic reflection profiling (the main stratigraphic horizons being also good reflectors).

Trusheim (1957, 1960) examined the growth of one salt dome, or stock. The mobilization of the salt and the formation of a salt pillow is reflected in the development of the Keuper (Fig. 15-9). During the early Cretaceous transgression, the salt pillow became a true penetrative diapir, and what Trusheim called the secondary rim syncline developed. By the end of the Tertiary, the salt pillow had been virtually replaced by a mature salt dome. Sannemann

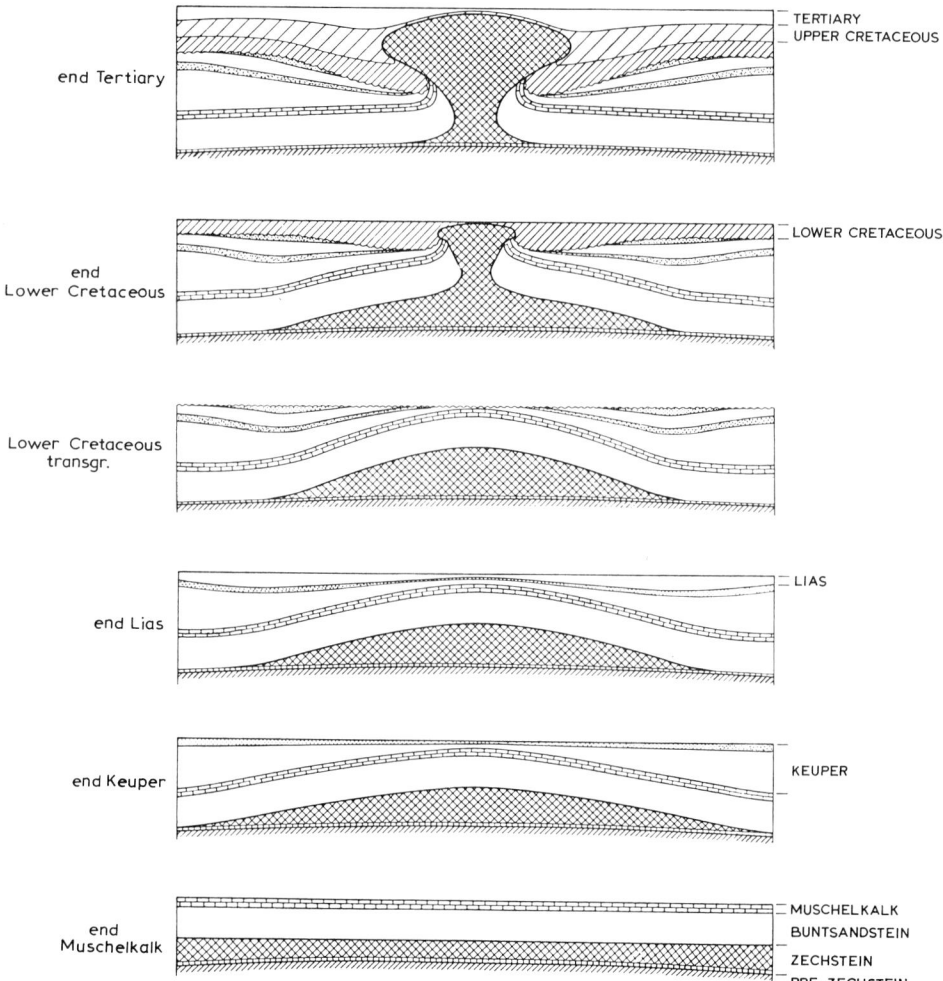

Fig. 15-9. Diagrammatic development of a single salt stock. (After Sannemann, in Trusheim, 1957: reproduced from Sannemann, 1968, p. 263, fig. 2, with permission.)

(1968) studied the stratigraphy around the salt domes and found that in groups of salt domes that he called salt-stock families, there was a sequential development from the "mother salt stock" outwards in space and time (Fig. 15-10), with progressively younger sedimentary rocks in the rim synclines. In this manner, he elucidated the development of the region (Fig. 15-11).

In general, the development of salt diapirs takes place during the accumulation and compaction of the overlying sedimentary rocks, and so is contemporaneous with fluid expulsion from them. The deformation of the sequence creates traps for any petroleum generated and expelled from the source rock; and ultimately such accumulations may be displaced to the flanks when penetration of the overburden occurs. Because there is no known causal relationship between salt and petroleum, we must regard these accumulations as coincidental, the deformation happening to take place while petroleum was being generated. This is not necessarily true of mudstone diapirism.

Mudstone diapirism has not received the same attention as salt diapirism, perhaps because it is not so distinctive; but it is probably more important in petroleum geology — particularly the geology of regressive sequences. This diapirism, from both observational and theoretical points of view, can begin soon after the more permeable, sandy, part of a regressive sequence begins to accumulate. Mud volcanism is mudstone diapirism at the surface, and incipient mudstone diapirism at depth is inferred for many parts of the world.

The apparent rarity of penetrative mudstone diapirs at depth (but see Gilreath, 1968; and Bishop, 1978) and their common occurrence at shallow depths is consistent with relative viscosity (and perhaps density) considerations because undercompaction is more pronounced at shallow depths. Mechanical instability in the sequence may remain until the mudstone is buried under 4

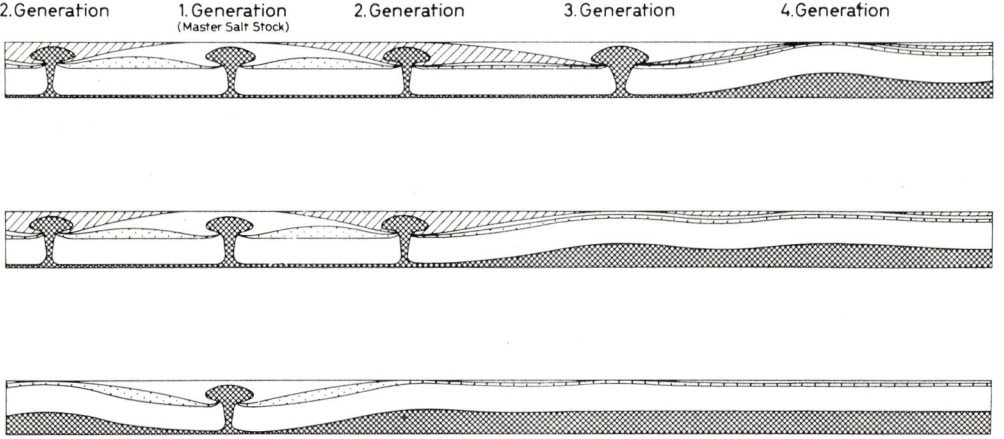

Fig. 15-10. Schematic diagram of development of a salt-stock family. (After Sannemann, 1968, p. 264, fig. 3.)

or 5 km of overburden, and so such instability is necessarily contemporaneous with the diagenesis of organic matter and with the expulsion of the bulk of the pore fluids in the mudstones (both in the overburden and below). If the mudstones are petroleum source rocks, in whole or in part, diapiric structures are contemporaneous with petroleum generation and so may form traps. The

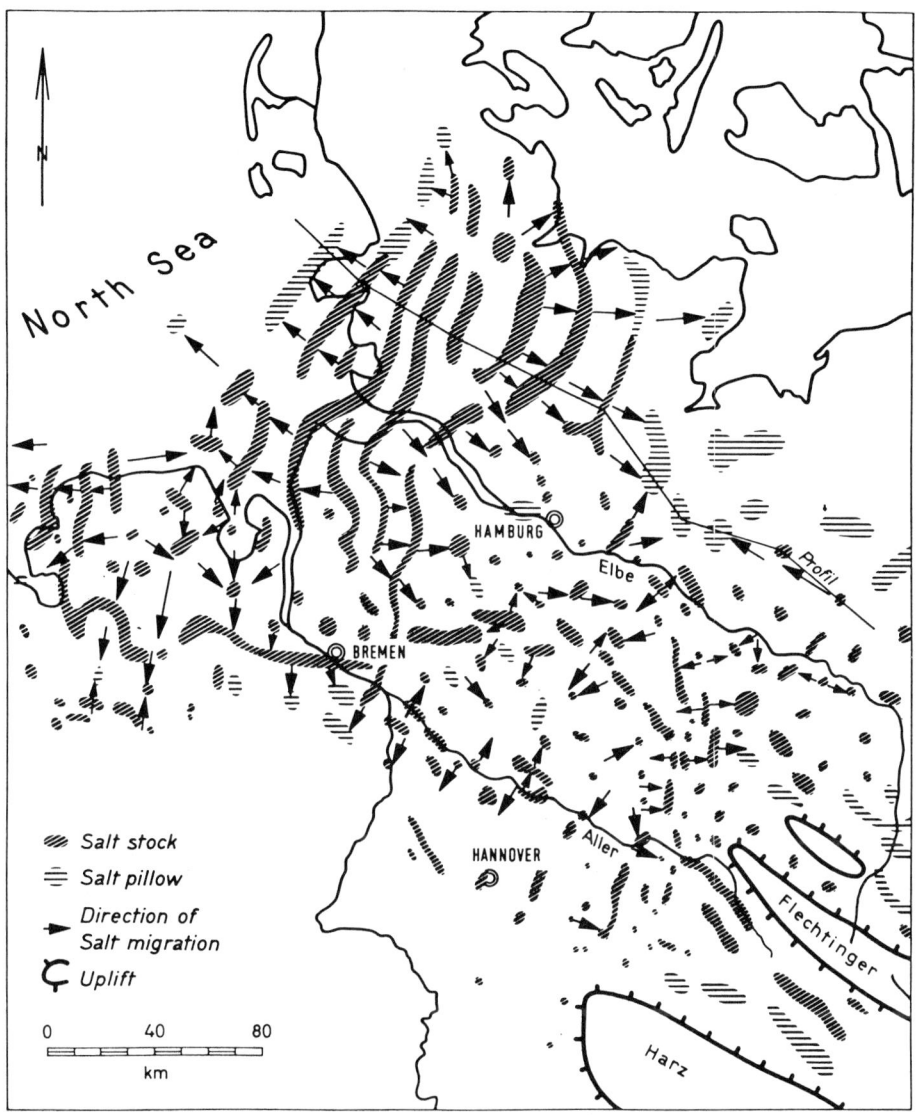

Fig. 15-11. Map showing the directions of salt migration, northwestern Germany (located on Fig. 15-8). (Reproduced from Sannemann, 1968, p. 269, fig. 9, with permission.)

association between mudstone diapirism and petroleum accumulations is inferred to be close and, from a practical point of view, causal. It is therefore of major importance in the geology of petroleum.

In its geological context, mechanical instability can develop whenever a sequence of greater density and equivalent viscosity accumulates on a sequence of smaller density and equivalent viscosity. There are examples of unstable sequences of carbonates on mudstones (see Chapman, 1981, pp. 162—168), but the dominant association is with regressive sequences.

GROWTH STRUCTURES AND INCIPIENT DIAPIRISM

Just as the development of a sedimentary basin is reflected in its stratigraphy, so are the details reflected in the local detailed stratigraphy (as Trusheim and Sannemann demonstrated for northern Germany, discussed above). Of particular importance are the variations of thicknesses of rock units due to contemporaneous developments of traps.

Petroleum occurs in dominantly regressive sequences in many parts of the world, and some of them are major producing areas: for example, California (Santa Barbara Channel region), the United States' Gulf Coast, Trinidad, the Niger delta, and many areas of South-east Asia. They have several features in common: (a) growth structures; (b) petroleum accumulations almost exclusively in structural traps; (c) underlying mudstones with abnormal pressures; (d) multiple sandstone reservoirs, with a tendency for the crude oils to be lighter in deeper reservoirs; and (e) Tertiary age.

Examples are taken from South-east Asia because they can be set in a sequence that illustrates the development of such traps and accumulations.

In Irian Jaya (W. New Guinea) in the early 1950s, field geologists used to say that the geology of the Tertiary basins was the geology of anticlines without synclines. The pattern of folding is not a wave-like sequence of anticlines and synclines, but rather of narrow, steep anticlines separating broad gentle synclines. This structural style is common in South-East Asia in onshore areas (with hints that it is less developed offshore). The anticlinal trends are commonly asymmetrical in section and sinuous in plan. They are not strictly parallel, but are generally parallel to the depositional strike. This structural style has been found in Irian Jaya (Visser and Hermes, 1962, pp. 140, 165—170, 227) and Papua New Guinea (Tallis, 1975, p. 59), and in Brunei, Sabah and Sarawak (Schaub and Jackson, 1958; Liechti et al., 1960, pp. 270—283). It also occurs onshore in Kalimantan, where it formed an important part of Van Bemmelen's hypothesis of *gravitational tectogenesis* (Van Bemmelen, 1949, e.g., p. 732). The steepness of the anticlines varies from gentle domes and elongated structures with dips to 30° in offshore and coastal areas, with some very steep anticlines with narrow, deformed cores, further inland. The steeper flank tends to be that on the seaward side, away from the source of the regression. The main petroleum accumulations are in the gentler anticlines.

Gesa anticline, Waropen coast, Irian Jaya

In the North Coast basin of Irian Jaya there is a very young regressive sequence under the Mamberamo delta (Fig. 15-12). The basin lies *south* of the present coast-line, where basement was found by drilling and geophysics at about 1200 m, rising to the north. A line of mud volcanoes occurs in the swampy plains of the Mamberamo delta. This line, about 35 km long (22 miles), forms a part of a general line about 100 km long (60 miles). The area has been described by Visser and Hermes (1962). Some of the mud volcanoes are on a narrow ridge, with outcrops of Plio-Pleistocene sandy and argillaceous rocks. Warm salt water and some methane are extruded with the mud. Further south, inland, there are some anticlines in which Plio-Pleistocene paralic sands with lignite beds and subordinate soft muds are exposed. The anticlines to the south tend to be steeper, with dips nearly vertical in their cores; but the first one to the south, the Gesa anticline, is a very gentle elongated dome with a sinuous axis roughly parallel (as are the others) to the line of mud volcanoes. Into this anticline, two boreholes were drilled between 1956 and

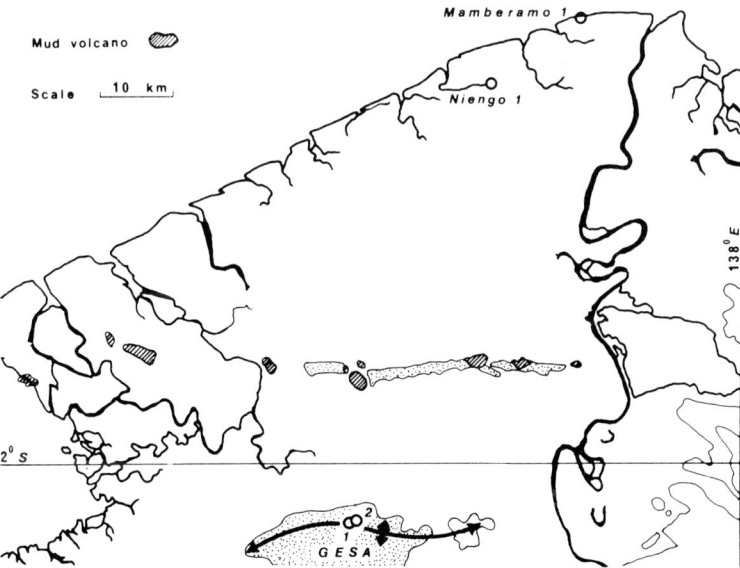

Fig. 15-12. Map of part of the Mamberamo delta, Irian Jaya. Mud volcanoes shown with diagonal lines. Relevant Plio-Pleistocene outcrop stippled. The Rombebai lake to the east of the line is probably evidence of local subsidence. (After Visser and Hermes, 1962, enclosure I-III.)

1958, both of which encountered abnormally high pore pressures in marine mudstones at depths of about 1300 m (4300 ft), underlying the sands. The ensuing drilling difficulties prevented their reaching their target, and no significant accumulation of petroleum was found.

The extent of the abnormality is not clear because the data given by Visser and Hermes (1962, fig. V-17, p. 230) cannot be taken at face value, and require careful analysis (after the correction of the pressure scale! An entertaining hour can be spent on this diagram: it illustrates well the difficulty of drawing valid conclusions from raw data). It seems quite certain that down to the depths penetrated the fluid pressure never exceeded the overburden pressure in spite of the mudweights used. It is inconceivable that the side-track to Gesa 1 (Gesa 1A) penetrated rocks with significantly higher pore pressures than those encountered in Gesa 1. So the troubles in Gesa 1A are to be attributed more to excessive mud weight and consequent loss of mud (as mentioned by Visser and Hermes on p. 228). The problems of Gesa 2, 500 m away, are likely to be analogous, with spuriously high pore pressures indicated by the mud weights. A value of λ of about 0.75 ($\delta = 0.54$) is indicated at a depth of about 1800 m (5900 ft) in Gesa 1.

The regional extent of this abnormal pressure is indicated not only by the mud volcanoes, but also by the seismic refraction survey, which detected a low-speed refractor at about 1500 m.

The Gesa anticline has therefore strong indications of a diapiric origin. It is clearly at a very early stage of development, and this may account for the lack of significant petroleum accumulation. Gas was reported in the *mudstones* only, so primary migration may be taken place.

Since Visser and Hermes mention that I suggested in 1954 that gravity sliding was the cause of these structures (1962, p. 171), it is appropriate to record here that the data obtained from drilling suggest that this hypothesis was wrong — at least for the Gesa anticline. This matter will be reconsidered when the evidence of other areas has been introduced.

Seria field, Brunei

The Seria field on the coast of Brunei (Fig. 15-13) is an asymmetric anticline of Tertiary sediments: it is about 20 km long and less than 5 km wide, with the axis parallel and close to the present-day coastline. The stratigraphic sequence is regressive, passing upwards from the neritic Setap Shale Formation (Oligo-Miocene), through the neritic mudstones and sandstones of the Miri Formation and the neritic sandstones and mudstones of the Seria Formation, to the paralic Liang Formation (Schaub and Jackson, 1958; Liechti et al., 1960). The regression is apparently still going on, towards the north or north-west.

The structure has no surface expression. It was revealed by shallow core-drilling under a flat, low-lying coastal swamp. The drilling of several hundred

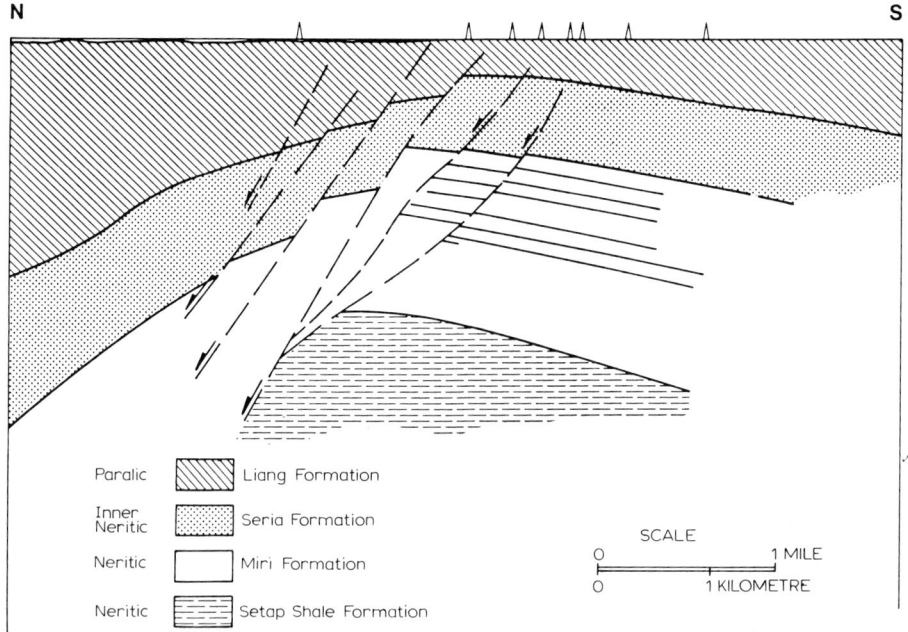

Fig. 15-13. Cross-section through Seria field, Brunei. (After Schaub and Jackson, 1958, p. 1333, fig. 3.)

wells has revealed beyond reasonable doubt that it is a growth anticline cut by growth faults, some of which cut the youngest sediments. So the anticline was formed in a stress field with a horizontal component of tension.

The petroleum is not distributed symmetrically in the anticline, but is restricted to the gentler south flank by growth faults. It should therefore be classed as a fault trap rather than as an anticlinal trap.

There are many sandstone reservoirs, and the crude oils tend to be lighter from the deeper reservoirs. The shallowest crudes are non-waxy 19°API (on average), followed by non-waxy 26°API, and the deepest are waxy 37°API crude oils. Some gas condensate occurs deeper. Insufficient data have been published to assess the source rocks for these reservoirs, but their variability and position in the south flank, trapped by faults, suggests interbedded source rocks and reservoirs with generation to the south.

The Setap Shale is abnormally pressured over a very wide area, and it is diachronous. It crops out in the core of the Jerudong anticline to the east, but it is still abnormally pressured below the Jerudong field on its flank (*Annual Report British Borneo Geol. Surv. for 1962*); and it forms mud volcanoes in the Jerudong-Limbang area in Brunei, and at Bulak Setap in Sarawak, from which it takes its name (Liechti et al., 1960, pp. 326—329).

Seria field is a clear example of what might be called the regressive syndrome: regressive stratigraphic sequence, abnormal pressures in the thick mudstone formation, growth structures and petroleum trapped in them, and lighter oil in deeper reservoirs. The evidence is strong, if not unambiguous, that deformation began early and continued through Miocene and Pliocene times at least. The fact that the structure was discovered by shallow coredrilling indicates that deformation has been continuing to the very recent past, if not the present. The evidence of growth faults is unambiguous that the deformation took place — is taking place — in a stress field in which the least compressive stress is horizontal, and it took place while the sequence in the area was subsiding. The younger growth faults (Fig. 15-13) lie north of the older, on their downthrown sides, in the direction of regression, affecting progressively younger sediments.

Schaub and Jackson (1958, p. 1335) noted the structural style of "wide, gentle synclines separated by relatively narrow and structurally complicated anticlinal zones" on one of which Seria is a culmination. They also noted that, in view of widespread evidence of diapirism, the Seria structure could "be due to, or have been modified by, a diapiric core of relatively plastic Setap shale".

We would go further than this: the internal evidence of stratigraphic sequence and stress field during contemporaneous deformation during subsidence indicates mechanical instability in the sequence, and the structural style should be regarded as the expression of the dominant wavelength of diapirism. A density inversion is unlikely now, but it seems an inevitable consequence of severe abnormal pressures that a density inversion did exist when the Setap shale was less deeply buried.

Miri field, Sarawak

Down the north coast of Borneo to the south-west from Seria field lies the Miri field. This field was discovered in 1910, and most of the development was done before modern logging techniques had been invented. According to Schaub and Jackson (1958), Miri lies on the same anticlinal trend as Seria, and contains a similar regressive sequence (Fig. 15-14). The cross-section shows that it is a steeper structure than Seria, and the shale/mudstone core is prominent in a fault block bounded on one side by a reverse fault. This reverse fault seems clearly caused by the shale core because the net throw across the block is small.

Miri differs from Seria in some respects. The oil is confined to the north flank of Miri, and there are two types of crude in the many reservoirs. The shallow crude is non-waxy 26°API, while the deeper crude is waxy 30°API.

The normal faults are growth faults that dip more steeply than usual on account of the dip of the beds they cut. The reverse fault is not to be interpreted as evidence of horizontal "compression" because its dip is consistent with a stress field with the greatest principal compressive stress vertical, not

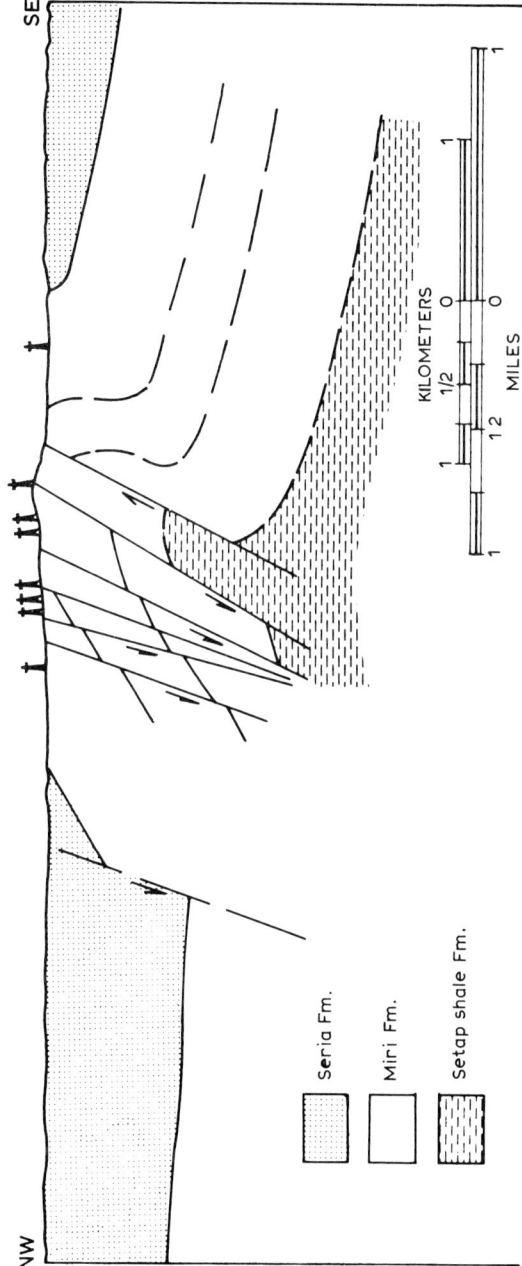

Fig. 15-14. Cross-section through Miri field, Sarawak. (After Schaub and Jackson, 1958, p. 1332, fig. 2.)

horizontal. It was perhaps a normal fault originally, with the movement subsequently reversed by mudstone diapirism acting on the original downthrown block.

Miri is perhaps at a more advanced stage of deformation than Seria. It is onshore, and the land surface has topographic relief.

It appears therefore that Gesa, Seria, and Miri represent stages in the development of growth anticlines in regressive sequences, and that this growth is vertical, due to mechanical instability in the stratigraphic sequence. The coexistence of growth faults in growth anticlines in Seria and Miri are evidence of a protracted period of extensional stress during subsidence. We shall pursue this line of reasoning in the last chapter.

It is clear that by its very nature, a regressive sequence tends to be "open-ended", and that the load on the abnormally pressured mudstones will tend to squeeze them out laterally (Fig. 15-15). Dailly (1976) has argued cogently for the reality of this on a large scale in the Niger delta; and, on a small scale, the evident lateral movements of folds and overthrusts under the mudlumps of the Mississippi delta (Morgan et al., 1968) is further support. This bulk flow may therefore be the additional process tending to flatten growth faults with depth in a regressive sequence and, indeed, may be the principal cause of growth faults. Hence Bruce's (1973, p. 884) observation that the abnormality of pore pressure is greater where the dip of the growth fault is flatter is probably ascribable to mass flow (rather than the direct mechanical effect that Bruce inferred).

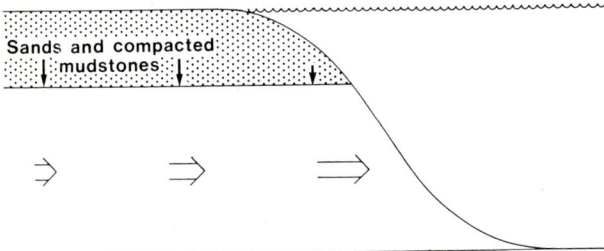

Fig. 15-15. A regressive sequence will tend to extrude the relatively plastic mudstone by virtue of the load (diagrammatic).

REFERENCES

Atwater, G.I. and Forman, M.J., 1959. Nature of growth of southern Louisiana salt domes and its effect on petroleum accumulation. *Bull. Am. Ass. Petrol. Geol.*, 43: 2592—2622.

Barton, D.C., 1931. Effect of salt domes on accumulation of petroleum. *Bull. Am. Ass. Petrol. Geol.*, 15: 61—66.

Barton, D.C., 1933. Mechanics of formation of salt domes with special reference to Gulf

Coast salt domes of Texas and Louisiana. *Bull. Am. Ass. Petrol. Geol.*, 17: 1025–1083.

Beck, R.H., 1972. The oceans, the new frontier in exploration. *J. Aust. Petrol. Explor. Ass.*, 12 (2): 7–28.

Beck, R.H. and Lehner, P., 1974. Oceans, new frontier in exploration. *Bull. Am. Ass. Petrol. Geol.*, 58: 376–395.

Biot, M.A. and Odé, H., 1965. Theory of gravity instability with variable overburden and compaction. *Geophysics*, 30: 213–227.

Bishop, R.S., 1977. Shale diapir emplacement in south Texas: Laward and Sheriff examples. *Trans. Gulf-Coast Ass. Geol. Socs.*, 27: 20–31.

Bishop, R.S., 1978. Mechanism for emplacement of piercement diapirs. *Bull. Am. Ass. Petrol. Geol.*, 62: 1561–1583.

Bornhauser, M., 1958. Gulf Coast tectonics. *Bull. Am. Ass. Petrol. Geol.*, 42: 339–370.

Braunstein, J. and O'Brien, G.D., 1968. Diapirism and diapirs. A symposium. *Mem. Am. Ass. Petrol. Geol.*, 8, 444 pp.

Bruce, C.H., 1973. Pressured shale and related sediment deformation: mechanism for development of regional contemporaneous faults. *Bull. Am. Ass. Petrol. Geol.*, 57: 878–886.

Carey, S.W., 1954. The rheid concept in geotectonics. *J. Geol. Soc. Aust.*, 1 (for 1953): 67–117.

Chapman, R.E., 1981. *Geology and water: an introduction to fluid mechanics for geologists*. Nijhoff/Junk, The Hague, 228 pp.

Dailly, G.C., 1976. A possible mechanism relating progradation, growth faulting, clay diapirism and overthrusting in a regressive sequence of sediments. *Bull. Can. Petrol. Geol.*, 24: 92–116.

Dickinson, G., 1953. Geological aspects of abnormal reservoir pressures in Gulf Coast Louisiana. *Bull. Am. Ass. Petrol. Geol.*, 37: 410–432.

Evans, C.R., McIvor, D.K. and Magara, K., 1975. Organic matter, compaction history and hydrocarbon occurrence — Mackenzie delta, Canada. *Proc. 9th World Petrol. Congress*, 2: 149–157.

Gansser, A., 1960. Über Schlammvulcane und Salzdome. *Vierteljahresschr. Naturforsch. Ges. Zürich*, 105: 1–46.

Gilreath, J.A., 1968. Electric-log characteristics of diapiric shale. In: J. Braunstein and G.D. O'Brien (Editors), Diapirism and diapirs. *Mem. Am. Ass. Petrol. Geol.*, 8: 137–144.

Hubbert, M.K., 1937. Theory of scale models as applied to the study of geologic structures. *Bull. Geol. Soc. Am.*, 48: 1459–1520.

Kent, P.E., 1979. The emergent Hormuz salt plugs of southern Iran. *J. Petrol. Geol.*, 2 (2): 117–144.

Lehner, P., 1969. Salt tectonics and Pleistocene stratigraphy on continental slope of northern Gulf of Mexico. *Bull. Am. Ass. Petrol. Geol.*, 53: 2431–2479.

Liechti, P., Roe, F.W. and Haile, N.S., 1960. The geology of Sarawak, Brunei and the western part of North Borneo. *Bull. Geol. Surv. Dept. Br. Terr. Borneo*, 3 (2 vols).

Lyell, C., 1867. *Principles of geology or the modern changes of the Earth and its inhabitants*, vol. 1 (10th ed.). Murray, London, 671 pp.

Mascle, J.R., Bornhold, B.D. and Renard, V., 1973. Diapiric structures off Niger delta. *Bull. Am. Ass. Petrol. Geol.*, 57: 1672–1678.

Morgan, J.P., Coleman, J.M. and Gagliano, S.M., 1968. Mudlumps: diapiric structures in Mississippi delta sediments. In: J. Braunstein and G.D. O'Brien (Editors), Diapirism and diapirs. *Mem. Am. Ass. Petrol. Geol.*, 8: 145–161.

Murray, G.E., 1968. Salt structures of Gulf of Mexico basin — a review. In: J. Braunstein and G.D. O'Brien (Editors), Diapirism and diapirs. *Mem. Am. Ass. Petrol. Geol.*, 8: 99–121.

Paraschiv, D. and Olteanu, Gh., 1970. Oil fields in Mio-Pliocene zone of eastern Carpathians (District of Ploiesti). In: M.T. Halbouty (Editor), Geology of giant petroleum fields. Mem. Am. Ass. Petrol. Geol., 14: 399—427.

Parker, T.J. and McDowell, A.N., 1955. Model studies of salt-dome tectonics. Bull. Am. Ass. Petrol. Geol., 39: 2384—2470.

Ramberg, H., 1981. *Gravity, deformation and the Earth's crust* (2nd ed.). Academic Press, New York, N.Y., 452 pp.

Sannemann, D., 1963. Über Salzstock-Familien in NW-Deutschland. *Erdöl Z. Bohr Fördertech.*, 79: 499—506.

Sannemann, D., 1968. Salt-stock families in northwestern Germany. In: J. Braunstein and G.D. O'Brien (Editors), Diapirism and diapirs. Mem. Am. Ass. Petrol. Geol., 8: 261—270.

Schaub, H.P. and Jackson, A., 1958. The northwestern oil basin of Borneo. In: L.G. Weeks (Editor), *Habitat of oil*. Am. Ass. Petrol. Geol., Tulsa, Okla., pp. 1330—1336.

Shepard, F.P., 1973. Sea floor off Magdalena delta and Santa Marta area, Colombia. Geol. Soc. Am. Bull., 84: 1955—1972.

Tallis, N.C., 1975. Development of the Tertiary offshore Papuan basin. J. Aust. Petrol. Explor. Ass., 15 (1): 55—60.

Tanner, W.F. and Williams, G.K., 1968. Model diapirs, plasticity, and tension. In: J. Braunstein and G.D. O'Brien (Editors), Diapirism and diapirs. Mem. Am. Ass. Petrol. Geol., 8: 10—15.

Trusheim, F., 1957. Über Halokinese und ihre Bedeutung für die strukturelle Entwicklung Norddeutschlands. Z. Dtsch. Geol. Ges., 109: 111—158.

Trusheim, F., 1960. Mechanism of salt migration in northern Germany. Bull. Am. Ass. Petrol. Geol., 44: 1519—1540.

Van Bemmelen, R.W., 1949. *The geology of Indonesia*. Vol. 1A. *General geology of Indonesia and adjacent archipelagoes*. Government Printing Office, The Hague, 732 pp.

Visser, W.A. and Hermes, J.J., 1962. Geological results of the exploration for oil in Netherlands New Guinea. Verh. K. Ned. Geol. Mijnbouwk. Genootsch. Nederland Koloniën, Geol. Ser., 20: 1—265.

CHAPTER 16

SYNTHESIS

SUMMARY

(1) The tendency for petroleum in transgressive sequences to be in stratigraphic traps and for petroleum in regressive sequences to be in structural traps suggests that sedimentary basins are not folded as a whole.

(2) The internal evidence of many fields in regressive sequences is that they were folded and faulted during subsidence, while sediment accumulated, in a stress field with a component of horizontal tension.

(3) An important class of sedimentary basin forms as part of broader crustal process that includes orogeny. Subsidence of the sedimentary basin proceeds concurrently with orogeny in a neighbouring area: the orogeny gradually forms mountains from which the sediment generated forms the regressive sequence. This regressive sequence is usually mechanically unstable.

(4) Petroleum source rocks are a facies of the sediments that accumulate in a sedimentary basin, and the products of their diagenesis normally migrate and accumulate in porous and permeable beds that are stratigraphically close. Primary migration is vertical, upwards and downwards: secondary migration is lateral along permeable beds, usually towards the land of the time, to be trapped in stratigraphic traps in transgressive sequences (because there are few structures) and structures in regressive sequences.

(5) Another important class of sedimentary basin is the rift basin, which leads to the accumulation of petroleum in unconformity traps and pre-unconformity fault traps. These are initiated by normal growth faulting, forming graben that accumulate continental and marine sediments in a dominantly regressive sequence (usually). A period of non-accumulation of sediment follows, with some erosion of higher surfaces, and fault movement ceases. Epeirogenic subsidence follows, and a transgressive sequence accumulates on the disconformity/unconformity surface, and a long period of subsidence without significant faulting follows. Petroleum source rocks may be in the pre-unconformity sequence or the post-unconformity sequence — commonly in the immediate post-unconformity sequence that seals the traps.

(6) Rift basins typically began in the Permian or Triassic and continued into the Tertiary, with the unconformity typically early Cretaceous in age, but ranging from late Jurassic to early Tertiary. The "Cretaceous transgression" is part of this world-wide, but not strictly synchronous, event.

(7) Sedimentary basins are dominantly extensional features of the Earth's crust, with vertical movements. They have characteristic features, world-wide;

and petroleum accumulations are part of the pattern. Much work remains to be done to reconcile the hypotheses of plate tectonics with the detailed evidence available from petroleum geology.

INTRODUCTION

Our purpose in this chapter is to pass in review the topics of earlier chapters and seek to relate them to broad concepts of sedimentary basin development and the occurrence of petroleum. These are geological concepts to which petroleum, as a natural substance, contributes significant data. Exploration for petroleum, and the development of petroleum fields, has provided us with regional and detailed geology of areas that would otherwise have waited decades before investigation. In particular, it has provided us with some knowledge of the geology of the continental shelves.

The 19th Century work on coal mines, railways and canals, brought great advances in geology; but some of the concepts developed from surface and near-surface work on land have been maintained despite contrary evidence from subsurface work offshore as well as on land during the last 30 years or so. It seems to have been generally accepted that the regressive sequences of sedimentary basins indicate uplift, and that their deformation was due to major tectonic events. Anticlines were (and perhaps still are) widely attributed to horizontally directed compressional stress. The evolution of sedimentary basins (and, to some extent, geosynclines) seems to have been regarded as involving three stages: subsidence, uplift and deformation, with the last two, perhaps, concurrent. While geologists were confined to the land, such a sequence of events was logical and consistent with the evidence available to them. But the discovery of anticlines offshore under the continental shelves has shown that they can occur in young sedimentary rocks without uplift and orogeny.

The world-wide search for petroleum has shown that just as a petroleum province has a character, so we have similar characters in petroleum provinces in different continents. The United States Gulf Coast province has much in common with the Niger delta, and many areas of South-East Asia: the Western Canada basin has much in common with northern Mexico and south-east Libya: the North Sea has much in common with the Gippsland basin of south-east Australia, the north-west shelf of Australia, and northern Alaska. Many of these similarities are so close that we feel entitled to believe that the principles of petroleum geology that apply to one, apply also to the others like it. This is not to deny individuality to fields, but to assert that we are concerned to discover the guiding principles of geology that govern the generation, migration, and accumulation of petroleum.

DEFORMATION OF SEDIMENTARY BASINS

It is one of the remarkable features of petroleum geology that some large petroleum provinces are virtually devoid of structure. Large areas of Silurian and Devonian reefs in North America, some within sight of the Rocky Mountains, have regional dips less than 5°. A geological history lasting 300—400 m.y. has taken them from sea level to depths of three kilometres or so with only the slightest deformation, and they were hardly affected by the orogeny that built the Rocky Mountains. Moreover, the physiographic basin within which the reefs grew remained favourable for them for 5—10 m.y. or 50 m.y. if one takes the combined Silurian and Devonian reef provinces of northern U.S.A. and western Canada. The geological picture is one of utmost tranquility.

Similarly, the Cretaceous reefs of northern Mexico, after 100 m.y. of geological history, lie almost horizontally at shallow depths down to about two kilometres. Were it not for the reefs that formed stratigraphic traps, these would not have become petroleum provinces.

By contrast, there are large areas of the world in which rocks less than 25 m.y. old have suffered substantial deformation, with folds and faults. These are areas that can be examined in outcrop and mapped. It is perhaps for this reason that we were taught by the collective experience of surface geologists that sedimentary basins were typically deformed by horizontally directed tectonic forces of orogeny after the sedimentary history had come to a close, or, alternatively, that these events terminated the sedimentary history of the basin.

In those areas, petroleum occurs in structural traps, and their study has revealed that much of this deformation took place during the sedimentary history of the basin, and that one must distinguish between pre-orogenic and orogenic deformation of such sedimentary basins. The offshore basins are no exception to these rules: the petroleum is in structural traps in parts of the basin that have not suffered orogeny, where deformation formed an integral part of the sedimentary history of the basin.

Since about 1970, yet another type of sedimentary basin has been recognized: the rift basin. It too has features that are shared by widely separated provinces, so that the petroleum geology of the North Sea is very similar to that of the north-west shelf of Australia, for example. The geological history of these basins consists typically of three stages. First, the formation of one or more graben (the bounding faults throwing perhaps 5—6 km eventually) provides a basin in which sediment accumulated. These graben contain numerous normal faults that moved while the sediment accumulated. A period of non-accumulation followed, with erosion on the higher parts and the upthrowing blocks of growth faults. During this period, the faults stopped moving, and the disconformity/unconformity was completed by epeirogenic subsidence, and the accumulation of fine-grained sediment followed. Faulting of the younger sequence is confined to minor movement of some faults. Petroleum in such basins is in unconformity traps that are commonly influenced by pre-unconformity faults and the surface of the disconformity/unconformity; and in stratigraphic and structural traps in the post-unconformity sequence.

Typically, but not exclusively, the graben began to form during the Permian or Triassic, the unconformity is late Jurassic or early Cretaceous, and the sequence above the unconformity consists of Upper Cretaceous and Tertiary sedimentary rocks that are virtually devoid of structure.

These three types of petroleum province are sufficiently well defined to require some general geological explanations that are in the nature of global, crustal processes. The undeformed transgressive sequences of North America were adjacent to a major orogenic event, yet did not suffer deformation. They are evidence of a prolonged neutral stress regime, unaffected by neighbouring events. The rift basins provide evidence of an extensional stress regime lasting perhaps 60—70 m.y., followed by a neutral regime lasting perhaps 100 m.y. Let us therefore review the salient features of the young deformed provinces with petroleum in regressive sequences.

REGRESSIVE SEQUENCES

Growth faults occur in both rift basins and the young regressive sequences that are now deformed to provide structural traps for petroleum, but we are only concerned here with the latter. The accumulation of sediment in both blocks of a growth fault is evidence that both blocks were subsiding, but that the downthrowing block was subsiding faster than the upthrowing block. These are almost invariably normal faults, so that we infer a stress field with the greatest principal stress vertical and the least horizontal and normal to the fault trace. We take this as unambiguous evidence that this deformation took place during subsidence in a stress field with a component of horizontal tension. Growth faults are typically younger in the direction of the regression, so we extend this conclusion to embrace the space and time involved in the regional deformation. In plan, growth faults are not strictly parallel, but they tend to be concave to the direction of regression, and to be en echelon with others. We therefore infer a prolonged extensional regime approximately normal to the depositional strike.

In section, growth faults in major regressive sequences flatten with depth, and appear to die out downwards in the massive mudstone unit that underlies the sandier part of the sequence. There is a strong stratigraphic association, and there is no evidence that these faults are caused by anything other than gravity acting on the prograding sequence. The pattern of growth faults in the Niger delta is particularly eloquent in this respect (Fig. 16-1).

Anticlines in regressive sequences are typically growth anticlines, which show thinning of the individual rock units onto the crest. They tend to form trends, or have their long axis roughly parallel to the depositional strike. They

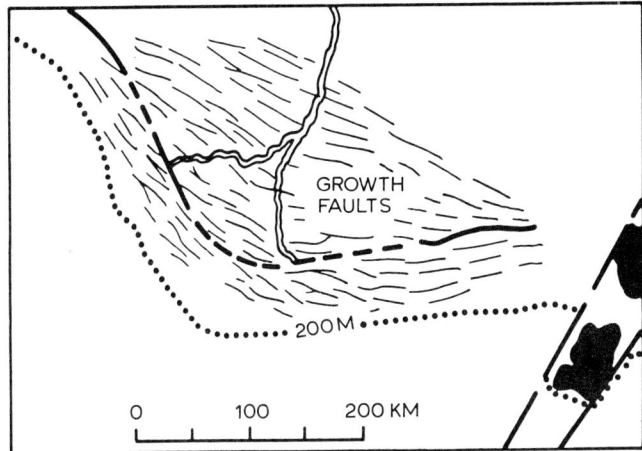

Fig. 16-1. Growth faults in the Niger delta reflect the growth of the delta. (After Weber, 1971, p. 560, fig. 2a.)

formed during the accumulation of sediment, and therefore also during subsidence and the movement of associated growth faults. We infer, therefore, that they too formed in a stress field with a component of horizontal tension.

It is part of the geological association in major young regressive sequences that the thick mudstone unit underlying the sandier part of the sequence is abnormally pressured. These abnormal pressures are found to be stratigraphically related, and closely related stratigraphically to the growth faults and growth anticlines. The abnormally pressured mudstones are typically undercompacted and less competent than the sandier, more compacted overburden.

Regionally, the anticlinal trends in such regressive sequences are found to be narrow, sinuous, and separated by broad, gentle synclines. Some anticlinal trends are steep, with complex cores.

We therefore infer that the deformation of regressive sequences of this sort is a gravity-induced phenomenon, with mechanical instability in the sequence leading to incipient diapirism and diapirism, and a tendency for the whole sequence to slide on the incompetent mudstones away from the land when unequal subsidence imposes a gentle slope on the *mechanical* interface at the top of abnormal pressures. The tendency to slide and form growth faults is near the distal limit of sand accumulation, but in the early stages of development, the slope on the mechanical surface is regionally down towards the land from which the regression is coming (it is roughly parallel to the diachronous surface separating the sandier sequence from the mudstones).

This deformation is seen as a contemporary deformation that results from the accumulation of the regressive sequence. It is only indirectly a consequence of orogeny. The role of orogeny is to create the mountains from which the sediment will be generated in sufficient quantities to accumulate in neighbour-

354

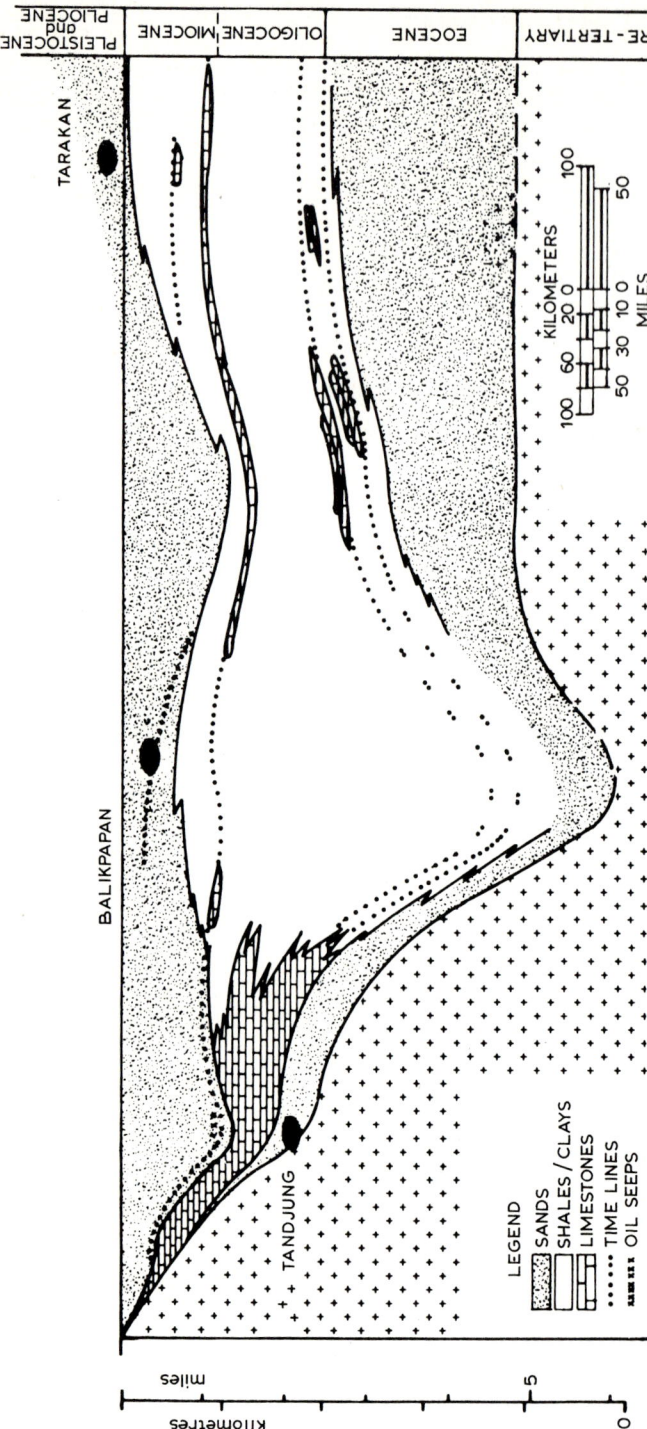

Fig. 16-2. Stratigraphic section through Kalimantan, Indonesia. (Redrawn from Weeda, 1958, pp. 1342–1343, fig. 3.)

ing sedimentary basins as a regressive sequence. The regressive sequence is the stratigraphic expression in a sedimentary basin of an orogeny outside the sedimentary basin. But just as the orogeny that created the Rocky Mountains had no effect on the Devonian reef province of the Western Canada basin (other than subsidence, perhaps), so the orogeny providing the sediment of a regressive sequence was a vertical event because the stress field in the sedimentary basin remained one with a component of horizontal tension.

In this respect, the island of Borneo is particularly interesting. The northwest coast, with Sarawak, Brunei and Sabah, is a Tertiary sedimentary basin that has had (and is having) a similar geological history to that in the east coast of Kalimantan. Petroleum occurs in both basins in numerous structural traps in the coastal region and offshore. A stratigraphic section down the east coast is shown in Fig. 16-2. The onshore areas have been successfully explored, and the area west of the Mahakam delta (Fig. 16-3), with the Samarinda anticlinorium, formed an important part of Van Bemmelen's (1949) hypothesis of gravitational tectogenesis. Long, narrow anticlinal trends with steep, sheared cores occur. These have the features of compressional structures, including thrust faults, and Van Bemmelen (1949, pp. 352 and 732) regarded these as the consequence of a tendency to slide into the "depressed

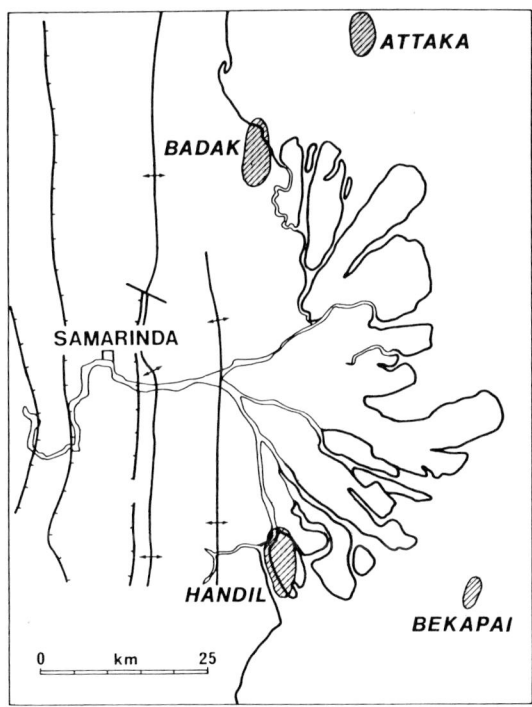

Fig. 16-3. Sketch map of Mahakam delta, Kalimantan, Indonesia.

areas" — the sedimentary basin. Orogeny, previously only supplying sediment, was now advancing and affecting the area directly.

Since Van Bemmelen's work, petroleum exploration moved offshore and several large oil and gas fields have been discovered. Four were found in and off the Mahakam delta: Badak (Gwinn et al., 1975; Huffington and Helmig, 1980) and Handil (Magnier and Samsu, 1976; Verdier et al., 1980) on one trend, and Attaka (Schwartz et al., 1974) and Bekapai (De Matharel et al., 1977; De Matharel et al., 1980) on another. These revealed an offshore structural style that contrasts strongly with that onshore. They are all growth anticlines; and Attaka and Bekapai on the seaward trend have growth faults. No faults are known in Badak, and there is one transverse fault in Handil. It therefore appears that the structures onshore are "compressional" while those offshore are extensional.

This paradox has not yet been resolved. Abnormal pressures exist in the massive mudstone formation at depth in the whole area, and it would be of great interest to know the shape of the top surface of them. The onshore structures could be the result of sliding following diapirism. Chapman (1974) postulated that the formation of diapiric anticlines in the pre-orogenic deformation of a regressive sequence would be modified by sliding if orogeny imposed an adequate slope on the top of abnormal pressures, and that the spacing of the resulting thrusts would be determined by the dominant wavelength of diapirism. He later demonstrated that the mechanics of subaerial sliding was about twice as efficient as that of submarine sliding, so that contrasting stress regimes could exist about sea-level (Chapman, 1979). In brief, the compressional tectonics onshore could be a superficial consequence of the stratigraphy. If the stress field changes from one in which compressional faults were generated to one in a neighbouring area in which extensional faults were generated, then mechanical theory would predict that there will be a substantial zone between the two in which there will be no faults because the magnitude of the principal stresses will be too similar. The lack of faulting in Badak and Handil may therefore be highly significant.

It is worth digressing here to note that the petroleum geology of these fields has other points of interest. Badak contains mainly gas; the others mainly oil, with some associated gas. The sands are generally lenticular, and difficult to correlate within the fields. The oil in both Handil and Bekapai in the south tends to be *heavier* with increasing depth to the reservoir. Both are regarded as immature from a petroleum source rock point of view, vitrinite reflectance in Bekapai being less than 0.5% and Handil's between 0.5 and 0.8%. On the grounds of immaturity of the mudstones within the productive sequence, Combaz and de Matharel (1978) postulated a source below 2600 m, with considerable vertical migration, possibly up faults. This is in contrast to the views expressed by Magnier in discussion at the 9th World Petroleum Congress in 1975, that the haphazard distribution of oil and gas in the Mahakam delta fields, and lack of faults, suggested source rocks adjacent

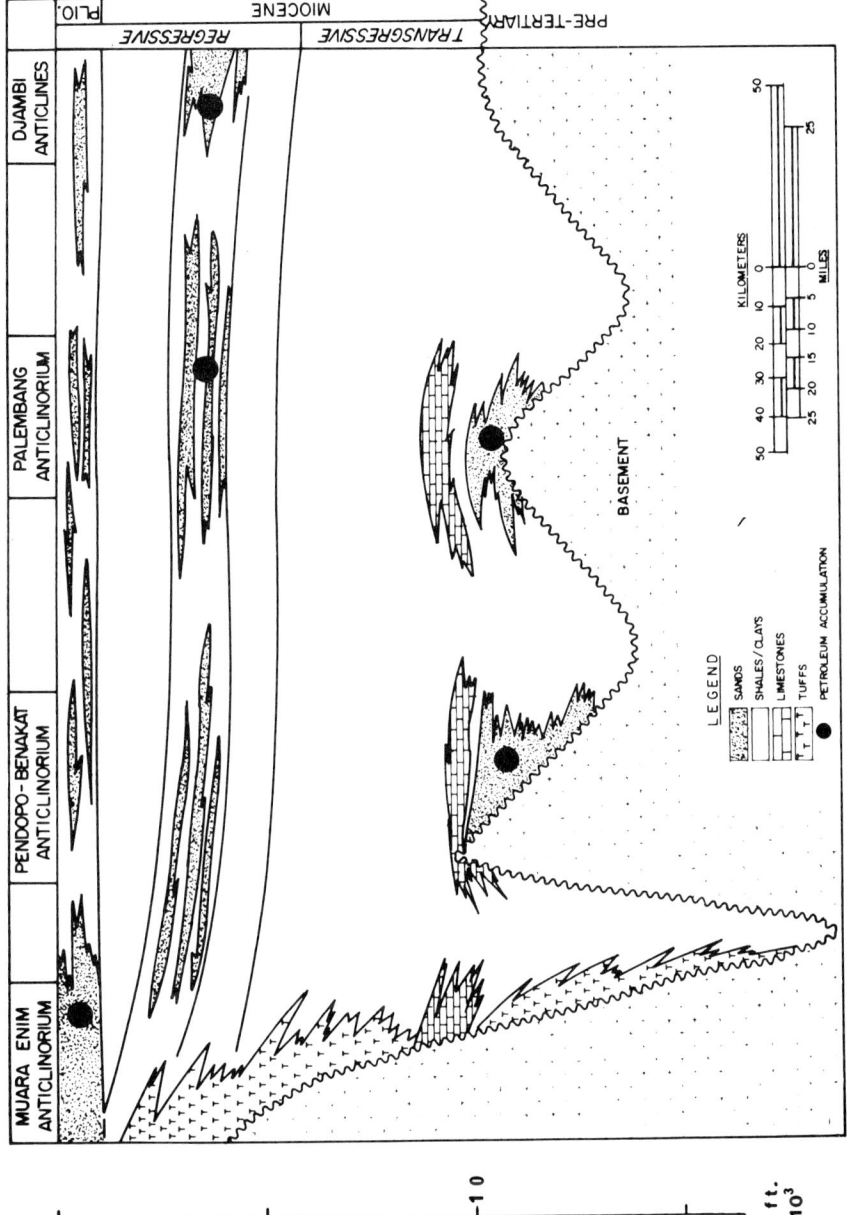

Fig. 16-4. Simplified section through South Sumatra basin. (Redrawn from Wennekers, 1958, p. 1356, fig. 7; petroleum has since been found in the transgressive sequence near "Djambi anticlines".)

to the reservoirs (see p. 243). These four fields appear to be a perfect natural laboratory for the study of all the problems discussed in Chapter 11 in an attempt at reconciling the geological, geochemical and hydraulic aspects of generation, migration and entrapment of oil and gas in these fields.

Returning to the regional geology of the island of Borneo, we see that structures of both Kalimantan in the east and Sabah, Brunei and Sarawak in the north-west, result from orogeny in the interior of Borneo, and the progress of this orogeny is recorded in the sedimentary basins around the island.

That Borneo is not a special case is seen in south Sumatra. The South Sumatra basin consists of an Eocene—Miocene—Pliocene sedimentary cycle (Fig. 16-4), its history beginning with a transgression across a pre-existing topography on pre-Tertiary metamorphic and igneous rocks (Wennekers, 1958; Adiwidjaja and De Coster, 1974; De Coster, 1975). The transgression led to palaeogeomorphic traps on the old topography: the regression led to anticlinal traps. Figure 16-5 shows how these two classes of accumulation are separated in space, suggesting that the folding of the regressive sequence was largely independent of basement relief.

The petroleum geology of basins with a simple sedimentary cycle, beginning with transgression and ending with regression, is dominated by the stratigraphy, which acquires its dominant characteristics from events outside the sedimentary basin. The sedimentary basin is the complement of orogeny, and its stratigraphy records the development of both. The inception of a

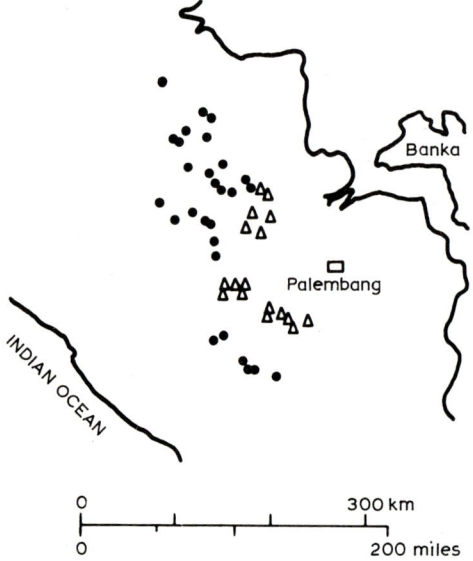

Fig. 16-5. Distribution of petroleum occurrences in South Sumatra basin. Triangles, in transgressive sequence; dots, in regressive sequence. (After Koesoemadinata, 1969, p. 2374, fig. 4.)

sedimentary basin in an area of low relief results in a carbonate—mudstone transgressive sequence, favourable for the growth of coral reefs. As subsidence in the basin and its complementary orogeny outside proceed, so the volume of sediment generated increases. Transgression gives way to regression; and as the regression develops, mechanical instability also develops in the regressive sequence. This is shown diagrammatically in Fig. 16-6.

Petroleum source rocks are a facies of the sedimentary basin. So are the carrier beds and reservoir rocks. Once petroleum is generated, it moves under physical laws to positions of lower energy. These are largely determined by the stratigraphy of the area: the superposition of permeable potential reservoirs and relatively impermeable source rocks and cap rocks normally restricts primary migration to short vertical paths to the nearest permeable bed with less energy in its pore fluids. From there it moves laterally until it is trapped — or dissipates at the surface.

In transgressive sequences, primary migration is usually downwards to the basal permeable unit, then lateral towards the land of the time, in the direction of the transgression. There is little structural relief on this surface, so trapping depends on a stratigraphic cause — palaeogeomorphic traps and fossil coral reefs being the most important. Secondary migration paths may be quite long, but both primary and secondary migration are confined to a diachronous sequence of lithologies, the source rock being coeval with the reservoir rock.

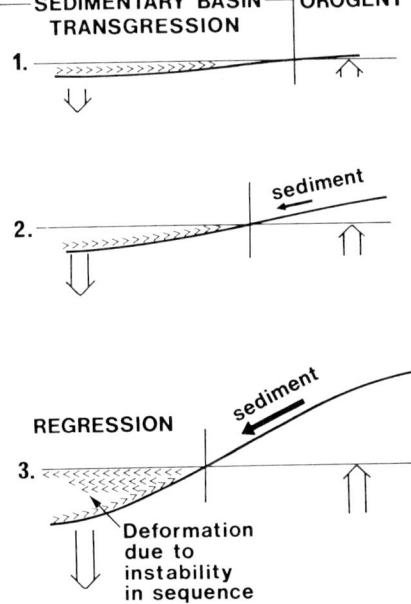

Fig. 16-6. Schematic diagrams showing development of sedimentary basin and nearby orogeny.

In regressive sequences, primary migration is vertical to the nearest land, then lateral towards the land of the time from which the regression comes. In a general sense, the source rocks tend to be below the massive sands; but here too the relationship is diachronous and the source rock is coeval with reservoir rock. Fluctuations of environment lead to the alternating sequence. Episodic transgressions and regressions within the dominant regression seem to favour the mudstone overlying the sandstone as the petroleum source rock because the more rapid accumulation of sediment favours the preservation of the organic matter. In detail, therefore, there will be important downward primary migration to the sandstone carrier beds, followed by lateral secondary migration to traps that formed concurrently with the accumulation of sediment.

The quality of the crude oil depends largely on the stratigraphy in the sense that it depends on the facies of the source rock and the conditions to which it was subjected during migration. The physical controls on migration are also related to the stratigraphy, so that petroleum normally accumulates in reservoirs that are stratigraphically close to their source rocks. Passage through growth faults may move this petroleum stratigraphically downwards as regards rock units (Fig. 16-7), but local migration may be in the reverse direction towards a growth anticline, and so pass through a growth fault to younger stratigraphic levels. The limits of this stratigraphic transfer are roughly the throws of the growth faults between source and accumulation.

Stratigraphy is more important than time or temperature in determining the quality of petroleum in a pool: there are too many examples of variations that bear no relationship to thermal gradients, but could be explained by facies changes. This is not to assert, of course, that temperature has no effect

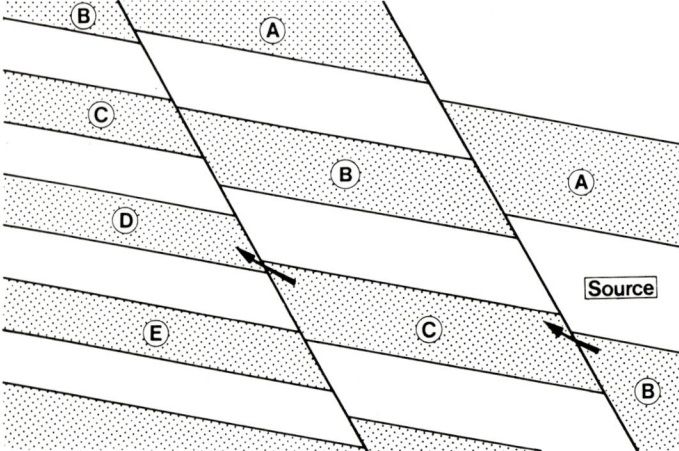

Fig. 16-7. Migration through faults results in accumulation in reservoirs stratigraphically removed from the source.

on the products of organic diagenesis, but that facies impose variations on these.

RIFT BASINS

Most of the principles discussed above apply equally to rift basins, but they deserve separate consideration because they provide evidence of a worldwide crustal process that is still incompletely understood. With little variation in detail, we find a sequence of geological events that spanned approximately 200 m.y. recorded in rift basins in the North Sea, and the western margin of Australia. Very similar events were recorded in the north of Alaska and in Arctic Canada, in the Atlantic margin of South America and around Africa. The data acquired in the exploration for petroleum has shown that the "Cretaceous" transgression is but an episode in a sequence of events of far greater importance that lasted from the Permian throughout the Mesozoic and into the Tertiary (Kent, 1977). It will be recalled from the discussion of stratigraphic traps in Chapter 13 that the main features of these events are these:

— Normal faulting that was contemporaneous with the accumulation of sediment during the Triassic and Jurassic.

— Erosion of these sediments on highs as deformation continued during the Jurassic and into the early Cretaceous, commonly resulting in two unconformities.

— Accumulation of fine-grained, low-energy mudstones, marls and silts, over the disconformity—unconformity surfaces.

— Faulting died out, but epeirogenic subsidence continued irregularly for the rest of the Cretaceous and into the Tertiary. Accumulation during the Tertiary was generally without faulting or folding, except in those areas where regression took place.

— The throw of the bounding fault or faults of the pre-unconformity sequence may be very large, to be measured in kilometres.

Just as the Cretaceous transgression has been found to be of rather different ages in different parts of the world, so the preceding deformation is also found to vary in age in different parts of the world — even different parts of a continent.

Along the western and north-western margins of Australia, numerous sedimentary basins (or sub-basins of a very large sedimentary basin) record the events listed above. Major structural trends were initiated at the end of the Permian by block faulting (Powell, 1976) that continued to be active in places throughout the Mesozoic. The main development of rift basins appears to have been late Triassic in the north. The phases were not contemporaneous over the whole area, but tended to migrate south with time. At the southern end of the north-west shelf, the main rifting phase, with the formation of horst and graben, took place at the end of middle Jurassic. It was followed by a late Jurassic transgression over a palaeotopographic surface. Regional subsidence led to a rapid early Cretaceous transgression, and open marine conditions replaced the restricted marine conditions — earlier in the north than the south. During late Jurassic and early Cretaceous, the major positive trends were enhanced prior to the marine transgression, and rifts formed (for example, between the Exmouth Plateau and the Rankin trend; Fig. 16-8).

The drilling of the Exmouth Plateau in water depths of about a kilometre, although very disappointing, revealed the nature of this part of the western margin of Australia (Barber, 1982) — and also illustrates the detail obtainable from modern seismic reflection surveys (Fig. 16-9). The style is virtually identical with the shelf areas, but the lack of Jurassic sediments was one of the few surprises (Fig. 16-10). It is interesting also for the close proximity to magnetic banding in the ocean crust, the oldest of which, in the Argo abyssal plain, is

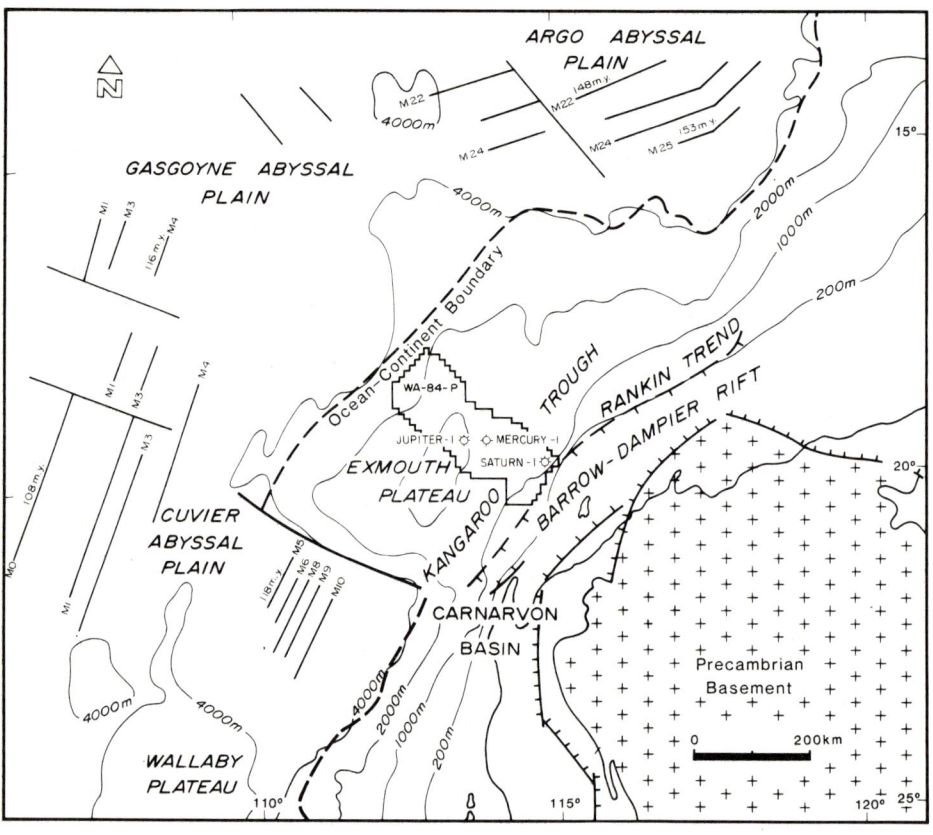

Fig. 16-8. The Exmouth Plateau, Western Australia. (Reproduced, with permission, from Barber, 1982, p. 132, fig. 1; courtesy of Phillips Australian Oil Co.)

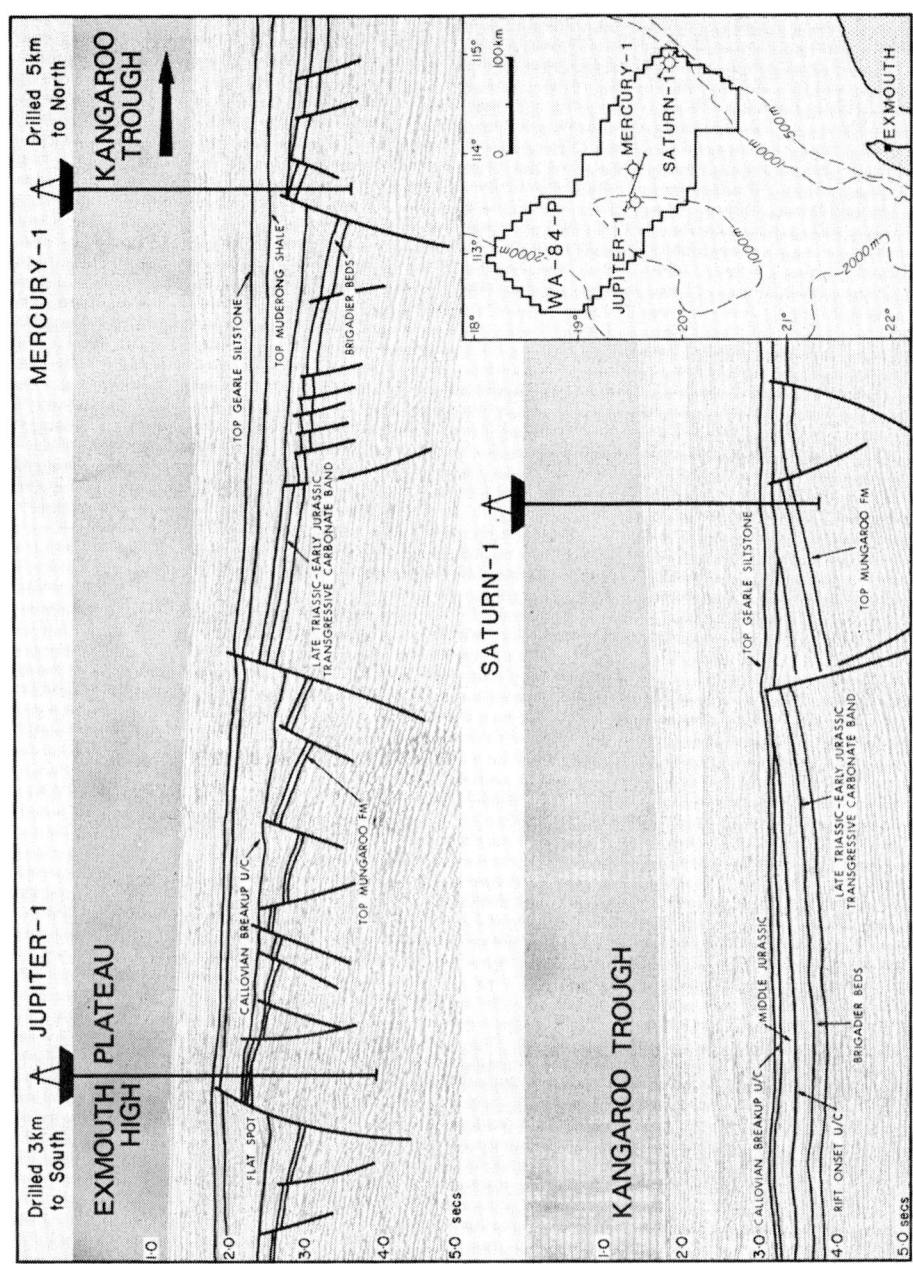

Fig. 16-9. Seismic profiles across the Exmouth Plateau through Jupiter 1, Saturn 1, and Mercury 1 wells. (Reproduced, with permission, from Barber, 1982, p. 133, fig. 2; courtesy of Phillips Australian Oil Co.)

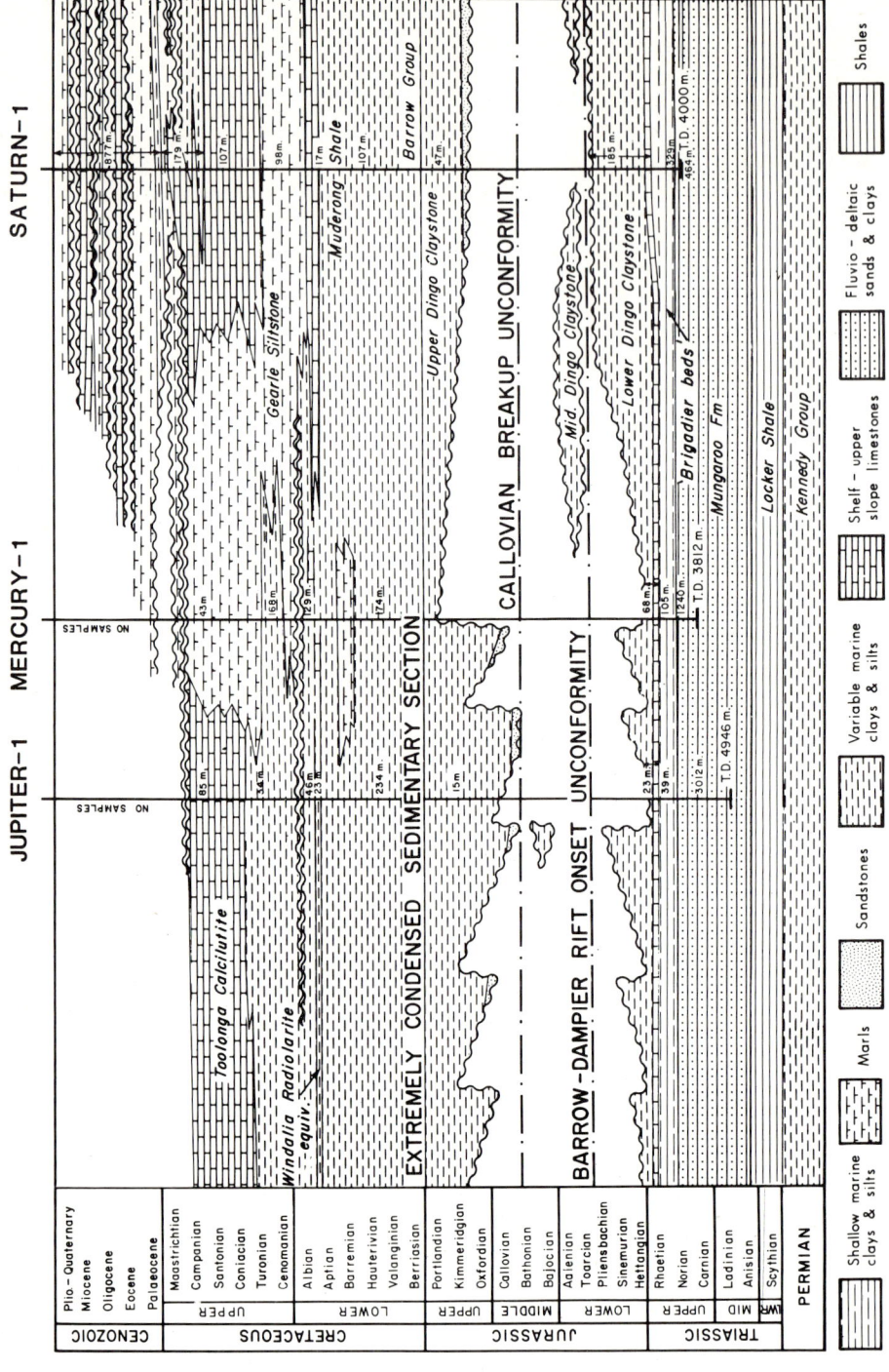

Fig. 16-10. Stratigraphic chart; Jupiter 1, Saturn 1, and Mercury 1. (Reproduced, with permission, from Barber, 1982, p. 135, fig. 4; courtesy of Phillips Australian Oil Co.)

dated at 153 m.y. (Fig. 16-8), slightly post-dating the Callovian unconformity*.

As in the North Sea, abnormal pressures are encountered in the north-west shelf of Australia in thick mudstones of Jurassic and Cretaceous age (Kyaw Nyein et al., 1977). Some of the large accumulations of gas and condensate are at normal hydrostatic pressures beneath the abnormally pressured mudstones.

Some of the record is on land. The Perth basin is half on land, half under the sea, dominating the south-west coast of Australia. It is a north-south trending graben and half-graben, 800 km long, bounded by the Darling fault on the east. Movement on the Darling fault since the early Permian has controlled the accumulation of at least 6 km of sedimentary rocks on its downthrown side, against the Precambrian shield to the east (Jones and Pearson, 1972; Jones, 1976). Marine and non-marine sediments accumulated from the early Permian. The rate of accumulation, controlled largely by movement on the Darling fault, peaked during the late Jurassic (4 km) and early Cretaceous (6 km) with intense fault activity. This ceased during the Neocomian, and tilted fault blocks were eroded. A marine transgression followed in late Cretaceous times, with only minor fault movement. Carbonates accumulated during the Tertiary. Thus the Perth basin acquired similar characteristics to the north-west shelf.

The southern Perth basin (Cope, 1972) is a graben in Precambrian basement, bounded to the east by the Darling fault, and to the west by the Dunsborough fault. In this graben accumulated about 6 km of Permian to lower Cretaceous sediments, and this is the throw on both bounding faults**.

The south coast of Australia, along the Great Australian Bight, has similar character to the northern Perth basin and north-west shelf, but rocks older than Cretaceous are not known (Pattinson et al., 1976). The Mesozoic is faulted by normal faults, but these barely affect the Tertiary sequence. Movements on faults from late Jurassic to late Cretaceous is inferred, with some moving later. In parts of the Bight, sediment accumulation was erratic from late Campanian to the Paleocene. This was followed by a middle Miocene transgression, then a prograding carbonate shelf.

Around the south-east corner of mainland Australia are three basins — the Otway, Bass and Gippsland — of which the easternmost is the main petroleum province of Australia (Ellenor, 1976; Brown, 1976; Threlfall et al., 1976)

---

* Another point of interest in Barber's paper is the report of geothermal gradients in a sedimentary column in deep water. In Jupiter 1 and Mercury 1, this was found to be 25.3°C/km from a surface temperature of 15.6°C; and in Saturn 1, it was 34°C/km.

** There is absolutely no evidence of a Mesozoic sequence of any thickness having accumulated on Precambrian rocks outside the graben — to the contrary. Such a sequence, if it had existed, would give the faults throws of more than 10 km and imply considerable depths of burial of the graben sequence. The lack of metamorphism and the space problem are evidence against this.

(Fig. 16-11). In the Otway basin, the post-unconformity transgression accumulated Oligo-Miocene calcareous sediments, and there is a mid-Cretaceous unconformity within a Cretaceous transgressive—regressive cycle, with marine mudstones in places on the unconformity and above. The Bass and Gippsland basins have much the same stratigraphy and style except that the Cretaceous in non-marine, with coal measures throughout.

The whole of the north-western, western and southern margin of Australia has a similar geological style although the events tended to be earlier in the north-west and progressively later anticlockwise round the continent. It seems reasonable to interpret the record of western Australia as the consequences of processes that resulted in the separation of India (or whichever continent lay to the north-west) and the record of the south coast as the consequences of processes that led to the separation of Antarctica — separation beginning in the north-west and progressing anticlockwise to the south-east. Tasmania then appears to be a rift remnant (*pace* Tasmanians!), with the Otway, Bass and Gippsland basins lying between it and the mainland. It is no doubt geologically significant that the petroleum provinces are at the ends of this margin, with little in the middle, but the reasons for this are not yet understood.

On the opposite side of the world lies the North Sea. The development of the North Sea basin (Ziegler, 1975, 1981; Kent, 1975a, b, 1977) is remarkably similar to that of the north-west shelf of Australia, with but two modifi-

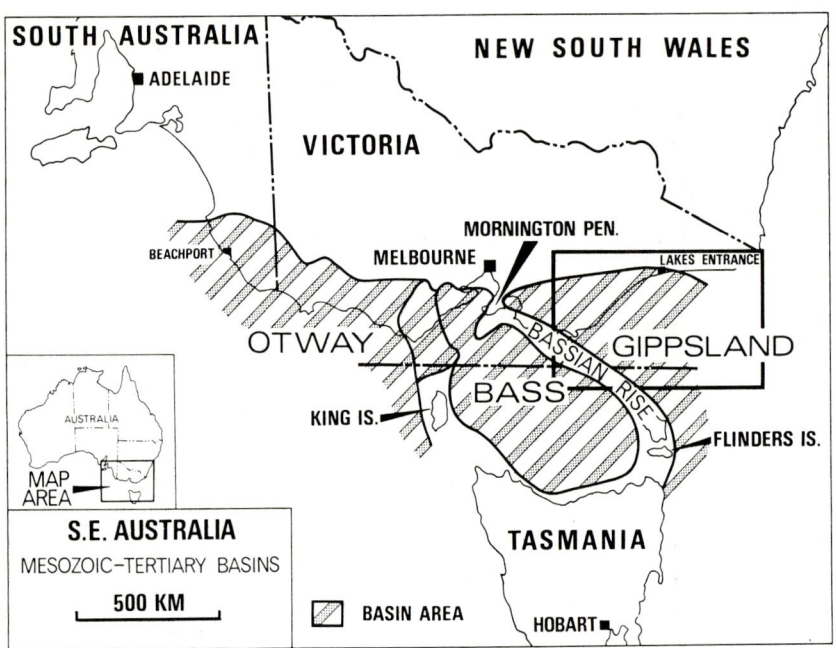

Fig. 16-11. The Otway, Bass and Gippsland basins of south-east Australia. (Courtesy of Esso Australia Ltd.)

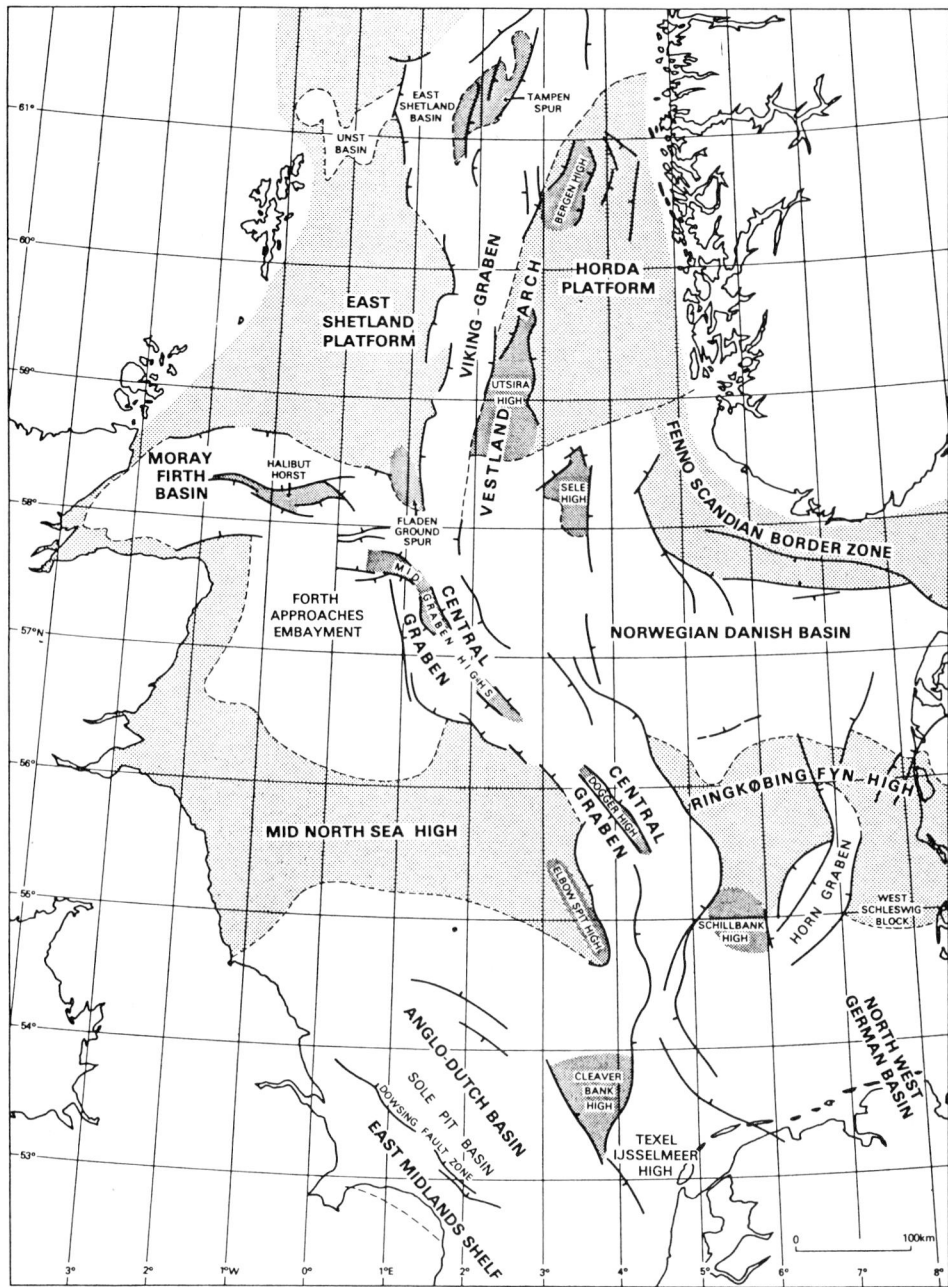

Fig. 16-12. The main structural units in the North Sea. (Reproduced, with permission, from Day et al., 1981, p. 78, fig. 1.)

cations: the North Sea is more of a graben, and Permian salt is diapiric in places. The main structural features are shown in Fig. 16-12.

Subsidence during the Permian and Triassic caused accumulation of sediments of various environments, and by the Triassic this subsidence and accumulation of sediment was accompanied by normal faulting that influenced the rate of subsidence and so the thicknesses accumulated. During the Cretaceous, fault movement became less important, and it had virtually ceased by the end of the Cretaceous. Thereafter, epeirogenic subsidence continued without faulting throughout the Tertiary, the rate of subsidence being variable in both space and time, and so controlling thicknesses. The physiography of the North Sea suggests that this subsidence is continuing.

In broad outline, the Permian period was continental in the North Sea area to begin with, followed by marine transgression of great extent in north-west Europe, and the accumulation of a sequence of evaporites that were to become diapiric in the southern North Sea and in north-west Germany. The continent persisted in the North Sea area, with conformable accumulation of cyclic sequences of evaporites (Brennand, 1975), with some marine influence in middle Triassic times. Marine conditions became more widespread by the Jurassic, and dominated thereafter as regards sediment accumulation.

But sediment accumulation was not continuous either in space or time. The North Sea basin consists of several sub-basins, the histories of which varied and are reflected in the petroleum accumulations. Early Jurassic sedimentary rocks, deformed by contemporaneous faulting, were eroded on highs; and in the northern North Sea there is sporadic evidence of the local accumulations of some of the products of this erosion, but in general it is not known where the products were carried. There was an episode of non-accumulation, followed by the accumulation of a thin sequence of Upper Jurassic shales and mudstones. Of these, the Kimmeridge Clay Formation appears to result from a transgression over the whole area with but minor exceptions, and the consistent facies and generally high organic content make this an important event in the evolution of the North Sea basin. The stratigraphic relationships about this late Jurassic non-accumulation are broadly of disconformity, but with local unconformity on eroded highs.

Still later in the Jurassic and into the early Cretaceous there was another, perhaps more important period of non-accumulation of sediment, especially in the north. The nature of this non-accumulation is not easily determined because the data by which rocks can be dated is only acquired from boreholes and these tend to be restricted to the highs on which erosion may have been deeper and transgression subsequently later. In the Danish North Sea, Childs and Reed (1975) report continuous shale accumulation from Late Jurassic into Early Cretaceous with but a shallowing of the seas (as indicated by sublittoral microfaunas). In the northern North Sea there is a hiatus that lasted typically from late Kimmeridge or Barremian times to the beginning of the Coniacian or, on top of the highs, even as late as the beginning of the Maastrichtian. The minimum hiatus *near highs* seems to be late Albian—Aptian to early Coniacian (Williams et al., 1975, p. 372, fig. 9). That faults were still moving at this time is shown in Piper, where Coniacian and Santonian marls accumulated on the downthrown side of a southern fault, but not everywhere on the upthrown side (as discussed in Chapter 13, pp. 279—302). This is evidence that here, for about 5 m.y., the upthrown block was not subsiding relative to base-level, but that the downthrown block subsided about 100 m (i.e., about 50,000 yrs/m!). So fine was the balance here that it is hard to escape the conclusion that accumulation was continuous in places (with all the reservations discussed in Chapter 1 and in Chapter 13, pp. 295—196).

By the end of the Cretaceous, beginning of the Tertiary, the North Sea basin accumulated sediment by virtue of subsidence without faulting. Extensive and thick mudstones with some sands accumulated, the thicker sequences in areas of greatest subsidence. At least some of the sands were to become petroleum reservoirs.

The present relief on the Jurassic—Cretaceous disconformities and unconformities, generally reported to be about 2000 m, has largely been induced by later events. Although some relief certainly existed on the erosion surfaces at the time, as shown by truncated strata on the present highs, it is unlikely to have been great. The general tendency towards greater thicknesses and more complete sequences in the lows is the result of a tendency to reduce relief by sediment accumulation.

The consistency of structure and stratigraphy in such widely separated areas as the North Sea and Australia, Arctic North America and South Africa, indicates, as Kent (1977) so clearly expounded, a world-wide sequence of events that is not to be explained simply by eustatic changes of sea level. Doubtless the broad crustal events changed the volume of the ocean basins so that there were concomitant eustatic sea-level changes, and there is evidence of this in the stratigraphy; but these were a consequence of more fundamental geological events that took place from late Palaeozoic times to the present in large areas around the world. The significant events seem to be the inception and cessation of normal faulting, because this is a more local phenomenon. Some of the periods of non-accumulation and erosion may have been due to eustatic sea-level changes caused by other events in the series in other parts of the world.

In general, continental margins of the "passive" or "aseismic" type were completed during the final post-unconformity stage of epeirogenic subsidence without faulting; but it is not a universal consequence of the whole process that continental margins were formed. The Sirte basin in Libya, for example, is a rift basin that was completed with transgressive Cretaceous and Tertiary, as we have seen. It is not a continental margin (and its geology must be taken into account in any hypotheses concerning northward relative movement of Africa against Europe). Nor was this great Mesozoic event necessarily the first of its kind. For example, the onshore Canning basin of north-western Australia (Horstman et al., 1976) appears to have had a pre-Permian rifting, or taphrogenic, phase followed unconformably by unfaulted Permian and (thin) Mesozoic. And the northern Perth basin onshore (Hosemann, 1971) contains a hint of Permian rifting with unfaulted transgressive Triassic on the unconformity.

From a petroleum geological point of view, we are concerned with the effect these events had on the stratigraphy and structure of the rift basins both in the taphrogenic phase and the following epeirogenic phase. In summary, they are these:

— *Taphrogenic phase:* subsidence of elongated regions, with normal faulting; brief transgression in places, followed by regression with, commonly, non-marine sediments, the accumulation of which was largely controlled by movement on the growth faults active at the time.

— *Epeirogenic phase:* a period of non-accumulation and erosion (at least locally) during which fault movement ceased; followed by marine transgression with fine-grained, commonly calcareous, sediments accumulating over a surface of generally *low* relief. The lithologies of the epeirogenic phase are commonly in strong contrast with those of the taphrogenic phase.

The taphrogenic—epeirogenic transition is the important event locally because this is a time of trap formation; but there are apparent eustatic events, both positive and negative, superimposed on the sequence as a consequence of events in other parts of the world. For example, the mid-Cretaceous hiatus in south-east Australia in non-marine sediments may be the consequence of an eustatic event caused by the taphrogenic-epeirogenic transition in the north-west shelf of Australia, the North Sea, and elsewhere. These too may form traps for petroleum, particularly when a relative fall of sea-level is followed by transgression and the accumulation of fine-grained sediment.

The most favourable conditions for accumulation of petroleum source rock were generally in the transgressions immediately following the end of the taphrogenic phase. These are the seals to the various unconformity traps, and petroleum will migrate to any permeable rock unit that underlies or subcrops against the source rock, and will accumulate if there is closure on the unconformity or disconformity surface. These fine-grained, undeformed post-unconformity sediments may also contain petroleum source rocks for palaeogeomorphic and reef traps formed during the transgression, and for submarine fans near the margins of the *physiographic* basin forming over the rift basin.

Pre-unconformity petroleum source rocks, both marine and non-marine, may accumulate at any time from the inception of rifting. A contributary favourable factor is the rapid rate of subsidence at times during the taphrogenic phase. However, the very rate of subsidence and accumulation of sediment may mean that early petroleum source rocks generate and lose their petroleum before large traps are formed or completed. Pre-unconformity regressive sands appear to accumulate without the thick mudstones of what might be termed orogenic regressive sequences (there was apparently no winnowing), so normal faulting is the main cause of traps.

If the post-unconformity sequence proceeds to regression, it acquires the characteristics of the normal transgressive—regressive sedimentary basin that were discussed early in this chapter, and petroleum may be trapped in growth structures in the "terminal" regressive sequences.

A remarkable feature of petroleum geology, when viewed in this broad perspective against the current hypotheses of plate tectonics, is that the major petroleum provinces are areas that were *extensional* over great spans of geological time, with not the slightest evidence for regionally-compressive structures, and the tectonics was *vertical* within, and in the immediate neighbourhood of, the sedimentary basins of many major petroleum provinces. Vertical movements were the dominant control on stratigraphy and structure. We do not yet understand the significance of these broad patterns of world geology and geological history, but we may rest assured that they are significant and that full understanding of the processes of the Earth can only follow understanding of these great areas. This will involve understanding the petroleum geology of these regions, because that is the economic incentive for their investigation.

REFERENCES

Adiwidjaja, P. and De Coster, G.L., 1974. Pre-Tertiary paleotopography and related sedimentation in south Sumatra. *Proc. Indones. Petrol. Ass.*, 2 (for 1973): 89—103.
Asmus, H.E. and Ponte, F.C., 1973. The Brazilian marginal basins. In: A.E.M. Nairn and F.G. Stehli (Editors), *The ocean basins and margins*, Vol. I, *the South Atlantic*. Plenum, New York, N.Y., pp. 87—133.
Barber, P.M., 1982. Palaeotectonic evolution and hydrocarbon genesis of the central Exmouth Plateau. *J. Aust. Petrol. Explor. Ass.*, 22 (1): 131—144.
Boeuf, M.G. and Doust, H., 1975. Structure and development of the southern margin of Australia. *J. Aust. Petrol. Explor. Ass.*, 15 (1): 33—43.
Brennand, T.P., 1975. The Triassic of the North Sea. In: A.W. Woodland (Editor), *Petroleum and the continental shelf of north-west Europe*, Vol. I. *Geology*. Applied Science Publishers/Institute of Petroleum, Barking, pp. 295—310.
Brown, B.R., 1976. Bass basin, some aspects of the petroleum geology. In: R.B. Leslie, H.J. Evans and C.L. Knight (Editors), Economic geology of Australia and Papua New Guinea. 3. Petroleum. *Australas. Inst. Min. Metall., Monogr. Ser.*, 7: 67—82.
Burk, C.A. and Drake, C.L., 1974. *The geology of continental margins*. Springer, New York, N.Y., 1009 pp.
Chapman, R.E., 1974. Clay diapirism and overthrust faulting. *Geol. Soc. Am. Bull.*, 85: 1597—1602.
Chapman, R.E., 1979. Mechanics of unlubricated sliding. *Geol. Soc. Am. Bull.*, 90: 19—28.
Childs, F.B. and Reed, P.E.C., 1975. Geology of the Dan field and the Danish North Sea. In: A.W. Woodland (Editor), *Petroleum and the continental shelf of north-west Europe*, Vol. 1. *Geology*. Applied Science Publishers/Institute of Petroleum, Barking, pp. 429—438.

Combaz, A. and de Matharel, M., 1978. Organic sedimentation and genesis of petroleum in Mahakam delta, Borneo. *Bull. Am. Ass. Petrol. Geol.*, 62: 1684—1695.

Cope, R.N., 1972. Tectonic style in the southern Perth basin. *Geol. Surv. W. Aust., Annu. Rep.* 1971, pp. 46—50.

Day, G.A., Cooper, B.A., Andersen, C., Burgers, W.F.J., Rønnevik, H.C. and Schöneich, H., 1981. Regional seismic structure maps of the North Sea. *In:* L.V. Illing and G.D. Hobson (Editors), *Petroleum geology of the continental shelf of north-west Europe.* Heyden/Institute of Petroleum, London, pp. 76—84.

De Coster, G.L., 1975. The geology of the Central and South Sumatra basins. *Proc. Indones. Petrol. Ass.*, 3 (for 1974): 77—110.

De Matharel, M., Klein, G. and Oki, T., 1977. Case history of the Bekapai field. *Proc. Indones. Petrol. Ass.*, 5 (for 1976): 69—93.

De Matharel, M., Lehmann, P. and Oki, T., 1980. Geology of the Bekapai field. *In:* M.T. Halbouty (Editor), Giant oil and gas fields of the decade 1968—1978. *Mem. Am. Ass. Petrol. Geol.*, 30: 459—469.

Douglas, J.G. and Ferguson, J.A., 1976. Geology of Victoria. *Geol. Soc. Aust., Spec. Publ.*, 5, 528 pp.

Ellenor, D.W., 1976. Otway basin. *In:* R.B. Leslie, H.J. Evans and C.L. Knight (Editors), Economic geology of Australia and Papua New Guinea. 3. Petroleum. *Australas. Inst. Min. Metall., Monogr. Ser.*, 7: 82—91.

Falvey, D.A., 1974. The development of continental margins in plate tectonic theory. *J. Aust. Petrol. Explor. Ass.*, 14: 95—106.

Gwinn, J.W., Helmig, H.M. and Kartaadipoetra, L.W., 1975. Geology of the Badak field east Kalimantan, Indonesia. *Proc. Indones. Petrol. Ass.*, 3 (for 1974): 311—332.

Hallam, A., 1980. A reassessment of the fit of Pangaea components and the time of their initial breakup. *In:* D.W. Strangway (Editor), The continental crust and its mineral deposits. *Geol. Ass. Can., Spec. Pap.*, 20: 375—387.

Horstman, E.L., Lyons, D.A., Nott, S.A. and Broad, D.S., 1976. Canning basin, on-shore. *In:* R.B. Leslie, H.J. Evans, and C.L. Knight (Editors), Economic geology of Australia and Papua New Guinea. 3. Petroleum. *Australas. Inst. Min. Metall., Monogr. Ser.*, 7: 170—184.

Hosemann, P., 1971. The stratigraphy of the basal Triassic sandstone, north Perth basin, Western Australia. *J. Aust. Petrol. Explor. Ass.*, 11 (1): 59—63.

Huffington, R.M. and Helmig, H.M., 1980. Discovery and development of the Badak field, east Kalimantan, Indonesia. *In:* M.T. Halbouty (Editor), Giant oil and gas fields of the decade 1968—1978. *Mem. Am. Ass. Petrol. Geol.*, 30: 441—458.

Jones, D.K., 1976. Perth basin. *In:* R.B. Leslie, H.J. Evans and C.L. Knight (Editors), Economic geology of Australia and Papua New Guinea. 3. Petroleum. *Australas. Inst. Min. Metall., Monogr. Ser.*, 7: 108—126.

Jones, D.K. and Pearson, G.R., 1972. The tectonic elements of the Perth basin. *J. Aust. Petrol. Explor. Ass.*, 12 (1): 17—22.

Kamerling, P., 1979. The geology and hydrocarbon habitat of the Bristol Channel basin. *J. Petrol. Geol.*, 2: 75—93.

Kent, P.E., 1975a. The tectonic development of Great Britain and the surrounding seas. *In:* A.W. Woodland (Editor), *Petroleum and the continental shelf of north-west Europe,* vol. 1. *Geology.* Applied Science Publishers/Institute of Petroleum, Barking, pp. 3—28.

Kent, P.E., 1975b. Review of North Sea basin development. *J. Geol. Soc. London,* 131: 435—468. (See also discussion, printed in "General index to volume 131" distributed with volume 132 (5) for September 1976 concerning fig. 18B).

Kent, P.E., 1976. Major synchronous events in continental shelves. *In:* M.H.P. Bott (Editor), Sedimentary basins of continental margins and cratons. *Tectonophysics,* 36: 87—91.

Kent, P.E., 1977. The Mesozoic development of aseismic continental margins. *J. Geol. Soc. London*, 134: 1—18.

Kent, P.E., 1978. Mesozoic vertical movements in Britain and the surrounding continental shelf. In: D.R. Bowes and B.E. Leake (Editors), Crustal evolution in northwestern Britain and adjacent regions. *Geol. J.* (Spec. Issue), 10: 309—324.

Kent, P.E., Bott, M.H.P., McKenzie, D.P. and Williams, C.A. (Editors), 1982. The evolution of sedimentary basins. *Philos. Trans. R. Soc. London*, Ser. A, 305: 1—338.

Koesoemadinata, R.P., 1969. Outline of geologic occurrences of oil in Tertiary basins of west Indonesia. *Bull. Am. Ass. Petrol. Geol.*, 53: 2368—2376.

Kyaw Nyein, R., McLean, L. and Warris, B.J., 1977. Occurrence, prediction and control of geopressures on the northwest shelf of Australia. *J. Aust. Petrol. Explor. Ass.*, 17 (1): 64—72.

Laws, R.A., and Brown, R.S., 1976. Bonaparte Gulf basin — south eastern part. In: R.B. Leslie, H.J. Evans and C.L. Knight (Editors), Economic geology of Australia and Papua New Guinea. 3. Petroleum. *Australas. Inst. Min. Metall., Monogr. Ser.*, 7: 200—208.

Lehner, P. and De Ruiter, P.A.C., 1977. Structural history of Atlantic margin of Africa. *Bull. Am. Ass. Petrol. Geol.*, 61: 961—981.

Magnier, P. and Samsu, B., 1976. The Handil oil field east Kalimantan. *Proc. Indones. Petrol. Ass.*, 4 (2) (for 1975): 41—61.

Pattinson, R., Watkins, G. and Van den Abeele, D., 1976. Great Artesian Bight basin, South Australia. In: R.B. Leslie, H.J. Evans and C.L. Knight (Editors), Economic geology of Australia and Papua New Guinea. 3. Petroleum. *Australas. Inst. Min. Metall., Monogr. Ser.*, 7: 98—104.

Pitman, W.C., 1978. Relationship between eustacy and stratigraphic sequences of passive margins. *Geol. Soc. Am. Bull.*, 89: 1389—1403.

Ponte, F.C., Fonseca, J. dos Reis and Morales, R.G., 1977. Petroleum geology of eastern Brazilian continental margin. *Bull. Am. Ass. Petrol. Geol.*, 61: 1470—1482.

Powell, D.E., 1976. The geological evolution of the continental margin off northwest Australia. *J. Aust. Petrol. Explor. Ass.*, 16 (1): 13—23.

Ridd, M.F., 1976. Papuan basin — on-shore. In: R.B. Leslie, H.J. Evans and C.L. Knight (Editors), Economic geology of Australia and Papua New Guinea. 3. Petroleum. *Australas. Inst. Min. Metall., Monogr. Ser.*, 7: 478—494.

Robson, D.A., 1971. The structure of the Gulf of Suez (Clysmic) rift, with special reference to the eastern side. *J. Geol. Soc. London*, 127: 247—276.

Schwartz, C.M., Laughbaum, G.H., Samsu, B.S. and Armstrong, J.D., 1974. Geology of the Attaka oil field east Kalimantan, Indonesia. *Proc. Indones. Petrol. Ass.*, 2 (for 1973): 195—216.

Threlfall, W.F., Brown, B.R. and Griffith, B.R., 1976. Gippsland basin, off-shore. In: R.B. Leslie, H.J. Evans, and C.L. Knight (Editors), Economic geology of Australia and Papua New Guinea. 3. Petroleum. *Australas. Inst. Min. Metall., Monogr. Ser.*, 7: 41—67.

Van Bemmelen, R.W., 1949. *The geology of Indonesia.* Vol. 1A. *General geology of Indonesia and adjacent archipelagoes.* Government Printing Office, The Hague, 732 pp.

Verdier, A.C., Oki, T. and Suardy, A., 1980. Geology of the Handil field (east Kalimantan — Indonesia). In: M.T. Halbouty (Editor), Giant oil and gas fields of the decade 1968—1978. *Mem. Am. Ass. Petrol. Geol.*, 30: 399—421.

Vidal, J., Joyes, R. and Van Veen, J., 1975. L'exploration pétrolière au Gabon et au Congo. *Proc. 9th World Petrol. Congress*, 3: 149—165.

Weber, K.J., 1971. Sedimentological aspects of oil fields in the Niger delta. *Geol. Mijnbouw*, 50: 559—576.

Weeda, J., 1958. Oil basin of East Borneo. *In:* L.G. Weeks (Editor), *Habitat of oil.* Am. Ass. Petrol. Geol., Tulsa, Okla., pp. 1337—1346.

Wennekers, J.H.L., 1958. South Sumatra basinal area. *In:* L.G. Weeks (Editor), *Habitat of oil.* Am. Ass. Petrol. Geol., Tulsa, Okla., pp. 1347—1358.

Whitten, E.H.T., 1977. Rapid Aptian—Albian subsidence rates in eastern United States. *Bull. Am. Ass. Petrol. Geol.*, 61: 1522—1524.

Williams, J.J., Conner, D.C. and Peterson, K.E., 1975. The Piper oil-field, UK North Sea: a fault-block structure with Upper Jurassic beach-bar reservoir sands. *In:* A.W. Woodland (Editor), *Petroleum and the continental shelf of north-west Europe*, Vol. 1. *Geology.* Applied Science Publishers/Institute of Petroleum, Barking, pp. 363—377.

Wise, R.A., 1976. The Papuan basin — off-shore. *In:* R.B. Leslie, H.J. Evans, and C.L. Knight (Editors), Economic geology of Australia and Papua New Guinea. 3. Petroleum. *Australas. Inst. Min. Metall., Monogr. Ser.*, 7: 494—499.

Ziegler, P.A., 1975. North Sea basin history in the tectonic framework of north-western Europe. *In:* A.W. Woodland (Editor), *Petroleum and the continental shelf of north-west Europe*, Vol. 1. *Geology.* Applied Science Publishers/Institute of Petroleum, Barking, pp. 131—148.

Ziegler, P.A., 1981. Evolution of sedimentary basins in north-west Europe. *In:* L.V. Illing and G.D. Hobson (Editors), *Petroleum geology of the continental shelf of north-west Europe.* Heyden/Institute of Petroleum, London, pp. 3—39.

APPENDIX I. CONSTRUCTION AND USE OF SHALE-TRANSIT-TIME PLOTS

(1) We require: (a) Sonic logs + caliper; (b) gamma-ray or S.P. logs for identifying mudstones; (c) mud log, giving mud weights and depths; (d) programmable pocket calculator; and (e) plotting sheets (see 4 below).

(2) Mark on the sonic log the mudstones, and tabulate the shortest transit times in each unit, concentrating on thinner units and the tops and bottoms of thicker units. These points will be used to determine the normal compaction curve and its parameters. Ignore calcareous mudstones and marls (or mark them clearly on the tabulation). Mark those where the hole is over-gauge because the transit time is probably too long. The tabulation requires six columns: depth below kelly bushing, depth below sea floor, Δt, δ, $\hat{p}$, and the total head h. Porosity and bulk wet density can be included.

(3) Examine the thicker mudstones and record trends of minimum transit time with depth within each mudstone unit. The points within a single mudstone bed will be joined by a line on the plot. Again, mark those where the hole is washed out or over-gauge, or irregular.

(4) *Plotting sheet* (Fig. A-1). Choose a scale for the logarithm of shale transit time that gives a good spread (most of those used are far too small to read properly). One cycle over 200 mm is recommended. The transit-time scale is

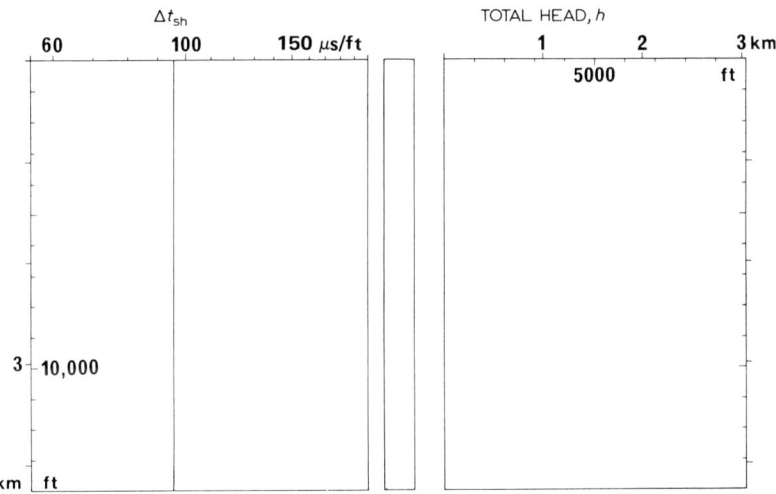

Fig. A-1. Plotting sheet for sonic log—hydraulic analysis of a well.

taken from 55 to 200 μs/ft, and this plot fits neatly onto the left-hand half of A4 paper (the rest will be used for plotting the total head).

The most suitable depth scale is 1 cm = 1000 *feet*. A similar scale in metres can be confusing, and it is easier to convert depths to feet.

It is not essential to use a logarithmic scale for transit time. A linear scale may be more appropriate when there is special interest in porosities or bulk densities because these will then also be linear. A suitable linear scale of transit time is 55—165 μs/ft over 11 cm (10 μs to 1 cm).

(5) Plot the data tabulated, joining all points in the same mudstone that formed the trends (Fig. A-2).

(6) Plot a generalized stratigraphic column close down the right-hand side of the transit-time plot. Mark the sea floor for offshore wells and record the elevation of log datum above sea floor.

(7) Determine the value of the scale length b, either: (a) by noting the depth below sea floor or ground level at which the shortest transit time is 95—96 μs/ft; or (b) by taking the deepest value of the transit time that could represent normal compaction, and solving the equation:

$$b = -z/\ln[(\Delta t - 55)/110] \qquad (A.1)$$

where z is the depth below sea floor, or below ground level onshore.

(8) Insert this value of b into the equation:

$$\Delta t = 110\, e^{-z/b} + 55 \text{ μs/ft} \qquad (A.2)$$

and check the minimum values of transit time at shallower depths.

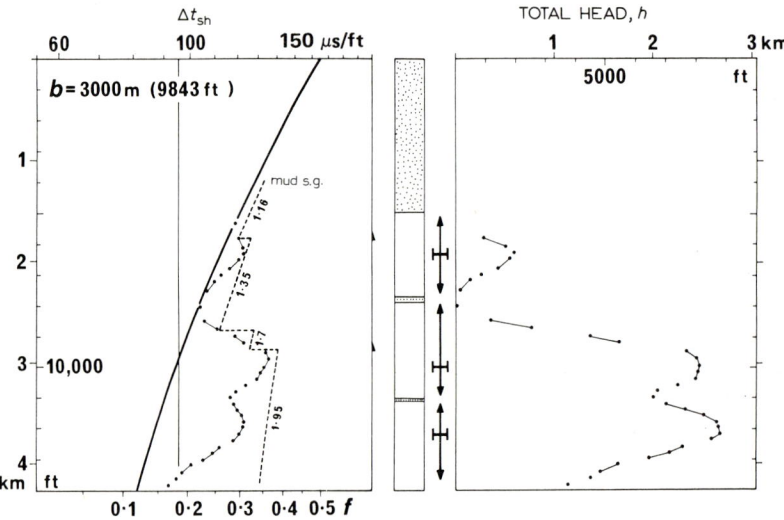

Fig. A-2. Plotting sheet with hypothetical example.

If these data points agree tolerably well with eq. A.2 (perfect correspondence is not to be expected), plot the curve. If there are transit times *shorter* than those predicted by the equation, recompute the scale length from the deepest of these, and plot the revised curve.

If all plots are made in a standard format, transparent overlays can be made from which the scale length can be estimated and the most suitable data point chosen for the estimation of the scale length (Fig. A-3).

(9) *The plot must now be checked* against the mud weights used during drilling, particularly those at which the well kicked. This is done from the formula:

$$\Delta t_{\text{mud equivalent}} = 110\, e^x + 55\ \mu s/ft \tag{A.3}$$

where $x = (z/b)$ (s.g. $-2.3)/1.3$, where s.g. is the specific gravity of the mud at depth z below sea floor or the ground surface.

The plot of mud equivalents should bound the data points on the side of longer transit time, in general, but slight imbalance may not be detected when drilling thinner or shallower mudstones.

(10) On the tabulation, compute the following:

$$\delta = -(b/z) \ln [(\Delta t - 55)/110] \tag{A.4}$$
$$\hat{p}_{\max} = \lambda \overline{\gamma}_{bw} z = (1 - 0.55\delta)\, 22.63 \times 10^{-3} z\ \text{MPa} \tag{A.5}$$
$$h_{\max} = (\hat{p}/\overline{\gamma}_w) - z = (\hat{p}/10.18 \times 10^3) - z\ \text{m} \tag{A.6}$$

where z is the depth below sea floor in *metres*, and b is also in metres. The value of $\overline{\gamma}_{bw}$ chosen, 22.63 kPa/m, is that that gives normal hydrostatic

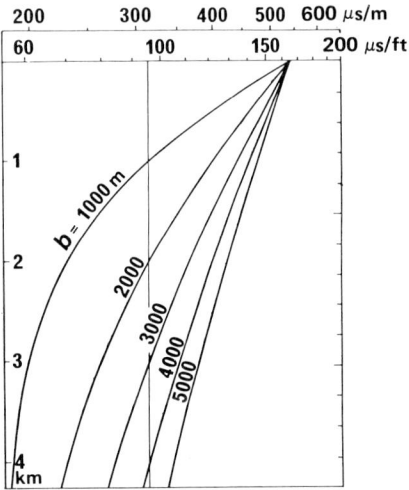

Fig. A-3. Curves of different scale lengths, for estimating the value of the scale length from mudstone—transit-time plots.

pressures for the density of water implied in λ in eq. A.5, which is $\rho = 1039$ kg m^{-3} or $\gamma_w = 10.18$ kPa/m — but if there is better data, use it and compute a better value of λ_e and λ. This gives zero total head (h) for normal hydrostatic pressures. For onshore wells, the elevation of land surface above sea level must be added to this h, or this z taken from sea level.

There are two great advantages in using total head rather than pressure. When pressure is estimated from eq. A.5 from sea floor, and used in eq. A.6, the value of total head obtained is for practical purposes the same as the value obtained by estimating the pressure from sea level. So the value of h is recorded relative to sea-level datum in offshore wells. The other advantage is that the total head is proportional to the energy of the pore water, and can therefore be used comparatively within a single borehole, and between boreholes.

(11) Plot the total head, h, against depth beside the stratigraphic column, and join points within a single mudstone bed that show trends.

(12) Mark and record the top(s) of abnormal pressures.

OPERATIONAL USES OF THE METHOD

(13) Before drilling, the construction of a synthetic sonic log from interval velocity data acquired from the seismic survey can be so plotted, and lines of equal mud weight drawn instead of lines of equal δ, using eq. A.3. From this plot, depths of abnormal pressures can be anticipated and the necessary mud-weights for their control estimated.

GEOLOGICAL INTERPRETATION

(14) Water flows from positions of higher energy to positions of lower energy. On the plot of total head against depth, therefore, all zones in which the total head increases with depth are zones of upward potential gradient and upward water flow. All zones of decreasing total head with depth are zones of downward water flow; and zones of constant total head are zones of no flow in a vertical direction.

Levels of minimum total head are therefore sinks, and water is being abstracted from the mudstones and then removed laterally along relatively permeable beds (check other logs for the lithology). Levels of maximum total head are levels separating upward flow above from downward flow below. The sequence penetrated by the borehole can therefore be divided into distinct hydraulic units, and the maxima of total head represent *total hydraulic and chemical barriers to the vertical migration of fluids*, as discussed in Chapter 9, pp. 181—182.

(15) Mark these barriers and sinks on the plot alongside the stratigraphic column. As a first assumption, the source of any accumulation and the accu-

mulation must both lie stratigraphically within a single hydraulic unit, and all sinks with source rocks in the same unit are prospective horizons. Where the sink is at normal hydrostatic pressures, with zero head, it is possible that the sink is connected with the overlying normally pressured units outside the immediate area of the well. Where the sink is abnormally pressured, such connexion is unlikely to be local.

(16) Correlate the barriers and sinks with other wells, and map the hydraulic units.

(17) *Accuracy and precision.* Transit times cannot be read better than ± 1 μs/ft, and tend to be longer than the true value. We are therefore estimating pressure no better than ± 1 MPa (in round numbers), or total head not better than ± 100 m, with a tendency to exaggerate both. Experience shows that consistent results can be obtained with careful work, as judged by the mud log. It is recommended that no rounding-off be performed in the sequential computations (i.e., program eqs. A.4, A.5 and A.6 as one), and that more importance should be attached to trends than to individual data points.

When estimating absolute pore pressures in mudstones from sonic transit times, two assumptions in the formulae must be remembered. These are, that the mean overburden pressure gradient, $\overline{\gamma}_{bw}$, is constant with depth, and that the value of λ_e ($= \overline{\gamma}_w / \overline{\gamma}_{bw}$) is constant at about 0.45. Neither assumption is strictly true, but eqs. A.5 and A.6, when used together, give zero total head for normal hydrostatic pressures (normal Δt and $\delta = 1$) and, when $\delta = 0$, the total head that corresponds with a mean overburden gradient of 22.63 kPa/m.

This means that when the mean overburden pressure gradient is some other

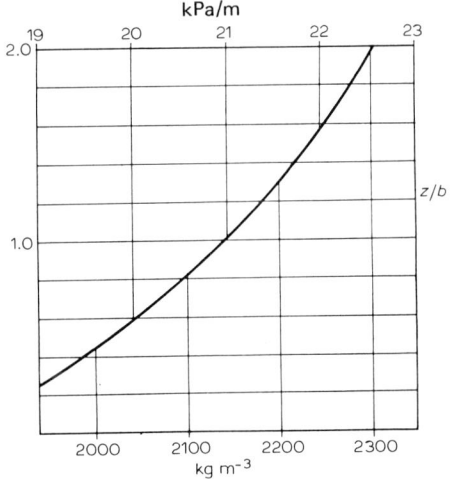

Fig. A-4. Plot of mean overburden pressure gradient (above) and mass density (below) as a function of the dimensionless depth, z/b, assuming $f_0 = 0.5$.

figure — almost invariably less in Tertiary sequences — the estimated pressures should be reduced proportionally, and the total head recomputed. In those areas where the normal mudstone compaction curve gives a good guide to the overburden density, $\overline{\gamma}_{bw}$ is a function of the scale length b as well as depth*. Figure A-4 shows the values of $\overline{\gamma}_{bw}$ and $\overline{p}_{bw}$ plotted against z/b, assuming $f_0 = 0.5$.

EXAMPLE OF PRESSURE ESTIMATION AND INTERPRETATION**

In the central North Sea, Tertiary mudstones are found from the surface down to the Paleocene sandstone or the Cretaceous chalk, and Herring (1973) used sonic log data to estimate pore pressures in these mudstones. Interpreting Herring's (1973, p. 76, fig. 4) normal shale transit time profile on the principles discussed on p. 50, using maximum values of transit time, we find the scale length b = 2700 m (8900 ft) for Δt_0 = 165 μs/ft. This determines the normal compaction—transit time curve (eq. A.2).

From the sonic log of a well drilled in 75 m of water, with log datum 100 m above sea floor, Table A-1 is constructed.

The pressure in the underlying oil reservoir is 48.3 MPa at 3050 m below *sea level* (these heads are computed from sea level, using the same value of 10.18 kPa/m for water as in the other computations), giving the water in the reservoir a total head of 1695 m.

We conclude from these data: (a) that the pore water in the mudstone overlying the reservoir is flowing *downwards* to the reservoir (note the present

---

* The mean relative porosity (f/f_0) in normally compacted mudstone from depth z_1 to z_2 is given by:

$$\overline{f}* = \int_{z_1}^{z_2} e^{-z/b}\, dz/(z_2 - z_1)$$

so the mean porosity from surface to depth z is given by:

$$\overline{f} = f_0 b\,[1 - e^{-z/b}]/z.$$

The mean bulk wet density can then be estimated by substituting this value into:

$$\overline{\gamma}_{bw} = \gamma_s - \overline{f}\,(\gamma_s - \gamma_w).$$

These equations with Hedberg's material constants (p. 47) lead to a predicted overburden pressure of 6070 psi at 6175 ft, whereas his estimate from measurements was 6050 psi (Hedberg, 1936, p. 254, table 1).

** This example is based on data published by Byrd (1975, p. 443, fig. 2) and Van den Bark and Thomas (1980, p. 2361, fig. 23) for Ekofisk oil field in the North Sea, the quantities having been read off their diagrams. It is given as an example of the application of the method, *not* as an interpretation of Ekofisk.

TABLE A-1

Tabulation of data and computed values

Depth (m)		Δt	δ	$\hat{p}_{max}$	h_{max}	Corrected*	
KB	SF			(MPa)	(m)	$\hat{p}$	h
1260	1160	137	0.68	16.4	449	14.7	288
2720	2620	150	0.15	54.4	2720	48.9	2186
3050	2950	120	0.48	49.1	1871	44.2	1389

* $\overline{\gamma}_{bw}$ could be as small as 20 kPa/m, giving a pressure correction factor of 0.9.

tense) from 2720 m below KB, but not through the oil accumulation (for reasons discussed in Chapter 8, p. 155); (b) that the energy of the pore water in the cap rock here is approximately equal to that of the oil in the accumulation at that level, probably a little less; (c) that if the cap rock is an oil source rock, oil will be undergoing primary migration above the accumulation, and its quality will be similar to that in the accumulation. It will only pass into the accumulation where its energy, less the capillary pressure energy, is greater than that of the accumulation, and it is not clear that this condition is satisfied in the position of this well; (d) that if the cap rock is not an oil source rock, no oil can enter the cap rock from the accumulation unless its energy exceeds that of the overlying water by an amount exceeding the injection energy required to overcome the capillary resistance; (e) that the lateral permeability of the reservoir rock is poor, whatever its value in the accumulation, because otherwise the abnormal pressures would have dissipated. This evidence supports the conclusion that migration of pore fluids from the mudstone is still in progress; and (f) that any petroleum generated in the mudstone below the barrier at 2720 m below KB is most unlikely to accumulate above that barrier over a wide area around the well (assuming the presence of a potential reservoir rock). Mapping of the barrier will determine the minimum area.

Conclusions based on the data of a single borehole can be dangerous, so they should be interpreted positively, to suggest where further prospects lie, rather than negatively.

REFERENCES

Byrd, W.D., 1975. Geology of the Ekofisk field, offshore Norway. In: A.W. Woodland (Editor), *Petroleum and the continental shelf of north-west Europe*, vol. 1. Geology. Applied Science Publishers/Institute of Petroleum, Barking, pp. 439—444.

Hedberg, H.D., 1936. Gravitational compaction of clays and shales. *Am. J. Sci.*, 31: 241—287.
Herring, E.A., 1973. North Sea abnormal pressures determined from logs. *Petrol. Engineer*, 45 (Nov.): 72—84.
Van den Bark, E. and Thomas, O.D., 1980. Ekofisk: first of the giant oil fields in western Europe. *In:* M.T. Halbouty (Editor), Giant oil and gas fields of the decade 1968—1978. *Mem. Am. Ass. Petrol. Geol.*, 30: 195—224.

GLOSSARY

Aliphatic — Organic compounds with molecules in which the carbon atoms are in open chains are called aliphatic compounds. Thus the alkanes (paraffins), olefins, carbohydrates and fats (from which the name is derived) are aliphatic compounds. Cf. *Aromatic*.

Appraisal — Drilling undertaken to determine the significance of a discovery is appraisal drilling. Development drilling follows if the appraisal indicates that the discovery is commercial.

Aromatic — Organic compounds with molecules in which the carbon atoms are in closed chains or rings are aromatic compounds. They normally contain a larger proportion of carbon than *aliphatic* compounds.

Assembly — The components of the drilling string below the drill pipe — bit, drill collars, reamers, stabilizers — are referred to collectively as the assembly. "To change the assembly" means to change the arrangement or to substitute reamers for stabilizers (for example).

Basement — The rocks that underlie, surround and pre-date those of a sedimentary basin. When basement is of sedimentary rocks, these have usually suffered deformation prior to the development of the sedimentary basin, and may have been metamorphosed.
Economic basement refers to rocks that have no economic prospects. In the context of petroleum geology, the two are commonly synonymous. It is unwise to regard igneous or metamorphic rocks as economic basement for petroleum, because they may (and do) form important reservoirs.

Billion — An ambiguous term that should only be used in a local context. 10^9 in the U.S.A.; 10^{12} in many European countries and Australasia — but 10^9 usually meant in the petroleum industry. Write 150×10^9 m^3, for example.

Block — A volume of rock bounded by one or more significant faults. A fault usually gives its name to the upthrown block in subsurface geology.

Blowout — An uncontrolled flow of fluid (gas, oil, or water) from a borehole that results when the pressure exerted by the mud is insufficient to contain the formation fluids *and* a human or mechanical failure prevents this excess pressure being contained at the surface by means of the blowout preventers (BOPs).

Bottom water — The water below an accumulation of petroleum in a reservoir, occupying the same general area (Fig. G-1).

Fig. G-1. Bottom water.

"Bottoms up", to circulate — To circulate mud through drill pipe or the drilling string in order to bring the mud at the bottom of the hole to the surface. This procedure is carried out when a change in the drilling performance suggests a significant change in the rocks and their contained fluids, e.g., after a *drilling break*.

Buttress sand — Sands that terminate against an *underlying* unconformity.

Capillary pressure — The difference in pressure across an interface between two immiscible fluids (e.g., water and oil). When oil and water, or gas and water, occupy the same pore space in a rock that is water-wet, the petroleum occupies the central position of minimum energy in the space. The pressure is higher on the concave side of the interface in oil than on the convex side in water, and the amount of this difference is related to the size of the pore and to the relative proportions of petroleum and water. If the petroleum is a discrete droplet in a pore, it can only be moved to another pore by exerting a displacement pressure that exceeds the capillary pressure. This increases as the size of the constriction decreases: it is a function of the radius of curvature of the interface. (See Hubbert, 1953, *Bull. Am. Ass. Petrol. Geol.*, 37: 1954—2026; and Levorsen, 1967, *Geology of Petroleum*, Freeman, San Francisco, Calif., for fuller explanations.)

Cap rock — (a) The rock that prevents petroleum from escaping from a reservoir rock. Clays, marls, mudstones and evaporites are common cap rocks. Although these rocks are relatively impermeable, they prevent the escape of petroleum by virtue of capillary forces associated with fine grain and small pore sizes.

(b) The material near the top of a salt plug, stock, or dome that is not salt. Usually gypsum and anhydrite.

(c) A hard, calcareous layer in mudstone above abnormal pressures.

Carbon, fixed — A curious term that relates to the matter in coal that is not volatile. It is an empirical measurement of considerable value. Coal consists of ash, moisture, volatile matter, and *fixed carbon*. It is simpler and quicker than full analysis, and reflects the rank of coal. Expressed as a percentage, the higher the fixed carbon the higher the rank of the coal, for volatiles decrease with increasing rank.

Carbon, total — *Total carbon, dry mineral matter free* relates to the total carbon in coals, including that in the volatile constituents.

Circulate — In rotary drilling, the mud *circulates* from the suction tank through the pumps, the drilling string, up the annulus between the drilling string and the wall of the hole, or the casing, and so through the shale-shaker and settling tanks back to the suction tank. When drilling is stopped, mud must be circulated to remove all cuttings before the bit is pulled from the hole.

Clastic ratio — A parameter used in some facies maps. It is the total thickness of conglomerate + sand, sandstone + clay, mudstone, shale divided by the total thickness of limestone + dolomite + evaporites.

Clastic/shale ratio — A parameter used in some facies maps. It is the total thickness of arenites (sandstones and calcarenites) divided by the total thickness of clay, mudstone, shale.

Coefficient of permeability — The coefficient K in Darcy's law when written in the form $q = K \Delta h/l$ (see p. 163). It is related to *intrinsic permeability* by $K = k\rho g/\eta$. It has the dimensions of a velocity, LT^{-1}. Synonym: *hydraulic conductivity*.

Completion — A well is *completed* on one or more petroleum reservoirs, after drilling and casing the hole, with a view to producing petroleum from it or them. If production is, or becomes, impossible or undesirable (due to high water cut, high gas/oil ratio, low yield, or other reason) the well may be *re-completed* on another reservoir or group of reservoirs.

Concurrent fault — Synonym for *growth fault*, over which it has priority (Tiddeman, 1890) but is rarely used.

Contemporaneous fault — Synonym for *growth fault* that is commonly found in the older literature.

Darcy — The darcy is the unit of intrinsic permeability based on incomplete expression of Darcy's law, $k = \eta q l/\Delta p$ (see p. 165). Common unit, millidarcy (md). One millidarcy is equal to 10^{-8} cm^2, or 1 μm^2, for practical purposes.

Depositional fault — Synonym for *growth fault* that is commonly found in the literature.

Diachronous — A rock unit is said to be diachronous when its age range varies from place to place. Diachronous rock units are the consequence of the migration of two contiguous environments during accumulation of sediment, so that isochronous surfaces are not strictly parallel to lithological surfaces. Synonym (undesirable): time-transgressive. Antonyms: isochronous, synchronous.

Diastem — A break in a stratigraphic sequence that is more often inferred than observed, but bedding planes within a lithological unit may be diastems. See Fig. 1-2. Defined by Barrell, 1917, *Bull. Geol. Soc. Am.*, 28, p. 794.

Differential sticking — See *wall sticking*.

Dog-leg — A sharp or sudden change of direction in a borehole. Dog-legs often have geological significance (change of lithology, fault, steep dip) but may also be caused by the driller. Obviously undesirable.

Down-time — Time lost during drilling due to repairs on drilling machinery, waiting on equipment or bad weather. Conditioning the hole, fishing, round-trips, and other operations that are integral to the drilling operation, are not strictly down-time.

Drilling break — A sudden increase in the rate of penetration. May be due to a change in lithology, but must be assumed to be due to an increase in the pore pressure relative to the hydrostatic.

Dry — A borehole that finds no accumulation of petroleum is said to be dry. Cf. *wet*.

Economic basement — See *Basement*.

Edge Water — The water below and lateral to a petroleum accumulation (Fig. G-2).

Fig. G-2. Edge water.

Equipotential surface, line — A surface or line on which points of equal fluid potential lie. Not synonymous with *potentiometric surface*. Synonym: isopotential, which is not to be preferred, but is too commonly used to be dropped now.

"Ethane plus" — Gases of the alkane (paraffin) series of higher molecular weight than methane (CH_4), i.e., ethane (C_2H_6), propane (C_3H_8), butane (C_4H_{10}) and pentane (C_5H_{12}).

Facies (singular and plural) — The character of a rock unit may change from place to place. The general aspect of a sedimentary rock is its facies; (a) in its lithological context (lithofacies), e.g. mudstone facies; (b) in its biological context (biofacies), e.g. graptolite facies; and (c) in its environmental context, e.g. littoral facies.

A *facies fauna* is a fossil fauna confined to a particular lithology.

Fish, fishing — When seeking to recover equipment accidentally left in a borehole, such as part of the drilling string, one is said to be *fishing*. The equipment accidentally lost in the hole is, of course, the *fish*.

Fluid potential — The mechanical potential energy of an element of fluid with respect to its physical environment. It is also (and equivalently) the amount of work required to move unit mass of fluid from the reference position and state (e.g., sea level and atmospheric pressure) to the point specified. Defined by Hubbert, 1940, *J. Geol.*, 48: 797—803. For liquids, the potential at a point is:

$$\Phi = gz + p/\rho$$

where g is the acceleration due to gravity, z the elevation of the point above (+) or below (—) the datum level (commonly sea level), p is the gauge pressure at the point, and ρ the mass density of the liquid.

The potential of a fluid can be determined for any point occupied by that fluid, or capable of being occupied by it (cf., potentiometric surface of an artesian aquifer).

Fluid potential gradient — When a fluid is at different potentials in two points in subsurface space, there is a fluid potential gradient between them, and fluid tends to flow from the point with higher potential towards that with lower potential:

$$\text{grad } \Phi = g \text{ grad } z + (1/\rho) \text{ grad } p.$$

More generally, fluid potential gradients exist in fluid that is not at rest (not in hydrostatic equilibrium).

Formation — (a) General term in petroleum industry for "bed", "reservoir", or any rock unit or group of rock units; e.g., formation water, producing formation, formation density log; (b) in strict geological usage, an objective subdivision of a sequence of rocks on the basis of lithology only. In subsurface geology, the basis may be electrical log characteristics.

Gas-cap drive — See *Water drive*.

Gas-cut mud (GCM) — The mud in circulation in rotary drilling is said to be gas-cut when it contains bubbles of gas at the shale shaker. Release from solution may be involved, but gas can be incorporated in the mud from the volume of rock drilled. The bubbles expand as they approach the surface due to the decreasing pressure: the mean density of the mud column is thus reduced, leading to a risk of a *blowout*. See also *Trip gas*.

Gas/oil ratio (GOR) — The ratio of a volume of gas to a volume of oil (not necessarily in the same units) at a specified temperature and pressure. The standard conditions of measurement vary from country to country, but are usually either 15°C and 760 mm of mercury or 60°F and 30 inches (760 mm) of mercury. Common units are: $m^3\ m^{-3}$, cubic feet of gas per barrel of oil.

Gauge pressure — Absolute pressure less atmospheric pressure; that is, the reading of a pressure gauge at atmospheric pressure is zero.

Genetic rock unit — Lithological units that are recognizable representatives of modern sedimentary environments that have been preserved in the stratigraphic record, e.g., point bar, channel fill, barrier bar, etc. It is unwise to suppose that all rock units may be recognizable in terms of genetic rock units.

Geostatic — Pressures and pressure gradients due to the gravitational load of the total overburden are called geostatic. Synonymous with overburden, which has priority. Undesirable synonym: *lithostatic*.

Growth fault — A fault that separates correlative sequences of which the thicker is on the downthrown side. More generally, it is used of faults that are inferred to have been moving during the accumulation of sediment in at least the downthrowing block. As a consequence of this thickness contrast, the throw tends to increase with depth. This is not a diagnostic criterion because antithetic faults may reduce the throw. Synonyms: *concurrent, contemporaneous, depositional, Gulf-Coast type, progressive, recurrent, synsedimentary*.

Halokinesis, halokinetic — "Salt tectonics". The deformation of sediments by a salt dome is halokinetic deformation, and the process is halokinesis. Used mainly in Europe.

Head — Refers to the vertical length of a column of fluid, usually liquid; ambiguous when not qualified. It is an energy per unit of weight, with dimensions of *length*, not pressure.

Elevation head (less desirably, *potential head*): the potential energy per unit weight due to elevation, i.e., the elevation of the point at which pressure p is measured, above (+) or below (−) an arbitrary horizontal datum (usually sea level).

Pressure head: the vertical length of a column of liquid supported, or capable of being supported, by pressure p at a point in that liquid; given by $p/\rho g$.

Velocity head: head due to the kinetic energy of the liquid. Negligibly small in most geological contexts, it is given by $V^2/2g$.

Total head: the algebraic sum of the pressure and elevation head, neglecting velocity head. $h = (p/\rho g) + z$.

Hiatus (plural, hiatus or hiatuses, not hiati) — A break in the sequence of accumulated sediments: a surface that represents the passage of time without the accumulation of sediment. See also *non-deposition*.

Homologue — Molecules of different substances that have the same general relationship between the atoms are said to be homologous. Ethane is a homologue of methane.

Hydraulic conductivity — Synonymous with *coefficient of permeability*.

Hydraulic gradient — The difference of *total head* (see *Head*) divided by the macroscopic length of porous material between the points where the total head is measured — properly, the steepest gradient at a point in the fluid. Note that the gradient of a *potentiometric surface* is not strictly the hydraulic gradient unless the aquifer is horizontal (but it may be a sufficiently close approximation). Hydraulic gradient is not the same as pressure gradient. Hydraulic gradient is dimensionless.

Interfacial tension — When two immiscible fluids are in contact in a capillary tube or fine-grained porous material, the interface between the two fluids acts as if it were an elastic membrane in a state of tension. Cf. *Surface tension*. Dimensions MT^{-2}.

Intrinsic permeability — The component of permeability that is ascribable to the porous material alone, independent of the physical properties of the fluid passing through it. Cf. *Coefficient of permeability, darcy*. Dimensions L^2.

Isochore — When the thickness of a rock unit is taken from borehole data without correction for dip or borehole deviation, and used on a map on which lines of equal apparent thickness are drawn, these lines are said to be isochores. They are *isopachs* uncorrected for dip or borehole deviation.

Isopotential surface, line — See *Equipotential surface, line.*

Isopycnic surface, line — A surface or line of equal density.

Jar — A tool by means of which an upward or a downward shock can be given to the drilling string beneath it, or to the fish. It is a means of freeing stuck pipe, and may be incorporated in the drilling string or in the fishing string.

Joint — A single length of pipe, or a single length of casing. The length is nominally 30 ft or 10 m, but their actual length is usually rather less than this. A drill collar is not a joint, viz., "two drill collars and three joints of drill pipe".

Juxtaposed — When two different lithological units are brought together across a fault, they are said to be juxtaposed.

Kelly bushing — Two half-sections that fit into the rotary table leaving a square or hexagonal opening through which the kelly passes. It is the means by which rotation is imparted to the kelly while leaving the latter free to be raised or lowered.

Kick — A borehole is said to "kick" if the pressure of the formation fluids comes to exceed that exerted by the mud column in the hole. A kick is the first stage of a blowout, but by no means all kicks lead to blowouts. Indeed, the safest way to drill to abnormal pressures is to drill with the lightest practicable mud so that the borehole kicks as soon as abnormal pressures are reached.

Liner — A string of casing that is not continuous to the surface. It is hung in a casing that has been cemented; and it is cemented through drill pipe that is later disconnected. It is an alternative to conventional casing as a production string.

Lithostatic — Used synonymously with *overburden* (adjective) and *geostatic*, the total pressure exerted by the overburden, solid and liquid, or its pressure gradient. Undesirable because both the other two have priority.

Marker — Any distinctive (but usually thin) part of a stratigraphic sequence that can be recognized over an area, or from borehole to borehole. In subsurface geology, it is commonly a point on the electrical log that is recognizable on the logs of different boreholes by correlation.

Non-deposition — Commonly refers to a *hiatus*. It is a phrase that has no merit. Non-deposition may mean the same as non-accumulation, but non-accumulation is factual and much to be preferred. An area of non-accumulation may well have been an area of deposition.

Offlap — Used in the opposite sense to *onlap*, an offlap sequence is one in which successively younger strata occur seaward of the older, as in some regressive sequences — but it is not strictly synonymous with regressive. It is doubtful whether this is a useful word or concept except in seismic stratigraphy.

Onlap — The accumulation of successively younger strata, each overlapping the other, in a transgressive sequence. It is not strictly synonymous with transgressive: it refers to the distribution of rock units relative to each other, and is thus a consequence of transgression.

Outpost, outstep — A borehole drilled during the development phase (see *Appraisal*) outside the area already drilled, but still to the same structure or trap, is an outpost or outstep well.

Overburden — In subsurface geology, the overburden is the total sequence above the layer of interest, solids and fluids. The pressures exerted by the overburden are overburden pressures (and they give rise to overburden pressure gradients). Synonym: *Geostatic* in the context of pressures.

Paralic — The general environment of the marine margin of a physiographic basin is paralic. It includes the littoral, lagoonal and shallow marine environments.

Permeability — A porous rock through which fluid can be passed is said to be permeable, and the rock has the property of permeability. A rock through which fluid can easily be passed is said to have high or good or large permeability: conversely, low, poor or small permeability. There are two distinct types of permeability: *intrinsic permeability*, which is a property of the rock independent of the fluid in the pores, and *coefficient of permeability*, which lumps together the properties of the rock and its contained fluids. Intrinsic permeability (k) has the dimensions of an area: coefficient of permeability (K) has the dimensions of a velocity. They are related by $K = k\rho g/\eta$. While rocks with large permeability tend also to have large porosity, porosity is but one of the components of permeability (see p. 165).

Effective permeability. When two immiscible fluids occupy the pore spaces of a rock, the movement of each is influenced by the other, and the saturations, and we speak of the effective permeability of that rock to oil, or to gas, or to water. It is in the nature of an intrinsic permeability.

Relative permeability is the ratio of effective permeability to oil, gas, or water, to the permeability when that material is saturated with that fluid only. Relative permeability is dimensionless.

For more detailed discussions, see Chapman, 1981, *Geology and water*, Nijhoff, The Hague, pp. 49—70; De Wiest, 1965, *Geohydrology*, Wiley, New York, N.Y., p. 161 et seq.; Hubbert, 1940, *J. Geol.*, 48: 785—944 (particularly pp. 785—819 and p. 915 et seq.; Levorsen, *Geology of petroleum*, Freeman, San Francisco, Calif., p. 104 et seq.

Permeability barrier — A reduction of permeability along a migration path that tends to inhibit further migration. It may be a facies change, or a fault. The force of capillarity is more significant than permeability in a barrier to petroleum migration.

Piezometric surface — See *Potentiometric surface*.

Pore ratio — See *Void ratio*.

Porosity — The ratio of pore volume to gross volume of rock, that is, the volume of pore space divided by the volume of pore space and the volume of solids. Symbol usually f. Usually spoken of as a percentage. Dimensionless. Related to *void ratio* by $f = \epsilon/(\epsilon + 1)$.

Potentiometric surface — If the *total head* (see *Head*) relative to the same datum surface is computed for various points in a body of fluid, points on a notional surface are obtained. The potentiometric-surface map is a map of total head. It can be contoured, and the contours are *equipotential* lines: fluid flow is in a direction normal to these contours, in the direction of decreasing head. *Piezometric surface* is a synonym not to be preferred because it implies a pressure-measuring surface, which it is not.

Progradation, prograde — The migration of facies seaward, as in a delta. Regression, progradation and offlap are associated.

Progressive fault — Synonym for *Growth fault*.

Recompletion — See *Completion*.

Recurrent fault — Synonym for *Growth fault*.

Regression — Strictly, a fall of sea level relative to the land, with a consequent extension of the land area at the expense of the sea. Sedimentary sequences that contain shallow-water sediments on deep-water sediments, or continental on marine, are called regressive. Most regressive sequences, however, are more readily understood in terms of an excess of sediment supply over that that can be distributed more widely by the energy available, so that this surplus accumulates and extends the land area, as in a delta, by *progradation*.

Round trip — In rotary drilling, pulling the bit and running in a new one.

Sand/shale ratio — A parameter commonly used in facies maps. It is the total thickness of sand, sandstone, and conglomerate divided by the total thickness of clay, shale and mudstone.

Saturation — The saturation of a rock with respect to a fluid is the proportion of the pore space filled with that fluid.

Secondary recovery — A petroleum reservoir is produced either by using the energy in the reservoir itself (a flowing well), or by pumping. Not all the petroleum in a reservoir can be produced in this way (primary recovery). Secondary recovery refers to methods of producing petroleum that would not be produced by primary recovery. These include water or gas injection to maintain the energy of the reservoir, steam to reduce the viscosity of shallow heavy oil, and solvents to remove the residual oil.

Sidetrack — If part of a borehole is lost while drilling (for example, on account of a *fish* that could not be recovered) the hole is plugged back with cement and side-tracked to drill past the hole lost, alongside it.

Slug — (a) A concentration of one fluid in another in a pipe or borehole, such as water slugs in an oil pipeline; and (b) unit of mass in American usage, as distinct from the pound, which is the unit of weight or force. A slug is the unit of mass that acquires an acceleration of 1 ft s^{-2} when acted on by a force of 1 pound. Thus, from Newton's second law of motion, unit acceleration is acquired by 32.174 lb mass, which equals one slug. For conversions, 1 slug = 14.594 kg (mass).

Sonde — The down-hole logging device that contains the moving electrodes for the electrical log, the source and detectors for radioactivity logs, and so on, each sonde being for a specific purpose. The sonde is attached to a cable within which pass the electrical cables that transmit signals to the surface equipment.

Specific surface — The surface area of solids divided by the bulk volume of the porous material. Dimensions L^{-1}.

Specific volume — The volume of unit mass of substance; the reciprocal of mass density. Dimensions $M^{-1}L^{-3}$.

Specific weight — See *Weight density*.

Stand — Rotary drilling. When pulling out of the hole, the drill pipe is stacked or stood in the derrick or mast in stands of three *joints*. A *short stand* consists of two joints: a *long stand*, of four.

Strain — The deformation of a material under stress.

Stress — A force per unit of area. It may be compressive, tending to shorten the dimension of the material under stress in the direction of the stress; or tensional.

A *principal stress* is a stress acting perpendicularly to a surface along which there is no shear stress. The principal stresses in a rock in the subsurface are considered to be as in Fig. G-3, σ_z being vertical by convention, σ_x and σ_y horizontal.

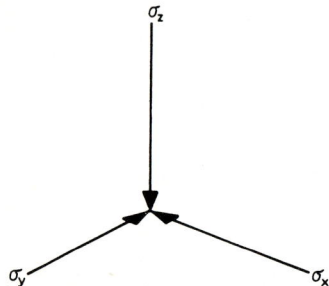

Fig. G-3. Principal stresses.

When the three principal stresses are equal, they are called hydrostatic (and care must obviously be used here). When they are unequal, they are labelled maximum, intermediate, and minimum principal stresses; and are designated σ_1, σ_2 and σ_3, respectively. Stresses may be resolved into components, such as that normal to a fault plane (normal stress, σ_n) and a shear stress, τ, along the fault plane.

See Hafner, 1951, *Bull. Geol. Soc. Am.*, 62: 373—398.

Subcrop — The area of a rock unit that is in contact with an overlying unconformity surface is the subcrop of that unit. A subcrop map may be regarded as a palaeogeological map at the time of the unconformity (strictly, the commencement of post-unconformity sediment accumulation).

Surface tension — The air/water interface in a capillary tube or porous rock acts as if it were an elastic membrane in a state of tension, supporting a pressure differential across it. This is surface tension. But it is defined in terms of the work required to separate unit area of the interface. Cf. *interfacial tension*. Dimensions MT^{-2}.

Synsedimentary — Occurring contemporaneously with sedimentation or sediment accumulation, e.g., synsedimentary fault, a synonym for *growth fault*.

T.D. (Total Depth) — The total depth of a borehole measured along the borehole from the derrick floor or rotary table. May be qualified by the addition of s.s. (sub-sea level) or s.d. (sub-datum). This figure may be very much larger than the *true vertical depth* (T.V.D.) in a deviated borehole. T.V.D. need not be qualified because it is normally below sea level or other datum.

Transgression — Strictly, a rise of sea level relative to the land, with a consequent reduction of land area. Sequences that contain deep-water sedimentary rocks on shallow-water sedimentary rocks, or marine on continental, are called transgressive. Transgression may also result from erosion of the coast without changes in relative sea level. Cf. *Regression*.

Trillion — 10^{12} in U.S.A.: 10^{18} in many other countries. Like *billion*, the word should only be used when there is no danger of misunderstanding. Natural gas volumes in trillion cubic feet (TCF) means 10^{12} standard cubic feet.

Trip gas — When resuming circulation (breaking circulation, in the jargon) after a round trip, the mud that was near the bottom of the hole is sometimes found to be *gas-cut* when it reaches the shale shaker. This is trip gas. It is distinguished from true gas-cutting of the mud by its short duration and its relationship to the round trip.

Tubing — Small diameter pipe run inside casing of a well for the production of petroleum.

T.V.D. (True Vertical Depth) — See *T.D.*

Twist-off — Rotary drilling. Failure of the drilling string is due to two main causes: excessive torque or excessive tension. The former is called a *twist-off*. When the latter happens, the drill string is said to *part*. Parting of the drill pipe and twist-offs (beware of pedantry!) entail a *fishing* job.

Viscosity — The internal resistance of fluid to flow; the larger the resistance, the larger the viscosity. There are two measures of viscosity.

Coefficient of viscosity, absolute, or *dynamic viscosity* is the ratio of shear stress to the rate of shear *strain:* $\eta = \tau/(dV/dz)$. It is the tangential force per unit area that maintains unit relative velocity between two parallel planes unit distance apart. Dimensions $ML^{-1}T^{-1}$.

Kinematic viscosity is the ratio of dynamic viscosity to mass density: $\nu = \eta/\rho$. It concerns motion without reference to force. Dimensions L^2T^{-1}.

Viscosity includes the dimension of time, so rocks in the context of geological time also have the property of viscosity. This is called *equivalent* or *effective* viscosity.

Void ratio — The ratio of pore volume to the volume of solids in a sedimentary rock. The symbol is usually ϵ or e. Thus, $\epsilon = f/(1-f)$. See *Porosity*. Engineers prefer void ratio to porosity; it has the merit that it is a ratio of a variable to a constant (in most practical geological contexts) and is useful in compaction studies.

Wall sticking — Rotary drilling. If the fluid potential of the mud in a borehole is much greater than that of the fluids in the permeable formations that have been penetrated, the mud cake is a site of large fluid potential gradient. If now the drill pipe impinges upon the mudcake, the pressure differential may be sufficient to hold it to the wall of the hole with considerable force (Fig. G-4). It is corrected by washing the well to water or lightening the mud, thus reducing the fluid potential gradient. There is a risk of wall sticking when drilling through depleted petroleum reservoirs, and when drilling into abnormally high pore pressures. Synonym: differential sticking.

Fig. G-4. Wall sticking (plan).

Water drive, gas-cap drive — When the volume of oil produced from a reservoir is replaced wholly or partly by expansion of the water in the reservoir rock unit, that reservoir is said to have water drive, and the oil/water contact tends to be displaced upwards. When the water drive is strong, loss of pressure in the reservoir is relatively small; when weak, it may be relatively large, and lead to a secondary gas cap.

When the energy of the reservoir comes largely from the expansion of a gas cap (primary or secondary), the reservoir is said to have gas-cap drive, and the gas/oil contact tends to be displaced downwards.

Weight density — The weight of a substance divided by its volume, i.e., ρg. This term is to be preferred over *specific weight*, but they are synonymous. Dimensions $ML^{-2}T^{-2}$. Symbol commonly γ.

Well — Strictly, a borehole that produces water, oil, or gas. In petroleum terminology, it is commonly used loosely for any borehole, whether or not it produces petroleum, e.g., outstep well.

Wet — When a borehole penetrates a formation that is a petroleum reservoir in the general area, the reservoir is said to be wet in that well if it contains water only. A dry hole may contain sands, for example, that are wet.

Wet, wetted, wetting — When two immiscible fluids saturate a porous material, one preferentially adheres to the solid surface, excluding (or tending to exclude) the other. The adhering fluid is the *wetting* fluid or wetting phase, the other is the non-wetting fluid or phase. Sands are usually water-wet.

Workover — A well is worked-over from time to time, to renew or clean tubing, clean casing, reperforate — or any operation that is to maintain production from the same reservoir. Cf. recompletion (see *Completion*), which involves a change of production interval.

AUTHOR INDEX

Achauer, C.W., 256, 276
Adiwidjaja, P. and De Coster, G.L., 301, 358, 371
Ager, D.V., 21
Alberding, H., see Renz, H.H. et al.
Allen, D.R., 177
Andersen, C., see Day, G.A. et al.
Andrichuk, J.M., 267, 268, 276
Archie, G.E., 114, 136
Armstrong, J.D., see Schwartz, C.M. et al.
Asmus, H.E. and Ponte, F.C., 371
Athy, L.F., 45, 47, 63
Atwater, G.I. and Forman, M.J., 346
Aubouin, J., 21

Bacon, M., see McQuillin, R. et al.
Bailey, N.J.L., Evans, C.R. and Milner, C.W.D., 226
Bailey, N.J.L., Jobson, A.M. and Rogers, M.A., 226
Bailey, N.J.L., Krouse, H.R., Evans, C.R. and Rogers, M.A., 222, 224, 226
Bailey, N.J.L., see Evans, C.R. et al.
Baker, E.G., 216, 226
Barber, P.M., 362, 363, 364, 371
Barclay, W., see McQuillin, R. et al.
Barker, C., 309, 321
Barker, C. and Horsfield, B., 310, 321
Barrell, J., 4, 5, 21, 89, 93, 386
Barss, D.L., Copland, A.B. and Ritchie, W.D., 16, 17, 21, 257, 263, 264, 265, 266, 276
Barsukov, O.A., Blinova, N.M., Vybornykh, S.F., Gulin, Yu.A., Dakhnov, V.N., Larionov, V.V. and Kholin, A.I., 136
Barton, D.C., 71, 93, 330, 346
Baugher, J.W., see Harkins, K.L. and Baugher, J.W.
Bebout, D.G., see Coogan, A.H. et al.
Beck, K.C., see Weaver, C.E. and Beck, K.C.
Beck, R.H., 326, 347
Beck, R.H. and Lehner, P., 326, 347

Beddoes, L.R., see Kenyon, C.S. and Beddoes, L.R.
Berry, F.A.F., 319, 321
Billings, M.P., 24, 38
Biot, M.A. and Odé, H., 332, 347
Bishop, R.S., 338, 347
Blinova, N.M., see Barsukov, O.A. et al.
Boatman, W.A., 63
Boeuf, M.G. and Doust, H., 371
Bonham, L.C., 215, 226
Bornhauser, M., 347
Bornhold, B.D., see Mascle, J.R. et al.
Botset, H.G., see Wyckoff, R.D. et al.
Bott, M.H.P., see Kent, P.E. et al.
Böttcher, H., 26, 38
Bottema, J.A., see Thomeer, J.H.M.A. and Bottema, J.A.
Bowen, J.M., 294, 301
Boyd, W.A., see Fowler, W.A. et al.
Brady, T.J., Campbell, N.D.J. and Maher, C.E., 273, 274, 275, 276
Braunstein, J. and O'Brien, G.D., 326, 347
Bredehoeft, J.D. and Hanshaw, B.B., 63, 183, 206, 218. 226
Bredehoeft, J.D., see Hanshaw, B.B. and Bredehoeft, J.D.
Brennand, T.P., 368, 371
British Petroleum Company Ltd., 93, 105
Broad, D.S., see Horstman, E.L. et al.
Brockett, L.D., see Jamison, H.C. et al.
Brooks, B.T., 212, 226
Brooks, J.D., 214, 226
Brooks, J.D. and Smith, J.W., 214, 227
Brown, B.R., 365, 371
Brown, B.R., see Threlfall, W.F. et al.
Brown, R.S., see Laws, R.A. and Brown, R.S.
Bruce, C.H., 346, 347
Bruce, W.A. and Welge, H.J., 156, 157, 178
Bubb, J.W., see Vail, P.R. et al.
Burgers, W.F.J., see Day, G.A. et al.
Burk, C.A. and Drake, C.L., 371

Burns, B.J., see Thornton, R.C.N. et al.
Burst, J.F., 63, 308, 318, 321
Byrd, W.D., 238, 239, 251, 380, 381

Campbell, I.R. and Smith, D.N., 200, 206
Campbell, N.D.J., see Brady, T.J. et al.
Carey, S.W., 330, 331, 347
Castillo-Tejero, C., see Viniegra-O., F. and Castillo-Tejero, C.
Chamberlin, T.C., 259, 276
Chapman, R.E., 33, 39, 43, 58, 63, 131, 136, 158, 178, 184, 192, 206, 217, 218, 227, 309, 310, 314, 320, 321, 340, 347, 356, 371, 392
Cheney, R.E. and Marsh, J.G., 21
Chevron Standard Limited, 269, 276
Childs, F.B. and Reed, P.E.C., 368, 371
Chilingarian, G.V., see Rieke, H.H. and Chilingarian, G.V.
Chin, G.E., see Langton, J.R. and Chin, G.E.
Cleland, R.L., see Debye, P. and Cleland, R.L.
Clifford, H.J., Grund, R. and Musrati, H., 301
Cloos, E., 39
Cobb, W.M., see Dowdle, W.L. and Cobb, W.M.
Colburn, W.A., see Hill, G.A. et al.
Cole, D.I., 136
Coleman, J.M., see Morgan, J.P. et al.
Combaz, A. and de Matharel, M., 356, 372
Combaz, A., see Tissot, B. et al.
Connan, J., 213, 227
Connan, J., Le Tran, K. and Van der Weide, B., 224, 227
Conner, D.C., see Williams, J.J. et al.
Coogan, A.H., Bebout, D.G. and Maggio, C., 271, 276
Cook, F.D., see Jobson, A. et al.
Cooper, B.A., see Day, G.A. et al.
Cope, R.N., 27, 39, 365, 372
Copland, A.B., see Barss, D.L. et al.
Cordell, R.J., 216, 227
Cordry, E.A., see Fränkl, E.J. and Cordry, E.A.
Cox, W.C., see Dickey, P.A. and Cox, W.C.
Currie, J.B., 25, 39

Dailly, G.C., 347

Dake, L.P., 173, 178
Dakhnov, V.N., see Barsukov, O.A. et al.
Dallmus, K.F., see Renz, H.H. et al.
Dalton, L.V., 93
Darcy, H., 163, 178
Daukoru, E., see Weber, K.J. and Daukoru, E.
Davis, J.B., 222, 227
Day, G.A., Cooper, B.A., Andersen, C., Burgers, W.F.J., Rønnevik, H.C. and Schöneich, H., 367, 372
Debye, P. and Cleland, R.L., 158, 178
De Coster, G.L., 301, 358, 372
De Coster, G.L., see Adiwidjaja, P. and De Coster, G.L.
De Matharel, M., Klein, G. and Oki, T., 356, 372
De Matharel, M., Lehmann, P. and Oki, T., 356, 372
De Matharel, M., see Combaz, A. and de Matharel, M.
De Mille, G., 263, 276
Dennis, J.G., 25, 39
Der, F., see Deroo, G. et al.
Deroo, G., Powell, T.G., Tissot, B. and McCrossan, R.G., 251, 269, 276
Deroo, G., Tissot, B., McCrossan, R.G. and Der, F., 223, 227
De Ruiter, P.A.C., see Lehner, P. and De Ruiter, P.A.C.
De Wiest, R.J.M., 121, 136
Dickey, P.A., 217, 227
Dickey, P.A. and Cox, W.C., 321
Dickey, P.A., Shriram, C.R. and Paine, W.R., 316, 321
Dickey, P.A., see Koinm, D.N. and Dickey, P.A.
Dickinson, G., 45, 56, 62, 63, 306, 316, 317, 321, 334, 347
Dobrin, M.B., 93
Dobryansky, A.F., 212, 227
Dohr, G., 93
Dott, R.H. and Reynolds, M.J., 93
Dott, R.H., see Halbouty, M.T. et al.
Douglas, J.G. and Ferguson, J.A., 372
Doust, H., see Boeuf, M.G. and Doust, H.
Dow, W.G., 215, 227
Dowdle, W.L. and Cobb, W.M., 134, 136
Drake, C.L., see Burk, C.A. and Drake, C.L.
Dron, R.W., 26, 39
Drong, H.J., see Philipp, W. et al.

Dufour, J., 227
Durand, B., see Tissot, B. et al.
Dutta, T.K., see Saikia, M.M. and Dutta, T.K.

Eicher, D.L., 21
Ekweozor, C.M. and Okoye, N.V., 243, 244, 251
Ellenor, D.W., 365, 372
Engler, C., 93
Espitalié, J., see Tissot, B. et al.
Esser, R.W., see Moody, J.D. and Esser, R.W.
Evamy, B.D., Harembourne, J., Kamerling, P., Knaap, W.A., Molloy, F.A. and Rowlands, P.H., 243, 244, 252
Evans, C.R., McIvor, D.K. and Magara, K., 347
Evans, C.R., Rogers, M.A. and Bailey, N.J.L., 234, 252
Evans, C.R., see Bailey, N.J.L. et al.
see also Milner, C.W.D. et al.
Evans, P.R., see James, E.A. and Evans, P.R.

Faber, J.M., see Hriskevich, M.E. et al.
Falvey, D.A., 372
Ferguson, J.A., see Douglas, J.G. and Ferguson, J.A.
Fertl, W.H., 63, 305, 307, 310, 311, 320, 321
Fitzgerald, T.A., 143, 146, 149, 153
Folinsbee, R.E., 143, 147, 153
Fonseca, J. dos Reis, see Ponte, F.C. et al.
Forman, M.J., see Atwater, G.I. and Forman, M.J.
Fowler, W.A., 308, 317, 321
Fowler, W.A., Boyd, W.A., Marshall, S.W. and Myers, R.L., 248, 252, 316, 321
Fränkl, E.J. and Cordry, E.A., 243, 244, 252
Fraser, W.W., 301
Frederick, W.S., 308, 321
Füchtbauer, H., 43, 64
Füchtbauer, H., see Philipp, W. et al.
Funkhouser, H.J., Sass, L.C. and Hedberg, H.D., 308, 321
Funkhouser, H.J., see Hedberg, H.D. et al.

Gagliano, S.M., see Morgan, J.P. et al.
Gaida, K.H., see Von Engelhardt, W. and Gaida, K.H.
Gansser, A., 347
Gardner, G.H.F., Gardner, L.W. and Gregory, A.R., 136
Gardner, G.H.F., see Wyllie, M.R.J. et al.
Gardner, L.W., see Gardner, G.H.F. et al.
see also Wyllie, M.R.J. et al.
Geertsma, J., 178
Gibbs, R.J., 219, 227
Gibson, H.S., 178
Gill, D., 203, 204, 205, 206, 232, 252, 259, 260, 261, 262, 276
Gillespie, J. and Sanford, R.M., 301
Gilreath, J.A., 338, 347
Glaessner, M.F. and Teichert, C., 21
Grames, L.R. and Reyner, R.R., 104, 105
Gregory, A.R., see Gardner, G.H.F. et al.
see also Wyllie, M.R.J. et al.
Gretener, P.E., 21, 64
Griffith, B.R. and Hodgson, E.A., 297, 301
Griffith, B.R., see Threlfall, W.F. et al.
Grund, R., see Clifford, H.J. et al.
Gulin, Yu.A., see Barsukov, O.A. et al.
Gussow, W.C., 202, 206, 232, 252, 269, 276
Gutenberg, B., 7, 11, 21
Gutjahr, C.C.M., 215, 227
Gutjahr, C.C.M., see Hood, A. et al.
Gwinn, J.W., Helmig, H.M. and Kartaadipoetra, L.W., 356, 372

Hackbarth, D.A. and Nastasa, N., 149, 153
Haddenhorst, H.-G., see Philipp, W. et al.
Haeberle, F.R., 72, 93
Hafner, 394
Haile, N.S., see Liechti, P. et al.
Halbouty, M.T., Meyerhoff, A.A., King, R.E., Dott, R.H., Klemme, H.D. and Shabad, T., 147, 148, 153, 198, 206, 272, 276, 281, 301
Hallam, A., 372
Hamblin, W.K., 39
Hancock, J.M. and Kauffman, E.G., 301
Hanshaw, B.B. and Bredehoeft, J.D., 64
Hanshaw, B.B. and Zen, E-an, 64
Hanshaw, B.B., see Bredehoeft, J.D. and Hanshaw, B.B.
Hardin, F.R. and Hardin, G.C., 25, 39

Hardin, G.C., see Hardin, F.R. and Hardin, G.C.
Haremboure, J., see Evamy, B.D. et al.
Harkins, K.L. and Baugher, J.W., 56, 64, 308, 316, 322
Hatlelid, W.G., see Vail, P.R. et al.
Haun, J.D., 153
Heacock, R.L., see Hood, A. et al.
Heatzig, G. and Michel, R., 272, 276
Hedberg, H.D., 44, 45, 46, 47, 52, 64, 93, 180, 184, 206, 220, 224, 227, 252, 320, 322, 380, 382
Hedberg, H.D., Sass, L.C. and Funkhouser, H.J., 27, 39, 233, 234, 235, 252
Hedberg, H.D., see Funkhouser, H.J. et al.
Helmig, H.M., see Gwinn, J.W. et al.
see also Huffington, R.M. and Helmig, H.M.
Hemphill, C.R., Smith, R.I. and Szabo, F., 16, 21, 257, 265, 266, 267, 276
Héritier, F.E., Lossel, P. and Wathne, E., 240, 241, 242, 252, 322
Hermes, J.J., see Visser, W.A. and Hermes, J.J.
Herring, E.A., 380, 382
Hewins, R., 270, 276
Hide, R. and Roberts, P.H., 15, 21
Hill, G.A., Colburn, W.A. and Knight, J.W., 319, 322
Hilt, C., 227
Hobson, G.D., 185, 206, 217, 227
Hobson, G.D. and Tiratsoo, E.N., 93
Hodgson, E.A., see Griffith, B.R. and Hodgson, E.A.
Höfer, H., 93
Holmes, A., 6, 21
Holser, W.T., see Kennedy, G.C. and Holser, W.T.
Hood, A., Gutjahr, C.C.M. and Heacock, R.L., 227
Horner, D.R., 174, 178
Horsfield, B., see Barker, C. and Horsfield, B.
Horstman, E.L., Lyons, D.A., Nott, S.A. and Broad, D.S., 370, 372
Hosemann, P., 27, 39, 370, 372
Hottmann, C.E. and Johnson, R.K., 50, 51, 64, 307, 322
Howarth, R.J., White, C.M. and Koch, G.S., 153

Hower, J., see Perry, E. and Hower, J.
Hriskevich, M.E., 264, 276
Hriskevich, M.E., Faber, J.M. and Langton, J.R., 257, 276
Hubbert, M.K., 121, 136, 164, 178, 180, 183, 184, 190, 191, 195, 206, 218, 227, 247, 252, 331, 347, 384, 387, 392
Hubbert, M.K. and Rubey, W.W., 57, 64
Hubbert, M.K., see Rubey, W.W. and Hubbert, M.K.
Huffington, R.M. and Helmig, H.M., 356, 372
Hull, C.E. and Warman, H.R., 201, 206, 248, 250, 252
Hunt, J.M., 72, 79, 93, 212, 213, 216, 220, 221, 223, 224, 227, 228, 237, 245, 252
Hunt, J.M., see Kidwell, A.L. and Hunt, J.M.

Illing, V.C., 80, 93, 206, 320, 322
Ingels, J.J.C., 17, 21, 260, 276

Jaboli, D., see Rocco, T. and Jaboli, D.
Jackson, A., see Schaub, H.P. and Jackson, A.
James, E.A. and Evans, P.R., 27, 39, 297, 301
Jamison, H.C., Brockett, L.D. and McIntosh, R.A., 286, 287, 288, 289, 301
Jankowsky, W., see Philipp, W. et al.
Janoschek, R., 27, 39
Jenkins, D.A.L. and Twombley, B.N., 301
Jobson, A., Cook, F.D. and Westlake, D.W.S., 224, 228
Jobson, A.M., see Bailey, N.J.L. et al.
Jodry, R.L., 260, 277
Johnson, R.K., see Hottmann, C.E. and Johnson, R.K.
Jones, D.K., 365, 372
Jones, D.K. and Pearson, G.R., 365, 372
Jones, H.P. and Speers, R.G., 287, 289, 290, 301
Jones, O.T., 21
Jones, R.W., 215, 216, 228
Joyes, R., see Vidal, J. et al.

Kamerling, P., 372
Kamerling, P., see Evamy, B.D. et al.
Kartaadipoetra, L.W., see Gwinn, J.W. et al.

Katz, H.R., 315, 322
Kauffman, E.G., see Hancock, J.M. and Kauffman, E.G.
Kawai, K. and Totani, S., 207
Kay, M., 20, 21
Keep, C.E. and Ward, H.L., 64
Kennedy, G.C. and Holser, W.T., 314, 322
Kent, P.E., 301, 347, 361, 366, 369, 372, 373
Kent, P.E., Bott, M.H.P., McKenzie, D.P. and Williams, C.A., 373
Kenyon, C.S. and Beddoes, L.R., 136, 233, 252
Kholin, A.I., see Barsukov, O.A. et al.
Khurana, A.K., see Thornton, R.C.N. et al.
Kidwell, A.L. and Hunt, J.M., 6, 21, 64, 79, 94, 315, 322
King, F.H., 180, 207
King, R.E., see Halbouty, M.T. et al.
Kirk, R.H., 290, 291, 292, 293, 301
Klein, G., see De Matharel, M. et al.
Klemme, H.D., see Halbouty, M.T. et al.
Klinkenberg, L.J., 178
Klosko, S.M., see Lerch, F.J. et al.
Klovan, J.E., 268, 277
Knaap, W.A., see Evamy, B.D. et al.
Knight, J.W., see Hill, G.A. et al.
Koch, G.S., see Howarth, R.J. et al.
Koesoemadinata, R.P., 358, 373
Koinm, D.N. and Dickey, P.A., 39
Kok, P.C. and Thomeer, J.H.M.A., 64
Kozeny, J., 43, 64
Krouse, H.R., see Bailey, N.J.L. et al.
Krumbein, W.C. and Sloss, L.L., 21
Kugler, H.G., 176, 178
Kündig, E., 20, 21
Kyaw Nyein, R., MacLean, L. and Warris, B.J., 365, 373

Lambert-Aikhionbare, D.O., 200, 207, 251, 252
Langton, J.R. and Chin, G.E., 264, 277
Langton, J.R., see Hriskevich, M.E. et al.
Laplante, R.E., 228
Larionov, V.V., see Barsukov, O.A. et al.
Laughbaum, G.H., see Schwartz, C.M. et al.
Laws, R.A. and Brown, R.S., 373
Lehmann, P., see De Matharel, M. et al.
Lehner, P., 347

Lehner, P. and De Ruiter, P.A.C., 373
Lehner, P., see Beck, R.H. and Lehner, P.
Lerch, F.J., Putney, B.H., Wagner, C.A. and Klosko, S.M., 22
Le Tran, K., see Connan, J. et al.
Levorsen, A.I., 94, 384, 392
Lewis, C.R. and Rose, S.C., 311, 322
Liechti, P., Roe, F.W. and Haile, N.S., 25, 27, 39, 340, 342, 343, 347
Lopatin, N.V., 213, 228
Lossel, P., see Héritier, F.E. et al.
Louis, M., 212, 228
Lowenstam, H.A., 16, 22, 256, 258, 259, 260, 277
Lyell, C., 11, 22, 329, 347
Lyons, D.A., see Horstman, E.L. et al.

MacLean, L., see Kyaw Nyein, R. et al.
Magara, K., 49, 51, 52, 61, 64, 218, 228, 249, 252, 311, 314, 318, 319, 322
Magara, K., see Evans, C.R. et al.
Maggio, C., see Coogan, A.H. et al.
Magnier, P. and Samsu, B., 356, 373
Maher, C.E., 294, 295, 302
Maher, C.E., see Brady, T.J. et al.
Marinoadi, see Mulhadiono and Marinoadi
Marsh, J.G., see Cheney, R.E. and Marsh, J.G.
Marshall, D.L., see Thornton, O.F. and Marshall, D.L.
Marshall, S.W., see Fowler, W.A. et al.
Mascle, J.R., Bornhold, B.D. and Renard, V., 347
Masson, P.H., see Winsauer, W.O. et al.
Masvall, J., see Renz, H.H. et al.
Matsumoto, T., 15, 22, 280, 302
Maximov, S.P., see Rodionova, K.F. and Maximov, S.P.
Maxwell, J.C., 43, 64
Mayuga, M.N., 175, 176, 178
McCrossan, R.G., 44, 64, 201, 207
McCrossan, R.G., see Deroo, G. et al.
McCulloh, T.D., 27, 39
McDermott, E., 212, 228
McDowell, A.N., see Parker, T.J. and McDowell, A.N.
McEvoy, D.I., 88
McIntosh, R.A., see Jamison, H.C. et al.
McIvor, D.K., see Evans, C.R. et al.
McKelvey, J.G. and Milne, I.H., 319, 322
McKenzie, D.P., see Kent, P.E. et al.

McQuillin, R., Bacon, M. and Barclay, W., 94
Meinschein, W.G., 216, 228
Meissner, F.F., 185, 207, 219, 228, 320, 322
Meyerhoff, A.A., 153
Meyerhoff, A.A., *see* Halbouty, M.T. et al.
Michel, R., *see* Heatzig, G. and Michel, R.
Milius, G., *see* Stäuble, A.J. and Milius, G.
Miller, E.G., 302
Milne, I.H., *see* McKelvey, J.G. and Milne, I.H.
Milner, C.W.D., Rogers, M.A. and Evans, C.R., 228
Milner, C.W.D., *see* Bailey, N.J.L. et al.
Minturn, L.W., XVII
Mitchum, R.M., *see* Vail, P.R. et al.
Moldowan, J.M., *see* Seifert, W.K. and Moldowan, J.M.
Molloy, F.A., *see* Evamy, B.D. et al.
Moody, J.D. and Esser, R.W., 148, 149, 153
Moody, J.D., Mooney, J.W. and Spivak, J., 142, 145, 146, 154
Mooney, J.W., *see* Moody, J.D. et al.
Moore, C.A., 136
Morales, R.G., *see* Ponte, F.C. et al.
Morgan, J.P., Coleman, J.M. and Gagliano, S.M., 329, 346, 347
Morgridge, D.L. and Smith, W.B., 287, 289, 302
Mörner, N.-A., 12, 15, 22
Morrow, N.R., 116, 136, 137, 160, 178
Mulhadiono and Marinoadi, 308, 322
Munn, M.J., 94, 207
Murray, G.E., 347
Murray, J.W., 265, 266, 277
Muskat, M., *see* Wyckoff, R.D. et al.
Musrati, H., *see* Clifford, H.J. et al.
Myers, R.L., *see* Fowler, W.A. et al.

Narain, K., *see* Smoot, T.W. and Narain, K.
Nastasa, N., *see* Hackbarth, D.A. and Nastasa, N.
Nordberg, M.E., 158, 178
Nott, S.A., *see* Horstman, E.L. et al.

O'Brien, G.D., *see* Braunstein, J. and O'Brien, G.D.
Ocamb, R.D., 25, 39

Odé, H., *see* Biot, M.A. and Odé, H.
Oki, T., *see* De Matharel, M. et al. *see also* Verdier, A.C. et al.
Okoye, N.V., *see* Ekweozor, C.M. and Okoye, N.V.
Olteanu, Gh., *see* Paraschiv, D. and Olteanu, Gh.
Owen, E.W., 270, 277
Owen, J.E., 137

Paine, W.R., *see* Dickey, P.A. et al.
P'an Chung-Hsiang, XVII
Paraschiv, D. and Olteanu, Gh., 348
Parker, T.J. and McDowell, A.N., 328, 348
Patterson, J.M., *see* Renz, H.H. et al.
Pattinson, R., Watkins, G. and Van den Abeele, D., 365, 373
Payton, C.E., 22, 89
Pearson, G.R., *see* Jones, D.K. and Pearson, G.R.
Perrodon, A., 94
Perry, E. and Hower, J., 64, 308, 318, 322
Peterson, K.E., *see* Williams, J.J. et al.
Philipp, W., Drong, H.J., Füchtbauer, H., Haddenhorst, G.-G. and Jankowsky, W., 200, 207, 252
Philippi, G.T., 79, 94, 228
Pirson, S.J., 178
Pitman, W.C., 373
Playford, P.E., 256, 257, 277
Ponte, F.C., Fonseca, J. dos Reis and Morales, R.G., 373
Ponte, F.C., *see* Asmus, H.E. and Ponte, F.C.
Porrenga, D.H., 219, 228
Postma, H., *see* Van Andel, Tj. and Postma, H.
Powell, D.E., 361, 373
Powell, T.G., *see* Deroo, G. et al.
Powers, M.C., 64, 308, 318, 322
Price, L.C., 215, 216, 228, 245, 252, 253
Putney, B.H., *see* Lerch, F.J. et al.

Quarles, M., 39

Ramberg, H., 332, 348
Reed, D.W., *see* Wyckoff, R.D. et al.
Reed, K.J., 220, 228, 244, 253
Reed, P.E.C., *see* Childs, F.B. and Reed, P.E.C.

Renard, V., see Mascle, J.R. et al.
Renz, H.H., Alberding, H., Dallmus, K.F., Patterson, J.M., Robie, R.H., Weisbord, N.E. and MasVall, J., 27, 39, 234, 253
Reyner, R.R., see Grames, L.R. and Reyner, R.R.
Reynolds, M.J., see Dott, R.H. and Reynolds, M.J.
Rich, J.L., 189, 207
Ridd, M.F., 373
Rieke, H.H. and Chilingarian, G.V., 64
Rigg, A.J., see Thornton, R.C.N. et al.
Ritchie, W.D., see Barss, D.L. et al.
Rittenhouse, G., 64
Roberts, J.M., 280, 283, 302
Roberts, P.H., see Hide, R. and Roberts, P.H.
Robie, R.H., see Renz, H.H. et al.
Robson, D.A., 373
Rocco, T. and Jaboli, D., 27, 39
Rodionova, K.F. and Maximov, S.P., 228
Roe, F.W., see Liechti, P. et al.
Rogers, M.A., see Bailey, N.J.L. et al.
see also Evans, C.R. et al., Milner, D.W.D. et al.
Rønnevik, H.C., see Day, G.A. et al.
Rose, S.C., see Lewis, C.R. and Rose, S.C.
Rose, W., 159, 178
Rowlands, P.H., see Evamy, B.D. et al.
Rubey, W.W. and Hubbert, M.K., 64
Rubey, W.W., see Hubbert, M.K. and Rubey, W.W.

Sadler, P.M., 22
Saikia, M.M. and Dutta, T.K., 228, 237, 253
Samsu, B., see Magnier, P. and Samsu, B.
see also Schwartz, C.M. et al.
Sanford, R.M., 302
Sanford, R.M., see Gillespie, J. and Sanford, R.M.
Sangree, J.B., see Vail, P.R. et al.
Sanneman, D., 336, 337, 338, 339, 340, 348
Sarkisyan, S.G., 253
Sarmiento, R., 137
Sass, L.C., see Funkhouser, H.J. et al.
see also Hedberg, H.D. et al.
Schaub, H.P. and Jackson, A., 340, 342, 343, 344, 345, 348

Schmidt, G.W., 319, 322
Schöneich, H., see Day, G.A. et al.
Schwartz, C.M., Laughbaum, G.H., Samsu, B.S. and Armstrong, J.D., 356, 373
Secor, D.T., 247, 253
Seifert, W.K. and Moldowan, J.M., 239, 253
Selley, R.C., 43, 65
Sengupta, S., 27, 39
Shabad, T., see Halbouty, M.T. et al.
Sharma, G.D., 262, 277
Shearin, H.M., see Winsauer, W.O. et al.
Shelton, J.W., 27, 39
Shepard, F.P., 348
Short, K.C. and Stäuble, A.J., 27, 39, 241, 243, 253
Shriram, C.R., see Dickey, P.A. et al.
Shrock, R.R., 258, 277
Skempton, A.W., 65
Sloss, L.L., see Krumbein, W.C. and Sloss, L.L.
Smith, D.A., 246, 247, 253
Smith, D.N., see Campbell, I.R. and Smith, D.N.
Smith, J.E., 65, 183, 207
Smith, J.W., see Brooks, J.D. and Smith, J.W.
Smith, N.E. and Thomas, H.G., 309, 323
Smith, R.I., see Hemphill, C.R. et al.
Smith, W.B., see Morgridge, D.L. and Smith, W.B.
Smoot, T.W. and Narain, K., 228, 253
Speers, R.G., see Jones, H.P. and Speers, R.G.
Spinks, R.B., 315, 323
Spivak, J., see Moody, J.D. et al.
Stanley, T.B., 39
Stäuble, A.J. and Milius, G., 214, 228
Stäuble, A.J., see Short, K.C. and Stäuble, A.J.
Steele, R.J., 94
Stephenson, L.P., 65
Stuart, M., 94, 207
Stutzer, O., 26, 39
Suardy, A., see Verdier, A.C. et al.
Suess, E., 11, 12, 22, 280, 302
Szabo, F., see Hemphill, C.R. et al.

Tallis, N.C., 340, 348
Tanner, W.F. and Williams, G.K., 348
Teichert, C., see Glaessner, M.F. and Teichert, C.

Teichmüller, M., 214, 229
Teichmüller, M. and Teichmüller, R., 229
Teichmüller, R., *see* Teichmüller, M. and Teichmüller, R.
Terry, C.E. and Williams, J.J., 17, 22, 272, 275, 277
Terzaghi, K., 57, 65
Thoesen, C.E., 316, 323
Thomas, B.M., 300, 302
Thomas, H.G., *see* Smith, N.E. and Thomas, H.G.
Thomas, O.D., *see* Van den Bark, E. and Thomas, O.D.
Thomeer, J.H.M.A. and Bottema, J.A., 65, 305, 323
Thomeer, J.H.M.A., *see* Kok, P.C. and Thomeer, J.H.M.A.
Thompson, S., *see* Vail, P.R. et al.
Thomson, W., 80, 94
Thornton, O.F. and Marshall, D.L., 156, 157, 178
Thornton, R.C.N., Burns, B.J., Khurana, A.K. and Rigg, A.J., 298, 299, 302
Thorsen, C.E., 30, 39
Threlfall, W.F., 302
Threlfall, W.F., Brown, B.R. and Griffith, B.R., 297, 302, 365, 373
Tiddeman, R.H., 25, 39, 385
Tiratsoo, E.N., *see* Hobson, G.D. and Tiratsoo, E.N.
Tissot, B.P. and Welte, D.H., 72, 74, 75, 94, 212, 213, 216, 220, 223, 229
Tissot, B., Durand, B., Espitalié, J. and Combaz, A., 211, 229
Tissot, B., *see* Deroo, G. et al.
Todd, R.G., *see* Vail, P.R. and Todd, R.G. *see also* Vail, P.R. et al.
Totani, S., *see* Kawai, K. and Totani, S.
Trend Exploration Technical Staff, 232, 253
Trusheim, F., 336, 337, 340, 348
Tunn, W., *see* Von Engelhardt, W. and Tunn, W.
Twombley, B.N., *see* Jenkins, D.A.L. and Twombley, B.N.

Vail, P.R., Mitchum, R.M., Todd, R.G., Widmier, J.M., Thompson, S., 'Sangree, J.B., Bubb, J.W. and Hatlelid, W.G., 15, 22, 89, 94
Vail, P.R. and Todd, R.G., 94, 302
Van Andel, Tj. and Postma, H., 219, 229

Van Bemmelen, R.W., 340, 348, 355, 373
Van den Abeele, D., *see* Pattinson, R. et al.
Van den Bark, E. and Thomas, O.D., 238, 239, 253, 380, 382
Van der Weide, B., *see* Connan, J. et al.
Van Hinte, J.E., 301
Van Veen, J., *see* Vidal, J. et al.
Verdier, A.C., Oki, T. and Suardy, A., 356, 373
Versluys, J., 159, 164, 178, 180, 207, 309, 323
Vidal, J., Joyes, R. and Van veen, J., 373
Viniegra-O, F., 277
Viniegra-O, F. and Castillo-Tejero, C., 269, 277
Visser, W.A. and Hermes, J.J., 320, 323, 340, 341, 342, 348
Von Engelhardt, W. and Gaida, K.H., 319, 323
Von Engelhardt, W. and Tunn, W., 178
Vybornykh, S.F., *see* Barsukov, O.A. et al.

Wagner, C.A., *see* Lerch, F.J. et al.
Walters, J.E., 40
Waples, D.W., 213, 229
Ward, H.L., *see* Keep, C.E. and Ward, H.L.
Warman, H.R., *see* Hull, C.E. and Warman, H.R.
Warris, B.J., *see* Kyaw Nyein, R. et al.
Wathne, E., *see* Héritier, F.E. et al.
Watkins, G., *see* Pattinson, R. et al.
Weaver, C.E. and Beck, K.C., 65, 309, 318, 323
Webb, J.E., 253
Weber, K.J., 40, 241, 242, 353, 373
Weber, K.J. and Daukoru, E., 243, 253
Weeda, J., 233, 253, 354, 374
Weeks, L.G., 19, 22, 253
Weisbord, N.E., *see* Renz, H.H. et al.
Weiss, A., 213, 229
Welge, H.J., *see* Bruce, W.A. and Welge, H.J.
Welte, D.H., *see* Tissot, B.P. and Welte, D.H.
Wennekers, J.H.L., 357, 358, 374
Westlake, D.W.S., *see* Jobson, A. et al.
White, C.M., *see* Howarth, R.J. et al.
White, D., 214, 229

Whitten, E.H.T., 374
Widmier, J.M., *see* Vail, P.R. et al.
Williams, C.A., *see* Kent, P.E. et al.
Williams, J.A., *see* Winters, J.C. and Williams, J.A.
Williams, G.K., *see* Tanner, W.F. and Williams, G.K.
Williams, J.J., 277, 280, 281, 282, 283, 302
Williams, J.J., Conner, D.C. and Peterson, K.E., 294, 302, 369, 374
Williams, J.J., *see* Terry, C.E. and Williams, J.J.
Williams, M., *see* Winsauer, W.O. et al.
Wilson, H.H., 253
Winsauer, W.O., Shearin, H.M., Masson, P.H. and Williams, M., 114, 115, 137
Winters, J.C. and Williams, J.A., 225, 229
Wise, R.A., 374
Woodland, A.W., 27, 40

Worthington, P.F., 137
Wyckoff, R.D., Botset, H.G., Muskat, M. and Reed, D.W., 167, 178
Wyllie, M.R.J., 137
Wyllie, M.R.J., Gregory, A.R. and Gardner, G.H.F., 131, 137
Wyllie, M.R.J., Gregory, A.R. and Gardner, L.W., 131, 137
Wyman, R.E., 137

Young, D., 270, 277
Youngquist, W., 236, 237, 253
Yurkova, R.M., 253

Zen, E-an, *see* Hanshaw, B.B. and Zen, E-an
Zhang Yi-gang, 229
Ziegler, P.A., 366, 374
Zipf, G.K., 143, 154
ZoBell, C.E., 211, 229

SUBJECT INDEX

Abnormal pore-fluid pressures, 57—63, 303—323, 329, 353, 365, 376—381
—, ages of, 62, 307
—, borehole log responses, 307—308, 310—311, 335
—, causes of, 303, 306, 317
—, —, diagenesis, 308—309, 318
—, —, osmosis, 319—320
—, —, petroleum generation, 320—321
—, —, reservoir geometry, 304, 318
—, —, tectonics, 319
—, —, thermal, 309, 312—314
—, depth of initiation, 59, 311—312, 315
—, drilling break, 59, 305, 310—311
—, drilling experience, 305—306
—, Ekofisk oil field, 239—240
—, formulae, 63
—, Frigg gas field, 242
—, geological contexts, 62, 306, 365
— and growth faults, 316
—, heaving shales, 305, 310
—, interpretation of, 59, 306ff
—, sand/shale ratios, 56, 308, 316
—, shallow, 315—316
—, Statfjord oil field, 294
—, stratigraphic control, 307, 353
—, terminology, 55
—, time of initiation, 311—312, 315
—, transition zone, 306—307, 310—313
—, variations 308, 316—317
Accumulation of sediment, see Sediment accumulation
Aerobic bacteria, 221—225
Alkane Series, 73
Alteration of crude oil, 221—225, 232, 244, 300
Amal field, Libya, 273, 280—281, 283
"Anticlines without synclines" see Regressive sequences, structural style
API density, 75
Archie's formula, 114
Aromatic Series, 74
Assam basin, 237
Asmari Limestone, Iron, Iraq, 201

Athabasca tar sands, 149
Attaka oil field, Kalimantan, 355—356
Augila oil field, Libya, 273, 279, 280—284
Australian continental margin, 296—301, 351, 361—366
—, abnormal pressures, 365
Authigenic minerals, 43, 162—163, 166—167, 251

Bacteria, 221—225
Badak gas field, Kalimantan, 355—356
Barracouta gas field, Australia, 297
Baselevel, 3—5, 7, 10, 28—29, 32—33, 37, 280, 295, 369
Basin, physiographic, 2—3, 7—8, 16, 18, 20—21, 370
—, —, sediments of, 2—3, 8
Basin, sedimentary, 1—22, 361—374 (see also Sedimentary basins)
—, —, age, 2
—, —, complement of orogeny, 358
—, —, concepts, 1—22
—, —, defined, 2
—, —, early deformation, 23—40, 351—359
—, —, geometry, 19
—, —, lithological associations, 15—18, 256
Bass basin, Australia, 366
Bekapai oil field, Kalimantan, 355—356
Bell Creek field, U.S.A., 225
Biodegradation of crude oil, 221—25, 244, 300
Bomu field, Nigeria, 244
Borehole logging, 107—137
—, electrical, see Electrical logging
—, radioactivity, 127—130
Brent oil field, North Sea, 294
Bunju oil field, Kalimantan, 233

Casing, 101—103
Capillary displacement/injection pressure, 185—186, 188, 217—218, 245—246

Catalysis, catalysts, 184, 211—212, 219—220
Clay-mineral diagenesis, 308—309
Clay-mineral facies, 219—220, 310
Coal, 213—215
Compaction, 31, 35, 41—65, 306, 312—313
—, carbonates, 44
—, effect on physical properties, 42
—, effect on thickness, 41, 53
—, growth structures, 31, 35
—, mudstone, shale, 44—62
—, retarded, 42, 46, 50, 51, 53—54, 57—62, 242
—, sandstone, 43—44
—, surface subsidence, 175—176
Compaction curves for mudstone, shale, 45—53, 376
—, Athy's, 45, 47—48
—, Hedberg's, 45—46
—, formulae, 46—48, 50—52, 63, 376—377
—, —, scale length, 47, 63
—, —, —, determination, 50—53, 376—377
—, interpretation, 45—46
—, from sonic transit times, 48—53, 63, 375—382
Compaction drive, 175—176
Compaction factor, 53
Continental margins, 279, 286—291, 296—301, 361—366, 369
Cores, 109—110
Craven fault, England, 25
Crude oil, 72—76
—, classification, 75
— density, 70—71, 75
— —, increases with depth, 71, 233—235, 356
— —, water salinity, 71, 233—235
—, microbial alteration of, 221—225, 244, 300
—, water-washing, 221—225, 232
—, waxy, 75, 210, 214, 220—221, 224, 233—235, 299, 343, 344
—, —, not in carbonates, 221
Crude oil reserves, countries, 142, 149—153, 270
—, fields, 142, 143, 145—149, 266, 269, 272, 274, 280, 281, 286—287, 291
—, world's ultimate recoverable, 147—149
cuttings, 96, 108
Cyclo-alkane Series, 74

Darcy's law, 121, 158, 163—168, 172, 196, 198
—, limits, 168
Darling fault, Western Australia, 365
δ (delta), defined, 58
Density of sediments, sedimentary rocks, 334—336, 379
—, formulae, 42, 334, 380
—, table, 43
Diachroneity, 62, 69, 218, 242—243, 279—282, 296, 307, 343, 353, 360
Diagenesis (see also Compaction), 200, 256, 262, 264, 267—268
—, inhibited by petroleum, 200, 251
—, organic, 210—211, 213—215
—, porosity in carbonates, 201
Diapirism, 38, 330—336, 353
—, dominant wavelength of, 332—333, 353, 356
—, models, 332—335
—, petroleum accumulation, 335
Diapirism and growth structures, 340—346
Diapirism and sliding, 356
Diapirs, 325—330, 353
—, in continental margins, 326
—, mudstone, 329—330, 338, 340
—, salt, 327—328
—, —, NW Germany, 336—338
Diastems, 5, 7, 20, 89
Differential entrapment, 202—206, 232, 262, 269
Dipmeter, 132—133
Disconformities/unconformities, see Unconformities/disconformities
Dominant wavelength of diapirism, 332—333, 353, 356
Drawdown, 97, 98, 171—172
Drilling, 95—105
—, deviated, directional, 103—105
— mud, 99, 108
— penetration rate, 100, 109, 305, 310—311

East Borneo basin, 354, 355—358
Eastern Venezuela, 233—236
Ekofisk oil field, North Sea, 238—240
Electrical logging, 110—127, 132—133
—, induction, 125
—, Microlog, 123—125
—, resistivity, 110—125
—, Spontaneous Potential (SP), 125—127

Equations, see Formula
Equilibrium compaction depth, 58, 313, 315
Evaporites, 15—16, 18, 204—205, 259—262, 264, 274
Exmouth Plateau, western Australia, 92, 93, 362—364

Faults and petroleum migration, 245—251
Fluid migration, see Migration of fluids
Fluid potential, 164, 171, 183, 190—191
Formation Density log, 129—130
Formation Resistivity Factor, 113—115, 124
—, formulae, 114
Formula, Archie's, 114
—, bulk density, 42, 334, 380
—, Darcy's law, 163, 164
—, δ (delta), 58
—, equilibrium compaction depth, 58
—, heads, 165, 173
—, Formation Resistivity Factor, 114
—, geostatic pressure, 55
—, Humble, 114
—, hydrostatic pressure, 54
—, λ (lambda), 57
—, liquid expulsion, 53
—, overburden pressure, 55
—, permeability, 165, 391
—, porosity—density, 130
—, porosity—depth, 47, 380
—, porosity—permeability, 165
—, porosity—tortuosity, 114
—, porosity— transit time, mudstones, shales, 49, 63
—, —, sandstones, carbonates, 131
—, potential (fluid), 171
—, scale length, 50, 63
—, specific surface, 158
—, temperature, bottom-hole, 134
—, Terzaghi's relationship, 57
—, tortuosity, 114
—, transit time—depth, mudstones, shales, 50, 63
Fortescue oil field, Australia, 299—300
Frigg gas field, North Sea, 240—242, 321

Gippsland basin, 87—88, 296—301, 365—366
Gamma-ray log, 127—128, 275
Gas, petroleum, nature of, 76

Gas chromatograms/chromatography, 223, 238
Geoid, surface of, 12—15
Geological inference, 68—72
Geophysics, 85—93
—, seismic record sections, 86—92
—, seismic stratigraphy, 89—93
—, velocity profiles, 88
Geosynclines, 20
Gesa anticline, Irian Jaya, 341—342
Giant oil fields, 79, 142—149
—, size distribution, 142—149
Great Artesian Basin, 191
Great Lakes area, N. America, Silurian reefs, 259—262
Groningen gas field, The Netherlands, 214
Growth anticlines, 34—38, 340—346
—, causes, 325—336
—, defined, 34
—, differential compaction, 35—36
—, effects on reservoirs, 37
—, geographical locations, 36—37
—, mechanical attenuation, 38
—, subsidence of, 37, 344
Growth faults, 25—34, 241, 294—296, 343—346, 351, 352—356
—, age, 30—31
—, causes, 31, 316
—, die out upwards, downwards, 28—29, 343
—, dip of, 31, 33
—, effect on reservoirs, 32
—, flattening of dip, 31, 33, 346
—, geographical locations, 26, 27, 235, 236, 241—242, 294—296, 343—346, 352—353, 365, 368—370
—, movement, 30—34
—, sediment accumulation, 28—29
—, sequences of, 29
—, shape, 26, 31
—, terminology, 24—25
Growth index, 30
Growth structures, 24—40, 241
—, axiom, 24
—, definitions, 24—25, 34—35
—, sand/shale ratio influence, 316
—, stress field, 352—353
growth structures and diapirism, 340—346, 371

Halibut oil field, Australia, 297—298, 300

Handil oil field, Kalimantan, 355—356
Heads, 190
—, elevation, 165, 173
—, pressure, 165, 173
—, total, 56, 63, 155, 165, 171, 173, 177
Horner plots, 134—135, 174
Hydraulic barriers to vertical migration, 182, 376, 378
Hydraulic gradient, 164, 173, 192, 201
Hydrodynamic trapping, 189—198
—, faults, 196—198

Idris field, see Intisar
Igneous rock reservoirs, 280—282
Intisar oil field, Libya, 202, 272—275, 279, 284
Isopach maps, 35—36, 38

Kerogen, 76, 211
Kingfish oil field, Australia, 87—88, 297

La Brea—Pariñas field, Peru, 236—237
λ (lambda), 248—249, 342
—, defined, 57
Leduc oil field, Canada, 263, 267
Libya, Paleocene reefs, 272—275
—, —, reserves, 272
Local petroleum geology not representative, 294
Logging, see Borehole logging

Mahakam delta, Kalimantan, 355—356, 358
—, source rocks, 243, 356, 358
Mackenzie delta, Canada, 18
Mamberamo delta, Irian Jaya, 320, 341
Marlin gas field, Australia, 297—299
Mexico, Cretaceous reefs, 269—272, 351
Midland field, U.S.A., 248—249, 316—317
Migration of fluids, 181—206
—, downwards, 61, 180, 182, 184, 239, 258, 285, 317
—, faults, 196—198, 245—251
—, lateral in permeable beds, 61—62, 317
—, stratigraphic control, 359—360
—, vertical, 61, 182
Migration of petroleum, 80—81, 180—207, 209—229
—, axioms, 181
—, in carbonates, 200—206
— and faults, 216, 239, 243—251

—, primary, 184—185
—, —, downwards, 61, 182
—, —, state during, 183
—, —, vertical, 61, 182
—, secondary, 185—188 (see also Secondary migration)
—, —, alterations during, 221—225
—, —, lateral, 62, 233, 317
—, —, rates, 198—199
—, —, in regressive sequences, 62
—, —, in transgressive sequences, 202—206
Miri field, Sarawak, 344—346
Mississippi delta, mudlumps, 329, 346
Molasse basin, S. Germany, 192, 320
Mudstone density formulae, 334
Mudstone permeability, 314
Mudvolcanoes, 329, 341, 343

Nafoora oil field, Libya, 273, 281, 283
Neutron log, 128—129
Niger delta, 27, 241, 242—244, 346, 353
—, source rocks, 243
North Sea basin, 290, 291—296, 351, 367—369
—, source rocks, 291—292, 320—321, 370

Ohm's law, 112
—, cf. Darcy's law, 121
Oil and gas fields
  Amal, 273, 280—281, 283
  Attaka, 355—356
  Augila, 273, 279, 280—284
  Badak, 355—356
  Barracouta, 297
  Bekapai, 355—356
  Bell Creek, 225
  Bomu, 244
  Brent, 294
  Bunju, 233
  Ekofisk, 238—240
  Fortescue, 299—300
  Frigg, 240—242, 321
  Groningen, 214
  Halibut, 297—298, 300
  Handil, 355—356
  Intisar, 202, 272—275, 279, 284
  Kingfish, 87—88, 297
  Kirkuk, 83
  La Brea—Pariñas, 236—237
  Leduc, 263, 267
  Marlin, 297—299

Midland, 248—249, 316—317
Miri, 344—346
Nafoora, 273, 281, 283
Piper, 294—296, 369
Prudhoe Bay, 286—291
Seria, 342—344
Statfjord, 291—294
Uzere, 241
W. Guara, 234
Wilmington, 175—176
Oil and gas fields, reserves, 139—154
—, giant, 142—149
—, —, size distribution, 142—149
Oil-rich countries, 149—152
—, reserves/production ratios, 150—153
Open fractures, 247—251
—, in Asmari Limestone, 248
Organic maturity, 211—215, 238—241, 243—245, 295, 356
Orinoco delta, early generation of hydrocarbons, 79
—, sediment accumulation rates, 6
Orogeny, 15
— and stratigraphy, 353—358
Otway basin, 366
Overburden pressure gradients, 379

Palaeogeomorphic, palaeotopographic traps, 85, 279—284
Pendular cement, 43
Pendular water, 159—160
Permafrost, 286, 290
Permeability, 61, 161—169, 391—392
Petroleum accumulation, 199—200
Petroleum accumulations, size distributions, 142—149
Petroleum, aqueous solubility, 215—216
Petroleum entrapment, 81—85
—, anticlinal, 81—83
—, fault, 83—84
—, stratigraphic, 84—85
Petroleum fields, see Oil and gas fields
Petroleum, nature of, 72—76
—, origin, generation, primary migration, 78—80, 179—182, 209—213
—, —, abnormal pressures, 219
Petroleum provinces, character, 350
Petroleum reserves, countries, 149—152, 270
—, fields, 139—154, 240, 266, 269, 272, 274, 280, 281, 286—287, 291
Petroleum reservoirs, nature of, 155—178

Petroleum source rocks, 181, 289—290
—, stratigraphic relationships, 72, 181, 221, 231—245, 262, 269, 280, 343
—, —, not contiguous with reservoir, 218—219
Pine Point lead—zinc, 264
Piper oil field, 294—296, 369
Pore water, composition, 76—78, 233—235
—, expulsion, 42, 44—46, 53—54, 319
—, —, formula, 53
—, —, from sonic log, 53
Porosity, mean, 380
Porosity—sonic transit time, 49—51, 131—132
—, formulae, 49—50, 131
—, inhibited diagenesis, 200
Potential, fluid, 164, 171, 183, 190—191, 387
Potentiometric surface, 171—173, 191, 192—197
Pressure, capillary displacement/injection, 185—186, 188, 217—218, 245—246
Pressures, pore-fluid, 54—63, 161
—, —, abnormal, see Abnormal . . .
—, —, mudstones, 56, 57—63, 182—183, 238, 242, 294, 305, 306—321, 329, 341—346, 356
—, —, —, estimation of, 59
—, —, —, —, assumptions, 379
—, —, sonic transit time, 50—53, 60
—, —, subnormal, 319—320
Pressures, overburden (geostatic), 379
—, —, estimation of, 379—380
Production mechanics, 169—176
Prudhoe Bay oil field, Alaska, 286—291

Radioactivity, Kimmeridge Clay, 321
Radioactivity logs, 127—130
Reefs, fossil organic, 16—18, 255—277
—, —, definition, 256
—, —, growth and extermination, 258, 260, 262, 264
—, —, not petroleum source rock, 258
—, —, Libya, 18, 272—275
—, —, Mexico, 18, 269—272, 351
—, —, northern U.S.A., 16—18, 259—262
—, —, Western Australia, 256—257
—, —, Western Canada, 16—18, 262—269, 351
Reef provinces, differential entrapment, 202—206, 232, 261—262, 269

Regional dip, 255, 262, 264, 266, 272, 351
Regression, 8—11, 15, 21, 28
—, defined, 9
Regressive sequences, abnormal pore-fluid pressures, 56, 62, 329, 341—346, 352—358
—, biodegradation of crude oil in, 224—225
—, diapirs, 329
—, mudvolcanoes, 329, 341, 343
—, orogeny, 353—354
—, structural style, 289, 340, 344, 353, 355
Relative permeability curves, 168—169, 187
Reservoir energy, 169, 171
Reservoir, pressure build-up, 294, 315
Reservoirs, abnormally pressured, 294, 315, 317
—, non-sedimentary, 280, 282
Resistivity index, 117
Reynolds number, 331
Rheidity, 331
Rift basins, 279—302, 361—371
Rim synclines, 327—328, 332
"Roll-over" anticlines, 26, 31, 241

Sand/shale ratios, abnormal pressures, 56, 308, 316
—, crude oil composition, 237
—, growth structures, 316
—, source rock, 236—238
Saturation, water, see Water saturation
Scale length, 47, 63
—, determination of, 50—53, 376
Scanning electron micrographs, 162—163, 166—167
Sea-level changes, 11—15, 32—33
—, eustatic, 15, 20, 32, 280
—, —, defined, 11
—, tide-gauge records, 7, 11
Secondary migration (see also Migration), 185—188
—, length of path, 232—233
—, in moving water, 189—198
—, rates, 198—199
Secondary recovery, 176—177
Sediment accumulation, 2—21, 28—29, 32—33, 284, 294, 296, 368—369
—, growth faults, 28—29, 32—33, 296, 369

—, lateral continuity, 8
—, net rates, 6
— from suspension, 6, 28
Sediment compaction, 31, 33, 36, 41—65
—, mechanical equilibrium, 57—58, 313, 315
—, —, interpretation, 59
Sediment supply, subsidence, 7—8, 15
Sediment transport, 28, 284, 295
Sedimentary basins
  Assam, 237
  Bass, 365—366
  East Borneo, 354—358
  East Coast, New Zealand, 315
  Eastern Venezuela, 71, 233—236
  Exmouth Plateau, 92—93, 362—364
  Gippsland, 87—88, 296—301, 365—366
  Great Artesian Basin, 191
  Mahakam delta, 355—356, 358
  Mackenzie delta, 18
  Mamberamo delta, 320, 341
  Michigan, 17, 203—206, 259—262
  Mississippi delta, 329
  Molasse, 192, 320
  Niger delta, 242—244
  North Sea, 290, 291—296, 320, 351, 367—369
  North-west Borneo, 342—346
  Orinoco delta, 6
  Otway, 366
  Sirte, 272—275, 279—284, 369
  South Sumatra, 358
  Trinidad, 305, 315
  Western Canada, 262—269, 351
  Williston, 215, 263, 320
Seria field, Brunei, 342—344
—, fault trap, 343
—, stress field, 344
Shale transit-time plots, 51—52, 60, 308, 335
—, construction of, 375—381
—, formulae, 376—377
Sirte basin, 272—275, 279—284, 369
Sonic log, 130—132
—, errors, 50
Sonic velocity, see Sonic transit time, Shale transit time
Source rock, facies, 181, 220—221, 238, 258, 262, 269, 280, 285, 356, 358, 359—360, 370
South Sumatra basin, 358

—, palaeogeomorphic traps, 358
Specific surface, 158
Statfjord oil field, North Sea, 291—294
Stratigraphic hiatus, 32—33
Stratigraphic influence on petroleum, 236—238, 360—361
Stratigraphic traps, 351
—, transgressive sequences, 256ff
Stratigraphy and orogeny, 353—358
Stratigraphy, trap associations, 359—361
Stress, effective, 57, 61
— field, 249—250, 352—353, 356
Structural style, regressive sequences, 289, 340, 344, 353, 355
Subsidence, 4, 7—8, 28, 37, 256, 283, 290, 309
— of faults, 28—29, 344
Temperature, subsurface, 133—136, 156, 216, 233, 244, 314, 365
Terzaghi's relationship, 57, 200, 212
Tortuosity, 43, 113—115, 165, 169, 218
Transgression, 8—11, 16—18, 21, 28
—, defined, 10
Transgression, "Cenomanian", "Cretaceous", 12, 280, 361—362
Transgressive sequences, 8—9, 10—11, 16—18, 255—302
—, lack of deformation, 351
Trinidad, 305, 315

Unconformities/disconformities, 5, 7, 20, 89, 361, 363—364, 368—370
—, nature of, 284, 295—296
—, relief on surface of, 284, 295—296, 368—370

Unconformity traps, 85, 284—301
—, source rocks for, 285
Uzere field, Nigeria, 241

Vitrinite reflectance, 214—215, 238—239, 356
Venezuela, E., oil density, water salinity, 71, 233—235
Void ratio, 44
VYCOR porous glass, 158, 217—218
—, Darcy's law, 158

Water, composition, 76—78, 233—235
—, connate, 77—78, 83, 156
—, immobile, 161
—, nature of, 76—78, 110, 112
Water-wet, wetting, 157, 188
Water saturation, 83, 115—117, 156—161
—, irreducible, 156—161, 169, 200, 202, 247
—, —, pendular ring geometry, 159
Water encroachment, Mexico, 270
Water density, subsurface, 71, 309
Water flow, directions, 60—62
—, downwards, 61—62
—, not through accumulations, 223—224
West Guara field, Venezuela, 234
Western Canada, Devonian reefs, 262—269, 351
White's Carbon Ratio Theory, 213—214
Williston basin, U.S.A., 215, 263

Zipf's law, 143—149